Chemistry of Phytopotentials: Health, Energy and Environmental Perspectives

L. D. Khemani, M. M. Srivastava, S. Srivastava
(Eds.)

Chemistry of Phytopotentials: Health, Energy and Environmental Perspectives

Springer

Editors
Prof. L D Khemani
Prof. M M Srivastava
Dr. Shalini Srivastava

ISBN 978-3-642-44177-6 ISBN 978-3-642-23394-4 (eBook)
DOI 10.1007/978-3-642-23394-4
Springer Heidelberg Dordrecht London New York

Softcover reprint of the hardcover 1st edition 2012

Printed on acid-free paper

Springer is part of Springer Science+Business Media (www.springer.com)

Preface

From the down of human civilization man is in close contact of nature and is still trying to find out solutions of their problems from natural sources. The plants have been considered as the most natural of all the other natural things and, therefore, attracted the attention of scientific community. There was a time not too long ago when most compounds came from plants. But beginning about 50 years ago, chemistry took over the charge from botany and started synthesizing the compounds. Infact, with increasing population, maintenance of our current standard of living and improvement in our quality of life forced the society to depend on the products of chemical industry. The 20th century has been highly successful in this regards. However, with advent of 21st century, a wave of environmental awareness and consciousness is developed regarding the side effects of used and generated hazardous chemical substances. An increasing concern is realized for using renewable natural resources in a manner which does not diminish their usefulness for sustainable development of future generations. Today, chemists, botanists, microbiologists, environmentalists, engineers and medicos have joined their hands for **greening the chemistry** and working for the search of remedies from natural resources.

The research all over the world on known and unknown plants has resulted in good amount of natural magic bullets. These researches have created interest and awareness among the people and they are changing their taste.

The picture of advertisements noticed these days demonstrates the unmistakable trends of popularity of natural green products.

Phytochemicals are classified as primary and secondary plant metabolites. Various primary metabolites like vegetative oils, fatty acids, carbohydrates, etc are often concentrated in seeds or vegetative storage organs and are generally required for the physiological development of the plant. The less abundant **secondary plant metabolites**, on the other hand, have apparently no function in plant metabolism and are often derived from primary metabolites as a result of the chemical adaptation to environmental stress. Thus, unlike compounds synthesized in the laboratory, secondary compounds from plants are virtually guaranteed to have biological activity.Plants are known to produce a wide range of secondary metabolites such as alkaloids, terpenoids, olyacetylenes flavanoids, quinones, phenyl propanoids, amino acids etcwhich have been proved to possess useful properties. Ten of thousands of secondary products of plants have been identified and there are estimates that hundreds of thousands of these compounds exist unexplored. These secondary metabolites represent a large reservoir of chemical structures with biological activity. With introduction of modern scientific methods of research, our knowledge in Plant Products has expanded vastly. Discoveries of physiological and pharmacological functions of medicinal plants, has initiated extensive research to utilize the properties of the plants in human needs and sufferings.

Presence of multiple active phytochemicals in plants offers exciting opportunity for the development of novel therapeutics, production of eco-friendly value added materials including agricultural, food products, enzymes, neutraceuticals, personal care products, herbal cosmetics, industrial products and sources of energy generations.

Our country has a long tradition of using plants derivatives for curing diseases. Rigveda and Atharveda describe various plant products used by our forefathers for various ailments. The varied climatic conditions have bestowed our country with a rich natural flora. Indian Material Medica shows that more than 90% of the drugs mentioned therein are of plant origin. A common Indian kitchen with onion, garlic, ginger, turmeric, tejpat, coriander, pepper, Ajowain, Jeera, tea, tulsi and neem leaves etc is actually a small herbal medical store.

Is it a fashion or mass hysteria which has gripped the world? Millions of people have started taking juice of roots; shoots, flowers and stem bark of the plants or incredibly dilute aqueous alcoholic solutions of Homeopathic drugs. **Herbalism** is in great demand and giving wake up call for conventional. Society is increasingly shopping for health, trying all the available options in magazines newspapers and on the Internet. Plants are the source of half the pharmaceutical in our modern medicine cabinet. Herbs could lead us away from synthetic bullets and towards a new generation of drugs. There are various health disorders from depression to multiple sclerosis for which no magic bullets are suitable.

Is crude extract more potent than isolated chemical? The issue is debatable and closely associated with the use of herbalism. Why to take a risk by swallowing something as unpredictable as plant material when modern science can isolate the active gradient and serve it to you straight. This approach has initiated intensive scientific research towards the isolation and characterization of bioactive principle of numerous plants for their respective pharmacological properties. While the Herbalists are of their views that as: mixtures are better than pure chemicals. Several biologically active compounds in a plant work together to produce greater effect then single chemical on its own. The mixture of chemicals found in herbs can be more potent than the single purified ingredient so beloved of drugs companies. **Chemical partnerships** explain why whole herbs can work better than single purified ingredients. In other words, the mixture has an effect greater than the sum of its parts. The synergism arises when two or more factors interact in such a way that outcome is not additive but multiplicative. The compound impact of the relationship can be so powerful that the result may be a whole order of magnitude greater than the simple sum of the components. The observation suggests that synergistic or antagonistic effect of various components of plant material in its crude natural state may enhance therapeutic effects and reduce side effects, which may not occur when one or more isolated chemical component are used alone in purified forms. Synthesizing the bioactive ingredients would inevitably reduce or eliminate that benefit. Anyway, herbal extract hopefully would delay resistance against diseases, while bioactive principles can become our therapeutic armamentarium.

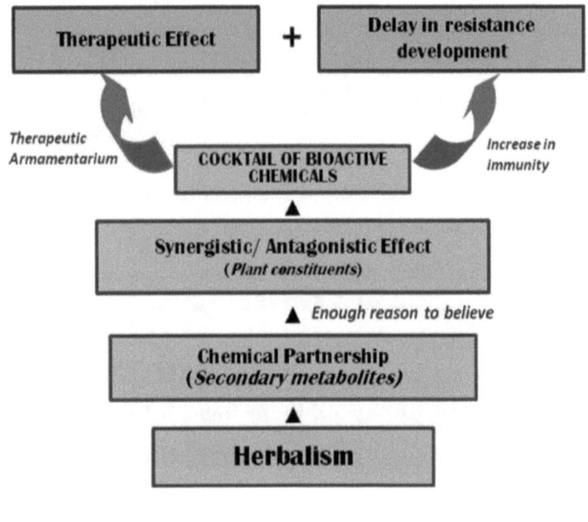

In recent years, research attention revolves around the trends of bringing technology into harmony with natural environment and to achieve the goals of protection of ecosystem from the potentially deleterious effects of human activity.Research findings have clearly raised strong doubts about the use of conventional methods based on the use of synthetic coagulants for water purification. Several serious drawbacks viz. Alzheimer's disease, health problems carcinogenic effects of alum lime, aluminum sulphate, polyaluminum chloride, polyaluminumsilico sulphate, iron hydroxide, iron chloride, soda ash, synthetic polymers and the reduction in pH of water resulting from such treatments have not been appreciated.

Phytoremediation involves processes that reduce overall treatment cost through the application of agricultural residues. This green process of remediation by plants lessen reliance on imported water treatment chemicals, negligible transportation requirements and offer genuine, localized and appropriate solutions to water quality problems. Regeneration of the plant biomass further increases the cost effectiveness of the process thus warranting its future success. Sorption using plant biomass thus has emerged as potential alternative to chemical techniques for the removal and recovery of metal ions. Structural modifications onto the biomaterials leading to the enhancement of binding capacity or selectivity are, therefore, in great demands. A special emphasis has been paid on chemical modifications resulting into tailored novel biomaterials improving its sorption efficiency and environmental stability making it liable for its commercial use as simple, fast, economical, ecofriendly green technologies for the removal of toxic metals from waste water particularly for rural and remote areas of the country.

Plants have also been explored for the generation of energy resources. The energy of sunlight has been harnessed through the process of photosynthesis not only to create the plant biomass on our planet today but also the fossil fuels. The overall efficiency of plant biomass formation, however, is low and cannot replace fossil fuels on a global scale and provide the huge amount of power needed to sustain the technological expectations of the world population now and in the future. However, the photosynthetic process is the highly efficient chemical reaction of water splitting, leading to the production of hydrogen equivalents and molecular oxygen. This new information provides a new dimension for scientists to seriously consider constructing catalysts that mimic the natural system and thus stimulate new technologies to address the energy/CO2 problem that humankind must solve. After all, there is no shortage of water for this cyclic non-polluting reaction and the energy content of sunlight falling on our planet well exceeds our needs.

India, with its rich floral wealth still needs intensive research on plants for their multidimensional uses. This resource is largely untapped for use. Several issues are to be resolved before such ideas can become a reality. No one expects these experiments to yield commercial benefits soon; there is growing awareness that basic studies implants biology may reap impressive and unusual harvest in the future and plants will be proved a dominant source of preventive and therapeutic safe drugs. Several plants' extracts have been characterized for various bioefficacies, but not many have reached to the level of **commercialization**. In fact, mainstream pharmaceutical industry is not really interested in herbs because they are difficult to patent. The marketing of herbal derivatives with patent protection are to be based on complete clinical trials. Manufacturers try to ensure the safety, along with the efficacy. The side effects must be taken into account for herbal preparation exhibiting any beneficial activity. Without the support of the pharmaceutical industry, herbs are likely to remain mired in uncertainty. There should be general worldwide guidelines for the registration of herbal products and special guidelines should be provided for natural products by various regulating agencies which will help in a long way in their promotion. It is time to think.

The present conference offers chemists from diverse areas to come to a common platform to share the knowledge and unveil the chemistry and magic potentials of phytoproducts leading to level of commercialization.

Conference Secretariat
Natural Products Research Laboratory
Dayalbagh Educational Institute, AGRA

Contents

About the Editors

Dr. LD Khemani, M.Sc. (Organic Chemistry, Jiwaji University, Gwalior, 1969), PhD (Chemistry, Agra University, 1977) is now Professor & Head in the Department of Chemistry of Dayalbagh Educational Institute, Agra, India and has experience of thirty five years of teaching and research in Environmental Toxicology and Medicinal Applications of Natural Products with reference to antioxidative, antidiabetic&antirenal failure bioefficacies. Prof. Khemani has 50 research papers in journals of repute. He has delivered lectures in various Universities of France, Spain and W.Germany. Prof. Khemani is member of American Diabetes Association, Washington U.S.A. and Society of Biological Chemists, New Delhi. He has extensive experience of various administrative positions of Chief Proctor, Student welfare and Discipline Committee; Board of Studies; Academic Council; Research Degree Committee; member of organizing committees of various National and International Conferences.

Dr. MM Srivastava, M.Sc. (Organic Chemistry, Agra University, 1976), M.Phil. (Organic Chemistry, H.P. University, Shimla, 1977), PhD (Chemistry, Agra University, 1983) is now Professor in the Department of Chemistry of Dayalbagh Educational Institute, Agra, India and has extensive experience of twenty six years of teaching and research in Analytical and Environmental Chemistry. Prof. Srivastava, currently, is engaged in the research under the domain of Green Chemistry working on Chemistry of Phytopotentials of indigenous plants with special reference to Anticancer activity and Green Nanotechnology. He has 90 research papers in journals of repute to his credit. Prof. Srivastava has delivered lectures in National Research Council, University of Alberta, Canada, University of Illinois, Chicago, Wisconsin, Maryland, USA and Basel, Switzerland. He has recently been elected as Fellow of Royal Society, London, UK (FRSC) and Fellow of Indian Society of Nuclear Techniques in Agriculture and Biology (FNAS). Prof. Srivastava has edited books on Recent Trends in Chemistry, Green Chemistry: Environmental Friendly Alternatives, Chemistry of Green Environment and HPTLC: fast separation technique with excellent hyphenation.

Dr. (Mrs.) Shalini Srivastava, M.Sc (Inorganic Chemistry, Agra University, 1979), Ph.D. (Chemistry, Agra University, 1983) is Associate Professor in the Department of Chemistry, Faculty of Science, Dayalbagh Educational Institute (Deemed University), Agra. Her major areas of research have been Fluoride Chemistry/Heavy Metal Interactions in Soil-Plant system/ Biological Pesticides. Currently, she is addressing the research problem of Phytoremediation of toxic metals under the domain of Green Chemistry. Dr. (Mrs.) Srivastava has 62 research papers in Journals, 72 presentations in Conferences of repute and is Member of various Scientific Societies. She has worked at Manchester University, UK in the area of Analytical Chemistry and also participated in the course WOMEN IN SCIENCE AND ENGINEERING (WISE), 1992 at Imperial College of Science and Technology, University of London, UK. Dr. (Mrs.) Srivastava has authored books on Recent trends in chemistry, DPH, New Delhi and Novel Biomaterials: Decontamination of toxic metals from wastewater, Springer, Germany. Dr. Srivastava has filed two patents on Green processes for the decontamination of toxic metal's polluted water using Agricultural wastes to her credit.

Section A Health Perspectives

1
Cruciferous Vegetables: Novel Cancer Killer and Guardians of Our Health

P. Bansal[1], M. Khoobchandani[1], Vijay Kumar[2] and M. M. Srivastava[1]
[1]Department of Chemistry, Faculty of Science Dayalbagh Educational Institute, Dayalbagh, Agra-282110
[2]Advisor, Medical and Health Care Committee, Dayalbagh, Agra-282110
Email: prachichemdraw@gmail.com

Abstract

Recent studies have shown that crucifers provide greater cancer protection than a diet high in a general mixture of fruits and vegetables. A diet rich in crucifers, such as Brussels sprouts and broccoli, is inversely associated with the risk of many common cancers. The high concentration of Glucosinolates (GLs) and their hydrolysis products (GLsHP) occurring in crucifers provide this protection through some mechanism. The present article describes the anticarcinogenic bioactivities of novel green bullets (Glucosinolates and their hydrolyzed products) and the mechanism of cancer protection.

Introduction

Glucosinolates are anionic, hydrophilic plant secondary metabolites. Toxic effects of GLs and their derivatives in humans have been described in animals. They are now less dramatic since new varieties of rape containing very low amounts of GLs have been bred. Nevertheless an ever increasing number of publications suggest a new potential of GLs-containing vegetables and are considered genuine candidates for protection against chemically induced cancer. Glucosinolates are found to play an important role in the prevention of cancer and other chronic and degenerative diseases. The intact Glucosinolates are capable of every carcinogen-metabolizing enzyme systems. Glucosinolates may breakdown to form isothiocyanates in plant material during processing by the diseases, especially cancers of various types. Recent researchers support that the chemopreventive effect of brassica vegetables and their constituents in various animal and clinical experiments. Such observations led the (American) Committee on Diet, Nutrition and Cancer to suggest that the consumption of cruciferous vegetables "was associated with a reduction in the incidence of cancer at several sites in humans".

Cruciferous are important sources of Glucosinolates (GLs) whose degenerated products like isothiocyanates were attributed to chemo-preventive activity. Vegetables of the *Brassica* genus (broccoli, cabbage, cauliflower, radish, mustard, etc.) have received much attention, because they are reported to have anticancer activity both *in vitro* and *in vivo*. Red cabbage (*Brassica oleraceae var rubra*) contains similar amounts of Glucosinolates like glucoraphanin, glucobrassicin, glucoiberin, progoitrin, sinigrin, gluconapin and glucoerucin. Broccoli sprouts are widely consumed in many parts of the world. A considerable number of epidemiological studies revealed an inverse relationship between consumption of *Brassica* vegetables (broccoli, red cabbage, Brussels sprout, kale, cauliflower, cabbage) and risk of cancer in various human organs. When brassica plant tissue is broken, GLs are hydrolyzed by the endogenous enzyme myrosinase (Myr), releasing many products including isothiocyanates (ITC). ITCs exert chemopreventive effects against chemically induced tumors in animals, modulating enzymes required for carcinogens activation/ detoxification and/or the induction of cell cycle arrest and apoptosis in tumor cell lines.

Crucifers

Vegetables of the *Cruciferae* family are in the botanical order *Capparales,* which includes the *Brassicas* genus. Crucifers contain a group of secondary meta-

M.M. Srivastava, L. D. Khemani, S. Srivastava, *Chemistry of Phytopotentials: Health, Energy and Environmental Perspectives*, DOI:10.1007/978–3-642–23394-4_1, © Springer-Verlag Berlin Heidelberg 2012

bolites called Glucosinolates (GLs) as well as numerous other bioactive compounds that play a role in cancer protection. The plant family *Cruciferae* (mustard family or *Brassicaceae*) includes broccoli, parsnip, Brussels sprouts, Chinese cabbage, radish, horseradish, wasabi, white mustard, watercress, and cauliflower. Crucifers also contain many other bioactive components including flavonoids. The chemopreventive effect of cruciferous vegetables is thought to be due to their relatively high content of Glucosinolates (β-thioglucoside *N*-hydroxysulfates), which distinguishes them from other vegetables.

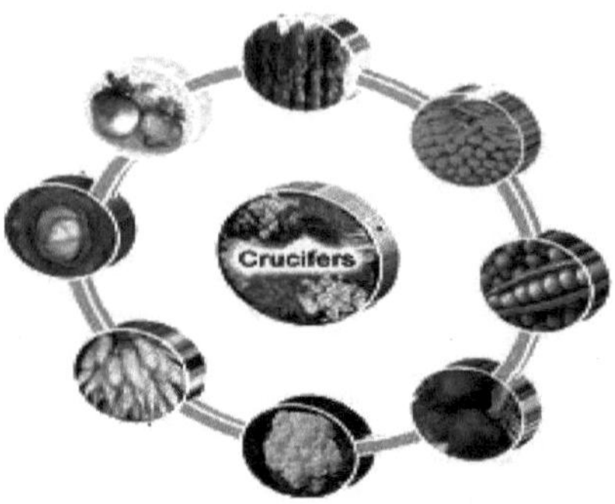

Fig. 1: Cruciferous Vegetables

Table 1: Vegetables and fruits of the family *Cruciferae*

Genus species (sub species)	Vegetable
Armoracia lapathifolia	Horseradish
Brassica camoestris (*rapifera*)	Turnip
Brassica camoestris (*oleifera*)	Rape
Brassica napus (*napobrassica*)	Swede
Brassica oleracea (*capitata*)	White/red cabbage
Brassica oleracea (*sabauda*)	Savoy cabbage
Brassica oleracea (*gemmifera*)	Brussels sprouts
Brassica oleracea (*cauliflora*)	Cauliflower
Brassica oleracea (*cymosa*)	Sprouting broccoli
Brassica oleracea (*laciniata*)	Curly kale
Brassica pekinensis	Chinese white cabbage
Lepidium sativum	Garden cress
Nasturtium officinale	Watercress
Raphanus sativus	Radish
Sinapis alba	White mustard
Carica papaya	Papaya

Among all of the cruciferous vegetables, broccoli sprouts have the highest level of the glucosinolates relevant to this enzymatic process. Just two or three tablespoons of broccoli sprouts a day provide a powerful dose of Glucosinolates. After broccoli sprouts, cauliflower sprouts are second highest in terms of containing the relevant Glucosinolates.

Glucosinolates

The Glucosinolates are a class of organic compounds that contain sulfur and nitrogen and are derived from glucose and an amino acid. They occur as secondary metabolites of almost all plants of the order *Brassicales*. The Glucosinolates are a class of secondary metabolites found in fifteen botanical families of dicotyledonous plants. So far about 100 Glucosinolates have been reported. Generally, levels in the seed are high (up to ten per cent of the dry weight). Studies have shown that myrosinases are localized in vacuoles of specialized plant cells, called myrosin cells. Thus the two components of the system are separated until autolysis or tissue damage brings them into contact.

Glucosinolate research has made significant progress, resulting in near-complete elucidation of the core biosynthetic pathway, identification of the first regulators of the pathway, metabolic engineering of specific Glucosinolate profiles to study function, as well as identification of evolutionary links to related pathways.

Hydrolysis of Glucosinolates

When crushed plant tissue or seeds containing Glucosinolates are added to water, myrosinases catalyze the hydrolytic cleavage of the thioglucosidic bond, giving D-glucose and a thiohydroximate-O-sulfonate (aglycone). The latter compound rearranges non enzymatically with release of sulfate to give one of several possible products. The predominant product is dependent on the structure of the Glucosinolate side chain and the presence of protein co-factors that modify the action of the enzyme. The most frequent fate of the unstable aglycone is to undergo rearrangement spontaneously via a proton independent Lossen rearrangement with a concerted loss of sulfate to yield an isothiocyanate, or a competing proton dependent desulfuration yielding a nitrile and elemental sulfur. Some Glucosinolates also give rise to the formation of thiocyanates.

Myrosinase is not properly identified as a single enzyme, but as a group of similar-acting enzymes. Multiple forms of the enzymes exist, both among species and within a single plant, and all perform a similar function. Myrosinases are fairly specific toward Glucosinolates. These enzymes cleave the sulfur-glucose bond regardless of either the enzyme or substrate source. Myrosinase is a cytosolic enzyme associated with membranes, perhaps surrounding a vacuole containing Glucosinolates. Glucosinolates are probably contained in vacuoles of various types of cells. In contrast, myrosinase is contained only within structures, called myrosin grains, of specialized myrosin cells that are distributed among other cells of the plant tissue. As Glucosinolate vacuoles do not appear to be present within myrosin cells, intercellular rather than intracellular separation occurs. Disrupting cellular tissues allows Glucosinolates and myrosinase to mix, resulting in the rapid release of Glucosinolate degradation products. Myrosinase activity and Glucosinolates are preserved in cold-pressed meal and are no longer physically separated. Thus, adding water immediately results in the production of the hydrolysis products, including isothiocyanate, without the need for additional tissue maceration.

Isothiocyanates

Glucosinolates are sulfur-containing molecules produced from amino acids by the secondary metabolites. Glucosinolates are not biologically active but are the precursor for the formation of a variety of potential allelochemicals, most important of these are Isothiocyanates (ITCs). They occur predominantly in various families: *Tovariaceae, Resedaceae, Capparaceae, Moringaceae and Brassicaceae.* Species belonging to these families are widely consumed or cooked as salad vegetables (cabbage, Brussels, sprouts, cauliflower, radish, water cress) or condiments (horseradish, mustard caper) cruciferous forages (kale, rape, turnip) and oilseed meals (rape, turnip rape) are used as foodstuffs for animals. Glucosinolates on enzymatic degradation by myrosinase enzyme in presence of water release isothiocyanates (ITCs), organic cyanides and ionic thiocyanates (SCN^-). Degradation also occurs thermally or by acid hydrolysis. Myrosinases are fairly specific towards Glucosinolates.

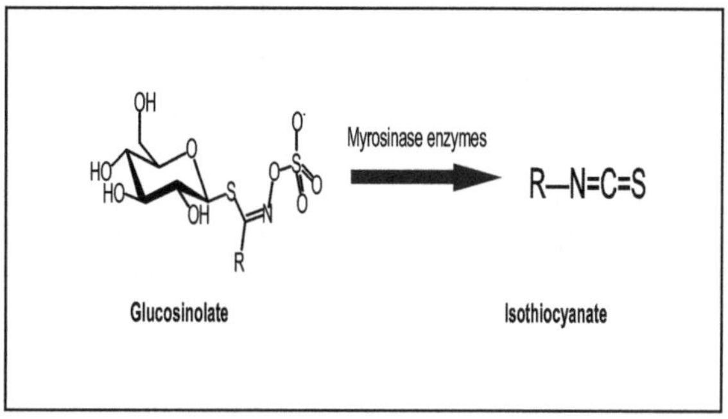

Fig. 2: Conversion of GLs into ITC

Isothiocyanates (ITCs) are found in many cruciferous vegetables, which are consumed widely. The flavor and odor peculiar to these vegetables are mainly ascribed to ITCs. They are classified as chemopreventive agents for cancer. Most studies on the cancer-preventive activities of crucifer-derived ITC have focused on those that occurs abundantly in common cruciferous vegetables which are frequently consumed by humans. ITC inhibits both the formation of cancer cells (anticarcinogenic activity) and the survival and proliferation of existing cancer cells. Such activities with each compound have been demonstrated in multiple organ sites of rodents. Considerable information on the molecular basis for both the anticarcinogenic and anticancer effects of ITC is available. It is now clear that ITC can target cancer in multiple directions, including inhibition of carcinogen-activating enzymes, induction of carcinogen-detoxifying enzymes, induction of apoptosis and arrest of cell cycle progression, as well as other mechanisms. It should be emphasized that ITC are dichotomous modulators of oxidative stress. While ITC transcriptionally stimulate many antioxidative enzymes and nonenzymatic proteins, leading to enhanced protection against oxidative stressors.

Table 2: Isothiocyanate structures and their efficacy

Isothiocyanate	Structure of ITCs	Efficacy
2-methylbutyl Isothiocyanate	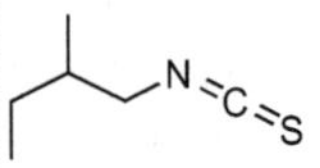	Determine Biogenetic pathway
4-hydroxy benzyl Isothiocyanate	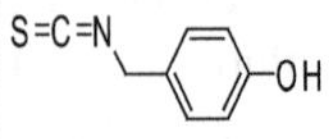	Essential oil, Food preservative
Benzyl Isothiocyanate	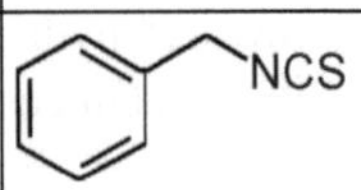	Cell cycle arrest in G_1-S phase, reduce atherosclerosis
2-Phenethyl-Isothiocyanate	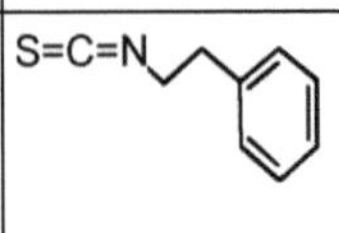	Anticarcinogenic Apoptosis induction, Antiinflammatory,
Sulforaphane	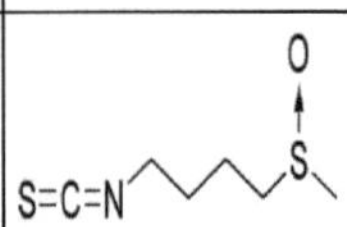	Cell cycle arrest Apoptosis, Anticarcinogenic activity, Antioxidant
Methyl Isothiocyanate	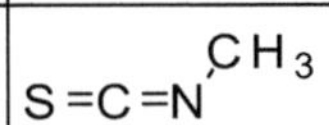	Biofumigation
1-methylpropyl Isothiocyanate	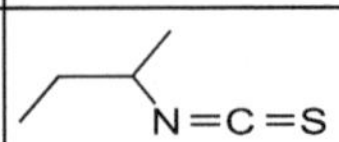	Herbicidal activity
3-butenyl Isothiocyanate	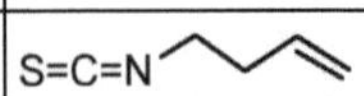	Cytotoxic and Antioxidant activity
4-(methylthio) butyl Isothiocyanate (Erucin)	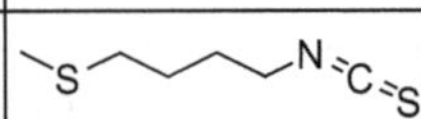	Antidiabetic, Antioxidant, Antiulcer, Antigenotoxic
Ethyl Isothiocyanate	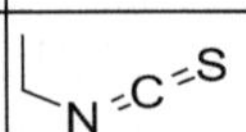	Apoptosis, Anticancer activity
4-pentenyl Isothiocyanate	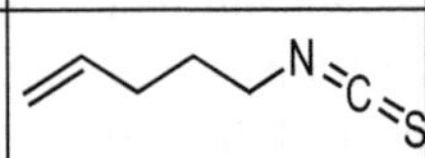	Antibacterial activity
Allyl Isothiocyanate	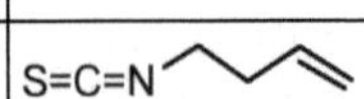	Anticarcinogenic activity, Apoptosis induction

They also directly alkylate and deplete cellular thiols, damage mitochondria, and elevate reactive oxygen species, leading to cellular stress. These paradoxical effects appear to occur in tandem: exposure of cells to ITC rapidly leads to an acute increase in stress, which is followed by a delayed but lasting increase in cellular protection against oxidants and carcinogens. Ironically, although ITC-induced stress may lead to oxidative damage, it has become increasingly clear that much of the chemopreventive activity of ITC stems from the response of cells to the stress induced by these compounds.

The most studied bioactive isothiocyanates are Sulforaphane, Phenyl ethyl isothiocyanate, Allyl isothiocyanate, but many other isothiocyanates present in lower quantities may contribute to the anticarcinogenic properties of crucifers. The isothiocyanates are strong inhibitors of phase I enzymes, particularly the cytochrome P_{450} enzymes. Another important activity of the isothiocyanates is induction of phase II detoxification enzymes including sulfotransferases, NAD(P)H quinone oxidoreductases, and N-acetyltransferases. Phase II enzymes catalyze the conjunction of carcinogens with endogenous ligands, resulting in the formation of hydrophilic conjugates, which are often less toxic and more easily excreted in the urine or bile. The isothiocyanates activate phase II enzymes and consequently reduce carcinogen titre within the body. The chemopreventive effects of the isothiocyanates were traditionally attributed to the enhancement of carcinogen detoxification by phase II induction and the blocking of carcinogen activation by phase I inhibition. Both of these actions explain the ability of the isothiocyanates to prevent tumorigenesis when administered prior to carcinogen exposure.

Protection against Oxidative Stress resulting from excessive exposure to environmental pollutants, ultraviolet light, or ionizing radiation may overwhelm the body's antioxidant system and result in oxidative damage to proteins and nuclear acids. This may lead to initiation of cancer and other degenerative diseases. Extracts of crucifers have direct free radical–scavenging properties ex vivo. Isothiocyanates may slow proliferation and increase apoptosis of cancer cells, resulting in a retardation of tumor growth. I3C arrests human breast cancer cells and prostate cancer cells in the G1-phase of the cell cycle. Cell cycle arrest is accompanied by abolished expression of cyclin-dependent kinase-6 and increased apoptosis. Sulforaphane arrests human colon cancer cells in G2/M-phase and increases expression of cyclin A and B, bax, and cell death by apoptosis.

Natural Products Research Laboratory, Department of Chemistry, Dayalbagh Education Institute, Dayalbagh, Agra is actively engaged in the research pertaining to extraction, isolation, structure elucidation and modification of bioactive principles of indigenous plants for antidiabetic, antioxidant and anticancer activities. The present focus is on the evaluation of cancer protective activity (antimelanoma, antimammary, anticolon) of isothiocyanate rich taramira *(Eruca sativa)* oil, addressing the role of stable conjugated and micro-encapsulated dietary isothiocyanates as promising cancer chemopreventing agent.

The article has been written as review paper and material is taken from sources that the authors have been directly involved with. Every effort has been made to acknowledge materials drawn from other sources.

Suggested Readings

1. K. K. Brown., Isothiocyanate induction of apoptosis in cells overexpressing Bcl-2, University of Canterbury, (2006).
2. L. Nugon – Baudon and S. Rabot., Glucosinolates and Glucosinolate derivatives: Implications for protection against chemical carcinogenesis *Nutrition research reviews*, 7, 205–231, (1994).
3. Committee on Diet, Nutrition and Cancer, National Research Council. *Diet, Nutrition and Cancer,* Washington DC: National Academy Press (1982).
4. G.R. Fenwick, R.K. Heaney, A.B. Hanley and E.A. Spinks., Glucosinolates in food plants. In *Food Research Institute, Norwich, Annual Report* (1986).
5. W.B. Jakoby., *Enzymatic Basis* of *Deroxication,* (1) London: Academic Press (1980).
6. J. Brown and M.J. Morra., Glucosinolate-Containing Seed Meal as a Soil Amendment to Control Plant Pests, National Renewable Energy Laboratory, *University of Idaho Moscow, Idaho,* (2000–02).
7. A. Bhattacharya et al., Inhibition of Bladder Cancer Development by Allyl Isothiocyanate, Carcinogenesis, (2009).
8. J. Fuhrman, Cruciferous vegetables: powerful anti-cancer foods, Community , (2010).
9. Studies on Antitumor Activity of Glucosinolates Hydrolytic Products *China's Outstanding Master's Theses Part C*, China Papers (2010).
10. D. Kiefer., Natural, Dual-Action Protection Against Deadly Cancers Cruciferous vegetables: Protective effects, LE Magazine, (2006)
11. J. Appleton., Vegetable extract prevents cervical cancer, Healthnotes newswire (2000).
12. A.P. Brown, J. Brown, J.B. Davis and D.A. Erickson., Intergeneric Hybridization between Yellow Mustard and Related Canola Species. American Society of Agronomy 86th Annual Meeting, (1994).

2

Synthesis of Bioactive Thiosemicarbazides: Antimicrobial Agents Against Drug Resistant Microbial Pathogens

M. Shukla[1], M. Dubey[2], H. Kulshrashtha[1] and D. S. Seth[1]
[1]Department of Chemistry, School of Chemical Sciences, St. John's College, Agra-282002, India
[2]Microbiology Research Lab, Department of Botany, RBS College Agra, India
Email: manisha_shukla12@rediffmail.com

Abstract

The synthesis of some new thiosemicarbazides derived from N-(substituted) phenyl malonamic acid hydrazide with 4-nitro phenyl isothiocyanate. All the new derivatives have been characterized by elemental analysis, IR, & NMR. The IR and NMR spectral data suggest the involvement of C=S, N-H, CH_2 N-N, CONH, N-C=O. Compounds have been synthesized in an open vessel under microwave irradiation (MWI) using a domestic microwave oven. The reaction time decreases from hours to minutes with improved yield as compared to conventional heating. The thiosemicarbazides have been tested in vitro against a number of microorganisms in order to assess their antimicrobial properties. The results indicate that the thiosemicarbazides possess antimicrobial properties.

Introduction

In the field of medicine the importance of thiosemicarbazides is well known. Thiosemicarbazides (—N=C=S group) have been known to show pronounced biological activities[1]. Thiosemicarbazides have shown activity against protozoa[2], small pox[3] and certain kinds of tumor[4]. The anticonvulsant activity of thiosemicarbazides has been reported in the isolated cerebral cortex preparation[5]. The influence of the thiosemicarbazides has also been on the electrical activity in the interior brain stem of the cat[6]. The anti-viral activity was tested of some thiosemicarbazides against the influenza virus (strain PR-8, type)[7,8]. Thiosemicarbazides have also been reported to posses hypoglycemic activity and usefulness in agriculture. Such types of compounds have been found to be useful as a large number of anticonvulsant, insecticides, rodenticides, anti-tubercular activity against *M. Tuberculosis* ($H_{37}R_v$), anti-viral, hypoglycemic, hypotensive as well as metabolic convulsants. The increasing application of microwave irradiation (MWI) in the synthesis of organic compounds has been receiving attention during recent years. Microwave heating has proved to be very useful tool to carry out certain organic transformations which not only excludes the use of hazardous non-eco friendly solvents but also enhances the reaction rates greatly. A much faster reaction under microwave makes it less expensive in terms of energy, yield and time compared to its thermal analogue. Also, reactions under this condition are very clean and no byproduct form even at high power irradiation. These features make microwave approach very compatible with the upcoming concept of "Green Chemistry".

Materials and Methodology

N-(substituted) phenyl malonamic acid hydrazide was prepared from N-(substituted) phenyl malonamate ester of various substituted aromatic amines. 4-nitro phenyl isothiocyanate used were of Sigma-Aldrich. Ethanol and other solvents of A.R. grade were used as received.

Synthesis of Thiosemicarbazides

Classical Heating Based Synthesis (Method A)

A mixture of N-(substituted) phenyl malonamic acid hydrazide (0.01mol) and 4-nitro phenyl isothiocyanate (0.01mol), dissolved in 10ml ethanol was re-

M.M. Srivastava, L.D. Khemani, S. Srivastava, *Chemistry of Phytopotentials: Health, Energy and Environmental Perspectives*, DOI:10.1007/978–3-642–23394-4_2, © Springer-Verlag Berlin Heidelberg 2012

fluxed for two hours. The solid obtained on cooling was recrystallized with hot absolute ethanol and was found to be N-(malon substituted anilic)-4-(4'-nitro phenyl) thiosemicarbazides.

Microwave "Jump Start" Synthesis (Method B)

A mixture of N-(substituted) phenyl malonamic acid hydrazide (0.01mol) and 4-nitro phenyl isothiocyanate (0.01mol), dissolved in 4ml ethanol and were exposed to microwave irradiation for 4–6 minutes.
The solid obtained on cooling was recrystallized with hot absolute ethanol and was found to be N-(malon substituted anilic)-4-(4'-nitro phenyl) thiosemicarbazides.

Physical Measurements and Analytical Data

Melting points were determined in open capillary tubes and are uncorrected (Table 1). The purity of the compound was checked by on TLC. The structures of the compounds are confirmed on the basis of their IR and ^1H NMR. All the compounds gave satisfactory microanalysis. Microwave irradiations were carried out in an unmodified IFB domestic microwave oven. All the chemicals were of analytical grade.

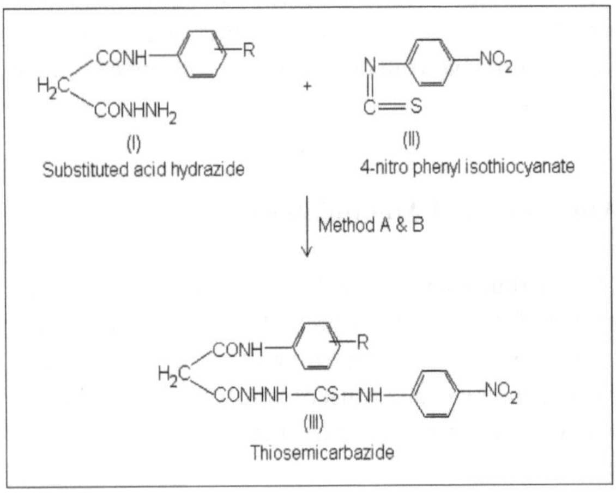

Fig. 1: Chemical reaction of N-(malon substituted anilic)-4-(4'-nitro phenyl) thiosemicarbazide

Antibacterial Activity

Antibacterial activity was evaluated by the paper disc method. The Müller-Hinton agar (beef infusion, casein hydrolyzate, starch, agar) and 5 mm diameter paper discs of whatman No. 1 were used. The compound was dissolved in DMSO. The filter paper discs were soaked in different solutions of the compounds, dried and then placed in the petriplates previously seeded with the test organisms *E. coli* and *S. aureus*. The plates were incubated for 24–30 hours at 28±2°C and the inhibition zone around each disc was measured[9].

Antifungal Screening

The antifungal activity of the compounds was evaluated against *Aspergillus niger* by the agar plate technique. The Sabouraud dextrose agar (dextrose, peptone, agar) and 5 mm diameter paper discs of whatman No. 1 were used. The compounds were dissolved in DMSO and then were mixed with in the medium. These petriplates were wrapped in the polythene bags containing a few drops of alcohol and were placed in an incubator at 25±2°C. The activity was determined after 96 hours of incubation at room temperature (25°C)[10].

Results and Discussion

Infrared Spectra

Infrared spectra of the substituted thiosemicarbazides show medium intensity bands at 3455–3168 cm^{-1} due to υ NH vibrations. A sharp bands found at 1245–1025 cm^{-1} due to υ C=S. υ N-N stretching bands in the thiosemicarbazides appeared at 980–1219 cm^{-1}. In the IR spectra of the substituted thiosemicarbazides the band appeared at 2997–1330 cm^{-1} due to the υ CH_2. υ CONH band appeared at 1620–1488 cm^{-1} in the compounds. A sharp and medium bands of υ N-C=O showed at 1529–1718 cm^{-1}.

^1H NMR

The bonding patterns of these compounds are further supported by the proton magnetic resonance spectral studies in DMSO-d_6. The compounds exhibit a singlet at δ 4.9–3.22 ppm due to NH. This compound shows multiplet in the region at δ 7.98–6.49 ppm attributable to the aromatic protons. Another singlet appearing at δ

4.34–3.33 due to the CH_2. A singlet due to the –CONH group appears around δ 11.20–8.61 ppm.

Antimicrobial Activity

The data in Table 2, showing zone of inhibition against the bacterium *S. aureus, E. coli* and fungus *Aspergillus niger* due to the different substituted thiosemicarbazides. G & H compound of thiosemicarbazides were found to be weak in activity against *E. coli* and compound D & F against *S. aureus*. Highest antimicrobial potential was observed with compound B & D against *E. coli* and compound C & G against *S. aureus*.

Compound A showed highest antifungal potential against *Aspergillus niger*.

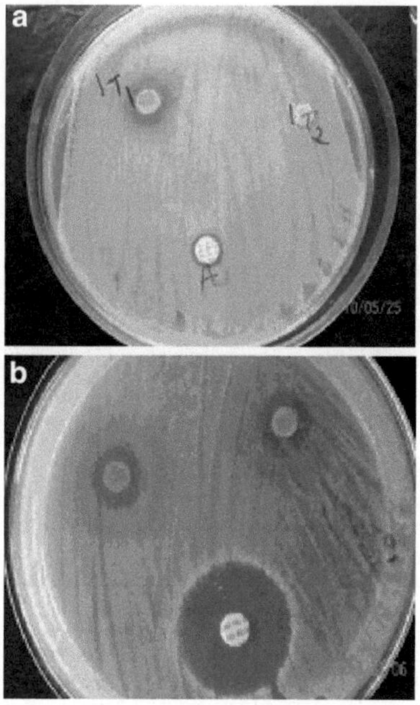

Fig. 2: Antibacterial activity of N-(malon substituted anilic)-4-(4'-nitro phenyl) thiosemicarbazides against **(a)** Escherichia coli and **(b)** Staphylococcus aureus

Table 1: Thiosemicarbazides obtained by the condensation of N-(substituted) phenyl malonamic acid hydrazide with 4-nitro phenyl isothiocyanate:

S. No	Substituted (R)	Mol. Formula	Color	M.P. (°C)	%Yield		C%	H%	O%	N %	S %
					A	B	Found/ Calc.	Found/ Calc.	Found/ Calc.	Found/ Calc.	Found/ Calc.
1	m-toluidine (A)	$C_{17}H_{19}O_4N_6S$	Yellow	146	80.51	82.38	50.64 (50.62)	4.70 (4.71)	15.86 (15.88)	20.80 (20.84)	7.91 (7.94)
2	2,4-di methyl (B)	$C_{18}H_{21}O_4N_6S$	Yellow	164	91.26	95.38	51.78 (51.79)	5.01 (5.03)	15.33 (15.34)	20.21 (20.14)	7.60 (7.67)
3	3,4-di methyl (C)	$C_{18}H_{21}O_4N_6S$	Yellow	152	73.26	75.65	51.81 (51.79)	5.02 (5.03)	15.31 (15.34)	20.23 (20.14)	7.69 (7.67)
4	3,5-di methyl (D)	$C_{18}H_{21}O_4N_6S$	Yellow	160	87.15	90.00	51.76 (51.79)	5.05 (5.03)	15.35 (15.34)	20.19 (20.14)	7.70 (7.67)
5	p-anisidine (E)	$C_{17}H_{19}O_5N_6S$	Yellow	160	84.76	89.50	48.67 (48.68)	4.51 (4.53)	19.10 (19.09)	20.02 (20.04)	7.59 (7.63)
6	p-chloro (F)	$C_{16}H_{16}O_4N_6SCl$	Yellow	140	82.42	85.16	45.40 (45.39)	3.76 (3.78)	15.14 (15.13)	19.79 (19.85)	7.58 (7.56)
7	2,4-di chloro (G)	$C_{16}H_{15}O_4N_6SCl_2$	Yellow	160	74.78	78.59	42.00 (42.01)	3.29 (3.28)	14.01 (14.00)	18.42 (18.38)	7.02 (7.00)
8	3,4-di chloro (H)	$C_{16}H_{15}O_4N_6SCl_2$	Yellow	152	75.65	80.00	42.02 (42.01)	3.26 (3.28)	14.02 (14.00)	18.36 (18.38)	7.05 (7.00)

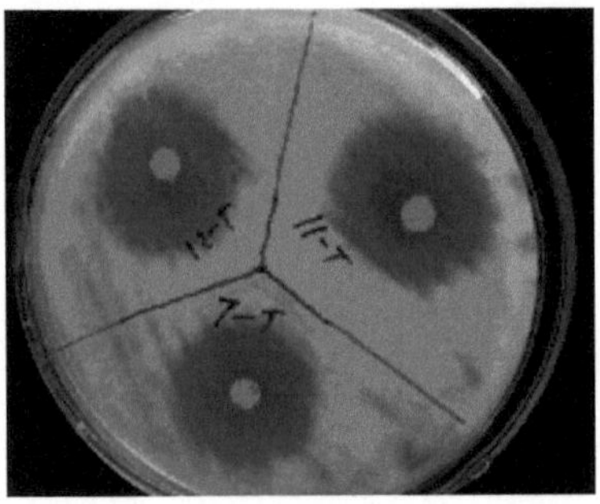

Fig. 3: Antifungal activity of N-(malon substituted anilic)-4-(4'-nitro phenyl) thiosemicarbazides against Aspergillus niger

Table 2: Antimicrobial Studies of N-(malon substituted anilic)-4-(4'-nitro phenyl) thiosemicarbazides

S. No	Compounds	Zone of inhibition (in mm)			
		E. coli	*S.aureus*	Positive control (Amikacin)	*Aspergillus niger* (Fungus)
1	(A)	10	11	25	30
2	(B)	14	12	25	25
3	(C)	10	15	23	22
4	(D)	15	9	25	26
5	(E)	11	11	25	22
6	(F)	12	10	25	23
7	(G)	8	14	25	22
8	(H)	9	11	24	24

Conclusion

In conclusion, from our point of view, microwave irradiation method has been proved here as a better method for the synthesis of thiosemicarbazides and increase in percentage (%) yield is in following order: **Method-1(Classical heating synthesis) < Method-2 (Microwave "jump start" synthesis).** N-(substituted) phenyl malonamic acid hydrazide with 4-nitro phenyl isothiocyanate were proved to have some antibacterial activity against Gram-negative *E. coli* & Gram-positive *Staphylococcus aureus* bacteria and these compound also showed highly antifungal activity against *Aspergillus niger.*

Acknowledgements

We are thankful to Central Drug Research Institute (CDRI), Lucknow for spectral and elemental analysis. We are also very grateful to Department Of Microbiology, R. B. S. College, Agra for antimicrobial screening.

References

1. G. Mazzone, F. Bonia, A. R. Reeina, G. Blandino, Farmaco, Ed. Sci., (1981) 36,181.
2. K. Butler, U. S. Pat., (1968) 3, 382, 266.
3. J.D. Bauer, L. St. Vincent, H.C. Kampe, W.A. Dowine, Lancet (1963) 494.
4. G.H. Peterling, H.H. Buskirk, E.G. Underwood, Cancer Res., (1964) 64, 367.
5. B.J. Preston, (Univ. Illinois, Coll. Med. Chicago) J. Pharmacol. Exptl. Therap. (1955) 115, 28–39.
6. Idem., Ibid, (1955) 115, 39–45.
7. P.N. Buu-Hoi, A. Bouffanais, P. Gley, D.N. Xuong, H.N. Nam, Experimentia (1956) 12, 73.
8. N.N. Orlova, A.V. Aksenova, A.D. Selidoukin, S.N. Bogdanova, N.G. Pershin, Farmakologiya. i. Toksikologiya (1968) 31, 725.
9. C. Saxena, D.K. Sharma, and R.V. Singh; Phosphorus, Sulfur and Silicon 85 (1993).
10. M. Jain, S. Nehra, P.C. Trivedi, and R.V. Singh; Heterocyclic Communications 9 (2003) 1.

3
Antineoplastic Properties of Parthenin Derivatives – The Other Faces of a Weed

A. Saxena[1], S. Bhusan[1], B. S. Sachin[3], R. R. Kessar[1], D. M. Reddy[2], H. M. S. Kumar[2], A. K. Saxena[1]

[1]Dept. of cancer pharmacology, Indian Institute of Integrative Medicine, Canal Road, Jammu, J&K, India 180001.
[2]Indian Institute of Chemical Technology, Uppal Road, Hyderabad, AP, India 500607
[3]Dept. of Chemical Technology, Babasaheb Ambedkar Marathwada University, Aurangabad, MH, India 431004.
Email: arpitaspak@gmail.com

Abstract

A spiro-isoxazolidine derivative of parthenin namely SLPAR13 was taken up for this study which induced cell death in three human cancer cell lines namely HL-60 (acute promyelocytic leukaemia), SiHa and HeLa (cervical carcinoma) with various inhibitory concentrations. The cytotoxicity test was also done on the normal cells hGF (primary human gingival fibroblast) and the inhibitory concentration was found to be more than 10 times higher than HL-60 cells. The cell death was confirmed by cell cycle arrest exhibited by the test compounds in a concentration dependent manner in HL-60 cells. The nuclear condensation and morphological changes induced by the test compounds further marked the HL-60 cell death which was confirmed to be apoptosis by DNA ladder which is hallmark of apoptosis by the formation of 180bp fragments.

Introduction

Parthenium hysterophorus (popularly known as Congress weeds, White top, Star weed, Carrot weed, Gajar ghas, Ramphool) is one of the ten worst weeds in the world. As a curse for the bio-diversity, this weed has always been criticized for its ill effects. Sesquiterpene lactones are the active constituents of a variety of medicinal plants used in traditional medicine. However, it has been found to be of interest due to its anti-cancer **[1, 2]** anti-bacterial **[3]**, anti-malarial **[4]** and allelopathic properties. The Spiro-isoxazolidine derivative of parthenin have been synthesized **[5]** and chosen for this study because of the fact that halogen substituted derivatives of most of the natural compounds show higher cytotoxicity **[6]**. Therefore this compounds has been identified for the present study to determine its potential as a novel anticancer therapeutic.

In cancer, the therapeutic goal is to trigger tumor-selective cell death. One of these events in cell deregulation is obligate compensatory suppression of apoptosis (programmed cell death), which provides support for neoplastic progression. Studies are being focused towards the induction of apoptosis in the cancer cells but being milder with the adjoining normal cells **[7]**. For the same reason natural and modifications of these natural compounds have become the centers of attraction of the oncologists and drug discovery groups **[8]**.

Aim

This study involves evaluation of anticancer potential of SLPAR13. Whatever the mechanisms involved, if the test compound induces apoptosis then that test compound may be the potential candidate for anticancer lead optimization.

Materials and Methods

Synthesis of SLPAR13

The synthesis of N-(phenyl)-C-(5-Bromo, 2-methoxy phenyl)-spiro-isoxazolidinyl parthenin (SLPAR13) (Fig.1) was done as described earlier **[5, 9]**.

M.M. Srivastava, L.D. Khemani, S. Srivastava, *Chemistry of Phytopotentials: Health, Energy and Environmental Perspectives*, DOI:10.1007/978–3-642–23394-4_3, © Springer-Verlag Berlin Heidelberg 2012

Cell Proliferation Assessment by MTT Assay

HL-60, Hela and SiHa cells were grown in suspension in T-75 flask and centrifuged at 100 g for 5 min. Cell pellet was suspended in RPMI medium and then 15×10^3 cells (HL-60) and 10×10^3 cells (HeLa, SiHa and hGF) were transferred to each well of Nunclon 96-well flat bottom plate and treated with SLPAR 13 and samples processed as described earlier **[10]**.

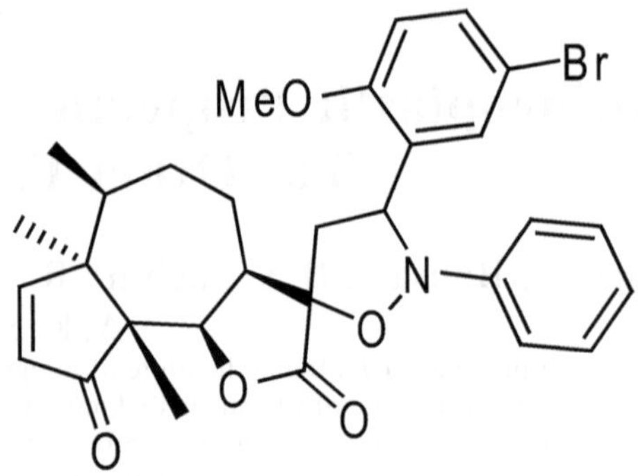

Fig. 1: Structure of SLPAR 13

DNA Content and Cell Cycle Phase Distribution

HL-60 cells (1×10^6/1.5 ml/well) treated with different concentrations of SLPAR 13 and incubated for 24h. The preparations were made as described earlier **[11]** then analyzed for DNA content using BD-FACS CALIBUR. Data were collected in list mode on 10,000 events for FL2-A vs. FL2-W.

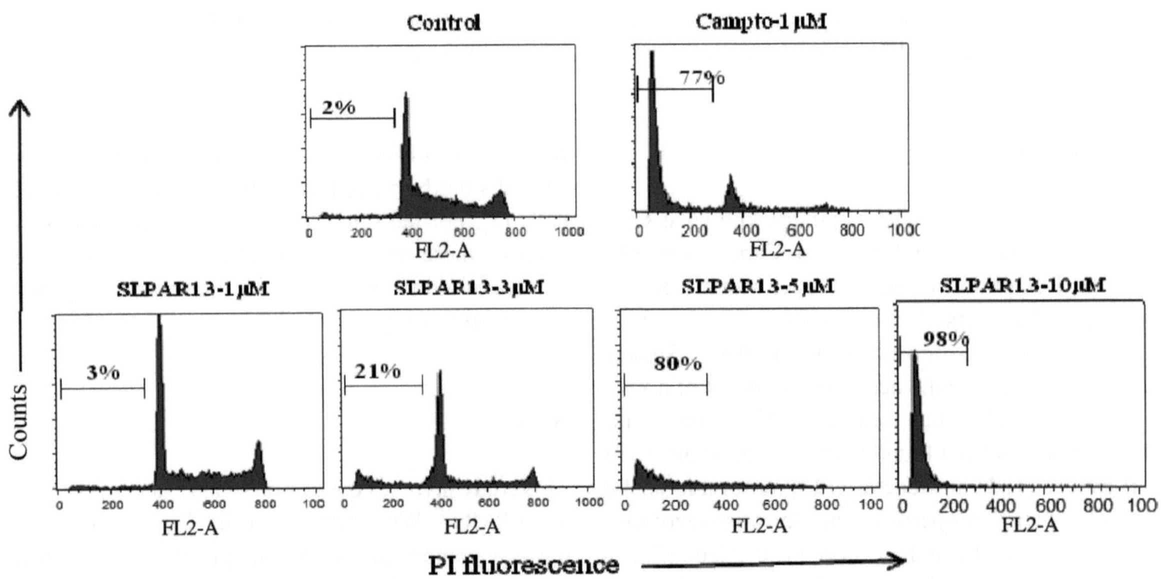

Fig. 2: HL-60 cells treated with SLPAR13, incubated for 24 h, were processed for acquisition in flow-cytometer as described in materials and methods. The compound inhibited cell cycle in a concentration dependent manner with maximum inhibition at 10 µM concentration

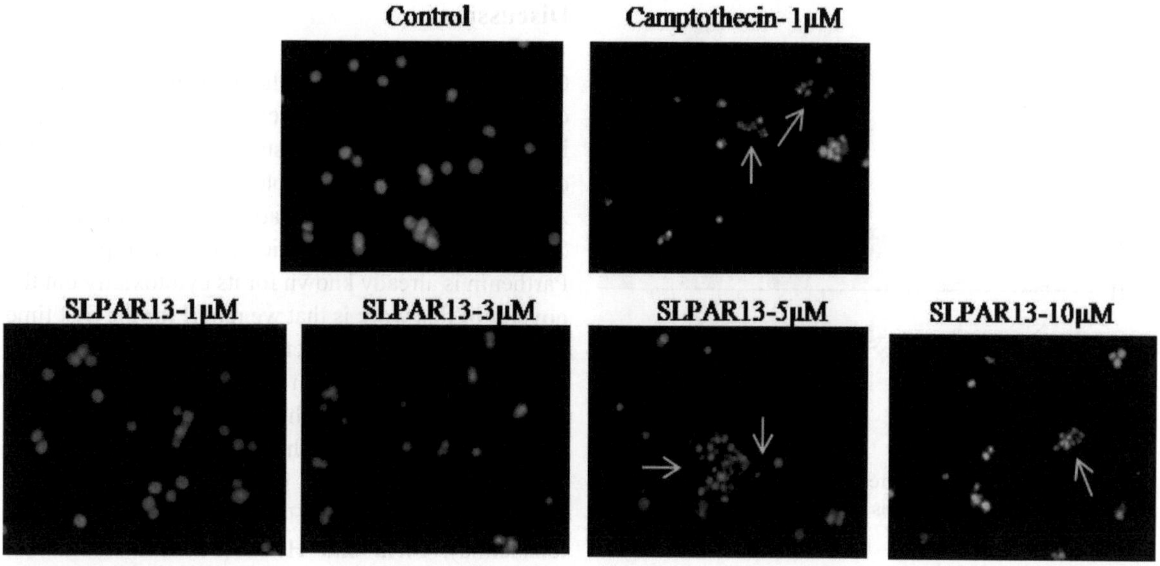

Fig. 3: SLPAR13 induced nuclear condensation stained by Hoechst dye. The nuclei of the control cells are round and uniform while the SLAPR13 treated cells, when stained with Hoecsht, exhibited condensed nuclei. The segregation and condensation of nuclei increased with increasing concentrations and was maximum at 10 µM concentration

Hoechst 33258 Staining of Cells for Nuclear Morphology

HL-60 cells (2×10^6 cells/3 ml/well) were treated with SLPAR 13 in a concentration dependent manner and incubated for 24 h. Cells were treated with Hoechst solution and spread on a clean slide and observed for any nuclear morphological alterations and apoptotic bodies under inverted fluorescence microscope (Olympus 1X70, magnification 60x) using UV excitation **[12]**.

Fragmentation of Genomic DNA

The genomic DNA was extracted from SLPAR 13 treated HL-60 cells. Cells (2×10^6/3 ml/well) after various treatments, incubated for 24h were centrifuged and processed for electrophoretic analysis as described earlier **[13]**.

Results

Cell Proliferation Assessment by MTT Assay

SLPAR13 inhibits cell proliferation in the three cancer cell lines namely HeLa, SiHa and HL-60 with the IC_{50} values 7.8, 9 and 0.7 µM respectively (Fig. 4). The IC_{50} of SLPAR13 was also calculated in normal cells hGF and was found out to be 14 µM.

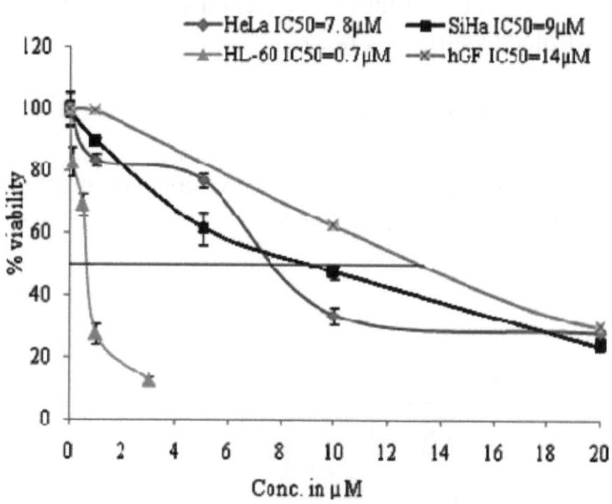

Fig. 4: The IC_{50} values of SLPAR13 were calculated by MTT assay against 3 different cancer cell lines HeLa, SiHa and HL-60 and one normal cell line hGF as described in materials and methods

DNA Content and Cell Cycle Phase Distribution

The peaks obtained by flowcytometry denoted arrest in cell cycle by SLPAR13 in a concentration dependent manner (Fig. 2). Higher concentration increased the extent of cell cycle arrest (Fig. 5).

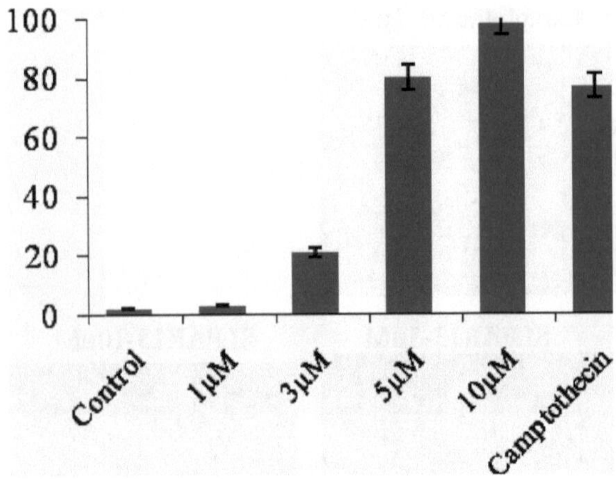

Fig. 5: Increasing cell cycle arrest activity of SLPAR13 against HL-60 cells. Camptothecin was taken as positive control

Hoechst 33258 Staining of Cells for Nuclear Morphology

Condensation of nuclei is observed in the treated HL-60 cells and the nuclear condensation increased with increasing concentration. The condensed nuclei are indicated by arrows (Fig.3).

Fragmentation of Genomic DNA

Fragments were obtained at 5 μM concentration of SLPAR13 as indicated (Fig. 6).

Fig. 6: Fragmentation of genomic DNA induced by SLPAR13. Other details are described in materials and methods

Discussion

Contemporary research in the anticancer drug development from plants has been focused on investigating the molecular mechanism by which an agent induces cytotoxicity and apoptosis in cancer cells **[13]**. A spiro-isoxazolidine derivative of parthenin namely SLPAR13 is a semi-synthetic derivative of parthenin. Parthenin is already known for its cytotoxicity but the novelty of this work is that we report for the first time the apoptotic inducing activity of a spiro-derivative of parthenin SLPAR13 in human leukemia and cervical cancer cell lines. The compound was tested in various models, one of them being the MTT assay, which induced cell death in three human cancer cell lines selectively namely HL-60 (acute promyelocytic leukaemia), SiHa and HeLa (cervical carcinoma) with various inhibitory concentrations. This indicated the potent cytotoxicity of the said compound against cancer cell lines at the same time being milder on normal cells. The IC_{50} of SLPAR13 in the hGF cells was found to be 14 μM. The therapeutic window was more than 10 times when the IC_{50} values of HL-60 and hGF were compared. The study demonstrated that SLPAR13 is a potential pro-apoptotic agent and hence can be developed into an important anti-cancer lead of therapeutic potential. This is evidenced from measurement of several biological end-points of the apoptosis such as appearance of apoptotic bodies, DNA fragmentation and increase in sub-G0DNA faction inHL-60 cells. The cell death was confirmed by cell cycle arrest exhibited by the test compound in a concentration dependent manner in HL-60 cells. The arrest marked the termination of series of events that takes place in a cell leading to its division and duplication (replication) which caused the cell death as a result of treatment of SLPAR13. The nuclear condensation and morphological changes induced by the test compound further marked the HL-60 cell death which was confirmed to be apoptosis by DNA ladder. The formation of fragments is hallmark of apoptosis due to the breaking of DNA strand into 180 bp fragments. This created a clear picture of apoptosis induced by SLAPR13 in the HL60 cells. Apoptotic cell death may involve intrinsic mitochondrial signaling pathway **[14, 15]** or extrinsic signaling cascade emanating through the activation of apical death receptors leading to caspase activation **[16]** and finally death of the cell. Successful drug treatment in human disease requires an adequate therapeutic index reflecting the treatment's specific effect on target cells and its lack of clinically significant toxic effect on the host **[17]**. Whatever the mechanisms involved, if the test compound induces apoptosis then that test compound may be the potential candidate for anti-cancer lead optimization. This study is expected to lead us to identify the active molecule that may have the potential for the treatment and management of cancer.

Conclusion

This study points towards the fact that natural compounds like parthenin and its halogenated derivatives induce death in human cancer cells. The mode of cell death was confirmed to be apoptosis which is a positive indication of these compounds being taken up for further studies as potential anticancer agents.

Acknowledgements

Arpita Saxena is a recipient of Indian Council for Medical Research, Senior Research Fellowship.

References

1. S.M. Kupchan, M.A. Eakin and A.M.J. Thomas; Med. Chem. 14 (1971) 1147–1152.
2. D. Mew, F. Balza, G.H.N. Towers and J.G. Levy; Plat. Med. 45 (1982) 23–27.
3. A.K. Picman and G.H.N. Towers; Bioch. Syst. Eco. 11 (1983) 321–327.
4. M. Hopper, G.C. Kirby, M.M. Kulkarni, S.N. Kulkarni, B.A. Nagasampagi, M.J. O'Neill, J.D. Philipson, S.R. Rojatkar, and D.C. Warhurs; Eur. J. Med. Chem. 25 (1990) 717–723.
5. M.S.K. Halmuthur, A.K. Saxena, S.C. Taneja, S.K. Singh, V.K. Sethi, S.D. Sawant, N.A. Qazi, M.R. Doma, A.H. Bandey, M. Verma,. and G.N. Qazi; US Patent (WO/2009/110007) (2009).
6. J.P. Begue and D.B. Delpon; J. Fl. Chem. 127 (2006) 992–1012.
7. W. Yeow, A. Baras, A. Chua, D. Nguyen, S.Sehgal, D. Schrump, D. Nguyen; *J. Tho. Card. Sur.* 132 (2006) 1356–1362.
8. F.M. Tadeusz, S.D. Doralyn, L.L. Sarah and P.S. Jonel; Nat. Rew. Dr. Dis. 8 (2009) 69–85.
9. D.M. Reddy, N.A. Qazi, S.D. Sawant, A.H. Bandey, J. Srinivas, M. Shankar, S.K. Singh, M. Verma, G. Chashoo, A. Saxena, D.M. Mondhe, A.K. Saxena, V.K.Sethi, S.C. Taneja, G.N. Qazi and H.M.S. Kumar; Eur. J. Med. Chem. (2011) (in press).
10. S. Bhushan, A. Kumar, F. Malik, S.S. Andotra, V.K. Sethi, I.P. Kaur, S.C. Taneja, G.N. Qazi, J. Singh; Apoptosis 12 (2007) 1911–1926.
11. A. Krishan; J. Cell Biol. 66 (1975) 188.
12. S. Bhushan, J. Singh, J.M. Rao, A.K. Saxena and G.N. Qazi; Nitric Oxide 14 (2006)72–88.
13. A. Kumar, F. Malik, S. Bhushan, V.K. Sethi, A.K Shahi, J. Kaur, S.C. Taneja, G.N. Qazi, J. Singh; Chem. Biol. Int. 171 (2008) 332–347.
14. S. Kumar; Cell Death Differ. 14 (2007) 32–43.
15. R.A. Kirkland, J.A. Windelborn, J.M. Kasprzak, J.L. Franklin; J. Neurosci. 22 (2002) 6480–90.
16. S. Nagata, P. Golstein; Science 267 (1995) 1449–56.
17. F. Bunz; Drug Discovery Today: Disease Mechanisms **2** (2005) 383–387.

4

In Vitro Antioxidant and Cytotoxicity Assay of *Pistia Stratiotes* L. Against B16F1 and B16F10 Melanoma Cell Lines

M. Jha[1], V. Sharma[2] and N. Ganesh[3]

[1,3] Department of Research, Jawaharlal Nehru Cancer Hospital & Research Center, Idgah Hills, Bhopal 462001 India.

[2] Department of Zoology, Dr. Hari Singh Gaur University, Sagar, M. P. India.

E mail:meghajhabtbpl@gmail.com

Abstract

In this study we investigated in vitro antioxidant activity and tumor growth inhibition by Pistia stratiotes L. on melanoma cell lines. The methanolic extract of Pistia stratiotes showed that percentage inhibition of DPPH increases with the increasing concentrations of test sample. The percentage inhibition of MEPS was (17.24–79.3 %) on DPPH against reference ascorbic acid (22.7–83.4 %) ranges (10–100 µg/ml). The effect of MEPS on the proliferation of B16F1 and B16F10 melanoma cell lines was determined by MTT and TBE bioassay. Among the two cell lines studied, the extract exhibited maximum anticancer activity with IC_{50} *(5.09). Structural elucidation of its bioactive principle is in progress.*

Introduction

Reactive oxygen species (ROS) capable of damaging DNA, proteins, carbohydrates and lipids are generated in aerobic organisms. These ROS include superoxide anion radical (O_2^-), hydrogen peroxide (H_2O_2), hydroxyl radical (OH^-), and single molecular oxygen.[1] Free radicals are associated with various physiological and pathological events such as inflammation, aging, mutagenicity and carcinogenicity. Cancer is one of the leading cause of the death worldwide. Among cancers melanoma is the most malignant skin cancer and its occurrence has remarkably increased during the past few decades due to increased UV-ray intensities and artificial skin tannings. [2] Melanoma now accounts for approximately 4 % of all cancers diagnosed in the United States. Studies on the pharmacological mechanisms and searching for chemical structures from herbal extract for new anticancer drug caught great interest. [3] Considering herbalism as an important strategy for cancer prevention, variety of animal experiments and cell lines culture have been carried out. [4]

Several research studies have demonstrated that herbal plants contain diverse classes of compounds such as steroids, polyphenols, alkaloids, tannins and carotenoids. [5] From the previous research it was found that *P. stratiotes L.* contains large amount of two di-C-glycosylflavones of the vicenin and lucenin and lesser amounts of the anthocyanin cyaniding-3-glucoside and a luteolin-7-glycoside, and traces of the mono-C-glycosyl flavones, vitexin and orientin. [6] With this background and abundant source of unique active components harbored in plant, the present study was taken up on this plant namely *Pistia stratiotes* belongs to the family Araceae. *P. stratiotes* is used in traditional medicine for its diuretic, antidiabetic, antidermetaphytic, antifungal and antimicrobial properties. [7] Research on relationships between antioxidants and prevention of non-communicable disease, such as cardiovascular disease, cancer and diabetes has been increasing sharply in recent year.[8] B16F10 murine melanoma cells have been widely used to elucidate the regulatory mechanisms of melanogenesis and pigment cell proliferation. B16F1 cell lines with melanin producing capability have an adherent growth patterns and fibroblast like morphology. Evidence for the utility of *in vitro* cytotoxicity tests has led many pharmaceuticals companies to screen compound libraries to remove potentially toxic compounds early

M.M. Srivastava, L. D. Khemani, S. Srivastava, *Chemistry of Phytopotentials: Health, Energy and Environmental Perspectives*, DOI:10.1007/978-3-642-23394-4_4,

in the drug discovery process. However, the property of this plant, especially its anticancer activity, has not yet been investigated. Therefore, this prompted us to investigate the inhibitory growth effect of this plant on two different melanoma cancer cell lines, B16F10 and B16F1.

Materials and Methods

Preparation of Plant Material

The *P. stratiotes* leaves were collected from upper lake, Bhopal (M.P.), India during the month of October. The collected plant material was dried under shade and then powdered with mechanical grinder. MeOH extract was prepared by macerating a powder with methanol/water (50/50, v/v) for 48 hr with constant stirring. Then it was filtered and the filtrate was evaporated in water bath at low temperature. The concentrated MeOH extract was then dried at 40°C in an oven and finally weighed.

Chemicals

(DPPH) 2, 2diphenyl-1 picrylhydrazyl-hydrate) reagent was purchased from Sigma chemical Co. Ascorbic acid were obtained from SD Fine Ltd, Baisar. All the other chemicals used were of analytical grade.

DPPH Assay[9]

The effect of methanolic extract of *P. stratiotes* (MEPS) leaves on DPPH radical was estimated using the method of Mensor et al. A solution of 0.3 mM DPPH in methanol was prepared. One ml of 0.3mM DPPH methanol solution was added to 2.5 ml of different dilutions of MEPS (10–100 µg/ml), and allowed to react at room temperature. After 30 min. the absorbance values were measured at 518 nm using UV-Spectrophotometer (VIS 260 Shimadzu, Japan). Methanol (2.5 ml) in DPPH solution (1 ml) was used as a control. Ascorbic acid was used as reference standard. The IC_{50} value is the concentrations of the sample required to scavenge 50 % DPPH free radical.

The percentage inhibition of DPPH assay was calculated using the formula-% Inhibition = $[(Abs_{(c)} - Abs_{(s)} / Abs_{(c)}) \times 100]$, where $Abs_{(c)}$ – Absorbance of blank, $Abs_{(s)}$ – Absorbance of sample.

In Vitro Antitumor Activity

Cell Lines and Culture

Melanoma cell line was obtained from National Cell Center of Science, Pune and maintained in Department of research, Jawaharlal Nehru Cancer Hospital and Research Center, Bhopal (M. P.). Cells were cultured in EMEM, supplemented with 10 %(v/v) fetal calf serum (FCS), 2 mM glutamine, streptomycin plus penicillin (100 µg/ml and 100 IU/ml, respectively). Cultures were maintained in a 5 % CO_2 humidified atmosphere at 37 °C until near confluence.

Determination of Inhibition of B16F10 and B16F1 Melanoma Cell Proliferation

Trypan Blue Exclusion Assay [10]

Cells (1×10^6/plate) were seeded in poly-l-lysine precoated tissue culture petri plates and allowed to adhere for 24 h in CO_2 incubator at 37 °C. The medium was replaced with incomplete EMEM medium containing dilution series of MEPS (10–100 µg/ml) again for 24 h in CO_2 incubator at 37 °C. 0.1 ml Trypan blue dye (0.4 % in water) was mixed with cell suspension, 15 min prior to completion of incubation period. At the end of incubation period, the petri plates were carefully taken out and 1.0 % Sodium dodecyl sulfate was added to each petri plates by pipetting up and down several times unless the contents get homogenized and the number of viable cells (not stained) counted using a hemocytometer. Viability was expressed as a percentage of control number of cells excluding Trypan blue dye. Although numbers of Trypan blue dye staining cells were not counted and it is recognized that these may be lost from the population relatively quickly.

Microculture Tetrazolium (MTT) Assay [11]

Cells (1×10^6/well) were seeded in poly-l-lysine precoated 96 well tissue culture plates and allowed to adhere for 24 h in CO_2 incubator at 37 °C. The medium was replaced with the serum free medium containing dilution series of MEPS (10–100 µg/ml) separately again for 24 h in CO_2 incubator at 37 °C. Tetrazolium bromide salt solution (10 µl/well) was added in cell suspension (100 µl), four hours prior to completion of incubation period. DMSO (200 µl) was added to each well and mixed the solution thoroughly to dissolve the crystals. Plate was placed in the dark for four hours at room temperature. The plates were kept on rocker shaker for 4 hr at room temperature and then read at 550 nm using Multiwell microplate reader (Synergy HT, Biotech, USA).

The average values were determined from triplicate readings and subtract from the average values of the blank. Percent of inhibition was calculated by using the formula: Percent of inhibition = (C – T)/C x 100, where C = Absorbance of control, T = Absorbance of Treatment.

Statistical Analysis

All experimental data were expresses in percent inhibition with respect to the control. The percentage inhibition was used to determine the IC_{50} values. The experiment was done in triplicate. The results are given as mean ±standard deviation. Significance of differences between the mean values was determined using student *t*-test. The IC_{50} value was calculated using probit analysis.

Results

DPPH Scavenging Activity of MEPS

Antioxidant react with DPPH, which is a nitrogen centered radical with a characteristics absorption at 518 nm and convert to 1, 1-diphenyl-2-picryl hydrazine due to its hydrogen accepting ability at a very rapid rate.

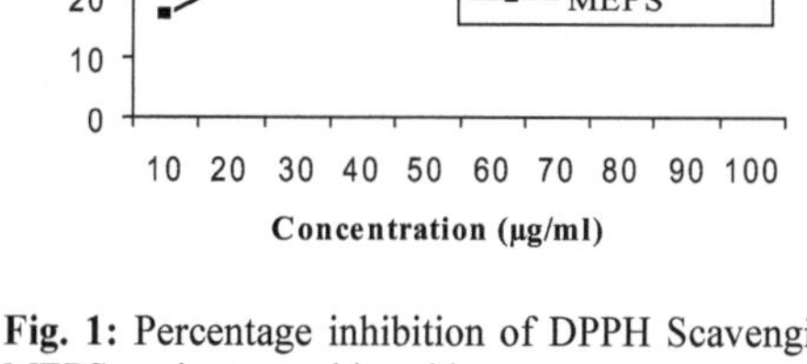

Fig. 1: Percentage inhibition of DPPH Scavenging Assay of MEPS against ascorbic acid

Since DPPH assay has been largely used as a quick, reliable, and reproducible parameter to search the *in vitro* general antioxidant activity of pure compounds as well as plant extracts. MEPS had significant scavenging effect on the DPPH radical which increased with increasing concentration in the 10–100 µg/ml range; the scavenging effect of MEPS was lower than that of Ascorbic acid. DPPH was reduced in the addition of the extract in concentration dependent manner. The MEPS indicated potencies of antioxidant by the discoloration of solution. The IC_{50} value of MEPS and ascorbic acid in DPPH radical scavenging activity was 5.74 µg/ml and 5.25 µg/ml.

Inhibitory Effect of MEPS on B16F1 and B16F10 Melanoma Cell Lines

Cytotoxicity activity of MEPS was screened against murine cell line B16F10 and B16F1 with ten increasing concentration (10–100 µg/ml) for 24hr first by the TBE and then followed by MTT bioassay. Percent inhibition of MEPS was calculated for B16F10 and B16F1 cell lines. The cytotoxicity of test sample varied with concentration level and the types of cell lines. The MEPS significantly inhibited the cell proliferation in a dose dependent manner in a range of 10–100 µg/ml Figure 2. The percentage of cytotoxicity observed shows an increasing pattern with increasing dosage. The maximum percent inhibition 83.3 % was achieved at 24hr exposure at the concentration level of 100 µg/ml by TBE assay while in MTT assay the growth of B16F1 cells was inhibited up to 85 % respectively at concentration level 100 µg/ml. Figure 3 indicate the noticeable percent inhibition of MEPS against B16F10 cell line by the TBE and MTT bioassay. Here also, in TBE assay the MEPS inhibit 65 % at 24 hr exposure at the concentration level 100 µg/ml. In MTT assay, the growth was inhibited up to 67.2 % at the same concentration. The percent inhibition for MEPS showed more pronounced efficacy against B16F1 compared to B16F10 cell lines. However, MEPS showed its best activity in the concentration level 100 µg/ml in B16F1 cell lines which was approximately similar to the activity of standard drug doxorubicin (Figure 4).

The IC_{50} values of MEPS calculated from MTT assay using probit analysis: B16F1 (5.09 µg/ml) and B16F10 (8.05 µg/ml). The regression constant and correlation coefficient were calculated for the MEPS (leaves) against B16F1 cell lines; Regression constant: (7.649x + 11.06) and Correlation coefficient (r): (0.986).

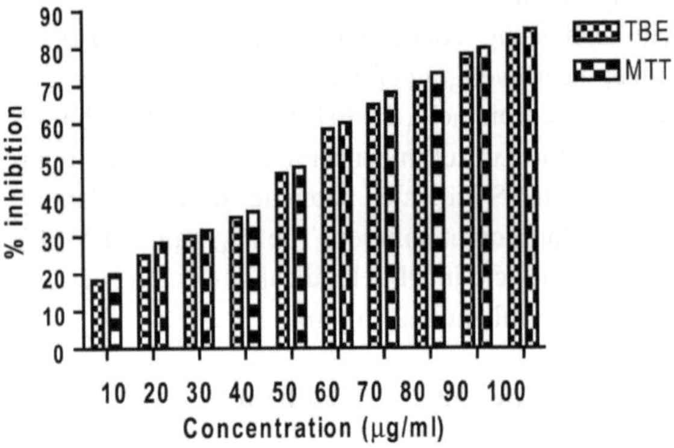

Fig. 2: Methanolic extract exhibited significant antiproliferative activity against cell line B16F1 showing maximum 83.3 % and 85.0 % inhibition in TBE and MTT bioassay at the concentration 100 µg/ml at 24 hrs. student t-test p>0.05

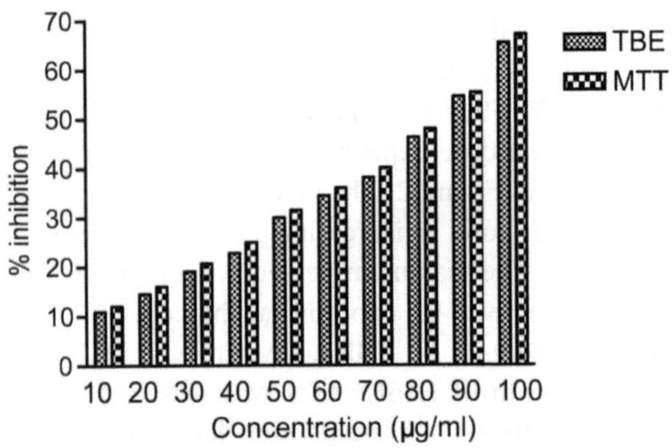

Fig. 3: Methanolic extract exhibited significant antiproliferative activity against cell line B16F10 showing maximum 65.4 % and 67.2 % inhibition in TBE and MTT bioassay at the concentration 100 µg/ml at 24 hrs. student t-test p>0.05

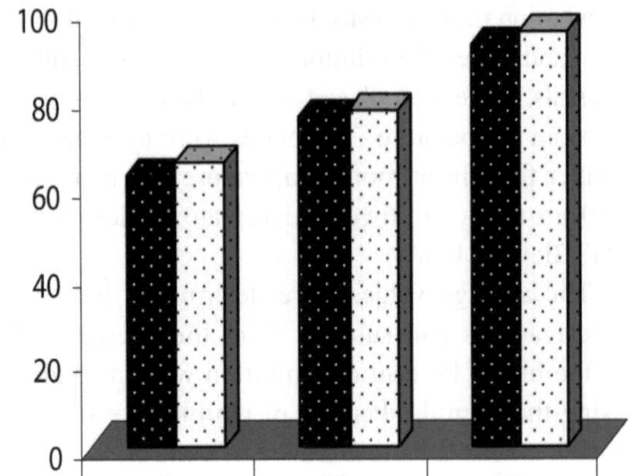

	5	10	15
■ TBE	62.11	75.68	92.16
□ MTT	65.68	77.67	95.58

Concentration (µg/ml)

Fig. 4: *In vitro* cytotoxicity of doxorubicin against Melanoma cell lines

Discussion

A positive correlation between the antioxidant potential and antitumor potential has been reported and shown that the high content of antioxidants is responsible for the inhibition of tumor cell proliferation. [12] The present study of the methanolic extract of *P. stratiotes* showed that the leaf possesses strong antiproliferative properties against the tested mouse tumor cell lines, and also showed antioxidant effects at certain concentrations. In this sense, new studies on this fraction are necessary for a better characterization of its possible biological application. Nowadays antioxidants have been at the centre of focus in chronic disease prevention research. The reduction of DPPH absorption is indicative of the capacity of the MEPS to scavenge free radicals, independently of any enzymatic activity and our results are in agreement with earlier reports on the ability of MEPS to scavenge free radicals and active oxygen species. [13] Therefore, we have evaluated the antioxidant potency through DPPH radical scavenging with the methanolic extract or ascorbic acid standard and the results indicated that the DPPH radical-scavenging activity of the extract enhanced with increasing concentration.

Melanoma is highly resistant to conventional chemotherapy; it is an invasive disease that shows preferential metastatis to the brain, lung, liver and skin. [14] Many naturally occurring agents have shown chemoprotective potential in a variety of bioassay systems and animal models. [15] MTT and TBE assay was used to study the antiproliferative activity of *P. stratiotes*. MTT is reduced to an insoluble purple formazan by mitochondrial dehydrogenase. Cell proliferative activity was measured by comparison of the purple color formation. Dead cells, on the other hand, did not form the purple formazan due to their lack of the enzyme. In TBE bioassay, dead cells uptake dye while the viable cells are excluded. The percent inhibition resulting from TBE and MTT assay demonstrated that MEPS is the efficient candidate as cytotoxic bioagent against these cell lines (B16F10 and B16F1). The cytotoxicity screening models provide important preliminary data to help select plant extracts with potential antitumoral properties for future studies. We demonstrate for the first time that MEPS has a strong dose- dependent antiproliferative activity on B16F1 and B16F10 cells as observed from the results of Trypan blue and MTT bioassay. This result is important because these cell lines are particularly resistant to cell death. In the study MEPS being potent, therefore it can be further use to study the time dependent (24–72hr) % inhibition at the dose of 10–100 μg/ml. The further study has to be extended for carrying out the *in-vivo* tumor potential of the MEPS extract using animal models in melanoma cancer. The efforts on the above lines are in progress.

Conclusion

In conclusion, our present *in vitro* study of the extract showed that MEPS possesses strong anti-proliferative effect against the tested melanoma cell lines, and also showed strong antioxidant potential through free radical scavenging ability in a concentration dependent manner. These observations also suggest that at least some of the flavonoids of this plant present in its methanolic extract are responsible for its anticancer property. More elaborative study in this plant with its pure compounds may lead to the development of natural antioxidant and alternative anticancer agent of clinical significance.

References

1. O.E. Ogunlana and O.O. Ogunlana; Research Journal of Agriculture and Biological Sciences 4 (2008) 666–671.
2. J.J. Sung, F.K. Chan, W.K. Leung, J.C. Wu, J.Y. Lau, J. Ching, K.F. To, Y.T. Lee, Y.W. Luk, N.N. Kung, S.P. Kwok, M.K. Li and S.C. Chung; Gastroenterology 124 (2003) 608–614.
3. A. Jemal, R. Siegel, E. Ward, T. Murray, J. Xu and M.J. Thun; Journal for Clinicians 57 (2007) 43–66.
4. G.M. Cragg and D.J. Newman; Cancer Lett 17 (1999) 153–62.
5. P. Nandi, G. Talukdar and A. Sharma; Pharmacotherapy 41 (1998) 53–55.
6. T.M. Zennie and J.W. McClure; Aquatic Botany 3 (1977) 49–54.
7. K.R. Kirtikar and B.D. Basu; Indian medicinal plants. Dehradun, Lalit Mohan Basu. 4 (1994) 2600–2602.
8. V. Patel, S. Shukla and S. Patel; Pharmacognosy Magazine. 5 (2009) 381–387.
9. L.L. Mensor, S.F. Menezes, G.G. Leitao, S.A. Reis, T.C. Santos dos, S.C. Coube and G. Leitao; Phytother. Re. 15 (2001) 127–130.
10. K.J. Pienta K.J. and J.E. Lehr; J. Ural. 49 (1993) 1622–1625.
11. S.E. Lee, E.M. Ju and J.H. Kim; Exp. Mol. Med. 34 (2002) 100–106.
12. W.Y. Li, W.W. Chan, D.J. Guo and P.H.F. Yu; Pharmaceu Biol 45 (2007) 541–546.
13. M. Jha, N. Ganesh and V. Sharma; International Journal of Chemtech Research 2 (2010) 180–84.
14. B. Gava, S. Zorzet, P. Spessotto, M. Cocchietto and G. Sava; J. Pharmacol. Exp. Ther. 317 (2006); 284–291.
15. C.L. Hsu, W.H. Lo and G.C. Yen.; J. Agric. Food Chem. 55 (2007) 7359–7365.

5
Synthesis, Characterization, Anti-Tumor and Anti-Microbial Activity of Fatty Acid Analogs of Propofol

A. Mohammad[1*], F. B. Faruqi[1] and J. Mustafa[2]
[1]Department of Applied Chemistry, Faculty of Engineering & Technology, Aligarh Muslim University, Aligarh, India
[2]Department of Pharmacognosy, King Saud University, Riyadh, Kingdom of Saudi Arabia
*Email: alimohammad08@gmail.com

Abstract

Derivatives of propofol (2, 6-diisopropylphenol) were prepared by coupling with 12-hydroxy-octadec-Z-9-enoic acid and Z-9-octadecenoic acid (oleic acid) with the C1-α-hydroxy function of propofol. Spectroscopic studies confirmed the formation of the desired product. The compounds were then investigated for its in-vitro anticancer activity against a panel of solid human tumor cell lines including human malignant melanoma, human leukemia cells. Their cytotoxicity was also determined against non-cancerous mammalian cells (VERO cells). The analogs were cytotoxic against all cancer cell lines whereas no effect was observed against normal cells. The compounds showed good antimicrobial activity against E. coli and S. albus.

Introduction

Synthesis and biological studies of short chain-length esters of propofol have been reported here. 3 to 8 μg/ml concentrations of propofol were reported to decrease the metastatic potential of human cancer cells, including HeLa, H71080, HOS and RPMI-7951 cells [1]. Siddiqui *et al.* [2] first reported the effect of two omega-3 fatty acids, combined with propofol on a breast cancer cell line in vitro. In view of the significance of long-chain FA in the treatment of cancer, we report here the synthesis and spectral studies of new propofol analogs containing two fatty acids12-hydroxy-octadec-Z-9-enoic acid and oleic acid along with their *in vitro* evaluation against a panel of human cancer cell lines including HeLa, SK-MEL, MCF-7 and HL-60 (human leukemia). Their cytotoxicity was also determined against non-cancerous mammalian cells (VERO cells). Ricinoleic acid (12-hydroxy-octadec-Z-9-enoic acid) is active component of castor oil (85 to 90%). (www.kristinasoil.com). Castor oil (Cremophor) is a chemomodulator and a MDR reversing agent used in anti-cancer drugs [3]. Oleic acid blocks the action of a cancer-causing oncogene called HER-2/neu which is found in about 30 percent of breast cancer patients. (www.oliveoilfarmer.com).

Experimental

Chemicals and Materials

A thin layer chromatographic applicator (Toshniwal, India), 20 × 3.5 cm glass plates and 24 × 6 cm glass jar were used for performing TLC. Silica Gel "G" (E. Merck, India) was used as a stationary phase. Petroleum ether and diethyl ether (1: 1, vol / vol) was used as a developing solvent. Reaction products on TLC plates were visualized by UV light and by exposure to iodine vapors. Column chromatographic separations were performed using silica gel "G" packing of particle size 60–120 mesh (petroleum ether/diethyl ether, 1: 1, v/v). ^{1}HNMR and ^{13}CNMR spectra were recorded on an Advance DRX-200 Bruker, (Switzerland) NMR Spectrometer. Mass spectra were obtained on a Jeol SX-102 (FAB) spectrometer (JEOL, Tokyo, Japan). FTIR Spectra were recorded in chloroform on a Spectrum RX-1 FTIR, Perkin Elmer Spectrometer. 2, 6-diisopropyl phenol, 4-dimethyl amino pyridine

M.M. Srivastava, L. D. Khemani, S. Srivastava, *Chemistry of Phytopotentials: Health, Energy and Environmental Perspectives*, DOI:10.1007/978–3-642–23394-4_5, © Springer-Verlag Berlin Heidelberg 2012

(DMAP) was procured from Acros chemicals. The coupling reagent-N, N-dicyclohexylcarbodiimide (DCC) was purchased from Fluka chemical corporation (New York). Oleic acid and – β-mercaptoethanol was purchased from Sigma Aldrich Chemicals and methylene chloride was purchased from CDH Chemicals (Mumbai, India). 12-Hydroxy-octadec-Z-9-enoic acid was isolated from *Ricinus communis* seed oil [4]. All solvents and reagents were of AR or HPLC-grade.

Synthesis of Compounds

Appropriate amounts of Fatty acid (1 mmol) and propofol (1 mmol) were dissolved in dry dichloromethane (5 mL), and DMAP (catalytic amount) was added to this solution. The reaction mixture was stirred at room temperature under nitrogen for 10 min before DCC (1 mmol) was added to it. The reaction mixture was allowed to stir at room temperature. Progress of reaction was monitored on TLC plates. Both coupling reactions showed the formation of single product and were completed in 12 h. The reaction mixture was filtered to remove solid dicyclohexylurea, and the filtrate was evaporated under reduced pressure at 20 °C. The semisolid mass was subjected to column chromatography (petroleum ether/diethyl ether, 1:1, v/v) on silica gel to purify the desired products.

Characterization of 12-Hydroxy-octadec-Z-9-enoic Acid (from Seed Oil of Ricinus Communis)

Viscous oil, $R_F = 0.2$, isolated yield, 95 %. *IR (CHCl$_3$, cm^{-1}):* 3418.0, 3013.0, 2930.4, 2858.4, 1710.8, 1640.0, 1460.5, 1216.9, 1104.4, 932.6, 763.5, and 668.8. MS-EI found [M+H] $^+$ 298.4638; $C_{18}H_{34}O_3$ [M+H] $^+$ requires 298.4659. *^{1}HNMR (CDCl$_3$, ∂H, ppm):* 0.89(t, J= 6 Hz, 3H of terminal –CH3), 3.39(s, 1H, CH-OH), 3.62(s, 1H, OH), 5.37(m, 1H, -CH=CH), 5.84(m, 1H, -CH=CH), 3.81–4.47(m, 4H), 2.13(m, 4H), 2.33–2.63(m, 6H), 1.31–1.83(m, 12H). *^{13}CNMR (CDCl$_3$, ∂c):*14.03, 22.53, 23.5, 24.6, 25.46, 27.14, 29.12, 31.83, 33.9, 35.86, 37.24, 42.8, 129.07, 130.63, and 179.46.

Spectral Studies of the Compound [1]-2, 6-diisopropyl-[1–12-hydroxy-octa-Z-9]-decenoate

Viscous oil, R_F = 0.5, (petroleum ether/diethyl ether, 1:1 v/v as a developer), isolated yield, 90 %. *IR (CHCl$_3$, cm^{-1}):* 3429.2, 3012.6, 2931.3, 2859.3, 1744.0, 1694.8, 1646.6, 1525.4, 1383.7, 1372, 1216, 1164, 1098.5, 930.7 and 754.7. MS-EI found [M+H] + 458.7287; $C_{30}H_{50}O_3$ [M+H] $^+$ requires 458.7297. *^{1}HNMR (CDCl$_3$, ∂H, ppm):* 0.88(t, J=6.4 Hz, 3H), 1.182(d, J=6.6 Hz, 6H), 1.27, 1.248(d, J=6.6 Hz, 6H), 2.615(m,6H), 2.40(m,3H), 2.92(m,2H), 3.20(m,2H), 3.615(m,1H), 3.90(m,2H), 4.22(m,2H), 5.38(m,1H), 5.673(m,1H) 6.894(m,1H), 7.042(d, J=7.5 Hz, 1H), 7.230(d, J=8.1 Hz, 1H), 1.308–2.14 (m, 12H). *^{13}CNMR (CDCl$_3$, ∂c):* 14.47, 22.72, 23.53, 24.95, 25.34, 26.29, 26.98, 27.44, 29.33, 31.46, 32.67, 34.10, 35.77, 37.35, 38.64, 49.65, 55.96, 68.08, 71.61, 73.60, 120.45, 123.30, 123.78, 126.34, 132.5, 149.97, 154.06, 172.34.

Spectral Studies of the Compound [2] – 2, 6-diisopropyl-[1-octa-Z-9]-decenoate

Viscous oil, R_F = 0.4, (petroleum ether/diethyl ether, 1:1 v/v as a developer), isolated yield, 90 %. *IR (CHCl$_3$, cm^{-1}):* 3012.6, 2931, 2859.3, 1744, 1628, 1645.2, 1512.1, 1362, 1243.6, 1160.4, 1089.6, 894.6, 752.6. MS-EI found [M+H] + 442.740; $C_{30}H_{50}O_2$ [M+H] $^+$ requires 442.747.*^{1}HNMR (CDCl$_3$, ∂H, ppm):*0.88(d, J=6.8 Hz, 6H), 1.194(d, J=6.8 Hz, 6H),1.22–1.55(m, 26H), 2.4(t, J=6.5 Hz, 3H), 2.615(t, J=6.5 Hz, 2H), 2.91(m, 1H), 3.197(m, 1H), 5.38(m, 1H, -CH=CH), 5.673(m, 1H, -CH=CH), 6.87(m, 1H), 7.06(d, J=6.8 Hz, 1H), 7.23(d, J=6.8 Hz, 1H).*^{13}CNMR (CDCl$_3$, ∂c):*22.72, 24.06, 24.95, 25.62, 26.32, 27.37, 28.75, 29.06, 34.02, 46.1, 53.39, 56, 114.1, 120.8, 123.3, 123.8, 126.34, 132.5, 140.2, 145.5, 150.0, 157.36, 172.32 and 173.6.

Assay for *in Vitro* Anti-cancer Activity

In vitro screening of new drug candidate against human cancer cell line panel is carried out and results are tabulated in Table 1. The assay is the same as we had done earlier [5] The number of viable cells was determined using modified Neutral Red assay [6] procedure. IC_{50} values were calculated from the dose curves generated by plotting % growth v/s the test concentration on a logarithmic scale.

Assay for Antimicrobial Activity

The *in vitro* antimicrobial activity was carried out against *E. coli, S. aureus* and *S. albus.* The assay is the same as we had done earlier [5]. To determine the zone of inhibition cup-plate method was employed [7].

Results and Discussion

After isolation of 12-Hydroxy-octadec-Z-9-enoic acid from seed oil of *Ricinus communis* it was characterized by various spectroscopic techniques. The IR spectra of the compound revealed strong absorption bands at 1710.8 cm^{-1} and 1216.9 cm^{-1} corresponding to C=O and C-O bonds respectively, indicating the presence of carbonyl carbon, showing carbon signal at ∂_C 179.46. Presence of hydroxyl group was confirmed by absorption band at 3418.0 cm^{-1} and its respective carbon signal appeared at ∂_C 71.77 with ∂_H 3.625ppm (s, 1H). IR spectra revealed a sharp band at 1640.0 cm^{-1} indicating the presence of double bond which is further related to chemical shifts at ∂_H 5.37(m, 1H) and 5.84(m, 1H) ppm for the two olefin protons 9H and 10H respectively with ∂_C 129.07 and 130.63. The bands at 2930.4 and 2858.4 cm^{-1} correspond to the aliphatic CH bonds. Some significant signals appeared at ∂_H 0.89(t, J= 6 Hz, 3H of terminal CH_3 group), 3.39(s, 1H, CH-OH) and 1.31–1.83(m, 12H) for the rest of the fatty acid chain length. The efficient synthesis of fatty acid conjugates of propofol is shown in (**Scheme 1**). DCC/DMAP was used to esterify the 1α-hydroxy group of propofol with the carboxylic acid.

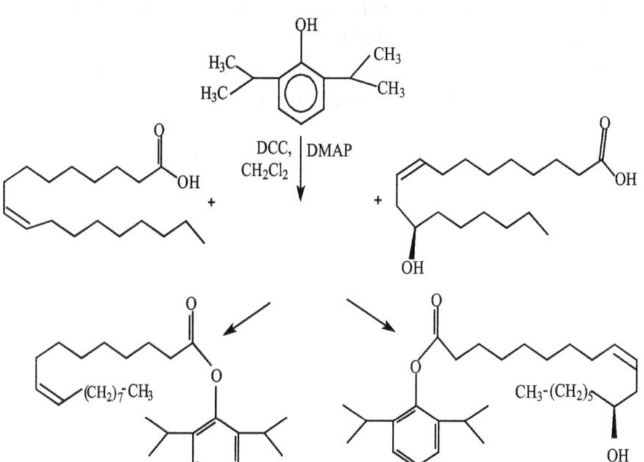

[2] [1]

(Scheme 1)

The IR spectrum of the compound [1] revealed broad, strong absorption bands at 1744.0 cm^{-1} and 1216 cm^{-1}which are attributable to C=O and C-O bonds ,respectively, and indicate the presence of an ester with their respective carbon signal at ∂_C 172.32. A strong band at 3429.2 cm^{-1} indicate the presence of hydroxyl group with ∂_H 3.615ppm, and its respective carbon signal appeared at ∂_C 71.61. The band at 3012.6 cm^{-1} is characteristic of an aromatic C-H (propofol) and the band at 2931.3 and 2859.3 cm^{-1} is characteristic of aliphatic C-H bonds. A distinct band at 1646.6 cm^{-1} shows the presence of alkene. The two olefin protons, 9'H and 10'H were observed at ∂_H 5.38ppm and 5.673 ppm and correlated with observations at ∂_C 126.34 and 132.5 respectively. The chemical shifts for aromatic protons are moved downfield at ∂_H 6.894 (m, 1H), 7.067 (d, J= 7.5 Hz, 1H), 7.230 (d, J=8.1 Hz, 1H) and their respective carbon signals appeared at ∂_C 120.45, 123.30, 123.78. For 12 protons of the two isopropyl groups, two doublets were observed at ∂_H 1.182(d, J=6.6 Hz, 6H) and 1.248 (d, J=6.6 Hz, 6H) and their respective carbon signals appeared at ∂_C 23.53 and 24.95. The broad and strong absorption bands at 1744 cm^{-1} and 1243.6 cm^{-1} of the compound [2] are attributable to C=0, C-O bands respectively, that indicate the presence of ester group with their respective carbon signals at ∂_C 172.32 and 173.6. The band at 3012.6 cm^{-1} is characteristic of an aromatic C-H and the bands at 2931.0 cm^{-1} and 2859.3 cm^{-1} for aliphatic C-H bonds. A distinct band at 1628 cm^{-1} show the presence of C=C of alkenes. The two olefin protons (terminal alkenes), 9'H and 10'H were observed at ∂_H 5.38 (m, 1H) and 5.673 (m, 1H) which are correlated with ∂_C 126.34 and 132.5 respectively. No O-H band was seen, indicating the absence of nonesterified propofol. The chemical shifts for three aromatic protons are moved downfield at ∂_H 6.87 (m, 1H), 7.06 (d, J= 6.8 Hz, 1H), 7.23 (d, J= 6.8 Hz, 1H) and their carbon signals appeared at 120.8, 123.3 and 123.53 ∂_C values. For 12 protons of the two isopropyl groups, two doublets were observed at ∂_H 0.88(d, J= 6.8 Hz, 6H) and 1.194 (d, J= 6.8 Hz, 6H) and their respective carbon signals appeared at ∂_C 24.06, 24.95.

The compounds were examined for their *in vitro* cytotoxicity against a panel of solid human tumor cell lines. Its cytotoxicity was also determined against non-cancerous mammalian cells (VERO cells) for comparison. The compounds [1] and [2] were cytotoxic against all cancer cell lines where as no effect was observed against normal cells (VERO cells) up to the highest concentration of 15 μM in the assay, thus demonstrating selectivity towards the tumor cells. The cytotoxic potency of compounds is expressed in terms of IC_{50} values as shown in Table 1. The significantly higher anti-cancer activity of [1] is attributed to the presence of a methylene interrupted 12-hydroxy and Z-9- monounsaturation in its C-18 fatty acid moiety. Compound [2] also show significant cytotoxicity against all the cancer cell lines especially HL-60 (human leukemia) because of Z-monounsaturation, but its anti-cancer activity is slightly lesser than that of [1], because of the presence of hydroxyl group in [1]- as it has been described earlier that ω-hydroxy and hydroxy fatty acids are potent anti-cancer agents.

Table 1: Anti-cancer activity of compounds

Compound[a]	Cell lines[b] (IC_{50}, μM) MCF-7	HeLa	HL-60	VERO
1	0.42	0.22	0.26	NA
2	0.54	0.30	0.24	NA

[a]The highest concentration tested was 15μM. [b]NA, not active; HeLa, Human cervical epitheloid carcinoma; MCF-7, Human breast adenocarcinoma; HL-60, Human leukemia; VERO, monkey kidney fibroblasts.

The compounds were also screened for their antimicrobial activity against *E. coli, S. aureus* and *S. albus.* Both of the compounds [1] and [2] exhibit significant antimicrobial activity against *E. coli* and *S. albus* while remain not active against S. aureus. Results of anti-microbial screening are reported in **Table 2.**

Table 2: Anti-microbial activity of compounds

	Diameter of zone of inhibition		
Compound	E.coli	S. aureus	S. albus
1	12mm	NA	11mm
2	15mm	NA	9mm
Chloromycetin	22mm	18mm	20mm

DMF used as the control; concentration = 100μg/ml of DMF; NA, not active.

Conclusion

These results suggest that the novel propofol-fatty acid conjugates reported here may be useful for the treatment of cancer as all of them show significant cytotoxicity against a panel of human solid tumor cell lines. Interestingly, none of them showed any cytotoxicity to normal cells. This feature places these products into the class of anticancer agents that possess selectivity toward cancer cells over normal cells. The conjugates also showed significant anti-microbial activity against *E. coli* and *S. albus.*

References

1. T. Mammoto, M. Mukai, A. Mammoto, Y. Yamanaka, Y. Hayashi, T. Mashimo, Y. Kishi and H. Nakamura; Cancer Lett. 184 (2002) 165.
2. R.A. Siddiqui, M. Zerouga, M. Wu, A. Castillo, K. Harvey, G.P. Zaloga and W. Stillwell; Breast Cancer Res. 7 (2005) 645.
3. D.D. Ross, P.J. Wooten, Y. Tong, B. Cornblatt, C. Levy, R. Sridhara, E.J. Lee and C.A. Schiffer; Blood. 83 (1994) 1337.
4. F.D. Gunstone; J. Chem. Soc. (1954) 1611.
5. A. Mohammad, F.B. Faruqi and J. Mustafa; J. Adv. Sci. Res. 1 (2010) 12.
6. E. Borenfreund, H. Babich and N. Martin-Alguacil; *In vitro* Dev. Cell. Biol. 26 (1990) 1030.
7. A.L. Barry, P.D. Hoeprich and M.A. Saubolle. Eds. 4th. LBS, Lea & Febiger, Philadelphia, (1976) p. 180–193.

6
Screening of Antioxidant Activity of Plant Extracts

H. Singh[1], R. Raturi[1], S. C. Sati[2], M. D. Sati[2] and P. P. Badoni[1]
[1]Department of Chemistry, HNB Garhwal Central University Campus Pauri Garhwal, India
[2]Department of Chemistry, HNB Garhwal Central University, Srinagar, Garhwal, India
Email: harpreetsngh08@gmail.com
Phone: +91–9997456808

Abstract

In the present study in vitro antioxidant activities of Salix babylonica and Triumfetta pillosa were carried out by using scavenging activity of DPPH (1,1 diphenyl-2-picrylhydrozyl) radical method. The plant extract showed remarkable antioxidant activity.

Introduction

In the past few years, there has been growing interest in the reactive oxygen species (ROS) due to their involvement in several pathological situations [1]. Reactive oxygen species (ROS) include, superoxide anion (O-2) and alkoxyl (RO.) radical, nitricoxide (NO), hydrogen peroxide (H2O2), peroxyl radical (ROO.) and hypochloride (HOCl). Superoxide anion radical(O-2) and hydrogen peroxide (H2O2) can interact in the presence of certain transition metal ions to yield a highly reactive oxidizing species, the hydroxyl radical (OH.) [2]. The oxidation induced by ROS may result in cell membrane disintegration, membrane protein damage and DNA mutations which play an important role in aging and can further initiate or propagate the development of many diseases, such as arteriosclerosis, cancer, diabetes mellitus, liver injury, inflammation, skin damages, coronary heart diseases and arthritis [3–4]. Although, the body possesses such defense mechanisms, as enzymes and antioxidant nutrients that arrest damaging properties of ROS [5] however, their prolonged exposure may lead to irreversible oxidative damage [4]. Therefore, antioxidants with free radical scavenging activities may have great relevance in the preservation and therapeutics of diseases involving oxidants or free radicals [6]. The antioxidants serve as a defensive factor against free radicals in the body. Enzymes such as superoxide dismutase, catalase and glutathione peroxidase are the main enzymes that oppose oxidation. If the free radical production becomes more than the capacity of enzymatic system to cope up with, then the second line of defense (vitamins) may come torescue. Vitamin A and C quench free radicals by oxidizing and inactivating them. The polyphenolic compounds commonly found in plants, mushrooms, and fungi have been reported to have multiple biological effects such as anti-inflammatory, antiarteriosclerotic, antitumor, antimutagenic, anticarcinogenic, antibacterial and cardioprotective actions including antioxidant activity [7]. *Salix babylonica* belongs to family salicaceae is a sub-deciduous or evergreen tree upto 15 meter high. Branches glaborous, drooping, leaves narrowly lanceolate commonly found along moist places and often cultivated as an ornamental tree [8]. *Triumfetta pillosa* belongs to family tiliaceae is an annual or perennial herb upto 2 meter high. Petals yellow, narrowly lanceolate, obtuse. Capsules tomentose, subglobose, commonly found in open waste places, forest edges and field terraces. The fruit juice of the plant is applied on cuts, its fruit infusion is given to women to facilitate delivery [9].The ethanolic extract of the rhizome of the plant showed significant antifungal activity. The ethanolic extract of the roots were analyzed for anticandidala activity [10].

Plant Material and Extract Preparation

The Plant materials of *Salix babylonica* and *Triumfetta pillosa* were collected from Bharsar, Pauri Garhwal, Uttrakhand, India in the month of August 2009 and identified from Taxonomy Laboratory, Department

M.M. Srivastava, L. D. Khemani, S. Srivastava, *Chemistry of Phytopotentials: Health, Energy and Environmental Perspectives*, DOI:10.1007/978–3-642–23394-4_6, © Springer-Verlag Berlin Heidelberg 2012

of Botany, H.N.B. Garhwal University Srinagar. A voucher specimens (GUH-8388, for *Salix babylonica* and GUH-8874, for *Triumfetta pillosa*) of the plants have been kept in the Departmental Herbarium for future records.

Determination of Antioxidant Activity

In order to measure antioxidant activity DPPH free radical scavenging assay was used. This assay measures the free radical scavenging capacity of the extract under investigation. DPPH is a molecule containing a stable free radical. In the presence of an antioxidant, which can donate an electron to DPPH, the purple color which is typical for free radical decays and the absorbance was measured at 517 nm using a double beam UV-VIS spectrophotometer [11]. The ethanolic extracts of the plants were re-dissolved in ethanol and various concentration (10, 20, 50 and 100 µg/ml) of extracts were used. The assay mixture contained in total volume of 1 ml, 500µl of extract, 125µl prepared DPPH and 375 µl solvent (methanol). After 30 min of incubation at 25°C, the decrease in absorbance was measured at 517 nm on spectrophotometer. The radical scavenging activity (RSA) was calculated as a percentage of DPPH using a discoloration using then equation

$$\% \text{ RSA} = [(A_0 - A_s)/A_0] \text{ x}100$$

Where A_0 and A_s are the absorbance of control and test sample respectively

Results and Discussion

The DPPH radical has been widely used to test the potential of compounds as free radical scavengers of hydrogen donor and to investigate the antioxidant activity of plant extracts [12]. The ethanolic extract of plants showed an effective free radical scavenging in DPPH (2, 2 diphenyl-1-picryl hydrazyl) assay (Table 1 and 2).

Table 1 and Table 2: Antioxidant activity of *Salix babylonica* and *Triumfetta pillosa* on DPPH free radical

Table 1

Concentration (µg/ml)	DPPH Free radical Scavenging activity (%)
10	10.12
20	21.9
50	46.67
100	88.00

Table 2

Concentration (µg/ml)	DPPH Free radical Scavenging activity (%)
50	3.12
100	6.15
150	10.69
200	23.27

The extract of *Salix babylonica* exhibited antioxidant effect at low concentration. When the extract of the plant was tested for DPPH radical scavenging activity, it was found that 50 µg/ml and 100 µg/ml of the extract lowered the DPPH radical levels above 46.7 % and 88 % respectively. Inhabitation of DPPH radicals 50 % considered as significant antioxidant properties of any compound [13]. The extract of plant *Triumfetta pillosa* also showed antioxidant property but at higher concentration which was found to be 10.69 % and 23.27 % at the concentration of 150 µg/ml and 200 µg/ml respectively. The results obtained in this study that the plant extract of *Salix babylonica* showed remarkable antioxidant activity in comparison to *Triumfetta pillosa* on DPPH free radical.

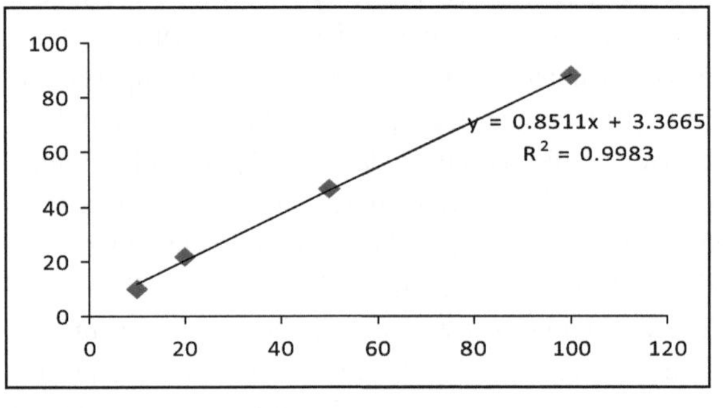

Fig. 1

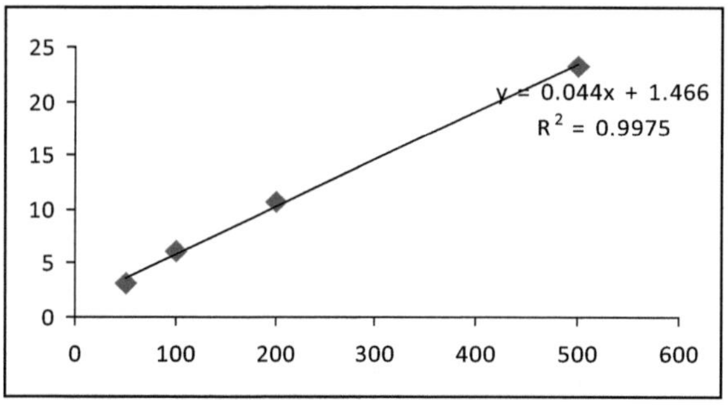

Fig. 2

Fig. 1 and Fig. 2: Antioxidant activity of *Salix babylonica* and *Triumfetta pillosa*

References

1. T.Y. Song and G.C. Yen; J. Agri. Foodchem. 50 (2002) 3322.
2. P. Valentao, E. Fernandes, F. Carvalho, P.B. Andrade, R.M. Seabra and M.L. Bastos; Biol. Pharm. Bull. 25 (2002) 1320.
3. B.S. Moon, I.J. Ryoo, B.S. Yun, K.S. Bae, K.D. Lee, I.D. Yoo and J.P. Kim; J. Antibiot., 59 (2006) 735.
4. O.P. Tiwari and Y. Tripathi; Food Chem., 100 (2007) 1170.
5. P. Valentao, E. Fernandes, F. Carvalho, P.B. Andrade, R.M. Seabra and M.L. Bastos; Biol. Pharm. Bull. 25 (2002) 1320.
6. G.R. Zhao, Z.J. Xiang, T.X. Ye, J.Y. Yaun and X.Z. Guo; Food Chem. 99 (2006) 767.
7. M.J. Miller, A.T. Diplock and C.A. Rice-Evans; J. Agric. Food Chem. 43 (1995) 1794.
8. R.D. Gaur; (1999) Flora of District Garhwal; Transmedia, Srinagar, Garhwal, 186.
9. R.D. Gaur, (1999) Flora of District Garhwal; Transmedia, Srinagar, Garhwal, 217.
10. W.W. Brand, H.E. Cuvelier and C. Berset; Food Sci. Technology 82 (1995) 25.
11. C.D. Hufford, S. Liu and A.M. Clark; J. Natural Prod. 51 (1988) 94.
12. C.D. Porto, S. Calligaris, E. Celloti and M.C. Nicoli; J. Agric. Food Chem. 48 (2000) 4241.
13. C. Sanchez-Moreno, J.A. Larrauri and F. Saura-Calixto; (1998). Bishen Singh Mahendra Pal Singh Publication, New Delhi, 123.

7
Andrographolide: A Renoprotective Diterpene from *Andrographis Paniculata (Burm. f.) Nees*

P. Singh[1], M. M. Srivastava[1], D. K. Hazra[2] and L. D. Khemani[1]
[1]Department of Chemistry, Dayalbagh Educational Institute (Deemed University), Agra, India.
[2]Nuclear Medicine Unit, S. N. Medical College, Agra, India.
Email: singh.prati@gmail.com

Abstract

Andrographis paniculata Nees, a well-known plant of Indian and Chinese traditional system of medicines, has drawn attention of researchers in recent times. An attempt was made to isolate the active principle from the plant using chromatographic methods. Andrographolide isolated from Andrographis paniculata (burm. f.) Nees (Acanthaceae) ameliorated the diabetes induced renal failure when treated for 28 days. The results demonstrated that andrographolide showed alleviation in terms of serum creatinine (54.73 %), serum urea (63.92 %), urinary proteins (32.35 %) at a dose of 90 µg/ml. The structure of the isolated compound was confirmed by various spectroscopic methods.

Introduction

Diabetic nephropathy is a common problem that is most likely to occur in patients who have worse glycemic control, hypertension, glomerular hyperfiltration or a genetic predisposition. The lifetime risk of nephropathy is roughly equivalent in type 1 and type 2 diabetes [1]. Diabetic nephropathy is one of the major complications of non-insulin dependent diabetes mellitus (NIDDM) which is a common cause of death in diabetic patients. The severity of renal disease in diabetic patients correlates with the levels of blood urea and serum creatinine [2]. Diabetic nephropathy accounts for considerably morbidity and mortality even in patients with well controlled blood sugar values [3]. Insulin therapy and oral hypoglycemic agents offer effective glycemic control; yet, their shortcomings limit their usage [4].

Plants are reputed in the indigenous systems of medicine for the treatment of various diseases [5]; the available literature shows that there are more than 800 plant species showing hypoglycemic activity [6]. The world health organization has also recommended the evaluation of the effectiveness of plants in conditions where we lack safe modern drugs [7]. Phytochemicals isolated from plant sources are used for the prevention and treatment of cancer, heart disease, diabetes mellitus and high blood pressure [8].

The hunt for complementary and alternative medicine is an ongoing process in the area of renal failure research. *Andrographis paniculata (Burm. f.) Nees* (Acanthaceae), a renowned plant in South-Asian traditional medicine is an established antidiabetic herb; hence attracts attention towards exploring the possible anti renal failure properties, especially against diabetic nephropathy. Andrographis paniculata is reported as a cold property herb in TCM and is used to get rid of body heat and to expel toxins. The plant is particularly known for its extremely bitter properties (often called king of bitters) and is used traditionally as a remedy against common cold, dysentery, fever, tonsillitis, diarrhea, liver diseases, inflammation, herpes, etc. [9–11]. The traditional uses and pharmacological aspects of A. paniculata have been well-documented in an extensive review recently [12]. A number of active principles are reported from the plant, which mainly include diterpene lactones, flavonoids and polyphenols [13–14]. However, the prime constituent andrographolide has been is mainly attributed for its therapeutic properties. Diterpenoid lactone andrographolide ($C_{20}H_{30}O_5$) is the principle compound found in A. paniculata, which is mainly concentrated in leaves and can be easily iso-

M.M. Srivastava, L. D. Khemani, S. Srivastava, *Chemistry of Phytopotentials: Health, Energy and Environmental Perspectives*, DOI:10.1007/978–3-642–23394-4_7,

lated from the crude plant extracts as crystalline solid [15–16]. The structure of the compound has been elucidated by X-Ray crystallographic analysis and the molecular stereochemistry, bond distances, bond angles, etc. all were determined [17]. Chemically designated as (3-[2-[decahydro-6-hydroxy-5-(hydroxymethyl) -5, 8-adimethyl-2-methylene-1-napthalenyl] ethylidene] dihydro-4-hydroxy-2(3H)-furanone),andrographolide (Fig. 1) exhibits extraordinarily vast range of biological activities [18–23]. In recent past, the compound is reported for its anti-tumor, anti-HIV and Cardioprotective properties [24–30]. In view of present literature and our previous experiments [31–32], an attempt was made to evaluate the effectiveness of isolated active principle 'Andrographolide' from *Andrographis paniculata* against stz-induced diabetes and renal damage in rats by studying the effect on blood urea and serum creatinine and urinary proteins levels.

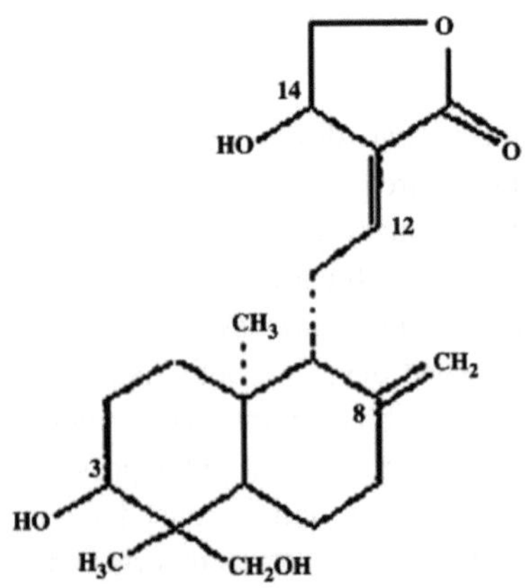

Fig. 1: Structure of the principle phytochemical compound of andrographolide

Materials and Methods

Plant Material

Plants were procured from NBRI, Lucknow (Accession No. AP-D02, Herbarium Voucher No. 445785) for the study. The fresh plants were washed quickly with the water to remove any foreign matter, shade dried and powdered and stored in airtight containers.

Isolation of Andrographolide

Andrographis paniculata (300 g) were macerated and extracted with methanol. The solvent was concentrated *in vacuo* to yield methanol extract (30 g), which was diluted with aqua distillate, and then partitioned by ethyl acetate, from which 10 g residue was obtained. The ethyl acetic fraction was subjected to silica column chromatography and gradiently eluted with chloroform methanol to afford andrographolide. The andrographolide crystal was recrystalized from hot methanol. Its IR, ^{1}H-NMR, ^{13}C-NMR and MS data were in accordance with those of andrographolide [33] and standard (Sigma, USA). Isolated andrographolide equivalent to dose 90 µg/ml was administered to rats.

Animals

Male Wistar albino rats weighing 250±50 g were housed in polypropylene cages and maintained at 24 ± 2 °C under 12 hour light/dark and 60 ± 5 % humidity. They were fed with Amrut Laboratory Animal Feed, manufactured by Nav Maharashtra Chakan Oil Mills Ltd., Pune, India. Water was provided *ad libitum*. The animals were acclimatized for a week under laboratory conditions. All experiments were performed according to the norms of the local ethical committee.

Experimental Design

Experimental animals were distributed randomly, in eight groups, containing six animals each.

a. Normal animals
Group I received vehicle only, throughout the experimental period.

b. Streptozotocin induced diabetic animals
Diabetes was induced in rats deprived of water for 24 hr followed by a single intraperitoneal injection of streptozotocin (STZ, 50 mg/kg) [34] dissolved in freshly prepared 0.01 M sodium citrate buffer, pH 4.5 [35] and the rats were left to develop diabetic nephropathy for a week. A week after the STZ injection, blood glucose levels were determined in blood samples collected from the tail vein, using Glucometer (Accu-Chek, one touch ultra). Animals were considered to be diabetic if they had elevated plasma glucose

concentrations > 250 mg/dl [36]. Then, these diabetic rats, referred to as STZ rats, were considered to show diabetic nephropathy if their blood urea and creatinine values were elevated. These animals further divided into two groups. Group II received vehicle only and group III received andrographolide, daily at a dose of 90 µg/ml (p. o.), respectively, for 28 consecutive days. Renoprotective activity against streptozotocin induced diabetic nephropathy, was screened in terms of alleviation in serum creatinine and serum urea.

Sample Collection

Individual rats belonging to different groups were placed in metabolic cages over a period of 24 h and urine was collected. At the end of 24 hours, rats were anesthetized with a combination of ketamine (60 mg/kg) and xylazine (5 mg/kg) given intraperitoneally. Blood samples were collected via retro orbital puncture in plain plastic tubes, left to stand at 4 °C for 1 hour, and centrifuged (900 × g for 15 min at 5 °C) to separate serum. The serum obtained was stored at – 5 °C until analysis.

Biochemical Analysis

Plasma and urine samples were assayed using standard diagnostic kits, viz. serum creatinine (Human, Germany), serum urea (Beacon Diagnostics, India) and urinary protein (ERBA Diagnostics, Germany).

Statistical Analysis

All values were expressed as mean ± standard error. Differences within groups were evaluated by paired *t*-test. One-way analysis of variance was used for examining differences among groups. Inter-group comparisons were made with Dunnet's multiple-comparison test. A p-value of < 0.05 was considered to indicate significance.

Results and Discussion

A. paniculata is cited in Ayurveda as a plant with diuretic, antioxidant and antidiabetic properties. Our previous study indicating renoprotective activity of *Andrographis paniculata* in gentamicin induced acute renal failure, thereby suggests this study as a complementary research for its anti renal failure effects against streptozotocin induced diabetic nephropathy.

Andrographolide (Figure 1) showed a significant alleviations in diabetes induced renal failure. Data presented in table 1 demonstrate that the isolated compound 'andrographolide' showed amelioration in terms of serum creatinine (54.73 %), serum urea (63.92 %), and urinary proteins (32.35 %) at a dose of 90 µg/ml. The isolated compound also diminished the blood glucose level. The currently available drug regimens for management of diabetes mellitus have certain drawbacks and therefore, there is a need to find safer and more effective antidiabetic drugs [36–38]. Diabetes mellitus of long duration is associated with several complications such as atherosclerosis, myocardial infarction, nephropathy etc. These complications have long been assumed to be related to chronically elevated glucose level in blood [39].

Table 1: Change in renal profile after the treatment of diabetic nephropathy rats with andrographolide at a dose of 90 µg/ml

Groups/Bio-markers	Andrographolide	
	0th Day	28th Day
Serum		
Glucose (mg/dl)	314.22±0.12	112.14±0.44 (–64.31)
Creatinine (mg/dl)	2.43±0.11	1.10±0.02 (–54.73)
Urea (mg/dl)	89.02±3.21	32.12±1.14 (–63.92)
Urine		
Proteins (mg/dl/day)	10.54±1.56	7.13±0.08 (–32.35)

Values are mean ± SE of 6 rats.
Figures in parenthesis indicate percent change with respect to 0 day.

p<0.0001 (28th day is compared with 0th day).

Renal disease is one of the most common and severe complications of diabetes. In diabetes mellitus, increased blood glucose, lipids and oxygen free radicals can induce glomerulosclerosis and chronic tubulointerstitial damage in the kidneys leading to diabetic nephropathy [39–43]. A progressive decline in the glomerular filtration rate due to loss of functioning nephrons and histological renal damage are common characteristics in the development of diabetic nephropathy [44].

Seven days after streptozotocin injection, the serum urea concentrations significantly elevated ($p<0.01$) in STZ induced diabetic nephropathy models. There were significant elevations in serum creatinine and urea levels when compared to that of the normal control animals, indicating impaired renal function of diabetic animals. In renal disease, the serum urea accumulates resulting in uremia, as the rate of urea production exceeds the rate of clearance [45]. Urea is the principal end product of protein catabolism and accumulates with renal failure. *A. paniculata* treatment showed a significant amelioration in all the renal biomarkers evaluated in the study, in a solvent-dependent manner. These results indicate that propolis can attenuate renal damage in diabetic rats. *A. paniculata* has a strong antioxidant and free radical scavenging effect [46]. This finding suggests that *A. paniculata* may improve the disturbed metabolism associated with diabetes. In this study, serum levels of glucose, was significantly elevated in diabetic rats compared to that of the normal control rats.

The present piece of work demonstrates that MeOH extract of the plant *A. paniculata* exhibited renoprotective activity against diabetic nephropathy. The chromatographic fractionation of the MeOH extract to its derived compound a diterpene lactone 'Andrographolide' has resulted into the amelioration with an effective concentration of 90 µg/ml and hence makes the plant metabolite interesting for further research.

References

1. E. Ritz, and S.R. Orth; The New England J. Med. 341 (1999) 1127.
2. G.P. Dubey, S.P. Dixit, and A. Singh; Indian J. Pharmacol. 26 (1994) 225.
3. Grenfel A: Clinical features and management of established diabetic nephropathy. In: *Textbook of Diabetes,* Vol. 2. Pickup JC, Williams G. Eds. Black Well Scientific Publications, London, (1991) p 677.
4. K. Anuradha, D. Hota, and P. Pandhi; Indian J. Exp. Biol. 42 (2004) 368.
5. R.O. Arise, S.O. Malomo, J.O. Adebayo, and A. Igunnu; J. Med. Plant Res. (2009) 77.
6. R.J. Marles, and N.R. Farnsworth; Phytomed. 2 (1995) 137.
7. World Health Organization. Second report of the WHO expert committee on diabetes mellitus. Technical Report Series (1980).
8. M.E. Waltner-Law, X.L. Wang, and B.K. Law; J. Biol. Chem. 277 (2002) 34933.
9. C. Patarapanich, S. Laungcholatan, N. Mahaverawat, et al; Thai. J. Pharm. Sci. 31 (2007) 91.
10. A. Panossian, T. Davtyan, N. Gukassyan, et al; Phytomed. 9 (2002) 598.
11. S.K. Mishra, N.S. Sangwan, and R.S. Sangwan; Pharmacog. Rev. 1 (2007) 283.
12. K. Jarukamjorn, and N. Nemoto; J. Health Sci. 54 (2008) 370.
13. Y.K. Rao, G. Vimalamma, C.V. Rao, et al; Phytochem. 65 (2004) 2317.
14. W. Li, X. Xu, H. Zhang, et al; Chem. Pharm. Bull. 55 (2007) 455.
15. M. Rajani, N. Shrivastava, and M.N. Ravishankara; Pharm. Biol. 38 (2000) 204.
16. L. Lomlim, N. Jirayupong, and A. Plubrukarn; Chem. Pharm. Bull. 51 (2003) 24.
17. A.B. Smith III, B.H. Toder, P.J. Carroll, et al; J. Crystall. Spec. Res. 12 (1982) 309.
18. X. Suo, H. Zhang, and Y. Wang; Biomed. Chromatogr. 21 (2007) 730.
19. K. Maiti, A. Gantait, K. Mukherjee, et al; J. Nat. Remed. 6 (2006) 1.
20. A.K. Gupta, and N. Tandon. Reviews on Indian Medicinal Plants. New Delhi: Indian Council of Medical Research, (2004).
21. B.C. Yu, C.R. Hung, W.C. Chen, J.T. Cheng; Planta Med. 69 (2003) 1075.
22. Y. Shen, C. Chen, and W. Chiou; Br. J. Pharmacol. 135 (2002) 399.
23. I.M. Liu, and J.T. Cheng; eCAM (2008) doi:10.1093/ecam/nen078.
24. A.Y.H. Woo, M.M.Y. Waye, S.K.W. Tsui, et al; J. Pharmacol. Exp. Ther. 325 (2008) 226.
25. F. Zhao, E.Q. He, L. Wang, et al; J. Asian Nat. Prod. Res. 10 (2008) 473.
26. A.P. Raina, A. Kumar, and S.K. Pareek. Ind J. Pharm. Sci. 69 (2007) 473.
27. S.R. Jada, G.S. Subur, C. Mattews, et al; Phytochem. 68 (2007) 904.
28. S. Rajagopal, R.A. Kumar, D.S. Deevi, et al; J. Exp. Ther. Oncol. 3 (2003) 147.
29. C. Calabrese, S.H. Berman, J.G. Babish, et al; Phytother. Res. 14 (2000) 333.
30. Y. Wang, J. Wang, Q. Fan, et al; Cell. Res. 17 (2007) 933.
31. P. Singh, M.M. Srivastava, and L.D. Khemani; Upsala J. Med. Sc. 11 (2009) 136.
32. P. Singh, M.M. Srivastava, and L.D. Khemani; Arch. Appl. Sc. Res. 1 (2009) 67.
33. T. Matsuda, M. Kuroyanagi, S. Sugiyama, et al; Chem. Pharm. Bull. 42 (1994) 1216.
34. J. Singh, S. Budhiraja, H. Lal, and B. Arora; Iranian J. Pharmacol. Therapeut. 5 (2006) 135.
35. G.H. Tesh, and T.J. Allen; Nephrol. 12 (2007) 261.

36. Y.J. Chen, and J. Quilley; J. Pharmacol. Exp. Therapeut. 324 (2008) 658.
37. J.K. Grover, S. Yadav, and V. Vats; J. Ethnopharmacol. 81 (2002) 81.
38. K. Rajagopal, and K. Sasikala; Singapore Med. J. 49 (2008) 137.
39. F.J. Alarcon-Aguilara, M. Jimenez-Estrada, R. Reyes-Chilpa, and R. Roman-Ramos; J. Ethnopharmacol. 72 (2000) 21.
40. X. Yin, Y. Zhang, H. Wu, et al; J. Pharmacol. Sc. 95 (2004) 256.
41. Y. Masumi, K. Hitoshi, K. Kouhei, and I. Mikio. Biol. Pharmaceut. Bull. 28 (2005) 2080.
42. P. Montilla, M. Barcos, M. Munoz, I. Castaneda, and I. Tunez; J. Biochem. Mol. Biol. 38 (2005) 539.
43. H. Okutan, N. Ozcelik, H.R. Yilmaz, and E. Uz; Clin. Biochem. 38 (2005) 191.
44. N. Yamabe, T. Yokozawa, T. Oya, and M. Kim; J. Pharmacol. Exp. Therapeut. 319 (2006) 228.
45. A.A. Adeneye, J.A. Olagunju, A.S. Benebo, et al; International J. Appl. Res. Nat. Prod. 1 (2008) 6.
46. B.L. Valadares, U. Graf, and A.M.A. Span; Food Chem. Toxicol. 46 (2008) 1103.

8

Enhanced Production of Antihypertensive Drug Ajmalicine in Transformed Hairy Root Culture of *Catharanthus Roseus* by Application of Stress Factors in Statistically Optimized Medium

D. Thakore[1], A. K. Srivastava[2] and A. Sinha[3]
[1, 2] Department of Biochemical Engineering and Biotechnology, Indian Institute of Technology Delhi, Hauz Khas, New Delhi-110016,
[3] Scientist, National Institute of Plant Genome Research, New Delhi
Email: dharathakore@gmail.com

Abstract

Ajmalicine is an important antihypertensive obtained from Catharanthus roseus. Transformed hairy roots were generated for the plant which had the ability to accumulate increased concentrations of ajmaciline. These roots can serve as the parent cell line for the economic mass in vitro production of ajmaciline, if cultivated in the appropriate media and right bioreactor configuration. The medium for hairy root cultivation was statistically optimized and thereafter the roots were subjected to different stress conditions primarily to influence the final metabolite accumulation. The objective of the present study was to investigate the effect of five stress factors mannitol, sodium chloride, potassium chloride, cadmium chloride and PVP polyvinyl pyrrolidone K-30 on growth and ajmalicine accumulation under in-vitro mass hairy root cultivation of Catharanthus roseus. Ajmalicine accumulation in actively growing hairy root was increased by addition of PVP to 2.53 mg/gDW(182 % increase than blank control) and by KCl to 4.09 mg/gDW(227 % increase). The maximum secretion of ajmalcine in the medium for mannitol, cadmium chloride, PVP and NaCl was 5.4 mg/l, 1.74 mg/l, 2.192 mg/l and 2.02 mg/l respectively as opposed to 1.32 mg/l in blank control. The mechanisms responsible for these stress effects are discussed herein.

Introduction

Plant hairy root culture has been a matter of active interest due to their distinct ability to grow actively and indefinitely in phytohormone free medium producing secondary metabolites at levels comparable to or even better than that of intact plants. *Catharanthus roseus* hairy root cultures have been widely investigated for the pharmaceutically important alkaloids. In the present investigation transformed hairy root line was used which had a distinct ability to overproduce antihypertensive ajmalicine. Mass scale bio-processing requires the understanding of the influence of external factors on the accumulation of secondary metabolites. One such yield enhancement strategy could be application of stress factors so that the content and/or rate of final metabolite can be enhanced. The present work focuses on the effect of osmotic salt and chemical stresses on *Catharanthus roseus* hairy root cultures cultivated in statistically optimized Gamborg's B5/2 medium.

Materials and Method

1. Maintenance of cell line and media optimization: The hairy roots of *Catharanthus roseus* (L.) G. Don cv. Prabal [4] were maintained in 40 ml of the independently statistically optimized media. Trial experimental protocols for concentration optimization were formulated using Design-Expert version 5.0.9 software (Stat-Ease Corporation, USA) using Gamborg's B5/2 medium (pH 5.8) at 25°C under 16/8 hours illumination conditions and 70 rpm. The optimal values of the selected key nutrients sucrose, potassium nitrate and sodium phosphate

M.M. Srivastava, L.D. Khemani, S. Srivastava, *Chemistry of Phytopotentials: Health, Energy and Environmental Perspectives*, DOI:10.1007/978-3-642-23394-4_8, © Springer-Verlag Berlin Heidelberg 2012

being 16 g/l, 2790 mg/l and 37.5 mg/L respectively, the rest of the components were kept similar to Gamborg's B5/2 medium.

2. Kinetics of growth: Knowledge about the growth kinetics of the hairy root culture in the optimized medium was a prerequisite for optimization of the various physical and chemical process parameters to achieve maximum ajmalicine volumetric productivity. Hence the hairy roots were cultivated in 250 ml Erlenmeyer flasks rotating on a gyratory shaker at 70 rpm containing 40 ml of the optimized medium. Temperature and pH were maintained at 25 °C and 5.8 respectively. The hairy root inoculum used for the experiment was 0.55 g/l of 15 days old culture on dry weight basis. The experiment was conducted in duplicate under 16/8 h L/D illumination conditions and the flasks (containing roots) were harvested at regular intervals for the analysis of biomass substrate and ajmaciline.
3. Analytical methods: Ajmalicine was extracted from the roots and quantified by High performance Liquid Chromatography[3]. The cell growth index was calculated by the formula: relative growth index = cell dry weighttreatment/cell dry weight-blank control. Cell viability rate was tested by TTC method. [2]
4. Addition of stress factors: With an aim to improve ajmalicine production several stress factors were applied. All chemicals used in the experiment were analytical grade. For salt stress treatment 2,3,4,6 g/l of NaCl and 3,4 and 5 g/l of KCl was used. For osmotic shock treatment,100mM, 200 mM, 400 mM and 600 mM mannitol, 0.1 %, 0.2 %, 0.4 %, 0.6 %, 0.8 % (w/v) PVP (polyvinyl pyrrolidone K-30 (PVP, MW: 40 000)) were used. Cadmium chloride concentrations used were 1,10,100,1000 µm. All the above trials were conducted in the statistically optimized medium and the pH was adjusted to 5.8 before autoclaving. The controls received the same amount of water.

Results and Discussions

1. Growth curve: After optimization of the medium the growth kinetics was established in shake flasks as shown in Fig 1. Each time point on the growth curve was an average of two replicates initiated with the same inocula from 15 day old cultures. After 30 days of cultivation the maximum biomass obtained was 42.71 g/l (fresh weight) equivalent to 5.025 g/l of dry weight. The stationery phase was reached at around 25th day. It was observed that after 30 days there was no further increase in growth. This may be due to the decrease in the limiting substrate by day 30.

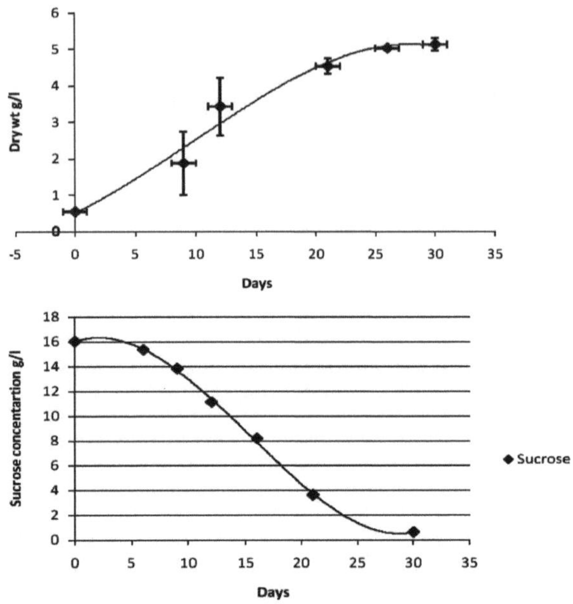

Fig. 1: Kinetics of growth

2. Osmotic Stress: The osmotic stress treatment by exposure to mannitol resulted in a maximum ajmalicine content of 1.27 mg/gDW at 400mM .The ajmalicine secreted in the medium was higher than the control for all treatments with the maximum being 5.4 mg/l at 200mM (Fig 2a.). However the root viability (Fig 2b.) as well as biomass concentration (Fig 2a.) decreased with the increase in the mannitol concentration. The viability was found to be 78 % for 100mM and decreased to 15.4 % for 600mM.
3. PVP treatment: As shown in Fig 3a. the PVP treatment did not result in any improvement in biomass concentration but the highest ajmalicine content in the roots (2.53 mg/gDW) was observed at 0.2 % w/v ajmalicine which is 182 % higher than the control. At 0.6 % w/v no ajmalicine was detected. The release of ajmalicine was stimulated at all concentrations with the highest being 2.192 mg/l at 0.4 %w/v; this yield was 66 % higher than the blank control. The viability of the roots was more or less same between 40–50 %, higher than that for the salt stress treatment though less than the control. (Fig 3b.)

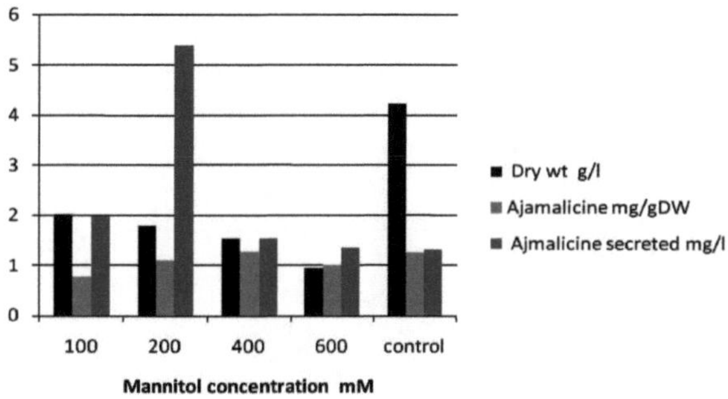

Fig. 2a: Effect of different concentrations of mannitol on growth and ajmalicine accumulation and secretion. Treatment was given on the day of inoculation and the roots were exposed to stress for 15 days

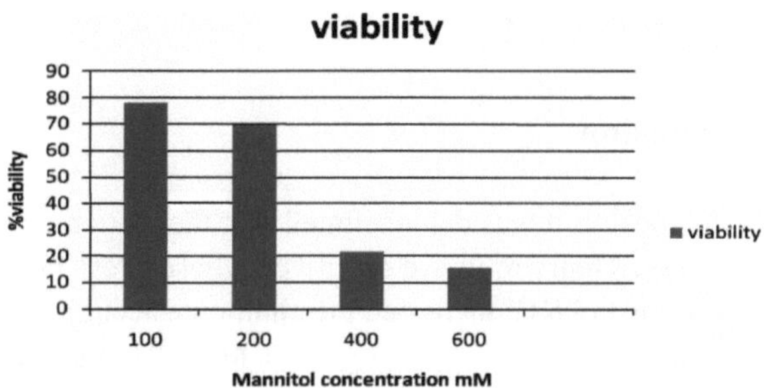

Fig. 2b: Effect of different concentrations of mannitol on viability. Treatment was given on the day of inoculation and the roots were exposed to stress for 15 days

4. Salt Stress: The hairy roots were exposed to different concentrations of NaCl and KCl (Fig 4a.b.). It was observed that for KCl treated hairy roots the ajmalicine concentration was higher at all concentrations with the maximum being 4.09 mg/gDW at 4 g/l as compared to the control for which the concentration was 1.25 mg/g DW (227 % increase). The maximum ajmalicine secreted was 1.33 mg/l at 5g/l KCl treatment. However for the case of NaCl the maximum ajmalicine content accumulated in the roots was 0.98mg/gDW at 4g/l whereas the maximum ajmalicine obtained in the medium was 2.02 mg/l at 6g/l NaCl, 53% higher than the control. The NaCl treatment resulted in less ajmalicine in the roots than KCl treatments

   The viability saw a decrease from 24.6 % to 18.2 % for KCl. The viability was same for all NaCl treatments. Reports on *Catharanthus roseus* cell cultures have shown similar observations wherein the ajmalcine content in the cells for NaCl treatment was more or less same as that in the control but the ajmalcine accumulated in the medium was higher. Also the report suggests that though the cell cultures became yellowish but the viability decreased for both salts.[3]

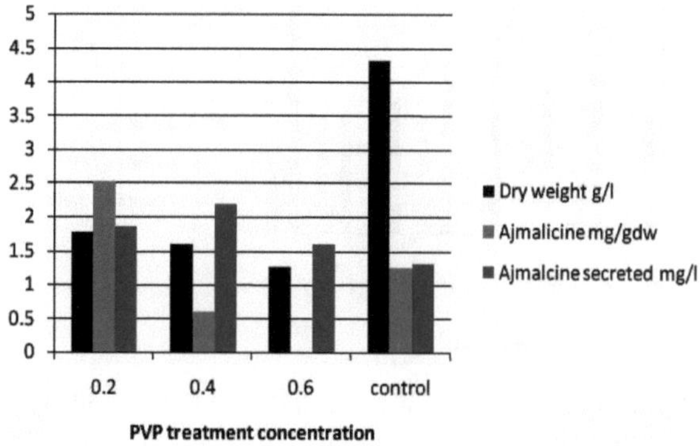

Fig. 3a: Effect of different concentrations of PVP on growth ajmalicine content. Treatment was given on the day of inoculation and the roots were exposed to stress for 15 days

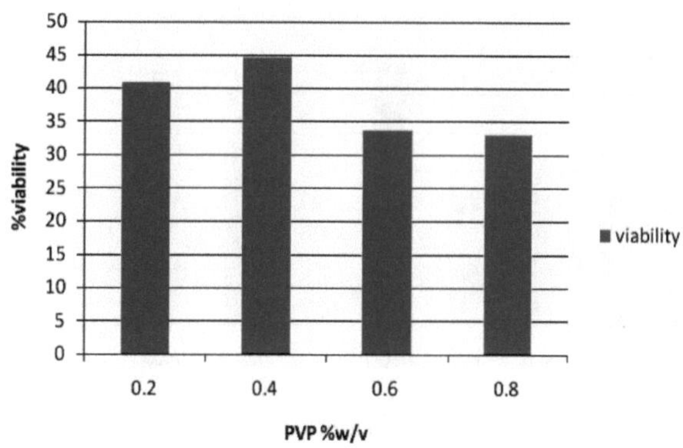

Fig. 3b: Effect of different concentrations of PVP on viability. Treatment was given on the day of inoculation and the roots were exposed to stress for 15 days

5. Cadmium Chloride Treatment: It was observed that the root viability was low at all concentrations (Fig 5b). The ajmalicine secreted in the medium increased with the increase in concentration of cadmium chloride (Fig 5a). The maximum ajmalicine secreted was 1.74 mg/l. Cadmium chloride treated cultures showed browning and the condition increased with increase in concentration with decreased viability which decreases with 20.2 % to 13.3 %. (Fig 5a.). This may be due to the exposure time to stress. Recent reports for *Catharanthus roseus* hairy roots and cell cultures of different ages (lag phase, mid exponential and stationery phase) treated with cadmium chloride suggest that treatment at the day of inoculation can lead to cell lysis whereas inoculating

during the early stationery phase enhances both the biomass and ajmalicine [1,5].

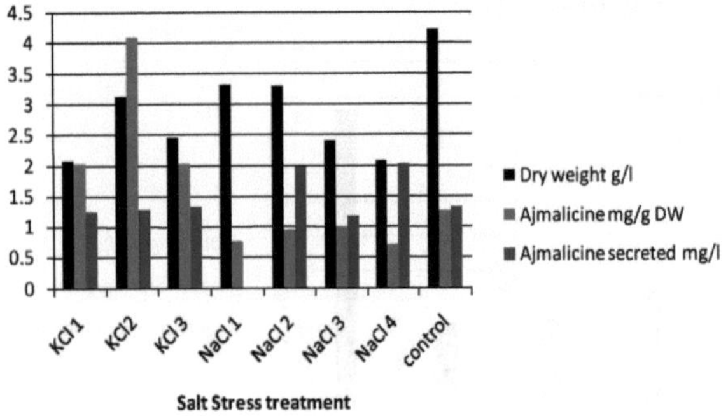

Fig. 4a: Effect of different concentrations of NaCl and KCl on growth ajmalicine accumulation. Treatment was given on the day of inoculation and the roots were exposed to stress for 15 days. KCl1, KCl2, KCl3 represents 3,4,6 g/l of KCl and NaCl 1, NaCl 2, NaCl 3, NaCl 4 represents 2,3,4,5 g/l of NaCl

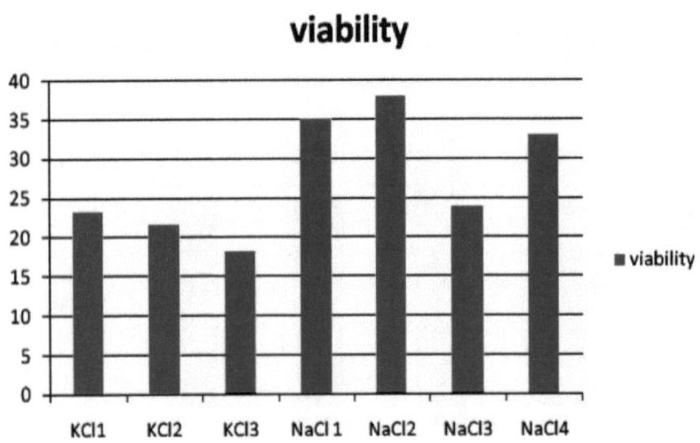

Fig. 4b: Effect of different concentrations of NaCl and KCl on growth ajmalicine accumulation and viability in hairy roots. Treatment was given on the day of inoculation and the roots were exposed to stress for 15 days. KCl1, KCl2, KCl3 represents 3,4,6 g/l of KCland NaCl1, NaCl2, NaCl3, NaCl4 represents 2,3,4,5 g/l of NaCl

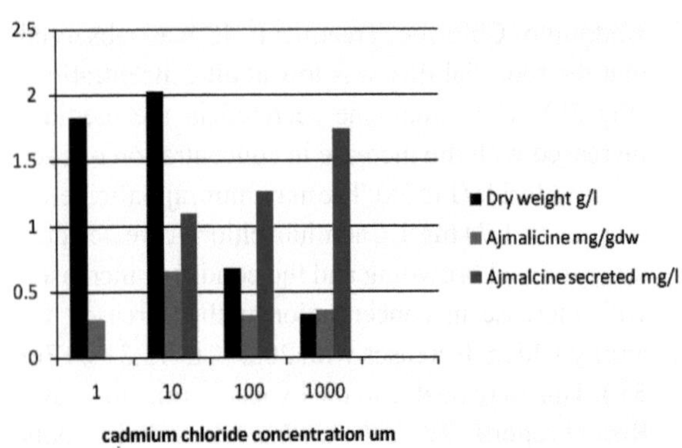

Fig. 5a: Effect of different concentrations of Cadmium Chloride on viability .Treatment was given on the day of inoculation and the roots were exposed to stress for15 days

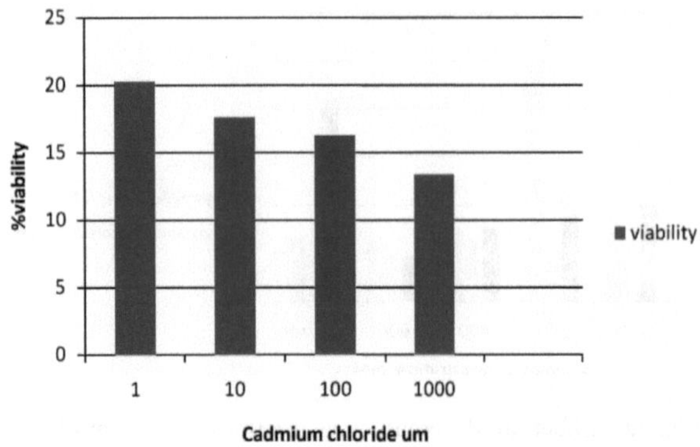

Fig. 5b: Effect of different concentrations of Cadmium Chloride on viability .Treatment was given on the day of inoculation and the roots were exposed to stress for15 days

Conclusion

In conclusion it was demonstrated that the *Catharanthus roseus* hairy roots are sensitive to these stress factors. PVP and KCl increased the ajmalcine accumulation in the roots to 2.53 mg/gDW and 4.09 mg/gDW respectively which was 182% and 227% increase with respect to additions of 0.2% w/v PVP and 4 g/l KCl respectively. Mannitol, NaCl and PVP strongly affected the release of ajmalicine in the medium with maximum being 5.4 mg/l, 2.02 mg/l and 2.192 mg/l respectively. Cadmium chloride treatment was found to be strongly dependent on the age of culture and time of exposure. The viability of the roots was low for salt stress treatment maximum being 24.6% for KCl and 35% for NaCl. The maximum viability for addition of stress factors, mannitol and PVP was 78% and 40.8% respectively which eventually decreased with increase in concentration(s).

References

1. A.Oscar, Moreno-Valenzuela Y. Minero-Garc'ıa, W. Chan, E. M. Geraldo, M. Victor, Loyola-Vargas, Biotechnol. Lett. 25 (2003)1345.
2. B.Ivano, R.Markus.Tree Physiology 23(2003) 257.
3. D.Singh, S. Pandey, S. Srivastava, S. K. Rai, R. Mishra S. Kumar, .J of AOAC International. 87 (2001) 1287.
4. J. Batra, A. Dutta, D. Singh, S. Kumar, J. Sen, Plant Cell Rep. 23 (2004) 148.
5. Z .Zheng, M. Wu, Plant Science, 106 (2004) 507.

9
Antioxidant Activity of Combined Extract of Some Medicinal Plants of Indian Origin

H. Ali[1] and S. Dixit[2]
[1]Department of Chemistry, Research scholar, MANIT, Bhopal
[2]Department of Chemistry, HOD, MANIT, Bhopal
Email: sana_soni26@yahoo.co.in, savitadixit1@yahoo.com

Abstract

Azdirachta indica, Tinospora cordifolia, Ocimum sactum, Triticum aestivum, Aloe barbandesis are Indian medicinal plants and has several medicinal properties. Antioxidants are essential substances which possess the ability to protect the body from damaged caused by free radical induced oxidative stress. A variety of free radical scavenging antioxidants exist with in the body which many of them derived from nutritional sources like food, vegetables. In this study the antioxidant activity of acetone and methanol (30:70) combine extract of selected plant materials, traditionally used by Indian population was evaluated against 2,2-diphenyl-1-picrylhydrazyl radical. This study suggests the possible mechanism of antioxidant activity due to the presence of flavonoids in each plant extract.

Introduction

About 80% of the world population depends absolutely on plants for their health and healing. Whereas in the developed world, confidence on surgery and pharmaceutical medicine is more usual but in the recent years, more and more people are complementing their treatment with natural supplements [1]. Furthermore, motivation of people towards herbs are increasing due to their concern about the side effect of drugs, those are prepared from synthetic materials. The people want to concern their own health rather than merely submitting themselves to impersonal health care system. Many botanical and some common dietary supplements are good sources of antioxidants compounds [2]. Plant materials containing phenolic constituents are increasingly of interest as they retard oxidative degradation of lipids and thereby improving quality and nutritional value of food [3, 4]. The importance of the antioxidant constituents of plant material is the maintained of health and protection from cancer [5]. They are very important substances that possess the ability to protect the body from damage caused by free radical induced oxidative stress [6, 7]. Several researches on the phenolic constituent and antioxidant activities in various plants have been conducted [8]. In the meantime, Galvez et al. [9] pointed out that there was a correlation between antioxidant capacity and phenolic content.

Recently, natural have received much attention as source of biological active substances including antioxidants. Numerous studies have been carried out on some plants, vegetables, fruits because they are rich sources of antioxidants such as vitamin C, vitamin A, vitamin E, carotenoids, polyphenolic compounds and flavonoids [10] which prevent free radical damage, reducing risk of various diseases. Thus the consumption of antioxidants from these resources is beneficial in preventing disease. The search for newer antioxidant compounds, especially of plant origin has ever since increased.

Materials and Methods

Collection of Plant Material

Fresh leaves of *Azadirachta indica, Ocimum sactum, Aloe barbandesis*, Stem of *Tinospora cordifolia*, Whole grass of *Triticum aestivum* were collected from their proper origin.

M.M. Srivastava, L.D. Khemani, S. Srivastava, *Chemistry of Phytopotentials: Health, Energy and Environmental Perspectives*, DOI:10.1007/978–3-642–23394-4_9, © Springer-Verlag Berlin Heidelberg 2012

Extraction of Plant Material

The plant materials (leaves of *Azadirachta indica, Ocimum sactum, Aloe barbandesis*, stem of *Tinospora cordifolia*, Whole grass of *Triticum aestivum* were air-dried at room temperature (26°C) for 2 weeks, after which it was grinded to a uniform powder. The mixture of acetone and methanol (30:70) extract were prepared by using 75 g each of the dry powdered plant materials in Soxhlet apparatus at 40°C for 48 h. The extract were filtered after 48 h, the extracts were concentrated using a rotary evaporator with the water bath set at 40°C. The percentage yield of extracts ranged from 7–17 %w/w. The extract of each plant material were mixed in equal proportion and make a single sample.

Determination of Antioxidant Activity

DPPH Scavenging Activity

1 mg of extract powder were dissolved in 1 ml of 90 % methanol solution to obtain 1000 µg/ml sample solutions were series diluted in to concentration ranging from 400–1000 µg/ml (i.e. 400, 500, 600, 700, 800, 900 and 1000 µg/ml). 200 µM solution of DPPH in methanol was prepared and 1.5 ml of this solution was added to 1.5 ml of methanol extract solution at different concentrations (400–1000 µg/ml). Mixture of methanol and DPPH were used as the standard control. Thirty minutes later, the absorbance was measured at 517 nm. The absorbance of DPPH solution decreases when kept in contact with antioxidant test sample and free radical scavenging activity is inversely proportional to the absorbance of DPPH solution [11, 12]. Percent inhibition of DPPH free radical scavenging activity was calculated using the following formula.

$$\text{DPPH Scavenged (\%)} = (A_{cont} - A_{test}) / A_{cont} \times 100$$

Where A_{cont} is the absorbance of the control reaction.

A_{test} is the absorbance in the presence of the sample of the extract.

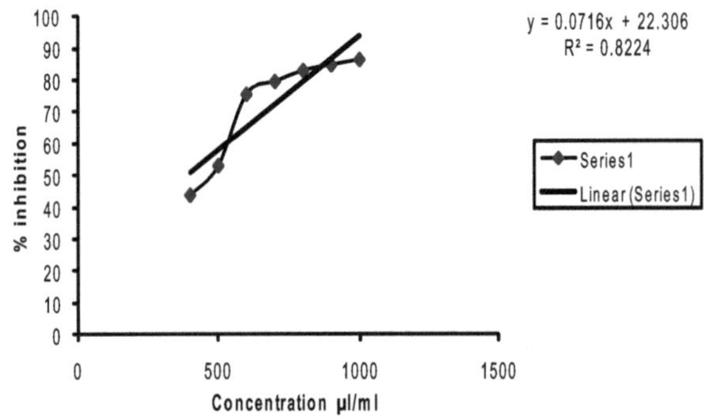

Figure 1: Inhibition of DPPH by the combine extract *Azadirachta indica, Ocimum sactum, Aloe barbandesis, Tinospora cordifolia, Triticum aestivum*

Table 1: Characteristics of the used medicinal plants

Extract yield %w/w	**Part used**	**Family name**	**Scientific name**	**Chemical constituent**
17.24	Leaves	Meliaceae	*Azadirachta indica*	Reducing sugar, Terpenoids, Flavonoids, Saponins, Alkaloids, Cardiac glycosides
16.09	Leaves	Lamicae	*Ocimum sactum*	Flavonoids, Tannins
10.08	Branch	Asphodelaceae	*Aloe barbandesis*	Flavonoids, Tannins, Alkaloids
9.26	Stem	Menispermaceae	*Tinospora cordifolia*	Reducing sugar, Flavonoids, Saponins, Alkaloids, Cardiac Glycosides, Tannins
7.94	Grass	Poaceae	*Triticum aestivum*	Flavonoids, Tannins, Alkaloids

Scavenging of Hydrogen Peroxide

Sample with different concentration (i.e. 400, 500, 600, 700, 800, 900 and 1000 µg/ml) were added to 0.1M phosphate buffer solution (pH 7.4, 3.4 ml) respectively and mixed with 43mM H_2O_2 solution (0.6 ml). After 10 minutes, the reaction mixture absorbance was determined at 230 nm. The reaction mixture without sample was used as the blank [13].

The % inhibition activity = $(A_{cont} - A_{test}) / A_{cont}$ X 100

Where A_{cont} is the absorbance of the control reaction.

A_{test} is the absorbance in the presence of the sample of the extract.

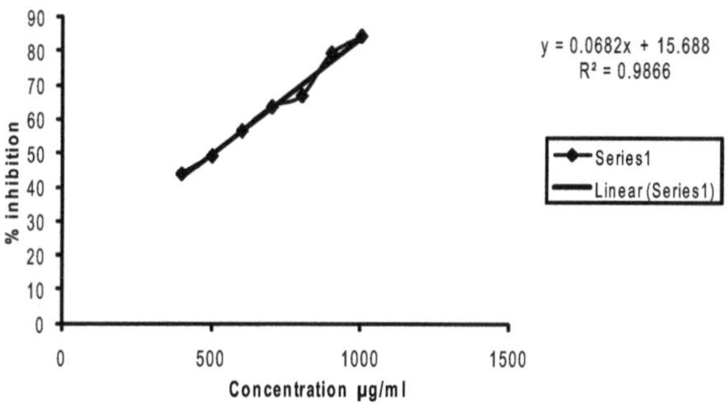

Figure 2: Inhibition of Hydrogen peroxide by the combine extract of *Azadirachta indica, Ocimum sactum, Aloe barbandesis, Tinospora cordifolia, Triticum aestivum*

Results and Discussion

The percentage yield of *Azadirachta indica, Ocimum sactum, Aloe barbandesis, Tinospora cordifolia, Triticum aestivum* was found to be 7–17 % w/w. Preliminary phytochemical screening of the extracts of individual plant showed the presence of flavonoids.

DPPH assay is based on the measurement of scavenging ability of antioxidant towards the stable DPPH radical. DPPH is relatively stable nitrogen centered free radical that can accept an electron or hydrogen radical to become a stable diamagnetic molecule. DPPH radicals react with suitable reducing agent as a result of which electron become paired off forming the corresponding hydrazine. The solution therefore loses color stoichiometrically at 517 nm [14]. From result it may be postulated that the combine extract have hydrogen donor thus scavenge free radical DPPH.

H_2O_2 is a weak oxidizing agent and can inactivate a few enzymes directly, usually by oxidation of essential thiol (SH) groups. Hydrogen peroxide can cross cell membrane rapidly. Once inside the cell, H_2O_2 can probably react with Fe^{2+} and possibly Cu^{2+} to form hydroxyl radical and this may be the origin of many of its toxic effects. It is therefore biologically advantageous for cell to control the amount of hydrogen peroxide that is allowed to accumulate. The decomposition of H_2O_2 by the combine extract may result from its antioxidant activity.

Results obtained in the present study indicate that the combine extract of *Azadirachta indica, Ocimum sactum, Aloe barbandesis, Tinospora cordifolia, Triticum aestivum* showed the maximum antioxidant activity with IC_{50} 386.7 µg/ml in DPPH and 504.7 µg/ml in hydrogen peroxide.

Conclusion

Results obtained in the present study indicate that the combine extract inhibit free radical scavenging activity. The overall antioxidant activity of extract might be attributed to its polyphenolic content and other phytochemical constituents. It could be a source of natural oxidant that could have greater importance as therapeutic agent in preventing various diseases.

References

1. E. Dursum, S. Otles and E. Akcicek ; Asian Pacific J. Cancer Prev. 5 (2004) 334–339.
2. M.Y. Khalil, A.A. Moustafa and N.Y. Naguib; World Journal of Agricultural Sciences. 3 (2007) 451–457.
3. L.G. Landry; Plant Physiol. 109 (1995) 1159.
4. C.A. Rice-Evans, N.J. Miller and P.G. Bolwell, P.M. Bramley and J.B. Prindham; Free Radical Res. 22 (1996) 375–383.
5. J. Lolinger,; Taylor and Francis, London. (1991) 129–150.
6. T. Yoshida, A.F. Ahmed and T. Okuda; Chemi and Pharmaco. Bull, Tokyo. 41 (1993) 672–679.
7. E. Souri, G. Amin, A.D. Sharifabadi, A. Nazifi and H. Farsam; Ir. J. Pharmac. Res. 3 (2004) 55–59.
8. I.V. Bol'shakova, E.L. Lozovskaia , I.I. Sapezhinskii; J. Agric. Food Chem. 47 (1999) 3954–3962.
9. M. Galvez, C. Martin-Cordera, P.J. Houghton and M.J. Ayuso; J. Agric. Food Chem. 53 (2005) 1927–1933.
10. T. Diplock, J.L. Charleux, G. Crozier-willi, F.J. Kok, C. Rice Evan, M. Roberfroid; Brit. J. Nutr. 80S (1998) S77.
11. P. Patil, V.V. Patil, V.R. Patil.; Pharmacologyonline. 2 (2009) 1344–1352.

12. Yerra Rajeshwar, G. P. Senthilkumar and Malwa Gupta; European bulletin of drug research. 1 (2005) 31–39.
13. N. Gupta, M. Agarwal, , V. Bhatia, S. K. Jha and J. Dinesh; Internal Journal of Pharmaceutical Sciences Review and Research. 6 (2011) 159–162.
14. K. L. Mukherji; Medical laboratory technology, 1st Edition. Tata McGraw Hill Publishing Company Limited, New Delhi (1989) 1124–1127.

10

Antioxidant and Antimutagenic Activities of Isothiocyanates Rich Seed Oil of *Eruca sativa* Plant

M. Khoobchandani[1], P. Bansal[1], S. Medhe[1], N. Ganesh[2], and M. M. Srivastava[1,*]
[1]Department of Chemistry, Dayalbagh Educational Institute, Agra 282110, India
[2]Jawaharlal Nehru Cancer Hospital and Research Center, Bhopal, 462001, India
Email : menkaresearch@gmail.com

Abstract

Eruca sativa seed oil has been explored for antioxidative and antimutagenic potential. Seed oil exhibited maximum percentage inhibition of hydroxyl (93.42%) and 2,2-diphenyl-1-picrylhydrazyl (92.02%) radicals at the concentration of 90 µg/ml. Treatment of mice with seed oil significantly ($p<0.01$) reduced oxidative stress and attenuated the altered changes of reduced glutathione. Seed oil treated group reduced the aberrant metaphase (47.50%) and micronucleus formation (85%) in the bone marrow of melanoma induced mice. The presence of isothiocyanates [allyl (40.3 µg/g), 3-butenyl (259.6 µg/g), 4-methylsulfinybutyl (743.1 µg/g), 2-phenylethyl (158.5 µg/g) and bis(isothiocyanatobutyl)disulphide (~5000 µg/g)] was confirmed by HS/SPME/GC–MS analysis of seed oil and found to be responsible for the observed bioefficacy. Our findings support the use of E. sativa seed oil (ignored edible oil) as health promoting food and its potential for clinical use.

Introduction

The involvement of reactive oxygen species (ROS) in degenerative diseases, including cancer, is now widely accepted [1]. An overproduction of ROS from disrupted metabolism referred as oxidative stress may cause damage through mutations terminating into cancer [2]. Mutations are the cause of innate metabolic defects in cellular system, triggering the morbidity and mortality in living organisms. A plethora of synthetic substances, apart from various genotoxic physical and biological agents, are known to act as mutagenic and or carcinogenic agents [3, 4]. However, these drugs are restricted by legislative rules because of doubts over their toxic and non-selective killing of the cells [5, 6]. Plants produce a great diversity of substances that could be of therapeutic significance in many areas of medicine. Dietary constituents and plant secondary metabolites viz., flavonoids, phenolic and glucosinolate compounds have been explored for significant antioxidant and anticarcinogenic properties [7, 8]. This in turn justifies the recent realization that a plant which has been explored for some particular pharmacological activities must be explored for other bona fide biological effects.

In continuation of our work on herbal screening for various pharmacological activities and chemical characterization of bioactive principle [9, 10], the present paper reports the antioxidant and antimutagenic activities of seed oil of *Eruca sativa* against melanoma. *Eruca sativa* Mill Thell (Cruciferae family) is biannual herb originated in the Mediterranean region but is presently found all over the world. It is consumed as salads in India and European countries. Traditionally, it is used as an astringent, diuretic, emollient, laxative, stimulant and rubefacient [11–13].

Materials and Methods

Authentication of the Plant

The plants were collected from Dayalbagh Agricultural Farms, Dayalbagh, Agra in March, 2008, and characterized as *Eruca sativa* Mill by Taxonomy Division, Department of Botany, Dayalbagh Educational Institute, Agra (India).

M.M. Srivastava, L.D. Khemani, S. Srivastava, *Chemistry of Phytopotentials: Health, Energy and Environmental Perspectives*, DOI:10.1007/978-3-642-23394-4_10, © Springer-Verlag Berlin Heidelberg 2012

Extraction of the Seed Oil

E. sativa seed oil was obtained by standard hot water extraction from (partly sprouted) ground seeds for 3–5 hours. Aqueous fraction was separated and dried over anhydrous Na_2SO_4. Seed oil was sealed under nitrogen and kept refrigerated (+4 °C) in the dark.

Fenton Assay

The hydroxyl radical scavenging activity of test samples was evaluated according to the Fenton reaction [14].

DPPH Assay

The DPPH radical scavenging activity of test samples was evaluated according to the DPPH assay [15].

α-tocoferol was used as reference antioxidant. All the tests were performed in triplicate and the percent inhibition of OH and DPPH radicals generation was calculated using the formula: [(C-T)/C]x100 where C was the absorbance of the control and T for the test samples.

Animals and Cell Line

C57BL/6 mice of 6–7 weeks old, weighing 25±5 gm were maintained in ventilated animal house of Department of Research, JNCH & RC, Bhopal (India). All mice were kept at controlled environmental condition (22±2 °C, 60±5 % humidity) with 12 hrs light/dark cycle. They were provided with standard pellet diet and water *ad libitum*. The animal experiments were performed according to the guidelines of Ethical Committee for the Purpose of Control and Supervision of Experiments on Animals (CPCSEA), Govt. India.

B16F10 mouse melanoma cells were maintained in DMEM supplemented with antibiotics, L-glutamine (2mM) and 10 % FBS and propagated by subcutaneous injection of 1×10^5 cells in the abdominal region of mice.

Animal Experiments, Treatment and Monitoring

To investigate the effects of test samples and reference anticancer drug (doxorubicin), mice were randomized and divided into 5 groups of 5 animals each. Group I was kept as normal control and Group 2–5 were injected with B16F10 melanoma cells subcutaneously into foot-pad on day zero. Mice were injected intraperitoneally with either Group II) saline; Group III) doxorubicin [16] (1 mg/kg body weight in 4 doses on the 1st, 5th 9th and 13th day of treatment); Group IV) seed oil (1 mg/kg body weight/day); Group V) seed oil (2 mg/kg body weight/day).

Mice in all groups were observed daily for survival and sacrificed after 21st day of the experimental schedule. The liver of each group of mice was dissected, weighed and stored at -80 °C until analyses were completed.

Assay for Reduced Glutathione (GSH)

The liver was homogenized in 154 mM saline (KCl) to give a 10 % homogenate (w/v). The crude homogenate was centrifuged at 2000 rpm for 15 min and supernatant was assayed for reduced glutathione activity [17]. The final mixture (5.0 ml) contained supernatant (1 ml), phosphate EDTA buffer (4.5 ml) and DTNB (0.5 ml) solution. The reaction mixture was incubated at room temperature for 20 min and monitored by the reduction of 5,5'-dithio-bis-(2-nitrobenzoic acid) to 5-thio-2-nitrobenzoate at 410 nm.

Chromosomal Aberrations Test

Cytogenetic damage in the bone marrow cells was studied by chromosomal aberration analysis at the end of experiment. All the animals were injected 0.025 % colchicines i.p and sacrificed 2 hr later to arrest the cells in metaphase by cervical dislocation. The femurs were dissected and cleaned to remove adherent muscles. Metaphase plates were prepared by the air drying method [18]. The bone marrow cells were flushed out, treated with pre-warmed 57 % hypotonic saline, fixed in Cornoy's fixative, stained with 4 % Giemsa and observed under a light microscope. A total of 400 metaphase spreads were scored per animal and the number of aberrations, namely chromosome and chromatid breaks, fragments and rings were scored.

Micronucleus Assay

The bone marrow was flushed out using minimum essential medium and the slides were prepared by the method of Schmid, 1975 [19]. Bone marrow was centrifuged and the pellet was resuspended in few drops of fetal bovine serum. Smears were prepared on pre-clean glass slides, stained with May-Grawnwald and followed by Gimsa stain and observed under a light microscope for micronuclei in polychromatic erythrocytes (PCEs) and normochromatic erythrocytes (NCEs).

Analytical Profile of Seed Oil

Chemistry Profile of Seed Oil

Head Space/Solid Phase Micro Extraction analysis of the crude oil resulted in the identification of isothiocyanates by GC-MS. Identification of ITCs was accomplished by comparison with NIST 05 MS-library (f-fit > 700; r-fit > 650) and was confirmed using authentic standards in all cases. Seed oil revealed the presence of significant amount of allyl-ITC (40.30 µg/gm), 3-butenyl-ITC (259.60 µg/gm), 2-phenylethyl-ITC (158.50 µg/gm), 4-methyl sulfinyl butyl isothiocyanate (743.10 µg/gm) and bis(4-isothiocyanatobutyl) disulphide (~5000 µg/gm) and traces of erucin [20].

Statistical analysis

All experimental data are given as mean ± SD. Statistical analysis was carried out using the one-way analysis of variances (ANOVA). Post Dunnett test was applied between control, reference and herbal test samples using Graph Pad Prism software. Probability values were found to be equal to or less than 0.05.

Results and Discussion

Free Radical Scavenging Activity

Seed oil was tested for their radical scavenging effect using known antioxidant (α-tocoferol) as a positive control in the concentrations range (10–100 µg/ml). Seed oil showed percent inhibition 95.55 % against hydroxyl radicals at the concentration 100 µg/ml while standard (α-tocoferol) antioxidant inhibited 99.46 %. A gradual increase in percent inhibition of DPPH˙ with ten increasing concentrations of test samples was also found to be dose dependent. The order of free radical scavenging in our experiment was as follows: α-tocoferol> seed oil at the concentration 90 µg/ml (Fig. 1).

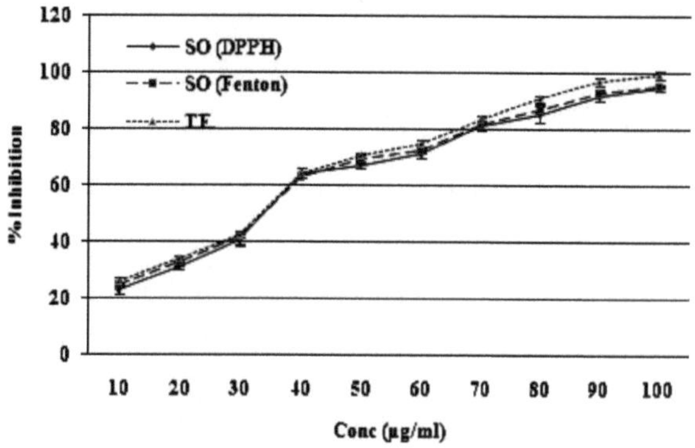

Fig. 1: Percentage inhibition of hydroxyl and DPPH radicals, concentration dependency of seed oil against standard α-tocoferol. Each value is mean ± SD (n=3). P>0.05 (SO) vs. α-tocoferol

Cancer Mediated Modulation of GSH Level

After induction of B16F10 melanoma cells, the weight and the level of reduced glutathione of liver tissues of experimental mice were recorded. The liver weight of seed oil treated mice was found similar to the normal control group mice (1.22 gm). The tumor control group shows lowering of reduced GSH (24.20 %) compared to normal control (51.54 %). Seed oil increased the level of down-regulation of reduced glutathione (49.54 %) at the concentration 2 mg/kg body weight. Seed oil also rendered significant protection against oxidative stress induced by melanoma in liver tissues in a dose dependent manner (Fig. 2).

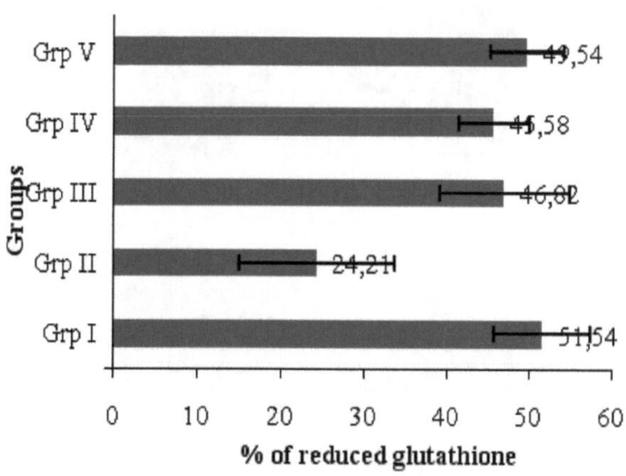

Fig. 2: Effect of *E. sativa* seed oil and doxorubicin drug on glutathione depletion in melanoma bearing mice

Table 1: Effect of *E. sativa* seed oil on percent of aberrant metaphases in the bone marrow of C57BL/6 mice after induction of melanoma tumor

S. No.	Group	CB	CR	FR	ACA	Pulvi	ICD	Abnormal metaphase
1	Normal Control	0.20±0.40	0.0±0	0.60±0.54	0.0±0	0.0±0	0.0±0	0.8±0.84
2	Tumor Control	12.80±2.28^{b}	6.20±0.83^{b}	23.80±2.49^{b}	34.20±1.92^{b}	4.20±1.09^{b}	4.40±3.20^{b}	82.60±2.49^{b}
3	Doxo + Tumor	10.60±3.36^{b}	5.60±1.14^{b}	17.40±2.30^{b}	33.60±2.88^{b}	3.20±1.92^{b}	5.40±1.94^{b}	77.80±3.76^{b}
4	SO 1 mg/kg + Tumor	8.60±1.82^{C}	2.60±1.14^{C}	11.24±1.51^{C}	28.92±3.12^{C}	0.0±0^{a}	2.60±1.14^{C}	51.20±5.32^{c}
5	SO 2 mg/kg + Tumor	8.20±1.09^{C}	2.40±0.70^{C}	10.30±1.93^{C}	23.35±3.42^{C}	0.0±0^{a}	3.60±1.14^{C}	47.50±2.95^{c}

Mean ± SE (n=5). Where; DOXO: Doxorubicin, SO: Seed Oil.
P value: ^{A}P>0.05 vs. Normal Control; ^{B}P<0.001 vs. Normal Control; ^{C}P<0.01 vs. Normal control and Tumor Control

Antimutagenic Activity

Antimutagenic activity of seed oil was observed in terms of chromosomal aberration and micronucleus assay by the induction of melanoma cells. Seed oil exhibited statistically significant (p<0.01) reduction in chromosomal aberration like chromatid breaks, centric rings, acrocentric association, acentric fragments, intercalary deletion, pulverization and total abnormal metaphases in bone marrow cells compared to tumor control and standard doxorubicin drug (Table 1). The decreasing order of chromosomal aberration in test sample treated mice was as follows: seed oil> doxorubicin drug> tumor control group (Fig. 3).

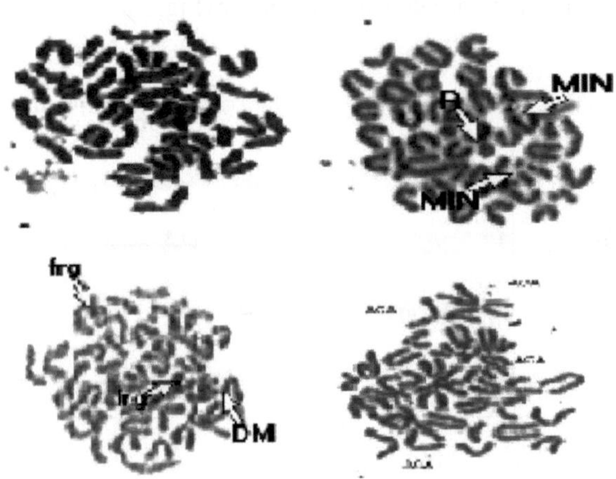

Fig. 3: Chromosomal aberration in (A) Normal control animal and (B) Tumor control animal (D) *E. sativa* seed oil treated animal (C) Doxorubicin treated animal after induction of B16F10 melanoma cells

The micronucleus assay has been used in cytogenetic studies to detect chromosomal changes such as acentric chromosome, chromatid fragments and chromosome lagging at anaphase. Micronuclei were considered an indication of a mutation effect [21]. Results of this study indicate that the seed oil reduced the frequency of micronuclei per polychromatic (PCEs) and normochromatic erythrocytes (NCEs) as compared to tumor control group (Fig. 4). The observed antimutagenic efficacy showed the similar trends of chromosomal aberration assay. However, ITCs-rich SO has been found to exhibit more potent inhibitory effect on melanoma growth in C57BL/6 mice without any toxicity and loss of body weight.

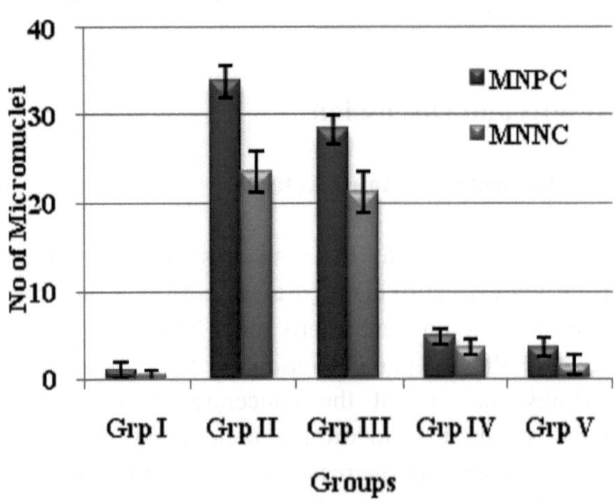

Fig. 4: Effect of the *E. sativa* seed oil and doxorubicin on frequency of micronuclei per PCEs and NCEs

Each value is mean ± SD (n=5). Grp I: Normal; Grp II: Tumor control; Grp III: Doxorubicin, Grp IV: Seed Oil (1 mg dose); Grp V: Seed oil (2 mg dose).

Conclusion

The present piece of work highlights the existence of significant antioxidant and antimutagenic bioefficacies in seed oil of *E. sativa* plant. ITCs-rich seed oil was notably capable of significant reduction of chromosomal aberration and micronuclei formation *in vivo* at 21st day, at doses only twice as large as those effective for the reference drug doxorubicin. It is suggested that this seed oil could be of potential use as dietary supplement as a naturally occurring antioxidant and antimutagenic bioagent.

Acknowledgements

The authors reciprocate Prof. V.G. Dass, Director, Dayalbagh Educational Institute, Dayalbagh, Agra, for providing necessary research facilities.

References

1. C. Keli, W.P. Geolf, N. Richard and Y.B. Bennett; J. Ethanopharmacol. 3 (2005) 201–205.
2. N.C.D. Nascimento, V. Fragoso, D.J. Moura, A.C.R. Silva, A.G. Fett-Neto and J. Saffi; Environ. Mol. Mutagen. 48 (2007) 728–734.
3. I. Shureiqi, P. Reddy and D.E. Brenner; Crit. Rev. Oncol. Hematol. 33 (2000) 157–167.
4. Z. Li, L.Z. Zhou, H.Z. Miao, L.P. Lin, H.J. Feng, L.J. Tong, J. Ding and Y.C. Li; J. Med. Chem. 52 (2009) 5115–5123.
5. Y. Yurekli, P. Unak, T. Ertay, Z. Biber, I. Medine and S. Teksoz; Annl. Nucl. Med. 19 (3) (2003) 197–200.
6. D. Shewach; American Chemical Society, Chemical Reviews, 7 (2009) 109.
7. B.B. Lee, M.R. Cha, S.Y. Kim, E. Park, H.R. Park and S.C. Lee; Plant Foods Hum. Nutr. 62 (2007) 79–84.
8. K.J. Jo, M.R. Cha, M.R. Lee, M.Y. Yoon and H.R. Park; Plant Foods Hum. Nutr. 63 (2008) 77–82.
9. M. Khoobchandani, B.K. Ojeswi, B. Sharma and M.M. Srivastava; Oxidative Medicine Cellular Longivity 2(3) (2009) 160–165.
10. B.K. Ojeswi, M. Khoobchandani, D.K. Hazra and M.M. Srivastava; Human Exp. Toxico. 29(5) (2010) 369–375.
11. L.M. Perry. Medicinal plants of east and Southeast Asia: attributed properties and uses; MIT Press: Cambridge, MA. 1978.
12. Z. Yaniv, D. Schafferman and Z. Amar; Ecol. Bot. 52 (1998) 394–400.
13. R.N. Bennett, F.A. Mellon, N.P. Botting, J. Eagles, E.A.S. Rosa and G. Willliamson; Phytochem. 61(1) (2002) 25–30.
14. B. Halliwell. JMC Gutteridge and C.E. Cross; J. Lab. Clin. Med. 119(6) (1992) 598–620.
15. K. Shimada, K. Fujikawa, K. Yahara and T. Nakamura; J. Agric. Food Chem. 40 (1992) 945–948.
16. W. Feleszko, I. Młynarczuk, D. Olszewska, A. Jalili, T. Grzela, W. Lasek, G. Hoser, G.K. Kowalska and M. Jakobisiak; Int. J. Cancer 100 (2002) 111–118.
17. E. Beutler. In Red Cell Metabolism. (Beultec) Grunne and Stratton (Eds), New York: 1984; pp 77–136.
18. J.R.K. Savage; Env. Mol. Mutagen. 22 (1993) 198–207.
19. W. Schmid; Mutat. Res. 31 (1975) 9–12.
20. M. Khoobchandani, B.K. Ojeswi, N. Ganesh, M.M. Srivastava, S. Gabbanini, R. Matera, R. Iori and L. Valgimigli; Food Chem. 120(1) (2010) 217–224.
21. C. Auerbach. Mutation: An introduction to research on mutagenesis Part I: Methods, Oliver and Boyed, Edinburgh, 1962.

11
Fungal Biosynthesis of Antimicrobial Nanosilver Solution: A Green Approach

M. Dubey[1], S. Sharma[2], S. Bhadauria[3], R. K. Gautam[4] and V. M.Katoch[5]
[1,2,3] Microbiology Research Laboratory, Department of Botany, R. B. S. College, Agra, India.
[4] School of Life Sciences, Dr. B. R. Ambedkar University, Khandhari campus, Agra, India
[5]Secretary to Govt. of India, Department of Health Research & Director General ICMR, New Delhi, India.
E. mail: dubeymanish22@rediffmail.com

Abstract

The biosynthesis of silver nanoparticles and their notable in vitro antibacterial activities against E. coli, a Gram-negative bacterial strain are reported here. Spherical shaped silver nanoparticles with a diameter of 20 nm are prepared by a fungus Penicillium expansum. The formation process of the silver nanoparticles is investigated by UV-visible (UV-vis) spectroscopy, X-ray diffraction (XRD) measurements and transmission electron microscopy (TEM). Bacteriological tests were performed in Luria–Bertani (LB) medium on solid agar plates and supplemented with 3.10 μg/ml concentration of nanosized silver particles. These particles were shown to be an effective bactericide. Scanning electron microscopy (SEM) was used to study the biocidal action of this nanoscale material. The results confirmed that the treated E. coli cells were damaged, showing deform morphology exhibits a significant increase in permeability, resulting in death of the cell.

Introduction

Nanotechnology collectively describes technology and science involving nano scale particles (nanoparticles) that increases the scope of investigating and regulating the interplay at cell level between synthetic materials and biological systems [1]. It can be employed as an efficient tool to explore the finest processes in biological processes [2] and biomedical Sciences. The enormous interest in the biosynthesis of NPs is due to their unusual optical, chemical, photochemical, electronic and magnetic [3] properties. NPs are either newly created via nanotechnology or are present naturally over the earth's crust.

Fabrication of NPs, through technology, can be undertaken using either chemical or biological systems [4]. The importance of biological synthesis is being emphasized globally at present because chemical methods are capital intensive, toxic, non ecofriendly. Utilizing potential biological systems from higher angiospermic plants or microbes, biosynthesis of NPs is currently under wide exploration [5]. The fungal mediated green chemistry approach towards the fabrication of NPs has many advantages. This includes easy and simple scale up method, economic viability, easy downstream processing and biomass handling, and recovery of large surface area with optimum growth of mycelia [6]. Aqueous Ag ions exposed to *Fusarium oxysporum* leads to the fabrication of extremely stable Ag hydrosol. The particles are in the 5–15 nm range and are stabilized in solution by the proteins excreted through the fungus [7]. Extracellular biosynthesis of Ag NPs in the 5–25 nm range using A*spergillus fumigatus* is found to be quite fast and manifested the production of dense fungal biomass [8]. White rot fungus, scientifically known as *Phaenerochaete chrysosporium* has also been used for biomimetics of Ag NPs. Fabrication of NPs, phytochelatin and NADPH dependent nitrate reductases for *in vitro* production of Ag NPs have been isolated from *Fusarium oxysporum* and been elucidated [9].

Ag-nanoparticle has been tested in various field of biological science Viz. drug delivery, wound treatment, binding with HIV gp-120 protein [10], in water treatment and an antibacterial compound against both Gram (+) and Gram (-) bacteria [11–12]. Most of the bacteria have yet developed resistance to antibiotics. Viewing all the above facts, it is future need to

M.M. Srivastava, L.D. Khemani, S. Srivastava, *Chemistry of Phytopotentials: Health, Energy and Environmental Perspectives*, DOI:10.1007/978-3-642-23394-4_11, © Springer-Verlag Berlin Heidelberg 2012

develop a substitute for antibiotics [13]. Ag-nanoparticles are attractive as these are non-toxic to human body at low concentration and having broad-spectrum antibacterial nature. Agnanoparticle inhibits the bacterial growth at very low concentration than antibiotics and as of now no side effects are reported [14].

Here we report a novel method for the synthesis and characterization of silver nanoparticles synthesized by fungus *Penicillium expansum* with their antibacterial properties against gram negate bacteria *E. coli*.

Experimental Details

Preparation of Silver Nanoparticles

The plant pathogenic fungus, *Penicillium expansum* was isolated from the fruits and maintained on potato-dextrose agar slants at 25 °C. The fungus was grown in 500 ml Erlenmeyer flasks each containing 100 ml MGYP medium, composed of malt extract (0.3 %), glucose (1.0 %), yeast extract (0.3 %), and peptone (0.5 %) at 25–28 °C under shaking condition (200 rpm) for 96 h. After 96 h of fermentation, mycelia were separated from the culture broth by centrifugation (5000 rpm) at 10 °C for 20 min and the settled mycelia were washed thrice with sterile distilled water. the carefully weighted 0.5 gm biomass was added to 100 ml of 1 mM aqueous AgNO 3 solution, in conical flasks of 250 ml content at room temperature and the reaction carried out for a period of 24 h. The biotransformation was routinely monitored by visual inspection of the biomass as well as measurement of the UV-vis spectra from the fungal cells.

UV-Vis Spectroscopic Studies

The bioreduction of Ag+ in aqueous solution was monitored by periodic sampling of aliquots (0.2 ml) of the suspension, then diluting the samples with 2 ml deionized water and subsequently measuring UV-Vis spectra of the resulting diluents. UV-Vis spectroscopy analyses of nanoparticles produced were carried out as a function of bioreduction time at room temperature on ELICO UV-Vis spectrophotometers at a resolution of 1 nm.

X-ray Diffraction Measurements

X-Ray diffraction (XRD) measurements of the bio-reduced silver nitrate solution drop-coated onto glass substrates were done for the determination of the formation of Ag by an X'Pert Pro analytical X-ray diffractometer instrument with X'Pert high score plus software operating at a voltage of 45 kV and a current of 40 mA with Cu Kα radiation.

TEM Observations

Samples of the aqueous suspension of silver nanoparticles were prepared by placing a drop of the centrifuged suspension on carbon-coated copper grids and allowing water to evaporate. TEM observations were performed on an H-600 electron microscope (Hitachi, Japan) operated at an accelerating voltage of 120 kV.

Antimicrobial Activity

Bactericidal effect of silver nanoparticles was studied against Gram-negative bacteria. The bacterial culture was obtained from *Microbial Type Culture Collection* and Gene Bank Institute of Microbial Technology, Chandigarh, India. Aqueous dispersions of silver nanoparticles of desired concentrations were made. An axenic culture of *E. coli M*TCC-443 was grown in liquid nutrient broth medium CM-01 (Oxoide, England) (containing (g/l): Lab lemco powder 1 g, NaCl 5 g, peptone 5 g and yeast extract 2 g). The experimental investigation, freshly grown bacterial inoculum (10^6 cells/ml) of *E. coli* was cultured on Mueller-Hinton agar plates. Empty sterile discs having a diameter of 6 mm were impregnated with silver nano solution at 3.10 μg/ml concentrations placed on inoculated surface of agar plate. These plates were incubate for 24 hour at 37 °C and measured the zone of inhibition in millimeter.

SEM Abservations

Interaction between bacteria and silver nanoparticles is shown in the scanning electron micrographs. For SEM analysis bacterial cells collected by centrifugation of liquid samples at 10000 r/min for 10 min. Collected bacterial cells that reacted with the silver nanoparticles were first fixed in 2.5 % glutaraldehyde and 2% paraformaldehyde (PF), in 0.1 M phosphate buffer (pH 7.4) for 6–12 hr at 4 °C. After being washed with 0.1 M phosphate buffer for 3 times, collection of the pellet by centrifugation and then dehydrolysis the sample by different ethanol volumes starting; 30%, 50%, 70%, 80%, 90% and 100% and for each ethanol volume incubated for 10 minutes. The cell biomass were fixed on the aluminum stubs and coated with a thin layer of gold for SEM analysis (ZEISS EVO 40 EP).

Results and Discussion

It is well known that silver nanoparticles exhibit reddish color in water [15]; this color arises due to excitation of surface plasmon vibrations in the metal nanoparticles [16]. The biomass incubated with deionized water retained its original color i.e. yellow-green, while the silver nitrate treated biomass turned dark red (as shown in Fig. 1a) after 6 hours due to the formation of silver nanoparticles extracellularly.

Fig. 1: Pictures show the color changes **(a)** before and **(b)** after the process of reduction of Ag+ to Ag nanoparticles

The surface plasmon resonance (SPR) band for spherical silver nanoparticles occurs in the range 380–440 nm. Graph shows the evolution of the absorbance spectra emanating from silver nanoparticles over time manifests increasingly sharp absorbance with increasing time of reaction at around 420 nm attributed to the surface plasmon resonance band (SPR) of the silver nanoparticles. After 24 h of incubation, no change in intensity at 420 nm was observed indicating the complete reduction of silver ions (as shown in Fig. 2).

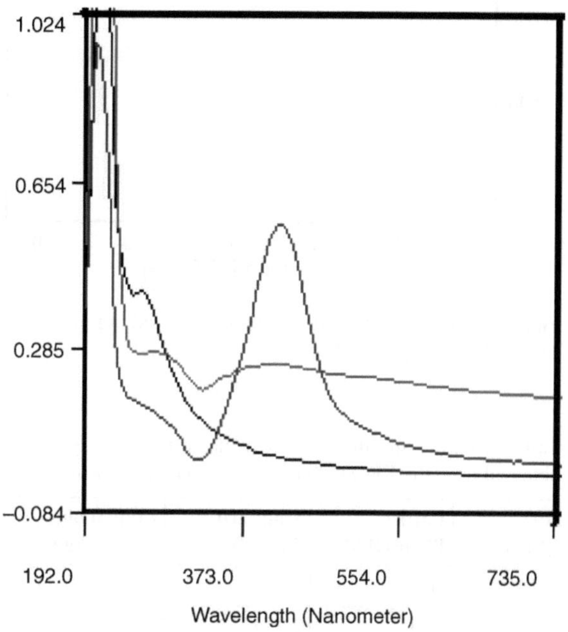

Fig. 2: UV-Vis spectra recorded as a function of time of reaction of 1 mM $AgNO_3$ aqueous solution with fungal biomass

Figure 3 shows the XRD pattern obtained for silver nanoparticles synthesized using fungal biomass represented by the curve. A number of Bragg reflections are observed for the silver nanoparticles, which may be indexed based on the fcc structure of silver and is shown in Table 3.1. In Fig. 3, a couple of Bragg's reflections are distinctly exhibited, which may be indexed on the basis of the face-centered cubic structure of silver. It exhibits a sharp and intense peak at ~38° and ~32° corresponding to diffraction from the (111) and (101) planes of silver with FCC lattice (JCPDS no. 04–0783). The XRD pattern thus clearly shows that the silver nanoparticles formed by the reduction of Ag+ ions by the fungal biomass are crystalline in nature. The XRD pattern of pure silver ions is known to display peaks at 2θ= 7.9°, 11.4°, 17.8°, 30°, 32°, 38°, and 44° [17]. This stimation confirmed the hypothesis of particle monocrystallinity. The sharpening of the peaks clearly indicates that the particles are in the nanoregime.

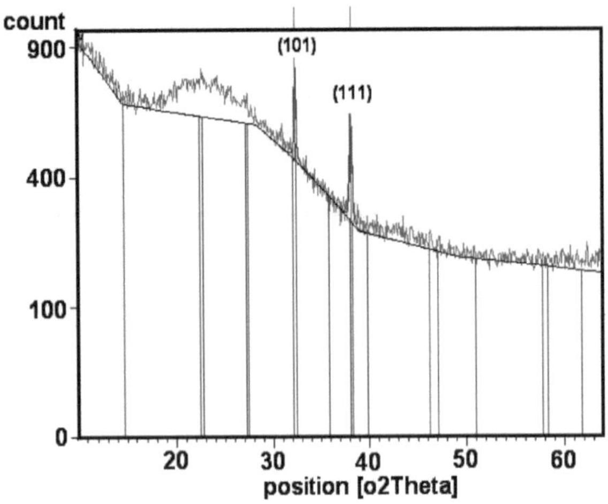

Fig. 3: XRD patterns recorded from drop-coated films of silver nanoparticles on glass substrates

Table 1: Lattice spacing values calculated from the 2θ values obtained from the XRD pattern of silver nanoparticles

Pos. [°2Th.]	Lattice Planes (hkl)	Standard Ag (A°)	Fungal bio-mass – Ag (A°)
32	(101)	2.815	2.81246
38	(111)	2.359	2.35291

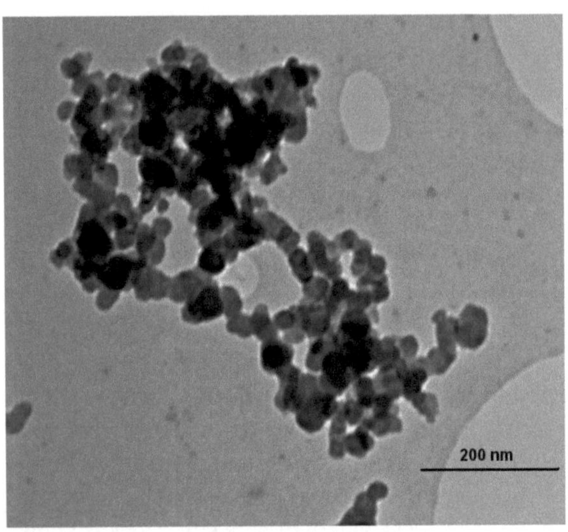

Fig. 4: TEM images of silver nanoparticles formed by reduction of Ag+ ions

Figures 4 show representative TEM images recorded for the Ag nanoparticles synthesized by treating $AgNO_3$ solution with fungal biomass for 24 h. The silver nanoparticles are observed to range in size from 10 to 30 nm with an average size of ca. 15 nm (Figure 3.11A). The morphology of the Ag nanoparticles is predominantly spherical. It is known that the shape of metal nanoparticles considerably changes their optical and electronic properties [18].

Finally, zone of inhibition test was done for identification of inhibition with 3.10 µg/ml Ag nanoparticle. It was found that 3.10 µg/ml concentration of Ag nanoparticle was able to inhibit bacterial growth and create a zone of 12 mm in *E. coli* (Figure 5). We found that increasing the Ag-nanoparticle concentration did not show a consistent increase in the zone size in *E. coli*, because the nanoparticles settled at the bottom of the wells in the plates and the particle solution was not able to diffuse properly in the agar medium.

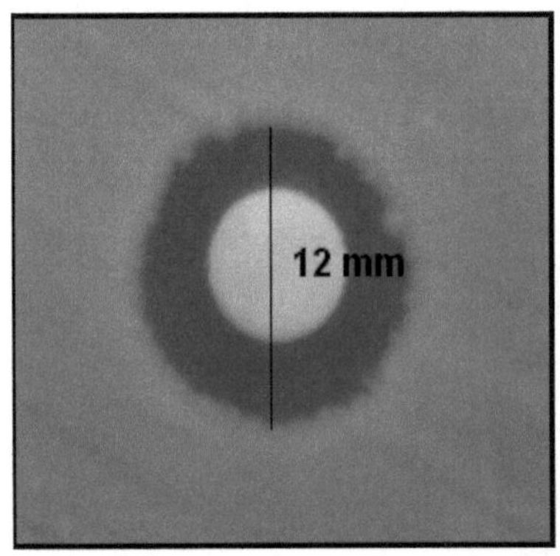

Fig. 5: Zone of inhibition of *E. coli* against silver nanoparticles

Interaction between Gram-negative *E. coli* and Ag nanoparticle is shown in the scanning electron photomicrographs (Figure 6). Reports on the mechanism of inhibitory action of silver ions on microorganisms show that upon Ag nano treatment, DNA loses its replication ability [19] and some other cellular proteins and enzymes essential to ATP production becomes inactivated [20]. In a previous report, silver has a greater tendency to react with sulfur- or phosphorus-containing soft bases, such as R-S-R, R-SH, RS or PR3. Thus, sulfur-containing proteins in the membrane or inside the cells and phosphorus-containing elements like DNA are likely to be the preferential sites for silver nanoparticle binding [21].

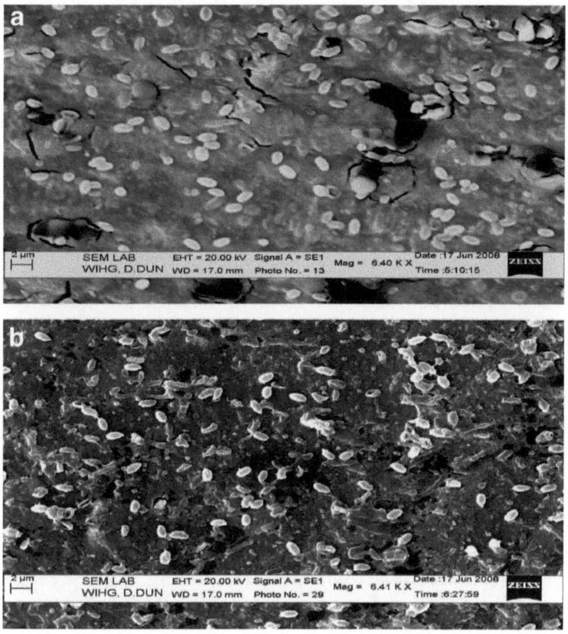

Fig. 6: SEM microphotographs of *E. coli.* **(a)** Untreated *E. coli* (magnification 6.40 KX). **(b)** *E. coli* supplemented with Ag nanoparticles indicates partial membrane damage (magnification 6.41 KX).

Conclusion

In conclusion, it has been demonstrated that the *Penicillium expansum* is capable of producing silver nanoparticles extracellularly and the silver nanoparticles are quite stable in solution. The antibacterial activity of the fungal synthesizes nanoparticles dispersion was measured by zone of inhibition method. The results of this study clearly demonstrated that the silver nanoparticles inhibited the growth and multiplication of the bacteria and the major mechanism through which silver nanoparticles manifested antibacterial properties was by anchoring to and penetrating the bacterial cell wall.

Acknowledgment

The authors gratefully acknowledge the financial support given to this research from the Indian Council of Medical Research, New Delhi, India under the IRIS (2009–08130). We are also grateful to the Dr. N.K. Saini, WIHG, Dehradun for providing XRD and SEM facility and Dr. Renu Pasricha for TEM.

References

1. L. Du, H. Jiang, X. Liu and E Wang; Electrochemistry Communications. 9 (2007)1165–70.
2. I. Sondi and B. Salopek-Sondi; Journal of colloid interface and Science. 275 (2004) 117–82.
3. J.H.P. Watson, D.C. Ellwood, A.K. Soper and J. Charnock; Journal of Magnetism and Magnetic Materials. 203 (1999) 69–72.
4. A. Panacek, L. Kvitek, R. Prucek, M. Kolar, R. Vecerova and N. Pizurova; Journal of Physical chemistry B. 110 (2006) 16248–53.
5. V. Bansal, D. Rautaray, A. Ahmad and M. Sastry; Journal of Materials Chemistry. 14 (2004) 3303–05.
6. M. Sastry, A. Ahmad, M.I. Khan and R. Kumar; Current Science. 85 (2003) 162–170.
7. A .Ahmad, P. Mukherjee, D. Mandal, S. Senapati, M.I. Khan, R. Kumar and M. Sastry; Colloids and Surfaces B: Biointerfaces. 28 (2003) 313–318.
8. K.C. Bhainsa and S.F. D'Souza; Colloids and Surfaces B: Biointerfaces. 47 (2006) 160- 164.
9. S.A. Kumar, M.K. Abyaneh, S.W. Gosavi, S.K. Kulkarni, R. Pasricha, A. Ahmad and M.I. Khan; Biotechnology Letters. 29 (2007) 439–445.
10. P. Jain and T. Pradeep; Biotech. Bioeng. 90 (2005) 59–63.
11. W.K. Son, J.H. Youk, T.S. Lee and W.H. Park; Macromol. Rapid. Commun. 25(18) (2004) 1632–1637.
12. P. Li, J. Li, C. Wu, Q. Wu and J. Li; Nanotechnology. 16 (2005) 1912–17.
13. C.N. Lok, C. Ho, R. Chen, Q.Y. He, W.Y. Yu, H. Sun, P.K. Tam, J. Chiu and C.W. Chi; J. Proteo. Res. 5 (2005) 916–924.
14. C. Baker, A. Pradhan, L. Pakstis, D.J. Pochan and S.I. Shah; J. Nanosci. Nanotec. 5 (2005) 244–249.
15. M. Sastry, V. Patil and S.R. Sainkar; J Phys Chem B. 102 (1998) 1404.
16. P. Mulvaney; Langmuir. 12 (1996) 788.
17. P. Gong, H. Li, X. He, K. Wang, J. Hu, W. Tan, S. Zhang and X. Yang; Nanotechnology. 18 (2007) 285604 (7pp).
18. H. Xu, M. Käll; J Nanosci. Nanotechnol. 4 (2002) 254.
19. Q.L. Feng, J. Wu, G.Q. Chen, F.Z. Cui, T.N. Kim and J.O. Kim; J. Biomed. Mater. Res. 52 (2000) 662–668.
20. M. Yamanaka, K. Hara and J. Kudo; Appl. Environ. Microbiol. 71 (2005) 7589–7593.
21. G. McDonnell and A.D. Russell; Clin Microbiol Rev. 12 (1999)147–179.

12
Natural Products as Inhibitory Agents of *Escherichia coli* and *Listeria monocytogenes*

P. Singh and A. Prakash
Department of Zoology, Faculty of Science, Dayalbagh Educational Institute, Agra-282110
Email: prakashdr.dei@gmail.com

Abstract

The potential application of the present research findings is of particular interest to search a natural preservative for milk products which could inhibit pathogenic Escherichia coli and Listeria monocytogenes. To achieve the aim pathogens were isolated and identified from cottage cheese samples, produced by the traditional methods using biochemical tests as well as sequencing of 16S rRNA gene sequence and their control was studied using natural products extracted in ethanol and water. In vitro studies shows that Mango seed and hime are more effective against E. coli whereas mausami leaves and peepal leaves inhibit L. monocytogenes more effectively.

Introduction

Milk and milk products represent an ideal growth medium for microorganisms. *Escherichia coli,* and *Listeria monocytogenes* are the most important pathogens which indicate the unhygienic conditions in milk and milk products. Most strains of *E. coli* are harmless but several strains are extremely pathogenic like *E. coli* 0157:H7 causing complications and death associated with hemorrhagic colitis and acute renal failure, especially in children. The infective dose of *E. coli* is estimated to be very low, about 10 cells: in contrast, the infective dose required for the *L. monocytogenes* is still unknown but it is believed that it varies with the specific strain of the bacterium and the susceptibility of the individuals. Regular FDA standards include ‚zero tolerance‘ for *L. monocytogenes* in all ready-to-eat products. *L. monocytogenes* is the only species in the genus *Listeria* that has been involved in known food-borne outbreaks of listeriosis, particularly in risk populations including neonates, immune-compromised hosts and pregnant women.

Food borne infections should be cured immediately after being diagnosed because they may lead to serious health problems and some time even death. Drugs used to cure these infections are known as ‚antibiotics‘. The clinical efficacy of many antibiotics is being threatened by the emergence of multidrug resistant pathogens. Natural products like Neem, Tulsi, and Garlic etc either as pure compounds or as standardized plant extracts provide unlimited opportunities for new drugs due to the unmatched availability of inherent chemical diversity of the natural products. Plant derived antimicrobial compounds may be of value as a novel means for controlling antibiotic resistant zoonotic pathogens which contaminate food animals and their products [1]. The urgent need to discover new antimicrobial compounds with diverse chemical structures and novel mechanism of action for new and re- emerging infectious diseases prompted us to take up the study on the inhibition of *E. coli and L. monocytogenes* by using various antibiotics and natural products. For this purpose we have isolated these microbes from cottage cheese and identified them by using the novel culture dependent methods that involve PCR amplification of bacterial small-subunit rDNA of all the two selected microbes using single set of universal primer.

Materials and Methods

Samples of cottage cheese were collected seasonally from in and around the different market areas of Agra city and examined for the presence of *E. coli and L. monocytogenes.* Standard strains of *E. coli* (MTCC-723) and *L. monocytogenes* (MTCC-1143) were procured from MTCC Chandigarh, India. Isolation and

M.M. Srivastava, L. D. Khemani, S. Srivastava, *Chemistry of Phytopotentials: Health, Energy and Environmental Perspectives*, DOI:10.1007/978-3-642-23394-4_12, © Springer-Verlag Berlin Heidelberg 2012

preliminary identification of the 100 isolates of each selected pathogenic bacteria were done using biochemical characterization (Singh and Prakash, 2008) and finally these were identified by 16S rRNA gene sequencing [2].

Extraction of DNA

Extraction of the template DNA was done as per the method of Tsai and Olson (3) with suitable modifications. 100 μl of 24 hrs pure culture was centrifuged in a micro-centrifuge at 4000 rpm for 12 minutes. The recovered pellet was suspended in 100 μl of sterilized DNase and RNase free water, heated in a boiling water bath for 10 min and then snap chilled in crushed ice. The obtained lysate was used as the DNA template. 50 μl Master mixture was prepared at a final concentration of 1X (10 X PCR buffer), 0.2mM (2mM dNTP mix), 2mM (25mM $MgCl_2$), 5pM (22pM each primer), 1.25U Taq DNA Polymerase, template DNA and MiliQ water. Amplification was done by using primer set 27F (5-AGAGTTTGATCCTGGCTCAG-3) and 1492R (5-TACGGTTACCTTGTTACGACTT-3) [4, 5]. The PCR mixture was subjected to thermal cycler (initial incubation 95°C for 5 min, 34 cycles of 30 s at 95°C for denaturation step, 30 s at 55°C for annealing step and 30 s at 72°C for extension step). 8μl of the reaction products were resolved by electrophoresis on a 1 % agarose gel containing 0.5 μg of ethidium bromide per ml in .5X Tris-borate-EDTA buffer at 7V/cm. A 100 bp DNA ladder (Bangalore genei) was included. The gel was visualized and photographed over the UV transilluminator (Zenith gel documentation system) and analyzed by gel doc software named UN-SCAN-IT gel 6.1.

Inhibition by Antibiotics and Natural Products

Inhibition under the influence of natural products including *Trechyspermum copticum (Ajwain), Gycyrrhiza glabar (Mulethi), Chebulic myrobalan* (Hime), *Piper chaba* (Choti pepper), *Mangifera indica* (Mango seed), *Ficus religiosa* (Peepal leaves), *Syzygium cumini* (Jamun seed) and *Citrus sinensis* (Mousami) was analyzed by agar well diffusion method. Extracts of natural products were prepared by soaking the product in ethanol and water in a ratio of 1:4 and 1:8 respectively, for 72 hrs, in sterile conical flasks at room temperature with uniform shaking. The extracts were then filtered and concentrated by evaporating to dryness at 45°C [6].

Results and Discussion

On the basis of Gram's staining and biochemical characterization (MR, VP, Indole, Nitrate, citrate utilization, H_2S production, fermentation of various sugars, hemolysis, and growth on chromogenic medium) 4 isolates have been confirmed as pathogenic *E. coli,* and 6 isolates as pathogenic *L. monocytogenes.* They were further confirmed to the species level by the amplification of 16S rDNA coding ~ 1400 bp 16S rRNA gene sequence using universal primer set 27F/1492R. Amplified products were submitted to the Institute of Molecular Medicine, New Delhi for sequencing. Obtained sequences were aligned through National centre for biotechnology Information (NCBI) database by using the Basic Local Alignment Tool (BLAST, 2.0 search programs) to determine their approximate phylogenetic affiliations. Sequence alignment with BLAST database Sequence determination by 16S rRNA gene confirmed the isolates as *E. coli* and *L. monocytogenes.*

Inhibition of two of the confirmed isolates, along with standard strain was studied in the presence of natural products (*Trechyspermum copticum (Ajwain), Gycyrrhiza glabar (Mulathi), Chebulic myrobalan* (Hime), *Piper chaba* (Choti pepper), *Mangifera indica* (Mango seed), *Ficus religiosa* (Peepal leaves), *Syzygium cumini* (Jamun seed) and *Citrus sinensis* (Mausami)) (Fig 1).

There are many reports available that prove antiviral, antibacterial, antifungal, antihelmintic, antimolluscal and anti-inflamatory properties of plants [7, 8]. For determining the MIC of the natural products the agar well diffusion method was employed (on Muller Hinton Agar, MHA). MIC of the natural products observed for *E. coli* shows that the least effective natural products were the extract of mausami leaves, jamun seed and ajwain while the most effective were the extracts of mango seed, followed by hime, choti peepal, and mulathi. (Fig 2).

Results obtained for the MIC of natural products against *L. monocytogenes* show that all the natural products were effective in inhibiting both the standard *L. monocytogenes* (MTCC 1143) and the two isolates,

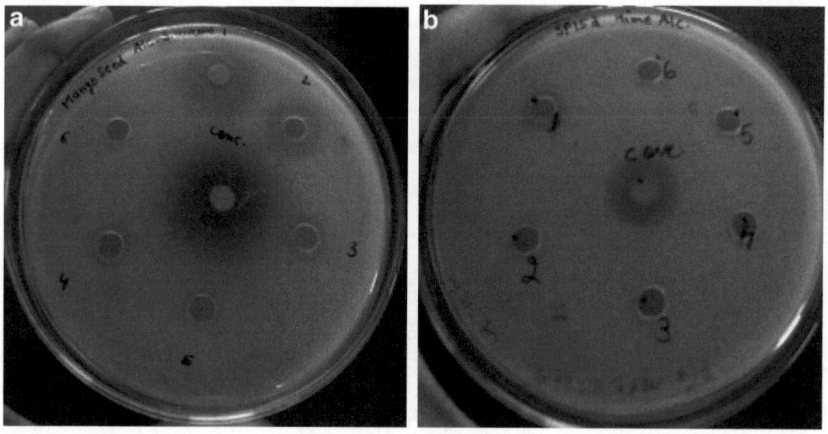

Fig. 1: Plate showing MIC **(a)** *E. coli* **(b)** *L. monocytogenes*

though the former was inhibited to a greater extent than the latter. The difference may be due to the differences in the virulent genes present in the standard *L. monocytogenes* (MTCC 1143) and its isolates from milk products. Of all the natural products the extracts of mausami leaves and peepal leaves were found to be most effective in inhibiting *L. monocytogenes* (Fig 3).

Ethanolic extracts were found to be more effective then aqueous extracts, this variation in the activity may be due to the better extraction of biologically active compounds (Alkaloids, flavonoids, essential oils, tennins etc.) which were enhanced in the presence of ethanol [9,10].

However, the present work suggests that if the active molecules are isolated from the crude extracts of the natural products they may prove to be more effective than the well known antibiotics currently in use for the control of these pathogens and can be easily incorporated in one's diet and exert no side effect as compared to antibiotics.

Besides antibiotic resistance, which poses an insurmountable problem in the cure of these bacterial infections could be taken care of by an increased use of natural products which possess a magical potential against these pathogens.

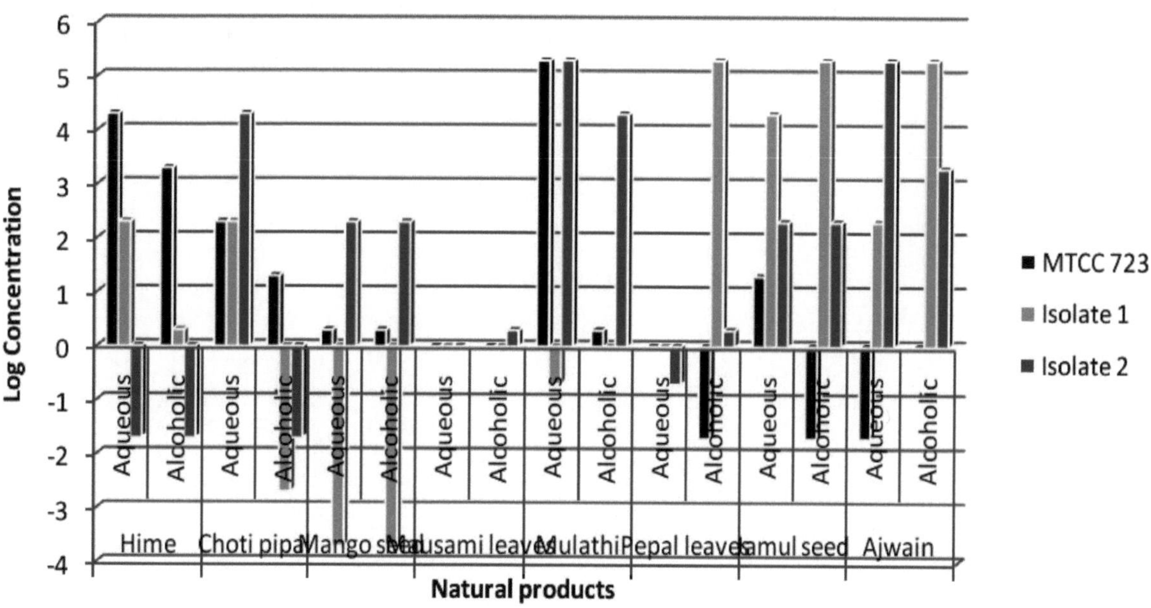

Fig. 2: Minimum Inhibitory Concentration of all the natural products against *E. coli*

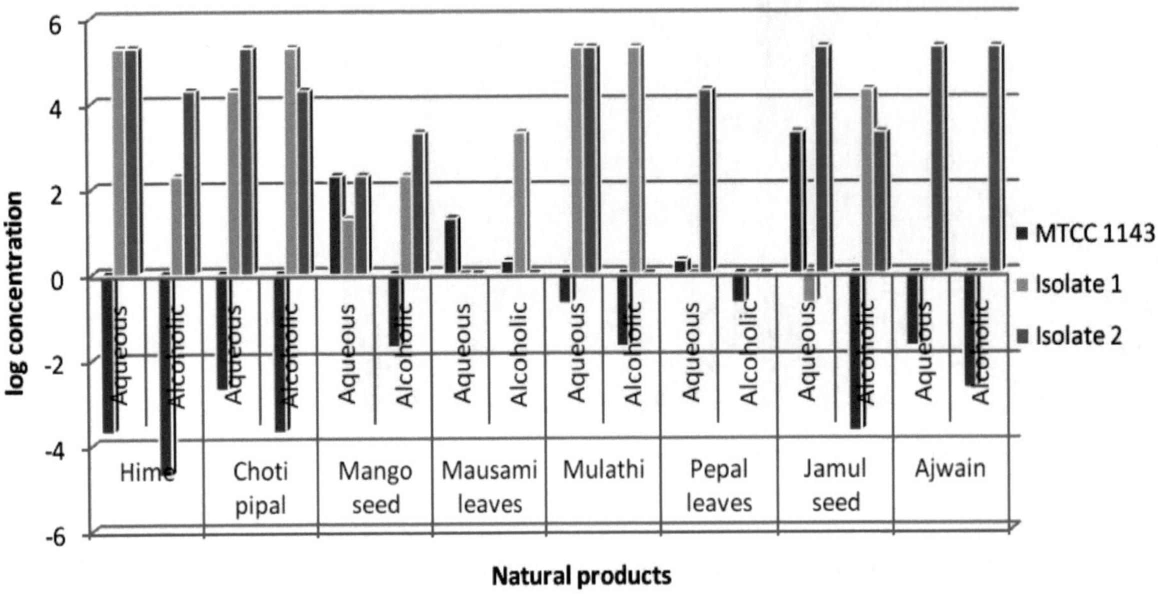

Fig. 3: Minimum Inhibitory Concentration of all the natural products against *L. monocytogenes*

References

1. R.A. Halley, and K. Palamappan; Int. J. Food Microbiol. 140 (2010) 164–168
2. P. Vandamme, B. Pot, M. Gills, P. De Vos, K. Kersters, and J. Swings; Microbiol. Rev. 60 (1996) 407–438
3. Y. L. Tsai, and B. H. Olson; Appl. Env. Microbiol. 57 (1991) 1070–1074
4. J.A. Frank, C.I. Reich, S. Sharma, J.S. Weisbaum, B.A. Wilson, and G.J. Olsen; Appl. Env. Microbiol. 74 (2008) 2461–2470
5. E. Acedo-Felix, and G. Perez-Martinez; Int. J. Syst. Evol. Microbiol. 53 (2003) 67–75
6. E.I. Mahmood, J.H. Doughari, and N. Landan; Afri. J. pharm. and pharmacol. 2 (2008) 89–94
7. S. Stepanovic, N. Antie, I. Dakic, and M. Svabicvlahovic; Microbiol. Res. 158 (2008) 353–357
8. S.K. Behera, and M.K. Misra; Microbiol. and Mol. Bio. Rev. 46(3) (2005) 242–280
9. A. Falodun, I. O. Okunrobo, and N. Uzoamaka; Afri. J. Biotechnol. 5 (2006) 529–53
10. A. Ghosh, B.K. Das, A. Roy, B. Mandal, and G. Chanda; J. Natur. Med. 62 (2008) 259–262

13
Wonders of Sesame: Nutraceutical Uses and Health Benefits

N. Shivhare and N. Satsangee
Department of Foundations of Education, Faculty of Education, Dayalbagh Educational Institute, Dayalbagh, Agra
Email: nehashivhare1@gmail.com

Abstract

The world today is witnessing a sharp rise in the number of people suffering from ailments such as hypertension, heart attacks, asthma, cancer, etc. One of the major causes of health related issues is our daily diet, the quality of which is reflected through our physique. There is a gamut of such natural edibles obtained from plants which can certainly enhance our physical health and also help in curing the ailments. Sesame is also one such plant which has magical potential to positively affect our health. The present paper highlights nutraceutical uses and health benefits of sesame seeds and sesame oil.

Introduction

Sesame or *til* (*Sesamum indicum,* L.) is one of the oldest products cultivated since the olden times, and is said to be one of the oldest seasonings used. It has long been used as a food, medicine, and for cosmetic applications. Sesame seed, composed of 50% lipid and 20% protein, is one of the important oil seed crops in the world. Many scientific studies have found health- promoting effects of sesame [1].

Sesame: General Characteristics

Sesame is a small annual plant growing up to 1 m high cultivated for its edible seeds, which grow in pods, and the oil which can be extracted from these. The species has a long history of cultivation, mostly for its yield of oil. The original area of domestication of sesame is obscure but it seems likely to have first been brought into cultivation in Asia or India [2].

The plant is usually 60 to 120 cm tall and the fruit is a dehiscent capsule held close to the stem. When ripe, the capsule shatters to release a number of small seeds. The seeds are protected by a fibrous 'hull' or skin, which may be whitish to brown or black depending on the variety. 1000 seeds weigh some 4–8 g. The seeds have a high oil content of 44–60% [2].

The plant is deep rooting and well adapted to withstand dry conditions. It will grow on relatively poor soils in climates generally unsuitable for other crops, and so it is widely valued for its nutritional and financial yield from otherwise inclement areas. It is well suited to smallholder farming with a relatively short harvest cycle of 90–140 days allowing other crops to be grown in the field. It is often intercropped with other grains [3].

Sesame is grown for its seeds, and the primary use of the sesame seed is as a source of oil for cooking. The young leaves may also be eaten in stews, and the dried stems may be burnt as fuel with the ash used for local soap making, but such uses are entirely subordinate to seed production.

Nutrients Found in the Sesame Seeds

Sesame seeds are a very good source of manganese, copper, calcium, magnesium, iron, phosphorus, vitamin B1, zinc and dietary fiber [2]. In addition to these important phytonutrients, sesame seeds contain two unique substances: *sesamin* and *sesamolin* [4]. Both of these substances belong to a group of special beneficial fibers called *lignans*, and have been shown to have a cholesterol-lowering effect in humans, and to prevent high blood pressure and increase vitamin E supplies in animals. Sesamin has also been found to protect the liver from oxidative damage [5]. An in-depth analysis of other nutrients present in sesame seeds are shown in Table I [6].

M.M. Srivastava, L.D. Khemani, S. Srivastava, *Chemistry of Phytopotentials: Health, Energy and Environmental Perspectives*, DOI:10.1007/978-3-642-23394-4_13, © Springer-Verlag Berlin Heidelberg 2012

Nutrients Found in the Sesame Seeds Oil

Like any oil, sesame oil is high in fat, much of which is the heart-healthy unsaturated variety [7]. The major components of sesame oil are shown in Table II [2]. Some of the valuable nutrients are discussed below.

Vitamin K: One cup of sesame oil provides 37 percent of the recommended dietary allowance for vitamin K, based on a 2,000-calorie diet.

Vitamin E: Sesame oil is also a good source of the antioxidant vitamin E, containing 75 % of the recommended daily allowance in 1 ounce of oil.

Omega-3 Fatty Acids: One tbsp. of sesame oil contains about 40 mg of omega-3 fatty acids.

Sesamin: Sesamin is a lignan, a plant compound that offers nutritional benefits to the body.

Lecithin: A molecule that plays a role in the prevention of vascular diseases, improves blood vessel elasticity, and hinders cholesterol buildup in the arteries.

Sesame oil also contains small amounts of magnesium, copper, calcium, iron, silicic acid, phosphorus and vitamins A and B.

Table 1: Nutritional Value per 36.00 gm of Sesame Seeds

Nutrient	Amount
calories	206.28
calories from fat	160.92
calories from saturated fat	22.56
protein	6.40 g
carbohydrates	8.44 g
dietary fiber	4.24 g
soluble fiber	0.88 g
insoluble fiber	3.36 g
sugar – total	0.40 g
disaccharides	0.32 g
other carbs	3.80 g
fat – total	17.88 g
saturated fat	2.52 g
mono fat	6.76 g
poly fat	7.84 g
trans fatty acids	0.00 g
cholesterol	0.00 mg
water	1.68 g
ash	1.60 g
Vitamins	
vitamin A IU	3.24 IU
vitamin A RE	0.36 RE
A – carotenoid	0.36 RE
A – retinol	0.00 RE
A – beta carotene	2.16 mcg
thiamin – B1	0.28 mg
riboflavin – B2	0.08 mg
niacin – B3	1.64 mg
niacin equiv	3.44 mg
vitamin B6	0.28 mg
vitamin B12	0.00 mcg
biotin	3.96 mcg
vitamin C	0.00 mg
vitamin D IU	0.00 IU
vitamin D mcg	0.00 mcg
vitamin E alpha equiv	0.80 mg
vitamin E IU	1.20 IU
vitamin E mg	0.80 mg
folate	34.80 mcg
vitamin K	0.00 mcg
pantothenic acid	0.00 mg
Minerals	
calcium	351.00 mg
chloride	3.60 mg
chromium	-- mcg
copper	1.48 mg
iron	5.24 mg
magnesium	126.36 mg
manganese	0.88 mg
molybdenum	10.64 mcg
phosphorus	226.44 mg
potassium	168.48 mg
selenium	2.04 mcg
sodium	3.96 mg
zinc	2.80 mg
Saturated Fats	
myristic	0.04 g
palmitic	1.60 g
stearic	0.76 g

arachidic	0.08 g
Mono Fats	
palmitol	0.04 g
oleic	6.68 g
eicosen	0.04 g
Poly Fats	
linoleic	7.68 g
linolenic	0.12 g
Other Fats	
omega 3 fatty acids	0.12 g
omega 6 fatty acids	7.72 g
Amino Acids	
alanine	0.28 g
arginine	0.84 g
aspartate	0.52 g
cystine	0.12 g
glutamate	1.24 g
glycine	0.40 g
histidine	0.16 g
isoleucine	0.24 g
leucine	0.44 g
lysine	0.20 g
methionine	0.20 g
phenylalanine	0.28 g
proline	0.24 g
serine	0.32 g
threonine	0.24 g
tryptophan	0.12 g
tyrosine	0.24 g
valine	0.32 g

Source: George Mateljan Foundation (www.whfooods.org) [6]

Nutraceutical Uses of Sesame

According to *Morris* (2002) [3], many nutraceutical uses have been discovered from sesame. Sesame lignans have antioxidant and health promoting activities [8].

High amounts of both sesamin and sesamolin have been identified in sesame [4]. Both sesamin and sesamolin were reported to increase both the hepatic mitochondrial and the peroxisomal fatty acid oxidation rate. Sesame seed consumption appears to increase plasma gamma-tocopherol and enhanced vitamin E activity which is believed to prevent cancer and heart disease [9]. Cephalin from sesame seed has hemostat activity and ranges from 133,168 to 233,856 ppm [10]. Historically, fiber is used as an antidiabetic, antitumor, antiulcer, cancer preventive, cardioprotective, and laxative. Fiber ranges from 27,100 ppm to 67,000 ppm in the seed with up to 166,000 ppm in the leaf. Sesame seed contains lecithin which has antioxidant and hepatoprotective activity and ranges from 58 ppm to 395 ppm [10]. Lecithin is also likely effective for reducing hepatic steatosis in long term parenteral nutrition patients and a successful treatment for dermatitis and dry skin [11]. Myristic acid has cancer preventive capability and is found in sesame seed ranging from 328 to 1,728 ppm [10]. The oil was used during the 4th century by the Chinese as a remedy for toothaches and gum disease. Sesame oil is known to reduce cholesterol due to the high polyunsaturated fat content in the oil. Other uses include the treatment of blurred vision, dizziness, and headaches. The Indians have used sesame oil as an antibacterial mouthwash, to relieve anxiety and insomnia [7]. In addition, sesame oil contains large amounts of linoleate in triglyceride form which selectively inhibited malignant melanoma growth [3].

Table 2: Nutritional Value per 100 gm of Sesame oil

Nutrient	**Amount**
Energy	3,699 kJ (884 kcal)
Carbohydrates	0.00 g
Fat	100.00 g
-saturated	14.200 g
-Monounsaturated	39.700 g
-Polyunsaturated	41.700 g
Vitamin C	1.40 mg (9 %)
Vitamin E	13.6 μg (0 %)
Vitamin K, Calcium, Iron, Magnesium, Sodium, Phosphorus	0 mg (0 %)

Percentages are relative to US recommendations for adults. Source: USDA Nutrient database Available on Wikipedia [3]

Health Benefits of Sesame Seeds

According to Takano [12] the rich assortment of minerals in sesame seeds translates into the following health benefits:

Copper Provides Relief for Rheumatoid Arthritis

Copper is known for its use in reducing some of the pain and swelling of rheumatoid arthritis. In addition, copper plays an important role in the activity of lysl oxidase, an enzyme needed for the cross-linking of collagen and elastin – the ground substances that provide structure, strength and elasticity to blood vessels, bones and joints.

Magnesium supports vascular and respiratory Health. Studies have supported magnesium's usefulness in: Preventing the airway spasm in asthma, Lowering high blood pressure (which is a contributing factor in heart attack, stroke, and diabetic heart disease), Restoring normal sleep patterns in women who are experiencing unpleasant symptoms associated with menopause.

Calcium helps prevent colon cancer, osteoporosis, migraine and PMS.

Phytosterols of Sesame Seeds Lower Cholesterol:

Phytosterols are compounds found in plants that have a chemical structure very similar to cholesterol, and when present in the diet in sufficient amounts, are believed to reduce blood levels of cholesterol, enhance the immune response and decrease the risk of certain cancers.

As an Antioxidant – A special element called ***"sesame-lignin"*** helps get rid of free radicals that cause aging and cancer, including fatty acid production.

Cholesterol reduction – Sesame seeds are effective for atherosclerosis as it has been confirmed by various experiments.

Beauty treatment – Sesame seed and sesame oil are now commonly talked about among beauty enthusiasts. Modern science discovered that sesame seed has been used traditionally for beauty. Initially, scientific findings show that the seed and its oil can keep the skin healthy and make the hair stronger.

Stress prevention – Scientific studies indicate that stress contributes to range of diseases. Because of sesame seed's chemical structure, it has a unique ability to reduce tension and stress when used regularly. It also helps nourish the nervous system, and relieves fatigue and insomnia.

Cancer prevention – Sesame seeds contain *'phytate'*, which is one of the most powerful antioxidants and one of the most potent natural anti-cancer substances. It helps inhibit the growth of various cancer cells.

Relieves constipation – Sesame seeds contain a lot of fiber that provides roughage to the stool, which softens the contents in the intestines making it easier to eliminate.

Prevention of anemia and osteoporosis

Iron and calcium contents of sesame seeds are valuable in prevention of anemia, osteoporosis, etc.

Health Benefits of Sesame Oil

Sesame oil is made from sesame seeds and proves to be a rich source of polyunsaturated fats, monounsaturated fats, antioxidants, and several vitamins and minerals. Clinical studies and anecdotal reports suggest that this oil offers a number of potential health benefits and indeed can help one maintain a normal body balance. Sesame oil is an excellent source of polyunsaturated fatty acids including omega-3, omega-6 and omega-9. Polyunsaturated fatty acids are necessary for growth and development and strong evidence supports their role in the prevention and treatment of chronic diseases such as coronary heart disease, hypertension, diabetes and arthritis. Sesame oil has been shown to help lower blood pressure, increase good cholesterol, decrease bad cholesterol levels and help maintain normal blood pressure levels. These effects have been primarily attributed to the naturally high polyunsaturated fat content found in sesame oil [13].

Consumption of polyunsaturated fatty acids may also help to prevent osteoporosis caused by estrogen deficiency, although the exact mechanism remains unknown. Another factor that renders sesame oil a healthful food is that sesame seeds and their oil are rich in antioxidants. Antioxidants counter the effects of molecules in the body that damage cells and accelerate the aging process, including bacteria, inflammation and viruses. Sesame oil contains significant amounts of sesame lignans: sesamin, episesamin, and sesamolin. Lignans are compounds that are found in plants and are partially responsible for the antihypertensive and antioxidant properties of sesame oil [13]. Vitamin E is cardio-protective and has been shown to reduce risk of coronary heart disease. Vitamin E may also reduce cancer risk and preliminary evidence supports the role of vitamin E in the prevention of Alzheimer's disease and cataracts. Because sesame oil increases the fluidity and flexibility of membranes throughout the body, sesame oil is often recommended to treat conditions related to dryness which include cough, constipation, arthritis, and dry skin. Anecdotal evidence suggests that regular consumption of sesame oil reduces anxiety, enhances circulation, prevents disorders of the nerves and bones, boosts the immune system and prevents bowel problems. Additional claims associated with sesame oil use include increased vitality, alertness, better sleep, reductions in chronic pain and muscle spasms, and even slowing of the aging process. A recent study in hypertensive diabetic patients showed that sesame oil supplementation for 45 days decreased systolic and diastolic blood pressure, body weight, body mass index, waist girth, hip girth, waist-to-hip ratio, glucose, glycosylated hemoglobin (a measure of long-term blood sugar control), total and LDL cholesterol and triglycerides. Sesame oil increases leptin (a hormone that regulates body weight) levels in the circulation, which may contribute to weight loss. Another recent study conducted at Louisiana State University reported that sesame oil consumption reduced total cholesterol, bad cholesterol and triglyceride levels [14].

Considerations

While a healthful addition to the diet, too much of a good thing can certainly backfire. According to the American Heart Association, not more than 10% of daily calorie intake should come from polyunsaturated fat. Therefore, despite the overwhelmingly positive health effects, sesame oil should be consumed in moderation as sesame oil is dense in calories. One tablespoon serving has 120 calories and 14 grams of fat. People suffering from digestive problems should avoid sesame seeds as they can intensify the disorder. Overconsumption of sesame oil may result in weight gain because of its high caloric content [15].

Conclusion

Although, the benefits of including sesame in daily diet have been acknowledged, there is still insufficient awareness and motivation among general masses regarding the usage of sesame oil. Hence, it is important to carry out further investigations in this direction and adopt appropriate methods for spreading the information for the benefit of society. Attention also needs to be given to the cultivation and processing aspects of sesame plant to get a better yield of it. To sum up it can be said that research based evidence of positive health effects of sesame oil suggests that sesame oil may play a future role in prevention, treatment of chronic diseases, and assuring a healthy life.

References

1. H. Mannan et al. (2008). Gamma Tocopherol Content of Iranian Sesame Seeds *In Iranian Journal of Pharmaceutical Research (2008),* 7 (2): 135–139
2. Sesame Oil from *Wikipedia*, the free encyclopedia. Retrieved on 26.04.2011 from http://en.wikipedia.org/wiki/Sesame_oil
3. J.B. Morris. (2002). Food, Industrial, Nutraceutical, and Pharmaceutical Uses of Sesame Genetic Resources Reprinted from: *Trends in new crops and new uses. 2002.* J. Janick and A. Whipkey (eds.). ASHS Press, Alexandria, VA.
4. S. Sirato-Yasumoto et al. (2001). Effect of sesame seeds rich in sesamin and sesamolin on fatty acid oxidation in rat liver. J. Agric. Food Chem., 49: 2647–2651. http://www.bioline.org.br/request?jb07431D.K.
5. K.D. Caulman, (2005). Whole Sesame Seed Is as Rich a Source of Mammalian Lignan Precursors as Whole Flaxseed *NUTRITION AND CANCER, 52*(2), 156–165
6. http://www.whfoods.com/genpage.php?tname=nutrientprofile&dbid=19
7. G. Annussek. (2001). Sesame oil. In *Gale encyclopedia of alternative medicine*. Gale Group and Looksmart.
8. M.J. Kato, A. Chu, L.B. Davin, N.G. Lewis. (1998). Biosynthesis of antioxidant lignans in *Sesamum indicum* seeds. Phytochemistry 47:583–591.
9. R.V. Cooney, L.J. Custer, L. Okinaka, and A. Franke. (2001). Effects of dietarysesame seeds on plasma tocopherol levels. Nutrition & Cancer 39:66–71.
10. S.M. Beckstrom-Sternberg and J.A. Duke. (1994). The Phytochemical Database. ars-genome.cornell.edu/cgi-bin/WebAce/webace?db= phytochemdb.
11. J.M. Jellin, P. Gregory, F. Batz, K. Hitchens et al. (2000). Pharmacist's letter/prescriber's letter natural medicines comprehensive database. 3rd ed Therapeutic Research Faculty, Stockton, CA. p. 1–1527.

12. J. Takano. Retrieved on 24.05.2011 from http://www.pyroenergen.com/articles10/sesameseeds.htm
13. L. Posch. ( ). Health Benefits of Sesame oil. Retrieved on 14–04–2011 from http://www.asianfoodrecipes.com/Health&Nutrition/Health_Benefits_of_Sesame_Oil.php
14. C. Bullock (2003). *Sesame oil helps reduce dose of blood pressure lowering medicine.* Retrieved on 16.05.2011 from http://eurekalert.org/pub_releases/2003–04/aha-soh042803.php
15. http://www.fatfreekitchen.com/cholesterol/cookingoil.html
16. D.E. Smith and J.W. Salerno. (2001). Essential Fatty Acids. 46:145–150.
17. http://www.buzzle.com/articles/sesame-oil benefits.html

14

Identification of Flavonoids in The Bark of *Alstonia Scholaris* by High Performance Liquid Chromatography-Electrospray Mass Spectrometry

Rahul Jain, S. Chaurasia, R. C. Saxena, and D. K. Jain
Pest control and Ayurvedic Drug Research Lab,
S. S. L. Jain P. G. College, Vidisha, M. P.
Email: rkjmb82@yahoo.com; ambrahul@gmail.com

Abstract

Alstonia scholaris is traditionally known for its medicinal properties. It contains many phenolic compounds that may have potential as antioxidant. High performance liquid chromatography coupled with electrospray mass spectrometry (LC-ESI/MS) was used for the identification of flavonoids in the bark of Alstonia scholaris. Mobile phase used in the chromatographic separation was 2mM ammonium acetate buffer (solvent-A, pH-2.5 with acetic acid) and acetonitrile (solvent-B) with gradient programming. The peaks were identified by the comparison of retention time, UV-vis spectroscopic and mass spectrometric data with authentic standards and/or literature data. The identified flavonoids included three anthocyanins (cyanidin 3-O-galactoside, cyanidin 3-O-glucoside and malvidin 3-O-glucoside) and three flavonol glycoside (quercetin 3-O-galactoside, quercetin 3-O-glucoside and kaempferol 3-O-glucoside).

Introduction

Alstonia schloris Linn is used extensively in various parts of the world against a wide range of ailment [1,2]. It is believed that the major active dietary constituents attributed to these protective effects are flavonoids [3,4]. The health related properties of flavonoids are due to their antioxidant activity.

During the study on the antioxidant and other protective activity of some medicinal plants, *Alstonia schloris* was found to have higher antioxidant activity. The objective of this research was to identify the flavonoids constituents in the bark of *Alstonia schloris* using High performance liquid chromatography-electrospray ionization mass spectrometry (LC-ESI/MS).

Experimental

All flavonoid standards were of HPLC grade. cyanidin 3-O-galactoside, cyanidin 3-O-glucoside and malvidin 3-O-glucoside were obtained from Polyphenol (Sandas, Norway). Quercetin 3-O-galactoside, quercetin 3-O-glucoside and kaempferol 3-O-glucoside was purchased from Indofine Chemical Co Inc (Hillsborough, NJ). Distilled and deionized water was further purified by a Milli-Q water system (Millipore ltd., Watford, U. K.) and used for all chromatographic analysis and sample standard preparations. All other solvents were of HPLC grade and were purchased from Qualigens fine chemicals.

The bark of Alstonia schloris Linn were collected from vidisha, MP, India during the month of October and identified by Dr. S. K. Jain dept of botany S. S. L. Jain P. G. College Vidisha.

Extraction and Isolation

The barks were dried in shade for a week. Then powdered and extracted with 90% alcohol and water by Soxhlet extraction (24hrs) to yield extract. The extract was filtered with filter paper under reduced pressure. The residue was resuspended in 100 ml of the same solvent and extracted for 5hrs. This extraction step was repeated twice. The combined alcoholic extract was concentrated with a rotary evaporator at 40 °C un-

M.M. Srivastava, L.D. Khemani, S. Srivastava, *Chemistry of Phytopotentials: Health, Energy and Environmental Perspectives*, DOI:10.1007/978-3-642-23394-4_14,

der reduced pressure to remove ethanol. The resulting solution was mixed with 200 ml of water. After filtration, the aqueous solution was applied to a 23 × 2.7 cm i. d. Diaion HP-20 absorption resin column (Supelco, Bellefone, PA), which was preconditioned by washing with ethanol and then equilibrated with water. Non phenolic impurities including sugars, amino acids, proteins and minerals were washed out with water (500 ml). Phenolic compounds were eluted from the resin with ethanol (200 ml), and the eluent was dried with a rotary evaporator under reduced pressure. The phenolic residue was redissolved in water (40mL) and freeze-dried to give a phenolic extract powder.

HPLC Analysis

HPLC analysis was performed using a Waters 2690 separation module system (Waters Associates, Milford, MA) equipped with an autosampler and a waters model 996 photodiode array detector. A phenomenex Luna C18 (2) analytical column (250 × 4.6 mm i. d., particle size, 5 µm) with a C18 guard column (Phenomenex, Torrance, CA) was used for separation. The binary mobile phase consisted of 2mM ammonium acetate buffer (solvent A, pH-250 with acetic acid) and acetonitrile (solvents B) and gradient program was as follows 0 %B to 15 %B in 45 min, 15 %B to 30 %B in 15 min, 30 %B to 0 %B in 5min. The flow rate was 1.0 ml/min for a total run time of 65min. The injection volume was 10µl for all samples. All standards except for anthocyanins were dissolved in methanol. The letter were dissolved in 1 %HCl in methanol. The detection were set at 280, 360 and 520 nm for simultaneous monitoring of the different group of phenolic compounds.

LC-ESI/MS Analysis

LC-ESI/MS analysis were performed with the same HPLC system as described above interfaced to a Waters micromass ZMD model mass spectrometer equipped with an ESI source, operated in both negative and positive ion modes.

Determination of Flavonoids

The *Alstonia schloris* bark powder was soaked in 80 % methanol containing 0.1 % HCl and kept in an airtight capped bottle at room temp for 2hrs. The suspension was then incubated at 45 °C in a water bath with continuous shaking for an additional 30min. after cooling to room temp the suspension was filtered by a syringe filter, and 10µl of the filtered extract was injected into HPLC analysis.

Compounds were tentatively identified by congruent retention times and UV-vis spectra with those of standards. Confirmation of identity was achieved by comparing the retention time and ESI-MS spectra of both standards and samples determined by LC-MS. All samples were prepared and analyzed in duplicate.

Results and Discussion

Once the LC-ESI/MS condition had been established for the compound studied, the bark extract was analyzed by the method in the full scan mode. Figure 1 shows the HPLC-UV-vis chromatograms of purified phenolic extract prepared from bark of *Alstonia schloris*. Identification of individual compounds was performed by comparison of LC retention time, photodiode array UV-vis spectroscopic, and ESI-MS spectrometric data (Table 1) with those of authentic standard or with published data. A total of 6 flavonoids were identified in the bark extract including three anthocyanins and three flavonol glycosides.

Anthocyanins

In the HPLC-vis chromatogram acquired at 520 nm three peaks (compound 1, 2 and 3) were detected. The UV-vis spectra of these compounds all showed strong absorption at 520 nm, which is characteristic of anthocyanin [5,6]. In mass chromatogram selected at m/z 449 and 479, three peaks (compounds 1, 2 and 3) were observed. The mass spectra of which showed their protonated aglycon ions $(A + H)^+$ to be *m/z* 287, 331 respectively corresponding to cyanidin and malvidin [7]. The protonated aglycon ion were all formed by loss of a sugar moiety with 162 units from their $(M+H)^+$, indicating that they are anthocyanidine monoglucoside. These suggest the presence of cyanidin 3-O-galactosides, cyanidin 3-O-glucosides and malvidin 3-O-glucosides in addition to comparison of their retention time, UV spectra and ESI mass spectra patterns with those of authentic standards.

Flavonol Glycoside

The HPLC-UV chromatogram acquired at 360 nm, three major peaks (Compound 4,5 and 6) with typical flavonol UV spectra was observed. The maximum absorption of these compounds were at about 265 nm and 360 nm, characteristic of flavonol compounds [7,8]. In mass chromatogram selected at m/z 463 and 447, three peaks (compounds 4, 5 and 6) were observed. The mass spectra of which showed their protonated aglycon ions $(A-H)^-$ to be m/z 301, 285 respectively corresponding to quercetin and kaempferol [9]. The protonated aglycon ion were all formed by loss of a sugar moiety with 162 units from their$(M-H)^-$, indicating that they are flavonol monoglucoside.. These suggest the presence of quercetin 3-O-galactosides, quercetin 3-O-glucosides and kaempferol 3-O-glucosides in addition to comparison of their retention time with those of authentic standards.

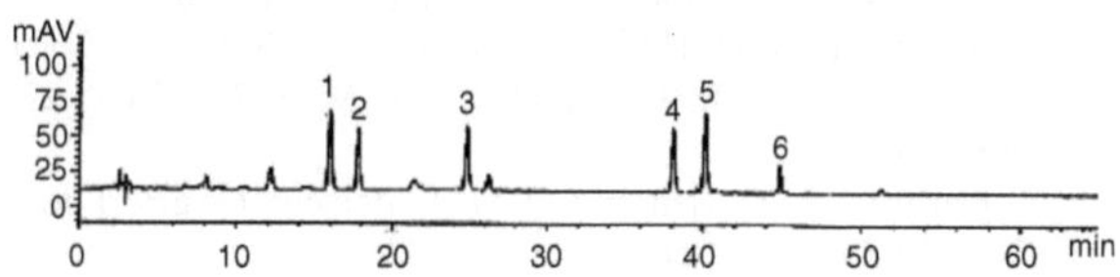

Fig. 1: HPLC chromatogram of purified phenolic extract of bark of *Alstonia scholaris*. 1, cyanidin 3-O-galactoside; 2, cyanidin 3-O-glucoside; 3, malvidin 3-O-glucosides; 4, quercetin 3-O-galactoside; 5, quercetin 3-O-glucoside; 6, kaempferol 3-O-glucoside.

Table 1: Identification of anthocyanins and flavonol glycosides in the bark extract of *Alstonia scholaris* based on HPLC retention time (t_R), UV-vis spectroscopic characteristics (λmax) and ESI/MS spectrometric pattern

Anthocyanins

Peak	t_R (min)	Identified compounds	λmax (nm)	Molecular ion $(M+H)^+$	Aglycon ion $(A+H)^+$	Compare with standard
1	15.92	cyanidin 3-O-galactoside	280, 519	449	287	Yes
2	17.95	cyanidin 3-O-glucoside	280, 514	449	287	Yes
3	24.92	Malvidin 3-O-glucosides	279, 526	493	331	Yes

Flavonol Glycoside

Peak	t_R (min)	Identified compounds	λmax (nm)	Molecular ion $(M-H)^-$	Aglycon ion $(A-H)^-$	Compare with standard
4	38.34	quercetin 3-O-galactoside	267, 357	463	301	Yes
5	40.21	quercetin 3-O-glucoside	267, 354	463	300.301	Yes
6	44.81	kaempferol 3-O-glucoside	265, 353	447	285	Yes

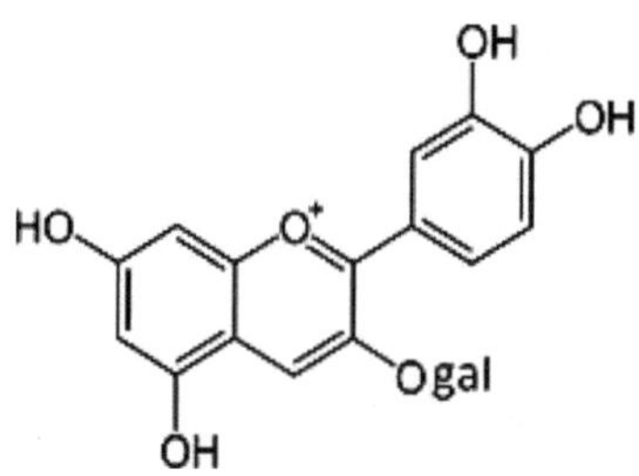

cyanidin 3-O-galactoside

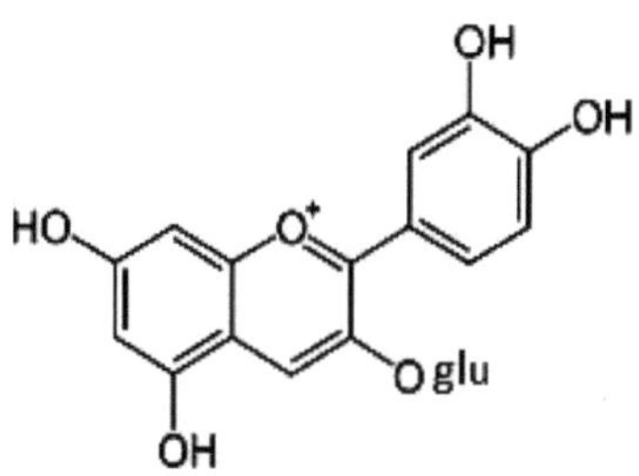

cyanidin 3-O-glucoside

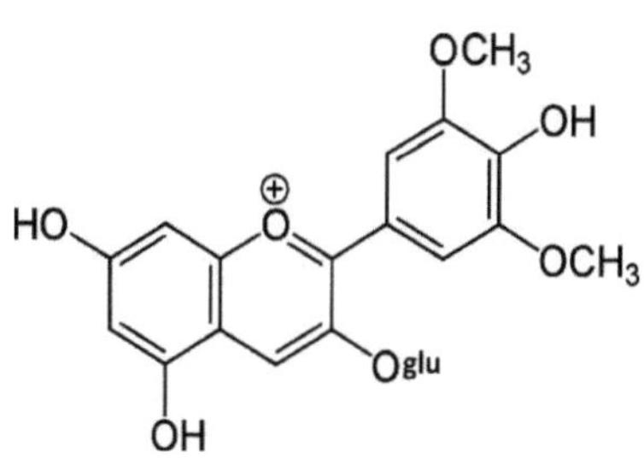

malvidin 3-O-glucoside

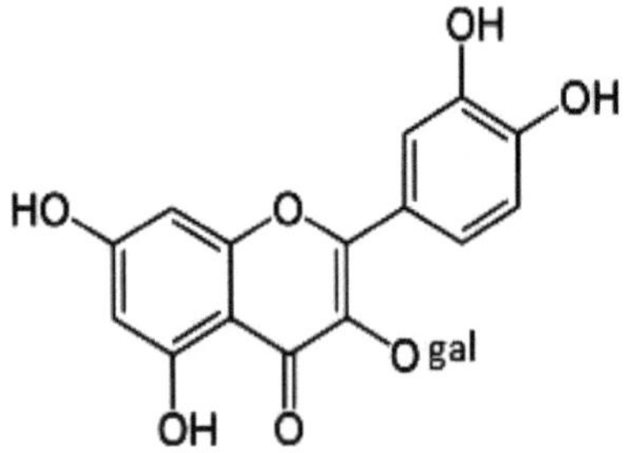

quercetin 3-O-galactoside

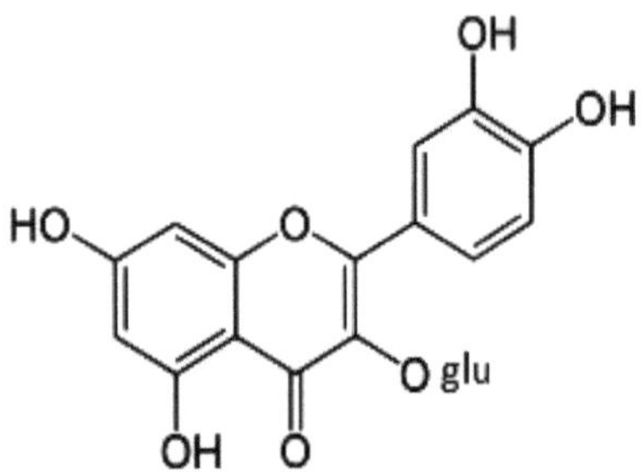

quercetin 3-O-glucoside

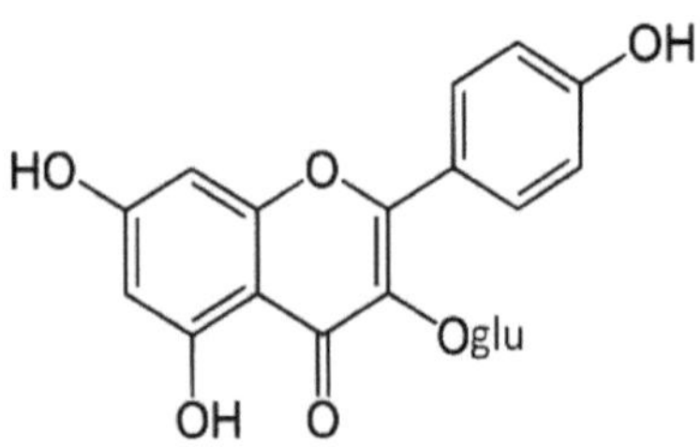

kaempferol 3-O-glucoside

Conclusion

Dietary intake of flavonoid antioxidant is shown to be related to various beneficial effects including risk cardiovascular diseases and certain forms of cancer. With regards of flavonoids in *Alstonia scholaris*, most studies focus on leaves of *Alstonia scholaris* [10,11]. Very limited information is available on other parts. This is the study on the composition of flavonoids present in the fruit of *Alstonia scholaris*. The high content of flavonoids present in the bark may contribute to high antioxidant activity observed for bark of *Alstonia scholaris*.

References

1. J.D. Hooker, *Flora of British India,* vol. 3, L. Reeve and Co, Ashford, Kent, (1882) p 641.
2. K.R. Kirtikar and B.D. Basu; *Indian Medicinal Plants,* vol. 2, L M Basu Publication, Allahabad, (1935) p 1565.
3. A. Sevanian and H. Hodis, Biofactors, 6 (1997) 385.
4. W. Ren, Z Qiao, H. Wang and L. Zhu, Med. Res. Rev., 23 (2003) 519.
5. L. Carbrita, T. Fossen and Q.M. Andersen; J Agric Food Chem., 68 (2000) 101.
6. E.B. Garcia, F. Cabello and E. Revilla; J Agric Food Chem., 51 (2003) 5622.
7. J.B. Harborne and C.A. Williams; Nat. Prod. Rep., 15 (1998) 631
8. L.Z. Lin, X.G. He, M. Lindenmarian, J.Yang, M. Cleary, S.X. Qui and G.A. Cordell; J Agric Food Chem., 48 (2000) 364.
9. R. Tsao, R. Yang, C. Young and H. Zhu; J Agric Food Chem., 51 (2003) 6947.
10. P. Versha, B. Ghosh, B. Anroop and M. Ramanjit; Indian Drugs, 40 (2003) 412.
11. A.P. Macabco, K. Krohn, D. Gehle, R.W. Read, J.J. Borphy and G.A. Cordell; Phytochemistry, 66 (2005) 1158.

15
Chemical Examination of *Morinda Pubescens* Var. Pubescens. (Rubiaceae) and Isolation of Crystalline Constituents

U.Viplava Prasad[1], B. Syamasunder[2], Anuradha. G[3] and J. Sree Kanth Kumar[4]
[1]Department of Organic chemistry, Andhra university, Visakhapatanam. A.P
[2]Principal, College of sciences, Acharya Nagarjuna University, Guntur. A.P
[3]Department of chemistry, Acharya Nagarjuna University, Guntur. A.P
[4] Department of Humanities and Sciences, Swarna Bharathi Institute of Science and Technology, Khamman, A. P
Email: viplav_31049@yahoo.com, profbsyamsundar@yahoo.co.in, anuradha.mythri@gmail.com, sreekanth.mythri@gmail.com

Abstract

Morinda pubescens var.pubescens (Rubiaceae) is a tropical plant whose leaves, bark, roots and fruits have been used as traditional remedy for various diseases. In our study we identified four Known compounds as two Pentacyclictriterpenoids Ursolic acid and Taraxerol, two sterols β-Sitosterol and Stigmasterol from Morinda pubescens var. pubescens (Rubiaceae). Solvent extraction, column chromatography were major techniques used for isolation of compounds, while structures were elucidated by integration of data from IR, UV, ^1H NMR AND ^{13}C NMR analysis. We are reporting presence of Taraxerol a pentacyclictriterpenoid first time from the leaves of Morinda pubescens var. pubescens of Rubiaceae family.

Introduction

The history of medicinal plants goes back to the history of human civilization. The ancient medical men through their trial and error mechanism identified herbs which are useful in healing various diseases and the same knowledge was passed on to the generations [1]. Even in recent times the plant products play very vital role in all the forms of medicine like Ayurveda, Sidda and Unnani. As the time proceeded the use of herbs transformed from raw form to purified form and extraction of chemical compounds leads to the synthesis of modern drugs [2]. The present study focuses on the Chemical Examination of Morinda pubescens var. pubsescens of Rubiaceae family and isolation of crystalline constituents.

Identification of the Plant Material

The plant was collected from Acharya Nagarjuna University Campus, Guntur. Andhra Pradesh, India. Based on Morphological and the Anatomical data [3] the plant was identified by B.S.I. Coimbatore as Morinda pubescens (J.E. Smith) var. pubescens of Rubiaceae family.

Chemical Composition of *Morinda* Species

The various chemical compounds were extracted from Morinda species like Anthraquinones [4, 10], Flavonoids [5], Iridoids [6, 9], Steroids [7] and Terpenoids [8].The molecular formula, Structure was elucidated based on melting point; I.R,U.V and ^{1}HNMR data in the present work.

Extraction Procedure Adopted

The leaves were dried at room temperature. Dried and powdered leaves are weighed. The weight of the material is 690 Grams. The powdered leaves are taken in a 5-liter Soxhlet apparatus and 4 liter of acetone is added. The mixture is kept for 24 hours. The first 3 liters cold fraction is collected into 5 liters round bot-

M.M. Srivastava, L.D. Khemani, S. Srivastava, *Chemistry of Phytopotentials: Health, Energy and Environmental Perspectives*, DOI:10.1007/978–3-642–23394-4_15, © Springer-Verlag Berlin Heidelberg 2012

tom flask. The acetone extract-I was kept for distillation. The remaining solvent about 2.5 liters collected after distillation of cold extract-I is again added to the Soxhlet jar. The cold acetone fraction–I after dryness under reduced pressure yielded a residue of 7.770 grams. The acetone cold extract-II was collected next day about 3 liters in round bottomed flask and kept for distillation under reduced pressure yield around 3 grams of residue.

The remaining material after cold extraction, taken into the Soxhlet apparatus under controlled temperature the hot acetone extract –I was collected after successive 12 siphons. The methanol hot extract also collected after 6 siphons. The hot acetone extract yield 1.200 gms and the hot methanol extract give 3gms residue.

The cold acetone extract I, II and hot acetone extract-I mixed up; the substance weighted around 11.97 grams. The column is setup with 200 grams column silica gel (acme) 100 – 200 mesh in 1250 ml of n-Hexane and Benzene in 9:1 ratio. The residue (around 12 grams) is taken and dissolved in acetone and impregnated in silica gel and the acetone is evaporated on water bath. The powdered substance with silica gel is further fine powdered (without granules). The fine powder is mixed with 9:1 Hexane: Benzene solvent. The column thus set is run with n-hexane, Benzene, ethyl acetate, ethyl alcohol (9:1 to 1:1) and finally with ethyl acetate.

Materials and Methods

General procedure: – TLC was carried out on percolated silica gel 60 F254 plates (Acme). Spots were detected under UV (254 and 366 nm) before and after spraying with 1% methanol sulfuric acid spray reagent, followed by heating the plate at 110°C for 5 min. Preparative TLC was performed on percolated silica gel 60 F254 plates, layer thickness 0.5 mm (Merck) unless indicated otherwise. Column chromatography was carried out on silica gel 60 (acme 100–200mesh) and sephadex LH20.

Chemical Examination of Morinda Pubescens Var. Pubescens Leaves:

The air-dried leaves powder was soxheleted with, n-hexane, Acetone and methanol. The hot and cold extracts of Acetone and n-hexane were found to be similar on TLC these extracts were combined and chromatographed extensively to yield eight compounds which are designated as AMP-1 to AMP-8 (Table 1) only four of them were isolated in a pure form for this present study. They are AMP-1 (Fig 1), AMP-4 (Fig 2), AMP-5 (Fig 3) and AMP-8 (Fig 4). The methanolic extract was in the form of greenish gummy substance. Hence the work was not pursued further.

Table 1: TLC of compounds AMP1–8

Compound Code	Yield	TLC Solvent System	R_f value
AMP – 1	70 mg	n-Hexane : Ethyl acetate 9:1	0.59
AMP – 2	30 mg	n-Hexane : Ethyl acetate 8:2	Long Streaks with no promising spots
AMP – 3	60 mg	n-Hexane : Ethyl acetate 8:2	No promising spot
AMP – 4	150 mg	n-Hexane : Ethyl acetate 4:1	0.78
AMP – 5	120 mg	n-Hexane : Ethyl acetate 4:1	0.71
AMP – 6	40 mg	n-Hexane : Ethyl acetate 1:1	0.61, Long streaks green color
AMP – 7	50 mg	n-Hexane : Ethyl acetate 1:1	0.66, Long streaks green color
AMP – 8	75 mg	n-Hexane : Ethyl acetate 9:1	0.59 with shot tailing

Examination of AMP – 1 :- (Urosolic Acid XCI)

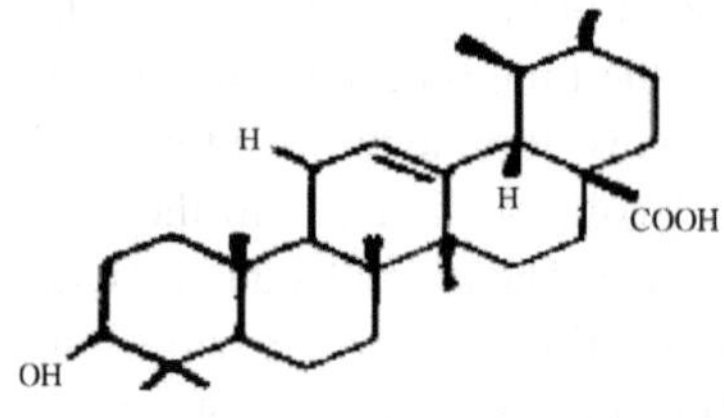

Fig. 1: structure of urosolic acid XCI

AMP-1 was colorless solid and recrystallized from ethylacetate gave a colorless compound. Indicated it is as triterpene. It give effervescences with $NaHCO_3$ and Na_2CO_3, $[\alpha]_D^{24}$ + 72.2°C (methanol) soluble in hot $CHCl_3$, MeOH, ethyl acetate. Melting point of the compound is 285.5°C. L. B. Test gave +ve result with pink to violet coloration. Molecular formula C_{30} H_{48}

O_3.Element analysis Found C, 83.72%, H, 11.78% Requires: C, 83.82%; H, 11.82%. I.R.: 3421, 1620 Cm^{-1} . ^1H (400, MHz, $CDCl_3$): δ0.8 (3H, S 24-Me), 0.84 (3H, S, 28-Me), 0.90 (6H, S,25, 29-Me), 0.95 (3H, S, 30-Me), 0.98 (3H, s, 23-Me), 1.20 (6H, S, 26, 27-M2), 3.15 (1H, M,H-3), 5.2 (1H dd, 15-H).

Examination of AMP – 4 :- (β-Sitosterol LXXXVI)

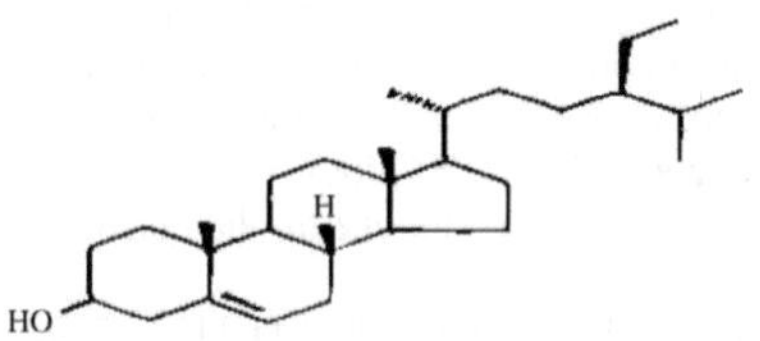

Fig. 2: structure of β-Sitosterol LXXXVI

AMP-4 isolated as white colored compound and recrystallized from methanol as colorless needles (150 mg) melting point is 135–136°C, $[\alpha]_D^{27}-35^0$, (C, 1.0, $CHCl_3$) positive L.B. test for sterols (violet → blue → green)

Molecular formula C_{29} H_{50} O. Element analysis Found : C, 83.86%, H, 11.90%Requires: C, 83.98%; H, 12.16% Rf : 0.67 (n-hexane, ethyl acetate-4:1).U.V.: No prominent absorption was noticed. I.R.:ν_{max}^{Nujol}: 3436, 1690, 1607, 1460, 1366, 985 Cm^{-1}.^1H NMR (400, MHz, $CDCl_3$) : δ0.6 (3H, S 18-Me), 0.81 (6H, d, J = 7 Hz, 26, 27-Me), 0.84 (3H, t, 29-Me), 0.89 (3H, d, J=6.3 Hz, 21-Me), 0.99 (3H, s, 19-Me), 3.2 (1H, m, 3_α-H], 5.3 (1H, m, 6-H),^{13}C NMR (125.70, MHz, $CDCl_3$) : δ36.1(C-1), 39.7 (C-2), 71.8 (C-3), 56.8 (C-4), 140.7 (C-5), 121.1 (C-6), 29.1 (C-7), 28.9 (C-8), 50.0 (C-9), 37.2 (C-10), 31.8 (C-11), 36.4 (C-12), 42.2 (C-13), 42.1 (C-14), 11.7 (C-15), 28.2 (C-16), 56.8 (C-17), 11.8 (C-18), 18.9 (C-19), 29.6 (C-20), 0.98 (C-21), 21.1 (C-22), 28.2 (C-23), 51.2 (C-24), 24.2 (C-25), 21.0 (C-26), 31.6 (C-27_, 19.4 (C-28), 19.3 (C-29).

Examination of AMP – 5 :- (Stigmasterol XCVII)

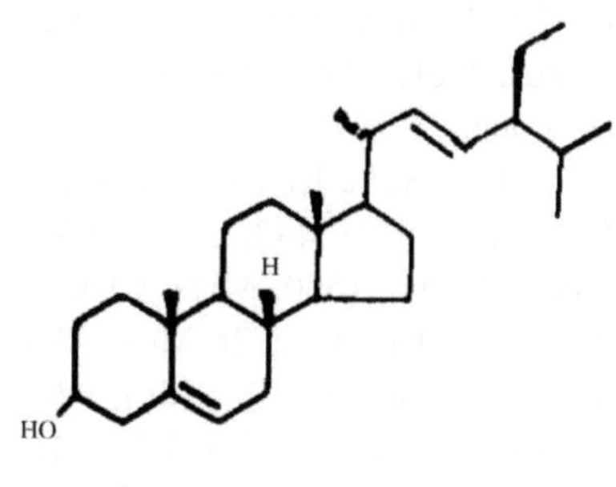

Fig. 3: Structure of Stigmasterol XCVII

AMP-5 was crystallized from chloroform-methanol as needles (120 mg). Melting point was 168 – 170°C, $[\alpha]_D$-45° (C, 1.2 $CHCl_3$). It showed pink color, which finally turned to green. It gave positive L.B. test for steroids. Molecular formula:C_{29} H_{48}O.Elementanalysis:Found:C,83.78%,H,11.60%Requires:C,84,40 %;H, 11.72% Rf: 0.65 (n-hexane, ethyl acetate-4:1) U.V.: No prominent absorption was noticed. I.R.: ν_{max}^{Nujol} 3442, 2928, 1693, 1457, 1387, 1031, 996 Cm^{-1}.^1H NMR (400, MHz, $CDCl_3$) : δ0.70, 0.80, 0.90, 1.00, 1.05, 1.30, (18H, 6xMe), 3.25 (1H, br, 3_α-H), 5.1 (2H, m, 22, 23-H), 5.3 (1H, m,6-H).^{13}C NMR (125.70 MHZ, $CDCl_3$):δ37.4 (C-1), 31.7 (C-2), 71.8 (C-3) 42.4 (C-4), 140.9 (C-5), 121.7 (C-6), 31.9 (C-7), 31.9 (C-8), 50.3 (C-9), 38.6 (C-10), 21.1 (C-11), 39.8 (C-12), 42.4 (C-13), 57.0 (C-14), 24.4 (C-15), 28.9 (C-16), 56.0 (C-17), 12.2 (C-18), 19.4 (C-19), 40.5 (C-20), 21.1 (C-21), 138.4 (C-22), 129.4 (C-23), 51.3 (C-24), 31.9 (C-25), 19.0 (C-26), 21.1 (C-27), 25.4 (C-28), 12.0 (C-29).

Examination of AMP – 8:- (Taraxerol XCVIII)

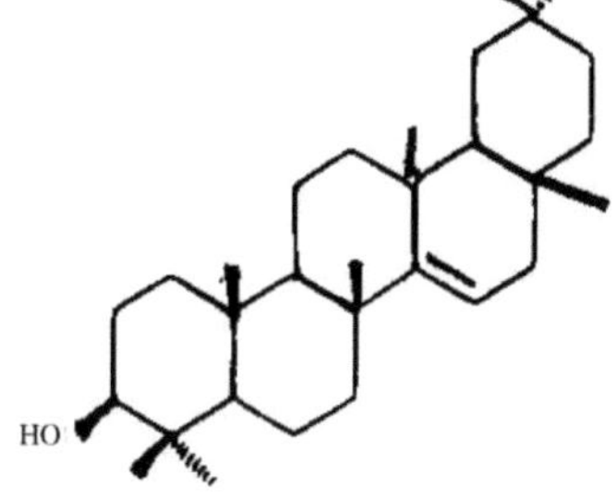

Fig. 4: Structure of Taraxerol XCVIII

AMP-8 was crystallized from chloroform-methanol, as crystalline solid (75 mg). Melting point 279 – 280°C, $[\alpha]_D^{20}$ + 0.72 (C, 0.97, $CHCl_3$). It showed pink color in L.B. test for triterpenes. Molecular formula: $C_{30}H_{50}O$. Element analysis: Found: C, 87.32 %, H, 11.63 % Requires: C, 84.45 %; H, 11.82 % Rf : 0.59 (n-hexane, ethyl acetate-9:1) I.R.: $\nu_{max}^{CHCL_3}$ 3442, 2933, 1694. 1458, 1376, 1033 Cm^{-1}. ^1H NMR (500, MHz, $CDCl_3$) : δ0.80 (3H, S, 24-Me), 0.84 (3H, S, 28-Me), 0.90 (6H, S, 25, 29 – Me), 0.95 (3H, S, 30 – Me), 0.98 (3H, S, 23-Me), 1.05 (6H, S, 26, 27 – Me), 3.15 (1H, m, H-3), 5.33 (1H, dd, J = 8.1 and 4.0 Hz, 15-H). ^{13}C NMR (150.8 MHz, $CDCl_3$): δ39.97 (C-1), 27.11 (C-2), 79.07 (C-3), 38.96 (C-4), 55.49 (C-5), 18.77 (C-6), 35.07 (C-7), 38.96 (C-8), 48.70 (C-9), 37.97 (C-10), 17.49 (C-11), 35.78 (C-12), 38.74 (C-13), 158.05 (C-14), 116.85 (C-15), 36.65 (C-16), 37.70 (C-17), 49.25 (C-18), 41.29 (C-19), 29.36 (C-20), 33.67 (C-21), 33.07 (C-22), 27.98 (C-23), 15.45 (C-24), 15.45 (C-25), 29.91 (C-26), 25.89 (C-27), 29.81 (C-28), 33.34 (C-29), 21.30 (C-30).

Acknowledgements

A special thanks to Prof. Y. L. N. Murthy, Dept. of Chemistry, Andhra university for his guidance and patience. Thanks to the IICT for providing spectral data.

References

1. P. Ranjan, Ethnobot. Studies in India, (1994) 5 148.
2. The wealth of India Raw materials. CSIR (1962), 423–425.
3. A. K. R. Kumari, D. Narasimhan, C. Livingstone and P.J; Raman, Phytomorphology, (2002) 52 (2,3) 207–215.
4. T. Akihisa, K. Matsumoto, H. Tokuda, K. Yasukawa, K. Seino, K. Nakamoto, H. Kuninaga, T. Suzuki, Y. Kimura, Jour. of Natu. Pro. (**2007**) 70 (5) 754–757.
5. T. Robak and R. J. Gryglewski, Bio chem. pharmacol. 37 (1988) 837 – 841.
6. S. Song, G. Lin Bioorg. Med. Chem. (2003) 11 2499 – 2502.
7. G. Mistra and N. Gupta, J. Inst. Chem. 54 (1982) 22.
8. S. M. Kupchan, H. Meshulam and A. T. Sneden, Phytochemistry 17 (1978) 767.
9. Bao-Ning Su, A. D. Pawlus, Hyun-Ah Jung, W. J. Keller, J. L. McLaughlin, A. Douglas Kinghorn, Jour. of Natu. Prod. 68 (4) (**2005**) 592–595.
10. Ye Deng, Y. Won Chin, H. Chai, W. J. Keller, A. D. Kinghorn .Jour. of Natu. Prod. 70 (**2007**) (12) 2049–2052.

16
Secretion of α-L-Rhamnosidase by Some Indigenous Fungal Strains Belonging to Penicillium Genera

S. Yadav[1], S. Yadava[2] and K. D. S. Yadav[3]
Department of Chemistry, DDU Gorakhpur University, Gorakhpur-273009
Email: dr_saritayadav@rediffmail.com

Abstract

Secretion of α-L-rhamnosidase by Penicillium purpurogenum MTCC-3010, Penicillium greoroseum MTCC-9424, Penicillium citrinum MTCC-3565, Penicillium corylophilum MTCC-2011, Penicillium brevicompactum MTCC-1999, Penicillium corylophilum MTCC-6492, Penicillium waksmanii MTCC-6480 in the liquid culture medium have been reported. The enzyme characteristics of α-L-rhamnosidase like K_m, pH and temperature optima have been determine using p-nitrophenyl α-L-rhamnopyranoside as substrate. The K_m values are found to be in the range 0.25–0.50 mM, pH optima values are found to be between 6.0–11.0 and temperature optima values are between 50–60 °C.

Introduction

α-L-rhamnosidase [EC.3.2.1.40] cleaves terminal α-L-rhamnose specifically from a large number of natural glycosides[1,2] such as naringin, rutin, hesperidin and quercetrin. The enzyme is widely distributed in nature and has been reported from animal tissues[3], yeasts[4], fungi[5] and bacteria[6].The enzyme has several biotechnological applications such as debittering of citrus fruit juices caused by naringin[7,8], hydrolysis of hesperidin to release hesperetin glucoside[9] which is an important precursor in sweetener production, preparation of many drugs and drug precursors by derhamnosylating the terminal L-rhamnose containing substrates[10] and preparation of prunin[11] and L-rhamnose. The enzyme has also been used in enhancing the aroma of wine[12] and derived beverages in grape juice[13].

Keeping in view, the biotechnological applications of this enzyme, we initiated studies on α-L-rhamnosidases secreted by some indigenous fungal strains belonging to *Penicillium* genera. Authors have reported the secretion of α-L-rhamnosidase by *Penicillium purpurogenum* MTCC-3010, *Penicillium greoroseum* MTCC-9424, *Penicillium citrinum* MTCC-3565, *Penicillium corylophilum* MTCC-2011, *Penicillium brevicompactum* MTCC-1999, *Penicillium corylophilum* MTCC-6492, *Penicillium waksmanii* MTCC-6480 in the liquid culture medium. The enzymatic characteristics of α-L-rhamnosidases like K_m, pH and temperature optima have been determine. Attempts have also been made to test the suitability of the enzymes for various applications.

Material and Methods

p-nitrophenyl-α-L-rhamnopyranoside, naringin, L-rhamnose, rutin, CM cellulose were purchased from Sigma Chemical Company, St. Louis (USA).All the chemicals including the protein molecular weight markers used in the polyacrylamide gel electrophoresis were procured from Bangalore GENEI Pvt. Limited Bangalore (India). All other chemicals were either from Merck Limited Mumbai (India) or from s. d. – fine CHEM limited Mumbai (India) and were used without further purifications.

The fungal strains *Penicillium purpurogenum* MTCC-3010, *Penicillium greoroseum* MTCC-9424, *Penicillium citrinum* MTCC-3565, *Penicillium corylophilum* MTCC-2011, *Penicillium revicompactum* MTCC-1999, *Penicillium corylophilum* MTCC-6492, *Penicillium waksmanii* MTCC-6480 were procured from MTCC Centre and Gene Bank, Institute of Microbial Technology, Chandigarh and were maintained in the laboratory on the agar slants as mentioned in

M.M. Srivastava, L. D. Khemani, S. Srivastava, *Chemistry of Phytopotentials: Health, Energy and Environmental Perspectives*, DOI:10.1007/978-3-642-23394-4_16, © Springer-Verlag Berlin Heidelberg 2012

the MTCC catalogue-2000.The fungal strains were screened for the secretion of the α-L-rhamnosidase in the liquid culture medium consisting of $CaCl_2$ 1.0 g, $MgSO_4.7H_2O$ 3.0 g, KH_2PO_4 20.0 g, $N(CH_2COONa)_3$ 1.5 g, $MnSO_4$ 1.0 g, $ZnSO_4.7H_2O$ 0.1 g, $CuSO_4.5H_2O$ 0.1 g, $FeSO_4.7H_2O$ 0.1 g, H_3BO_3 10.0 mg, sucrose 40.0 g, ammonium tartarate 8.0 g. water (Milli-Q)1000 ml. The sterilized liquid culture medium was improved by adding separately sterilized 2.0 g rice grain or 1.0 g corn cob or 1.0 g bagasse particles or 2.0 g orange peel or 0.005 g naringin or 0.005 g rhamnose or 0.002 g rutin or 0.002 g hesperidin. The fungal growth was allowed to take place in BOD incubator at different growth temperature under stationary culture conditions.

0.5 ml aliquots of culture media were withdrawn at the regular intervals of 24 hrs and analyzed for activity of α-L- rhamnosidases. 1.0 ml assay solution consists of 0.4 ml of 1 mM substrate in buffer (0.1 m citric acid–NaH_2PO_4 (McIlvaine buffer pH 3–7, Clarks and Lubs solution pH 8–10 and phosphate buffer solution pH 11–13), 0.1 ml crude enzyme and 0.5 ml same buffer at 55 °C temperature. The activity of α-L-rhamnosidase was assayed by using p-nitrophenyl-α-L-rhamnopyranoside as the substrate and monitoring the libration of p- nitrophenol spectrophotometrically at λ= 400 nm using the molar extinction coefficient value of 21.44 $mM^{-1} cm^{-1}$.

ture medium through four layers of cheese cloth and then centrifuged using Sigma refrigerated centrifuge model 3K30 at 8,000 rpm for 20 minutes at 4 °C to remove the particles.

The Michaelis-Menten behavior of the purified enzyme for the substrate p-nitrophenyl-α-L-rhamnopyranoside was determined by measuring the steady state velocity of the enzyme catalyzed reaction at different concentrations of p-nitrophenyl-α-L-rhamnopyranoside (0.05 mM to 1.5 mM) using the reported method. The K_m and V_{max} value were determined by linear regression analysis of the data points of the double reciprocal plot. The pH optimum of the purified enzyme was determined by using p-nitrophenyl-α-L-rhamnopyranoside as the substrate and measuring the steady state velocity of the enzyme catalyzed reaction in solutions of varying pH in the range 5.0 to 13.0. The steady state velocity was plotted against pH of the reaction solutions and pH optimum was calculated from the graph. The temperature optimum was determined by measuring the steady state velocity of the enzyme catalyzed reaction in solutions of varying temperatures (40–70 °C) using p-nitrophenyl- α-L-rhamnopyranoside as the substrate. The steady state velocity of the enzyme catalyzed reaction was plotted against the temperature of the reaction solution and temperature optimum was calculated from the graph.

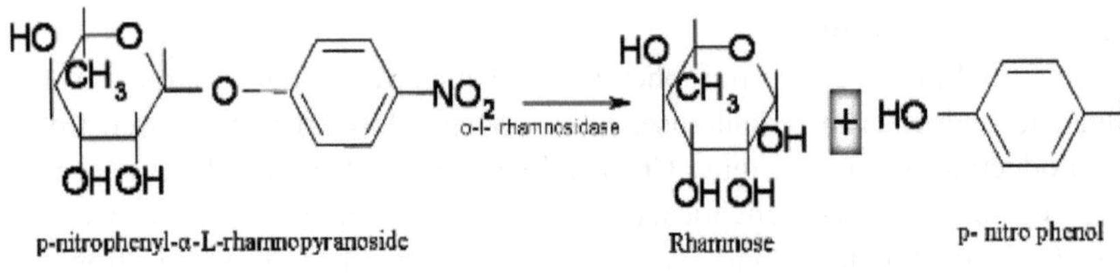

In order to maximize the secretion of α-L- rhamnosidase in the liquid culture medium the appearance of enzyme in the growth medium was followed with time using growth medium amended by the addition of rhamnose, rutin, naringin, hesperidin, corncob, bagasse, rice grain and orange peel. The maximum activity of the enzyme appeared between 3 to 11d after inoculation of spores for different fungal strains. The fungal strains were grown in fifteen sterilized 100 ml culture flasks each containing 25 ml liquid culture medium amended with different inducers and harvesting the culture on the day of maximum enzyme activity. The fungal mycelia were removed by filtering the cul-

Results and Discussion

The secretion of α-L-rhamnosidase the liquid culture media amended with different inducers for the different fungal strains of genera *Penicillium* are shown in Figures 1–7. The growth conditions and the fungal strains are described. Figure 1 shows the secretion of α-L-rhamnosidase in the growth medium of MTCC-3010. It is obvious from the figure that the enzyme secretion was maximum in the liquid culture medium amended with rutin it occurred on the 3rd d after the inoculation of the fungal spores and was 0.215 IU/ml of the culture medium. In the case of MTCC-9424,

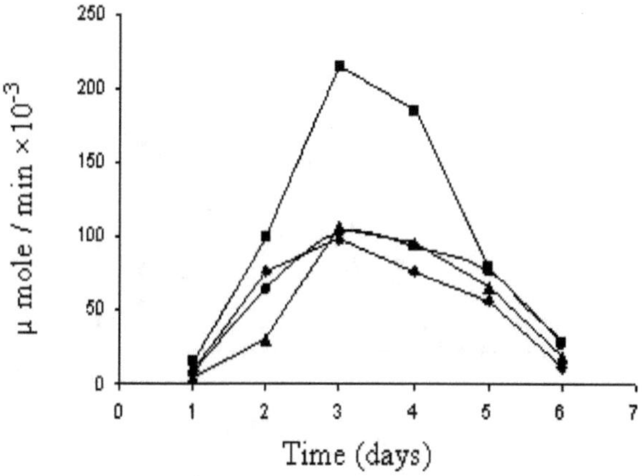

Fig. 1: Secretion of α-L-rhamnosidase in the growth medium of MTCC-3010. Medium blank (●), medium + rutin (■), medium + rhamnose (▲), medium + naringin (◆)

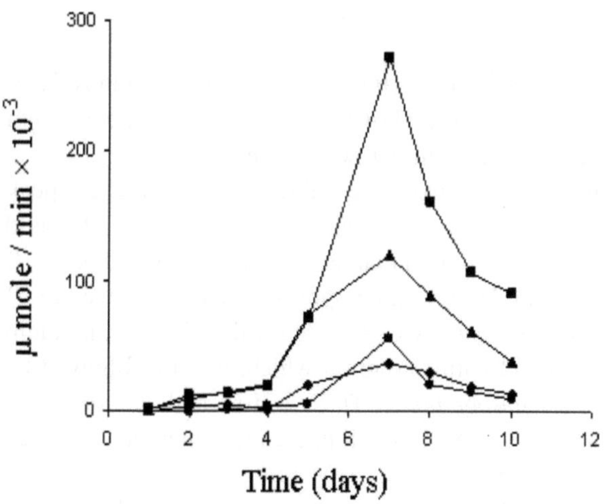

Fig. 2: Secretion of α-L-rhamnosidase in the growth medium of MTCC-9424. Medium blank (◆), medium + rice grain (■), medium + corncob (▲), medium + orange peel (●)

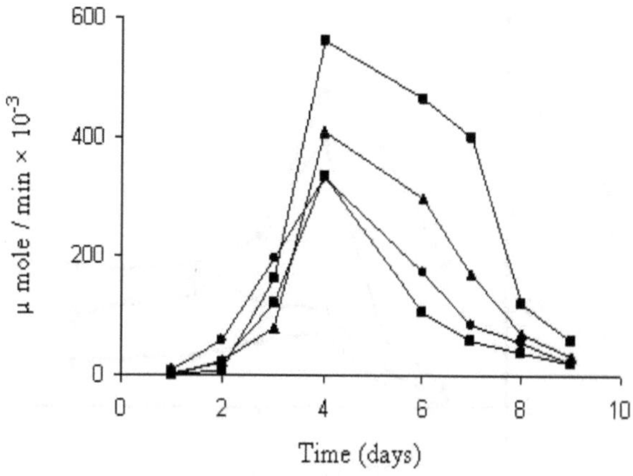

Fig. 3: Secretion of α-L-rhamnosidase in the growth medium of MTCC-3565. Medium blank (◆), medium + corn cob (■), medium + orange peel (▲), medium + rhamnose (●)

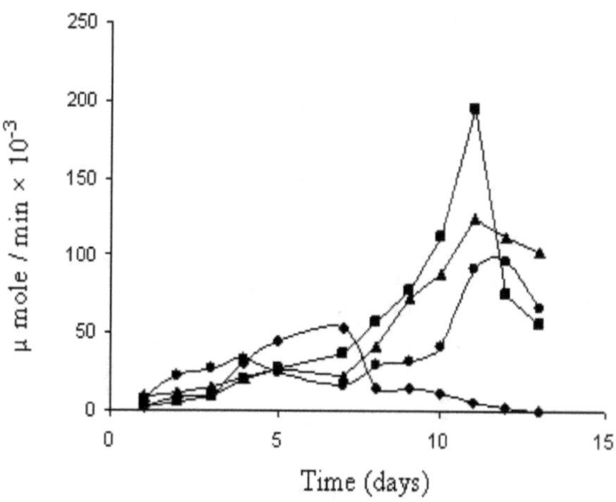

Fig. 4: Secretion of α-L-rhamnosidase in the growth medium of MTCC-1999. Medium blank (◆), medium + bagasse (■), medium + corncob (▲), medium + rice grain (●)

also shown in the Figure 2 the maximum secretion α-L-rhamnosidase occurred in the liquid culture medium amended with rice grain on 7th d after the inoculation of the fungal spores and was 0.271 IU/ml of the culture medium. In case of MTCC-3565 shown in figure 3 the maximum secretion of the enzyme took place on the 4th d after the inoculation of the fungal spores and 0.563 IU/ml occurred in the liquid culture medium amended with corn cob. In the case of MTCC-1999 the maximum secretion of enzyme was in the liquid culture medium amended with bagasse particles and occurred 0.195 IU/ml on the 11th d after the inoculation of the fungal spores in the liquid culture medium which is shown in the figure 4. The maximum secretion of α-L- rhamnosidase in the case of MTCC-2011, shown in the figure 5, took place in the liquid culture medium amended with corn cob and occurred on the 3rd d after the inoculation of the fungal spores and was 0.128 IU/ml of the culture medium. The enzyme secretion in the case of MTCC-6492, is shown in the figure 6, it was in the liquid culture medium on the 4th d after inoculation of the fungal spores and was 0.139 IU/ml of the culture medium. Figure 7 shows the secretion of enzyme in the liquid culture medium of MTCC-6480. It is obvious from the figure that the maximum enzyme secretion was in the growth medium amended with rutin on the 8th d after the inoculation of the fungal spores and was 0.211 IU/ml.

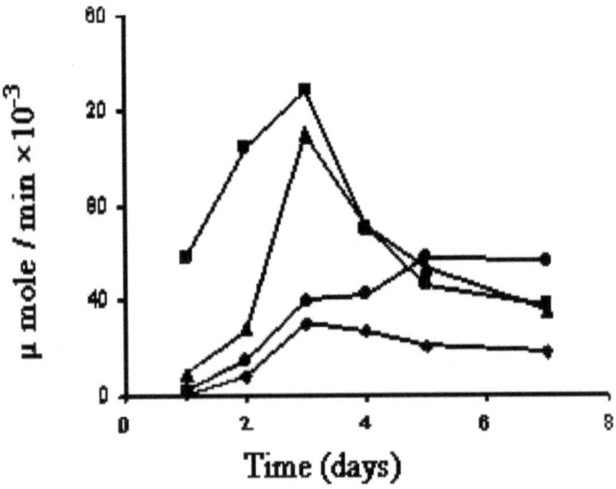

Fig. 5: Secretion of α-L-rhamnosidese in the growth medium of MTCC-2011. Medium blank (●), medium + corncob (■), medium + bagasse (▲), medium + naringin (◆)

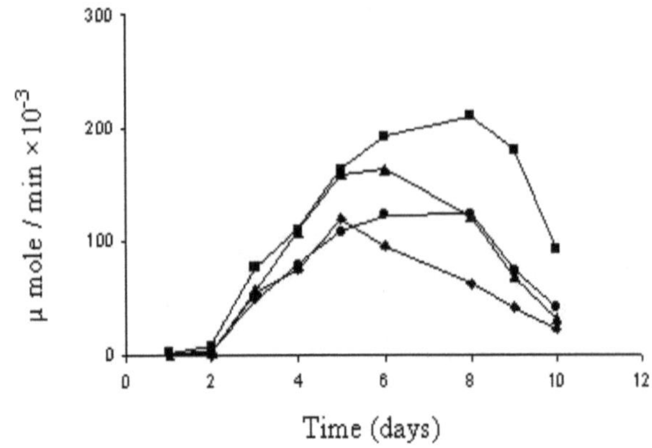

Fig. 7: Secretion of α-L-rhamnosidase in the growth medium of MTCC-6480. Medium blank (▲), medium + naringin (◆), medium + rutin (■), medium + hesperidin (●)

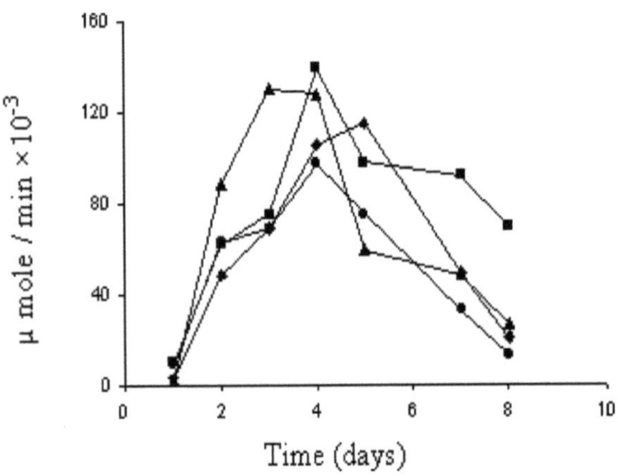

Fig. 6: Secretion of α-L-rhamnosidase in the growth medium of **MTCC-6492.** Medium blank (■), medium + naringin (◆), medium + bagasse (▲), medium + corn cob (●)

The enzyme characteristics of α-L-rhamnosidases secreted by the above fungal strains of Penicillium genera were summarized in Table 1.

It is obvious from the Table 1 that the highest peak value of enzyme production is in the case of liquid culture medium amended with corn cob containing *P. citrinum* MTCC-3565 spores which is followed by *P. greoroseum* MTCC-9424 in the liquid culture medium amended with rice grain which, in turn, followed by *P. purpurogenum* MTCC – 3010 and *P. waksmanii* MTCC – 6480 in the liquid culture medium amended with rutin.

The K_m value using p-nitrophenyl-α-L-rhamnopyranoside at the optimum pH and the optimum temperature mentioned in the table for α-L-rhamnosidase from different fungal strains varies from 0.25 to 0.5 mM. The optimum pH values of α-L-rhamnosidase

Table 1: Enzyme characteristics of α-L-rhamnosidese

Fungal Strains		Growth Temperature (°C)	Peak Value of enzyme produced IU/ml	K_m mM	pH	Temperature (°C)
P. purpurogenum	MTCC-3010	27	0.215	0.33	8.5	50.0
P. brevicompactum	MTCC-1999	25	0.195	0.48	10.0	60.0
P. corylophilum	MTCC-2011	25	0.128	0.50	11.0	55.0
P. corylophilum	MTCC-6492	30	0.139	0.43	8.0	57.0
P. citrinum	MTCC-3565	30	0.563	0.36	8.5	50.0
P. greoroseum	MTCC-9424	30	0.271	0.28	6.0	53.0
P. waksmanii	MTCC-6480	25	0.211	0.25	11.0	57.5

from different fungal strains varies from 6.0-11.0 where as temperature optima are between 50–60°C.

Though the secretion of α-L-rhamnosidase in the liquid culture medium containing different *Penicillium* strains are not very high, with the development in the area of molecular biology it would be possible to isolate the gene of the above enzymes and over express them in suitable vector so that the amount of enzyme needed for commercial applications could be produced. The reported studies in this communication will be useful in achieving the above objective.

Acknowledgements

Authors acknowledge the financial support of Department of Science and Technology, Government of India, New Delhi in the form of a Women Scientist-A Fellowship no. SR/WOS-A/CS-21/2009 awarded to Dr. Sarita Yadav, under which this research work has been done.

References

1. M. Puri and U.C. Banerjee; Biotechnology Advances.18 (2000) 207–217.
2. V. Yadav, P.K. Yadav, S. Yadav and K.D.S. Yadav; Process Biochem. 45 (2010), 1226–1235.
3. Y. Kurosawa, K. Ikeda and F. Egami; J. Biochem, 73 (1973) 31–37.
4. S. Qian, H. Yu, C. Zhang, Lu, M.H. Wang and F. Jin; Chem. Pharm. Bull. 53 (8) (2005) 911–914.
5. P. Manzanares, M. Orejas, E. Ibanez, S. Valles and D. Ramon ; Lett Appl Microbiol, 31(2000) 198–202.
6. I.S. Jang and D.H. Kim; Biol Pharm Bull. 19 (1996) 1546–1549.
7. G.M. Gray and A.C. Olson; J. Agric. Food Chem. 29 (6) (1981). 1299–1301.
8. S. Yadav and K.D.S. Yadav; J. of Scientific & Industrial Research. 59 (2000) 1032–1037.
9. T. Chase; Adv. Chem. Ser. 136(1974) 241–266.
10. A.A. Elujoba and R. Hardman; Fitoterapia. 58 (1987) 299–303.
11. M. Roitner, T.H. Schalkhammer, and F. Pittner; J. .Applied Biochem. & Biotech. 9 (1984) 483–488.
12. C. Caldiny, F. Bonomi, P.G. Pifferi, G. Langirini and Y.M. Alante; Enzyme Microb. Technol. 16 (1994) 286–291.
13. Z. Gunata, S. Bitteur, J.L Brillouet, C.Bayonove and R. Cordonnier; Carbohydrate Res. 184 (1988) 139–149.

17

Collection, Establishment, Acclimatization and Quantification of Shatavarin IV in the Medicinally Important Plant – *Asparagus racemosus* Willd

J. Chaudhary and P. K. Dantu
Department of Botany,
Faculty of Science, Dayalbagh Educational Institute (Deemed University),
Dayalbagh, Agra 282110
Email: premdantu@dei.ac.in

Abstract

Medicinal plants constitute an important group of non-wood forest products. Approximately 80 % of the world population uses plants as a source of medicine for healthcare. Domestication and large scale cultivation is a viable alternative and offers the opportunity to overcome the problems that are inherent with herbal drugs-misidentification, genetic and phenotypic variability, extract variability and instability, toxic components and contaminants. Harvesting from the wild, the main source of raw material, results in loss of genetic diversity and habitat destruction. Asparagus racemosus being a rejuvenating herb, its restorative action is beneficial in women's complaints and is mainly known for its phytoestrogenic properties. The tuberous roots of A. racemosus are rich in the saponins – Shatavarin I to IV and used in several diseases. Use of this plant is ever increasing and has become 'critically endangered'. However, production of this crop has remained consistently low to fulfill the higher market demands. The present paper discusses the strategy of in situ conservation by way of collection, establishment and acclimatization of A. racemosus germplasm collected from different parts of India. All plants grown in this study originated from different climatic zones. The plants that were acclimatized for over three years in the Botanic Garden of the Institute contained about 0.25 % and 0.31 % of Shatavarin IV by dry weight of roots in thick and thin roots, respectively.

Introduction

Medicinal plants are a vital source of natural drugs and thus constitute an important group of non-wood forest products. Approximately 80 % of the world population uses plants as a source of medicine for healthcare [1]. The economic value of plant based drugs is over US $ 60 billion and is likely to rise to US $ 5 trillion by the year 2050 [2]. In India, the medicinal plant sector is part of a time-honored culture that has traditionally occupied an important position in the socio-cultural, spiritual and medicinal arena of rural and tribal populations. The increased demand for natural herbs is putting pressure on the natural resources resulting in overexploitation, habitat destruction and loss of genetic diversity. There is an urgent need to conserve the biodiversity and meet the genuine needs through commercial cultivation. Domestication and large scale cultivation is the only viable alternative and offers the opportunity to overcome the problems that are inherent with herbal drugs collected from the wild – misidentification, genetic and phenotypic variability, extract variability, instability, toxic components and contaminants [3].

Asparagus belongs to the family Asparagaceae of the monocot order Asparagales. The genus *Asparagus* consists of herbs, shrubs and vines that are widespread all over the world and is a highly valuable plant species having therapeutic and nutraceutical importance in addition to being consumed as food [4]. It includes 150 species distributed throughout Tropical Asia, Africa and Australia. 17 species are reported from India from the Sal and mixed forests of Madhya Pradesh, Chattisgarh and Jharkhand [5]. *Asparagus racemosus* (locally known as Shatavar) is one of the important medicinal plants extensively used by the traditional

M.M. Srivastava, L.D. Khemani, S. Srivastava, *Chemistry of Phytopotentials: Health, Energy and Environmental Perspectives*, DOI:10.1007/978–3-642–23394-4_17, © Springer-Verlag Berlin Heidelberg 2012

practitioners in India for its medicinal value. Origin of *Asparagus racemosus* is uncertain but it is believed to be native to the East Mediterranean and the Middle East.

The tuberous roots of *A. racemosus* are medicinally important in several diseases. *A. racemosus* is distributed throughout tropical and sub-tropical parts of India up to an altitude of 1500 m [6]. It is straggling or scandent, much branched perennial, spinose, under shrub with fusiform, tuberous, fasciculated root stock. Tuberous roots contain four saponins, Shatavarin I to IV. The healing qualities of Shatavar are useful in a wide array of ailments. Being a rasayana or rejuvenating herb, its restorative action is beneficial in women's complaints. *A. racemosus* is mainly known for its phytoestrogenic properties. With increasing realization that hormone replacement therapy with synthetic estrogens is neither as safe nor as effective as previously envisaged, the interest in plant-derived estrogens has increased tremendously making *A. racemosus* particularly important [7, 8]. Roots of *A. racemosus* were found to possess antioxidant and anti-ADH [9], antitumor and anticancer [10, 11] anti-ulcerogenic [12] and antimicrobial activities [13]. The aim of the present study was to obtain information on suitability of the semi-arid climatic conditions of Agra, growth and biomass production, as well as quantitative and qualitative aspects of secondary metabolite produced under the adopted conditions with an ultimate aim of conserving and developing suitable cultivation practices for *A. racemosus*. To the best of our knowledge no such work has been undertaken so far for this region for this plant.

Materials and Methods

Germplasm Collection: The germplasm of *A. racemosus* was collected from different parts of India. The different land races were established in the net house of Botanic Garden at Dayalbagh Educational Institute, Agra. IC471921 and IC471923 were originally collected from Nauni Forest, Solan, Himachal Pradesh and IC471924 was collected from Ochlaghat, Solan, Himachal Pradesh. IC471910 was collected from Mandala, Madhya Pradesh. The landraces of *A. racemosus* used in the present study and their sources is given in Table 1.

Establishment: The crowns of *Asparagus* were planted in 6 to 8 inches deep trench in rows 4 to 5 feet apart, and 12 to 18 inches between crowns. Roots were spread out in trench with buds pointing upward covered with soil and irrigated. Subsequent irrigations were done at 4–6 day intervals until a month and thereafter at weekly intervals. Initially crowns were planted in the net house with 70 % shading net in 2007. Growth rate was slow in the net house, therefore, the rooted plants were transplanted to the open field in 2009. The plants were periodically treated with urea. As the crop is a climber it required support for its proper growth, which was provided by 4–6 feet long iron stands with circular frames at the top.

Acclimatization: After transfer to field spears sprouted from established plants. These spears developed into healthy plants with cladophylls at nodes.

Table 1: Collection sites of different landraces of *A. racemosus*

Name of the Line	Cultivar source
FFDC	Fragrance and Flavor Development Centre, Kannauj
SG1	Forest Department, Jabalpur
KAU	Thrissur (KAU), Kerala, India
CDH	Haryana Forest Department, Chandigarh
IC471921	NBPGR New Delhi
IC471923	NBPGR New Delhi
IC471924	NBPGR New Delhi
IC471927	NBPGR New Delhi
IC471911	NBPGR New Delhi
IC471910	NBPGR New Delhi
IC471909	NBPGR New Delhi
IC471908	NBPGR New Delhi

Shatavarin Quantification: Roots of accession no. IC471911 were harvested 18 months after establishment dried and ground to a fine powder. 8 g of root powder was weighed and extracted overnight in 50 ml of Ethyl Acetate:Methanol (1:1) on a shaker [14]. Filtrate was collected on Whattman filter paper and residue was re-extracted with 50 ml of Ethyl Acetate: Methanol (1:1). Both the filtrates were combined. The filtrate was concentrated in a Rotary evaporator (Eyela, Japan) and Freeze Drier (Allied Frost, India) and samples sent to M.F.P. Processing & Research Centre, Barkhera Pathani, Bhopal, India for HPLC analysis.

Results and Discussion

It is true that most plants can be propagated from root suckers but the success of this is dependent on factors such as the type of soil, rainfall and pests. To prevent root rot and fungal diseases lines were treated with dithane M45 periodically. The plants were healthy and produced several spears during the growing season (Fig. 1a). Two accessions IC471927 and IC471910 showed comparatively poor growth. Color of foliage and the time of flowering is given in Table 2. The plants flowered in the last week of October and first week of November while fruiting occurred in late November and continued till December (Fig. 1b).

Table 2: Variation in time of flowering and color of foliage in established lines of *Asparagus racemosus* in the Botanic garden

Name of the Line	Color of Foliage	Time of Flowering
FFDC	Medium Green	Mid November
SG1	Medium Green	Late October
KAU	Light Green	Late September
CDH	Dark Green	Late October
IC471921	Dark Green	Late October
IC471923	Dark Green	Late November
IC471924	Dark Green	Late November
IC471927	Dark Green	Late November
IC471911	Medium Green	Mid November
IC471910	Dark Green	Late November
IC471909	Medium Green	Mid November
IC471908	Medium Green	Late November

The landraces varied in terms of foliage color, spear pigmentation and berry color at maturity. There was variation in the time of flowering in different races. Those collected from Kerala region flowered earlier than other lines while some lines collected from FFDC, Kannauj flowered late in mid November.

Table 3: Qualitative and Quantitative analysis of the thick and thin roots of *Asparagus racemosus*

Tests	Thick Roots	Thin Roots
Length (cm)	26.8	20.9
Diameter (cm)	4.44	2.89
Moisture, % loss on drying	81.24	70.76
Saponins, %	4.62*	5.11*
Shatavarin IV, %	0.25*	0.31*

*Value expressed on dry weight basis.

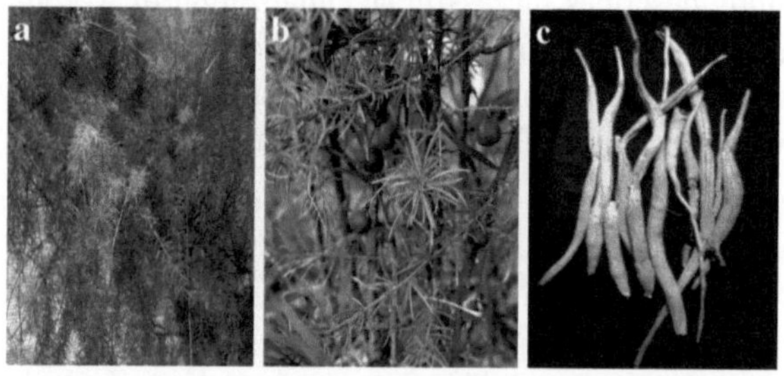

Fig. 1: **(a)** Healthy Plants after acclimatization in Botanic Garden of the Institute. **(b)** Fruiting in the month of November. **(c)** Roots harvested in the month March

Roots were harvested from accession no. IC471911 and separated into two distinct groups on the basis of diameter (Fig. 1c). Both the thick and thin roots were sent for saponin analysis. Interestingly, the thin roots showed greater content of the saponin and Shatavarin IV. The thick roots had 4.62% saponins and 0.25% Shatavarin IV, while the thin roots had 5.11% saponins and 0.31% Shatavarin IV (Table 3). Shatavarin IV as a marker has been identified in *A. racemosus* by HPLC with ELS detector [14], by HPTLC [15] and also has been identified in *A. gonoclados* [16]. Steroidal saponins have also been investigated in fruits of *A. racemosus* [17]. There are few reports on structural revision of Shatavarin I and IV from roots of *A. racemosus* [18, 19]. The quantity of both saponins and Shatavarin IV was much less than the average reported in the literature [15]. This low level of saponins and Shatavarin IV in the present study could be attributed to the fact that this was the first root harvest since establishment of the plant in Agra and the Shatavarin IV amount might increase as the plant stabilizes in the adopted home.

Introduction of new plant species with different economical value into the new environment and agronomic practices during the history have always had great importance. The introduction and acclimatization of any new plant species into a new environment is a long-term research. All plants grown in this study originated from different climatic zones characterized by differences in soil conditions, temperature and relative humidity. The cultivation system of *A. racemosus* needs additional research. Further studies on molecular modeling and structural studies of secondary metabolites is required. In conclusion, the present study describes successful acclimatization of *A. racemosus* collected from different climatic conditions in the semi-arid region of Agra may result in increased in vivo production of Shatavarin IV in roots.

Acknowledgements

The authors wish to thank the Director of the Institute for providing the laboratory facilities for carrying out this work. Funding by University Grants Commission, New Delhi, through a major project is gratefully acknowledged by PKD. JC wishes to thank the UGC, New Delhi, for the Rajiv Gandhi Fellowship. National Bureau of Plant Genetic Resources, New Delhi is gratefully acknowledged for providing the germplasm of *A. racemosus*.

References

1. Anonymous. Compendium on Phyto-medicine. Council of Development of Rural Areas. Gramin Chhetriya Vikas Parishad, Delhi 1997
2. Anonymous. Medicinal plants: Cure of 21st Century (Biodiversity Conservation and utilization of Medicinal Plants). Faculty of Forestry, University of Putra, Malaysia. 1998
3. P.H. Canter, H. Thomas and E. Ernst; Trends in Biotech. 23 (2005) 180
4. A.K. Shasnay, M.P. Darokar, D. Sakia, S. Rajkumar, V. Sundaresan and S.P.S. Khanuja; J. Med. Arom. Plant Sci. 25 (2003) 698
5. Anonymous. The Wealth of India, Raw materials, Publication and Information Directorate, CSIR, N. Delhi. 1987
6. S. Velvan, K.R. Nagulendran, R. Mahesh and V.V. Hazeena Begum; Pharmacognosy Rev. 1 (2007) 350
7. D. Grady, T. Gebretsadik, K. Kerlikowske, V. Ernster and D. Petitti; Obstetrics and Gynae 85 (1995) 304
8. E. Barrett-Connor; Ann Rev of Pub Health. 19 (1998) 55
9. J.P. Kamat, K.K. Boloor, T.P.A. Devasagayam and S.R. Venkatachalam; J. Ethnopharmacol. 71 (2000) 425
10. S. Diwanay, D. Chitre and B. Patwardhan; J. Ethnopharmacol. 90 (2004) 49
11. Y. Shao, C.K. Chin, T. Ho-Chi, W. Ma, S.A. Garrison and M.T. Huang; Cancer Lett. 104 (1996) 31
12. G.K. Datta., K. Sairam, S. Priyambada, P.K. Debnath and R.K. Goel; Indian J. Exp. Biol. 40 (2002) 1173
13. S.C. Mandal, C.K.A. Kumar, M. Lakshmi, S. Sinha, T. Murugesan, B.P. Saha and M. Pal; Fitoterapia 71 (2000) 686
14. S. Penumajji, V. Bobbarala and C.K. Naidu; J. Pharm. Res. 3 (2010) 159
15. N.K. Satti, K.A. Suri, P. Dutt, O.P. Suri, M. Amina, G.N. Qazi and A. Rauf; J. Liq. Chrom. & Related Technol. 29 (2006) 219
16. V. Madhavan, R.D. Tijare, R. Mythrey, M.R. Gurudeva and S.N. Yoganarasimhan; Ind. J. Nat. Products and Resources. 1 (2010) 57
17. D. Mandal, S. Banerjee, N.B. Mondal, A.K. Chakravarty and N.P. Sahu; Phytochem. 67 (2006) 1316
18. P.Y. Hayes, A.H. Jahidin, R. Lehmann, K. Penman, W. Kitching and J.J. De Voss; Tetrahedron Lett. 47 (2006) 6965
19. P.Y. Hayes, A.H. Jahidin, R. Lehmann, K. Penman, W. Kitching and J.J. De Voss; Tetrahedron Lett. 47 (2006) 8683

18
Chemical Composition and Biological Activities of Essential Oils of *Cinnamomum Tamala, Cinnamomum Zeylenicum* and *Cinnamomum Camphora* Growing in Uttarakhand

R. Agarwal[1], A. K. Pant[2] and O. Prakash[3]
Department of Chemistry, College of Basic Science & Humanities,
G. B. Pant University of Agriculture & Technology,
Pantnagar-263 145, Uttarakhand.
Email: ruchiagarwal19.ru@gmail.co

Abstract

Studies were undertaken to chemically examine the essential oils of different species of Cinnamomum growing in Kumaon region of Uttarakhand and screening of their antibacterial, and antioxidant activities. Eight samples of Cinnamomum tamala, two samples of Cinnamomum zeylenicum and one sample of Cinnamomum camphora were collected. Among all these oils, four chemotypes of C. tamala, two of C. zeylenicum and one of C. camphora are introduced by GC-MS analysis. Antibacterial activity of essential oils of Cinnamomum species were tested against three pathogenic bacteria viz. Pasturella multocida, Escherichia coli and Salmonella enterica enterica by disc diffusion method and compared with Gentamicine. Antioxidant activities of the essential oils were evaluated by three different methods viz. reducing power activity, DPPH radical scavenging activity and effect on the chelating activity of Fe (II) ions. Results showed that the essential oils of Cinnamomum species have effective antimicrobial and antioxidant activities. Therefore they could be used as food preservatives and as medicines.

Introduction

Spices are dried seeds, fruits, roots, barks, leaves and vegetative substances used for nutritive purposes in insignificant quantities as food additive for the purpose of flavor, color, or as preservatives. Many of these substances are also used for other purposes such as medicines, in religious rituals, as cosmetics, in perfumery or as vegetables [19].

Family Lauraceae or Laurel family comprises a group of flowering plants included in the order Laurales [28]. Himalayan region particularly Uttarakhand region represent 10 genera and 23 species of the plant belonging to family lauraceae [8]. The genus *Cinnamomum* contains over 300 species, distributed in tropical and subtropical regions of North America, Central America, South America, Asia, Oceania and Australasia. The genus *Cinnamomum* contains 250 species in Indo-Malaysia and south-east Asia and 16 species in India [23]. The species of *Cinnamomum* have aromatic oils and terpenoids in their leaves and bark.

Cinnamomum tamala Nees & Eberm also known as Indian *Cassia,* a native species of India [3] is widely distribution in Himalayan region (900–2000 mt) also known as “Tejpat”, popular among the north Indians as flavoring ingredient in various dishes. *Cinnamomum tamala* is mainly used in pharmaceuticals preparations because of its hypoglycemic, stimulant and casuistic properties. It is also used in therapeutics, colic, diarrhea, enlargement of spleen and snake bite. Leaves are commonly used as spice and are reported to possess antioxygenic, antibacterial and antifungal properties. Essential oil of *Cinnamomum tamala* is important for spices and perfumes [4, 30]. *Cinnamomum camphora* is commonly known as camphor. It has medicinal importance as Anti-inflammatory, antiseptic, antiviral, bactericidal, counterirritant, diuretic, expectorant, stimulant, rubefacient, vermifuge. *Cinnamomum zeylanicum* Blume. (cinnamon) is known

M.M. Srivastava, L.D. Khemani, S. Srivastava, *Chemistry of Phytopotentials: Health, Energy and Environmental Perspectives*, DOI:10.1007/978–3-642–23394-4_18, © Springer-Verlag Berlin Heidelberg 2012

as ‚Dalchini' in India. Cinnamon bark is widely used as a spice. It is principally employed in cookery as a condiment and flavoring material [4, 30].

Materials and Methods

Isolation and Chemical Investigation

Plant materials were collected from different natural habitats of the Kumaon region of Uttarakhand during March to May 2009. Eight samples of *Cinnamomum tamala* from Munsyari, Pithoragarh, Lohaghat, Champawat, Pantnagar, Dogaon, Nepal and Tanakpur, two samples of *Cinnamomum zeylenicum* from south India and Pantnagar and One sample of *Cinnamomum camphora* from Pantnagar were collected. 1 Kg. each of the fresh aerial part of the plant were hydrodistilled in Clevenger's type apparatus for 8 hours. The oils were extracted in diethyl ether. Removal of solvent under vacuum and drying over anhydrous Na_2SO_4 yielded varies amount of essential oils. The oils were kept in refrigerator for further study. The chemical analysis of volatile oils was undertaken by Gas chromatography-mass spectrometry (GC-MS) technique. The GC-MS data were obtained on Thermo Quest, Trace 2000, GC coupled with Finnigan Mat Polaris Q MS. In GC-MS analysis column was RtX- 5 (Res tex corp.) fused silica capillary column (30 m×0.25 mm, 0.25 µm) film coating, helium used as carrier gas with 1 ml/ min flow rate, EI (70 eV) mode and 210°C injection temperature. Components of essential oils (Table 1) were identified by matching their Mass spectra and GC retention indices with those in NIST- MS Wiley Library, comparing with literature reports and published data [24].

Antibacterial Activity

Antibacterial activity was screened against three gram negative bacteria viz., *Salmonella enterica enterica*, *Escherichia coli* and *Pasturella multocida*. These were collected from Department of Virology, Indian Veterinary Research Institute (Bareilly) and maintained in the laboratory by regular subculturing on to nutrient agar. Antibacterial screening of the essential oils against test bacteria was done by Disc diffusion method as reported by [14] with slight modification. 5µl, 10µl, 15µl, 20µl concentrations of essential oils were applied on each sterilized paper disc. The antibacterial screening was measured by zone of inhibition against bacteria. The plates were incubated at 37°C for overnight to observe the zone of inhibitions around the disc, which were compared with the standard antibiotics (Gentamicin).

Antioxidant Activity

Antioxidant potential of the essential oils were evaluated in terms of 2,2'-diphenyl-1-picrylhydrazyl (DPPH) radical scavenging ability, effect on the chelating activity on Fe^{+2} and reducing power in comparison with the synthetic and natural antioxidants. Butylated hydroxyl toluene (BHT), catechin and gallic acid were taken as standards.

Reducing Power Activity

The reducing power of the essential oils was determined by the method reported earlier [14–16]. Different amount of essential oils (5µL, 10 µL, 15 µL and 20 µL) were mixed with 2.5mL of the phosphate buffer (200 mM, pH 6.6) and 2.5mL of 1% potassium ferricyanide, $K_3[Fe(CN)_6]$. The mixtures were incubated at 50°C. After incubation, 2.5mL of 10% trichloroacetic acid was added to the mixtures, followed by centrifugation at 650 rpm for 10min. The upper layer (5mL) was mixed with 5mL of distilled water and 1mL of 0.1% ferric chloride and absorbance of the resultant solution were measured at 700 nm using UV-Vis spectrophotometer (Visiscan-167, Indian).

DPPH Radical Scavenging Activity

The scavenging effect on DPPH radical was determined according to the methods developed earlier [14–16, 21]. Various amounts of essential oils (5µL, 10 µL, 15 µL and 20 µL) were mixed with 5mL of 0.004% methanolic solution of DPPH. Each mixture was placed for 30min. in the dark and the absorbance of the samples was read at 517 nm using UV-Vis spectrophotometer (Visiscan-167, Indian). The percentage (%) radical scavenging activity was determined according to the Equation [1].

Effect on the Chelating Activity of Fe^{2+}

The chelating activity of essential oils on ferrous ions (Fe^{2+}) was measured with slight modification of the method reported earlier [12]. Different amounts of essential oils (5µL, 10 µL, 15 µL and 20 µL) were first mixed in 1mL of methanol. Then mixtures were left for reaction with ferrous chloride (2mM, 0.1mL) and ferrozine (5mM, 0.2mL) for 10min at room temperature and, absorbance was measured at 562nm in UV-Vis spectrophotometer (Visiscan-167, Indian). A lower absorbance indicates a higher chelating power. The chelating activity on Fe^{2+} of the oil was compared with that of EDTA (0.01mM) and Citric acid (0.025M) and percentage (%) chelating activity was calculated according to the Equation [2].

Results and Discussion

GC-MS Analysis of Essential Oils

The present investigation shows several chemotypes of *Cinnamomum tamala* which have mainly linalool, 1,8-cineol, cinnamyl acetate, *E*-cinnamaldehyde and eugenol were presented in major quantities. The essential oil of *Cinnamomum tamala* collected from Munsyari has linalool (52.5%) and E-cinnamaldehyde (26.4%) as major components another major component is 1,8-cineol (4.2%). 31 compounds have been identified in the oil which contributes to 98.4% of the oil. The oil is rich in monoterpenoids contributing of the total oil 94.12% of which 7.44% were hydrocarbons and 86.68% were oxygenated monoterpenoids. Only 3.68% sesquiterpenoids could be identified. Essential oil of *Cinnamomum tamala* collected from Nepal has similar pattern of essential oil composition. It has 53.2% linalool, 25.0% *E*-cinnamyl acetate and 16.1% 1,8-cineol. In oil 34 components were identified in the oil and the oil was rich in monoterpenoids (92.60%). The sesquiterpenes identified were epi-cubenol (0.6%) and caryophyllene oxide (0.6%). The above two species are linalool-*E*-cinnamaldehyde type. No camphor was detected in these samples. The essential oil of *Cinnamomum tamala* collected from Lohaghat showed the presence of linalool (29.8%), camphor (44.0%) and E-cinnamaldehyde (14.3%) along with other constituents. Its constituents contributing 96.6% of the oil were identified. Similar pattern of essential oil components was formed in the essential oil collected from Champawat collection. It has linalool (24.7%), camphor (25.5%) and *E*-cinnamaldehyde (30.4%). In all 25 constituents were identified in the oil which contribute to 96.3% of the oil. The above two collections are linalool-camphor-cinnamaldehyde type. The *Cinnamomum tamala* essential oils collected from Pithoragarh and Tanakpur showed similar pattern of the major constituents. Samples contains linalool (22.2%) and (38.0%), *E*-cinnamaldehyde (44.6%) and (25.0%) and cinnamyl acetate (15.1%) and (3.5%) respectively. Both the oils also contain borneol. The oil of *Cinnamomum tamala* collected from Dogaon also showed similar pattern in its linalool (27.2%), borneol (2.2%), cinnamaldehyde (42.5%) and E-cinnamyl acetate (1.8%). Essential oil of *Cinnamomum tamala* collected from MRDC Pannagar has eugenol (65.0%) as major constituent. It also contains *E*-cinnamaldehyde (3.8%) and *E*-cinnamyl acetate (2.6%) along with epi-cubenol (2.9%) and caryophyllene oxide (4.1%) hence it is eugenol type [1–2, 25–27].

Analysis of two sample of *Cinnamomum zeylenicum* cultivated at MRDC Pantnagar and CIMAP (introduced from South India) showed quit difference in the chemical makeup. The essential oil of *Cinnamomum zeylenicum* of Pantnagar contains linalool (7.4%), *E*-cinnamaldehyde (10.9%) and *E*-cinnamyl acetate 58.5% and 92.3% of the oil contributed by 25 compounds, identified by GC-MS. The essential oil of *Cinnamomum zeylenicum,* south Indian collection showed dominance of eugenol (74.1%). Unlike other oils, this was also rich in sesquiterpenoids. β-caryophyllene (1.2%), γ-elemene (1.9%), aromadendrene (0.1%), γ-gurjunene (2.3%), spathulenol (0.5%) and caryophyllene oxide (9.6%) were among the identified sesquiterpenoids [11, 18, 20, 31]. GC analysis of essential oil of *Cinnamomum camphora* cultivated at MRDC revealed the presence of a camphor single major constituent (82.4%). However other chemotypes of *C. camphora* are also reported in literature [5, 7, 9–10, 17, 22, 29, 32].

Antibacterial Investigation

All the oils showed activity against *Pasturella multocida, Escherichia coli* and *Salmonella enterica enterica* in comparison to gentamicine which was used as standard. Antibacterial activity of essential oils of the essential oil of *Cinnamomum tamala* collected from Pithoragarh, Lohaghat and Champawat showed zone of inhibition against *Pasturella multocida* but minimum activity against *Salmonella enterica enteric. Cinnamomum tamala* collected from Dogaon, Tanakpur and Munsyari were found to show maximum zone of inhibition against *Salmonella enterica enterica, Escherichia coli* and *Pasturella multocida. Cinnamomum camphora* oil collected from Pantnagar was effective against *Pasturella multocida* but not against *Salmonella enterica enterica* and *Escherichia coli.* Concluding all these results, it can be said that the volatile oils of *Cinnamomum* species posses excellent antibacterial properties even at very low concentration against tested bacterial strains due to their major components. The results demonstrate that the essential oils of *Cinnamomum* species posses antibacterial activity against *Pasturella multocida, Escherichia coli* and *Salmonella enterica enterica* which are pathogenic bacteria which are responsible for diseases in animals and human beings.

Antioxidant Investigation

Antioxidant activities of the essential oils were evaluated by three different methods viz. reducing power activity, DPPH radical scavenging activity and effect on the chelating activity of Fe (II) ions. In presence of chelating agents, the complex formation between ferrous and ferrozine is disturbed, resulting in decrease the color of the complex. Measurement of color reduction therefore allows estimating the metal chelating activity of the coexisting chelator. Among all the essential oils tested essential oil of *C.zeylenicum* of south India (A_{700}=1.685 to 2.396 (±0.000 to 0.002)), *C. tamala* from Pantnagar (A_{700}= 3.812 to 4.000 (±0.020 to 0.000)) and from Pithoragarh (A_{700}=1.565 to 1.895 (±0.003 to 0.005)) possess maximum reducing power, followed by essential oil of *Cinnamomum tamala* from Tanakpur (A_{700}=1.243 to 1.693 (±0.001 to 0.000)). The reducing powers of essential oils were significantly lower compared to the BHT, gallic acid and catechin (Fig 1).

For the determination of radical scavenging activity, DPPH radical method is a very fast method to evaluate the antiradical power of an antioxidant activity by measurement of the decrease in absorbance of the DPPH radical at 517 nm. In the radical form this molecule had an absorbance at 517 nm which disappeared after acceptance of an electron or hydrogen radical from an antioxidant compound to become a stable diamagnetic molecule. The DPPH radical scavenging activity was tested for the essential oil of *C.zeylenicum* of Pantnagar (47.11±0.000%) and *C. tamala* of Tanakpur (39.64±0.006%) exhibited maximum DPPH radical scavenging activity among all collections which is followed by sample of Champawat (78.08±0.468%) and Lohaghat (75.14±0.573%) (Fig 2).

All the essential oils showed dose dependent chelating activity on Fe (II) ions. Among the transition metals, iron is known as the most important lipid oxidation pro-oxidant due to its high reactivity. The ferrous state of iron accelerates lipid oxidation by breaking down hydrogen and lipid peroxide to reactive free radicals like Fenton type reaction. Among all the essential oil, the oil of *Cinnamomum tamala* cultivated at Pantnagar (62.48±0.126%) showed maximum chelating activity, followed by oil of *Cinnamomum tamala* from Munshyari (57.29±0.165%) and oil of *Cinnamomum zeylenicum* collected from South India (54.33±0.044%) (Fig 3). Therefore it can be concluded that the essential oils are an effective chelating agent and could effort protection against oxidizing agent. With the increase in the amount of essential oils, increase in chelating activity was observed for all the oils. Therefore we assume that the observed variation in antioxidant potential is related to differential chemical composition and concentration level. Therefore we can use the essential oils as natural antioxidant in place of synthetic antioxidant.

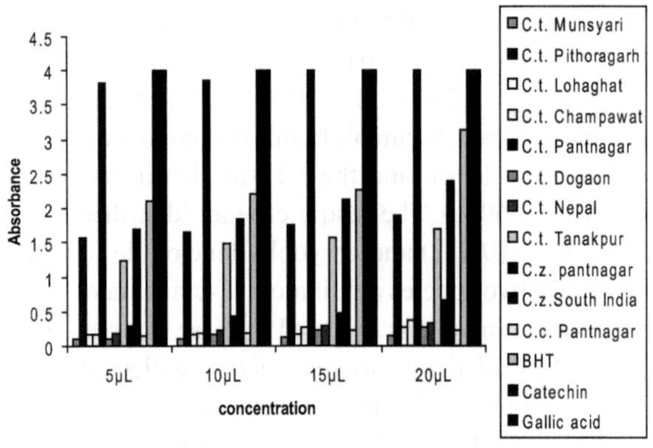

Fig. 1: Reducing power activity of essential oil of *Cinnamomum* species

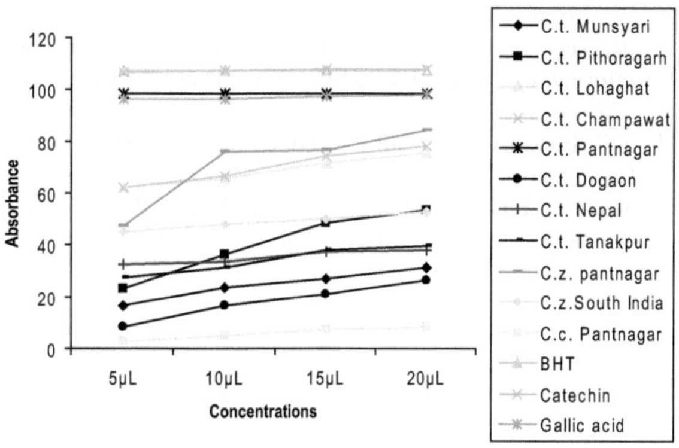

Fig. 2: Radical scavenging activity of essential oils of *Cinnamomum* species on DPPH

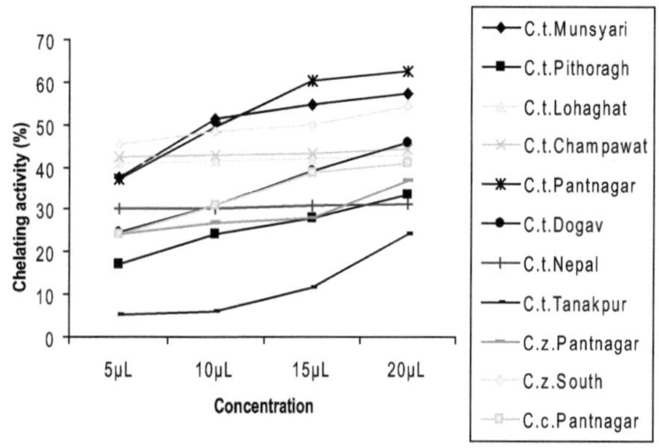

Fig. 3: Effect of the essential oils of *Cinnamomum* species on the chelating activity of Fe (II) ions

Equations

Equation [1]

$$\text{DPPH Radical Scavanging Activity (\%)} = [1 - (A_t/A_o)] \times 100$$

(Where A_t is the absorbance of the sample at 517 nm and A_o is the absorbance of the control at 517 nm).

Equation [2]

$$\text{Chelating activity (\%)} = [1 - (A_t/A_o)] \times 100$$

(Where A_t is the absorbance of the sample at 562 nm and A_o is the absorbance of the control at 562 nm).

References

1. A. Ahmed, M.I. Choudhary, A. Farooq, B. Dermirci, F. Dermirci and K.H.C. Baser; Flavour and Fragrance Journal 15(6) (2000) 388–390.
2. A. Baruah, S.C. Nath and A.K. Hazarika; IndianPerfumer 51(3) (2007) 50–52.
3. A. Goswami; MEP-News 11(2) (2001) 12–13.
4. A. Hussain, O.P. Virmani, S.P. Popli, L.N. Mishra and A.K. Gupta; *Dictionary of Indian Medicinal Plants*. CIMAP, Lucknow (1980).
5. A.K.S. Baruah, S.D. Bhagat, J.N. Hazarika and B.K. Saikia; Indian Journal of Pharmacy 37 (2) (1975) 39–41.
6. J.E. Angmor, D.M. Dicks, W.C. Evans and D.K. Santra; Planta Medica 21(4) (1972) 416 – 420.
7. B.J. Stubbs, A. Specht and D. Brushett; Journal of Essential Oil Research16(3) (2004) 200–205.
8. B.P. Uniyal, J.R. Sharma, U. Chaudhary and D.K. Singh; *Flowering Plants of Uttarakhand* (2007) 210–217.
9. C.D. Frizzo, A.C. Santos, N. Paroul, L.A. Serafini, E. Dellacassa, D. Lorenzo and P. Moyna; Brazilian Archives of Biology and Technology43(3) (2000) 313–316.
10. C.H. Liu, A.K. Mishra, B. He and R.X. Tan; International Pest Control 43(2) (2001)72–74.
11. E. Schmidt, L. Jirovetz, G. Buchbauer, G.A. Eller, I. Stoilova, A. Krastanov, A. Stoyanova and M. Geissler; Journal of Essential Oil Bearing Plants 9(2) (2006) 170–182.
12. E.A. Decker and B. Welch; Journal of Agri. Food Chem 38(1990) 674–677.
13. E.F. Gilman and D.G. Watson; Cinnamomum Camphora, Fact Sheet ST-167(1993).
14. G. Singh, P. Marimuthu, H.S. Murali and A.S. Bawa; Journal of Food Safety25 (2005) 130–145.
15. G. Singh, S. Maurya, P. Marimuthu, H.S. Murali and A.S. Bawa; Natural Product Radiance 6(2) (2007) 114–121.
16. G.C. Yen, and H.Y. Duh; Journal of American Oil Chemist's Society 70(1993) 383–386.
17. G.F. Zhang, C. Chen, Z.P. Chen, R.Y. Chen and X.S. Lin; Journal of Plant Resources and Environment 17(1) (2008) 24–27.
18. G.R. Mallavarapu, S. Ramesh, R.S. Chandrasekhara, B.R.R. Rao, P.N. Kaul and A.K. Bhattacharya; Flavour and Fragrance Journal 10(4)(1995) 239–242.
19. J.I. Miller; *The Spice Trade of the Roman Empire*. Oxford: Oxford UP (1969).
20. J.C. Chalchat and I. Valade; Journal of Essential Oil Research 12(5) (2000) 537–540.
21. M. Cuendt, K. Hostettmank and O. Potterat; Helv. Chem. Acta. 80 (1997) 1144–1152.
22. N.X. Dung, P.V. Khien, T.C. Ho and P.A. Leclercq; Journal of Essential Oil Research 5(4) (1993) 451–453.
23. R.D. Gaur; *Flora of the District Garhwal North-west Himalaya* (with ethnobotanical notes), Transmedia, Srinagar Garhwal (1999).
24. R.P. Adams; *Identification of essential oil component by gas chromatography/ mass spectroscopy,* carol, stream, Illinois USA. Allured Publications (1995).
25. R.P. Sood, C.D. Padha, Y.P. Talwar, R.K. Jamwal, M.M. Chopra and P.R. Rao; Indian Perfumer 23(2) (1979) 75–78.
26. S.C. Joshi, D.C. Bisht. R.C. Padalia. K.K. Singh and C.S. Mathela; Journal of Essential Oil Bearing Plants11(3) (2008) 278–283.

27. S.C. Nath, A.K. Hazarika and R.S. Singh; Journal of Spices and Aromatic Crops 3(1) (1994) 33–35.
28. W.S. Judd, C. Christopher, K. Elizabeth, S. Peter, and D. Michael; *Plant Systematics a Phylogenetic Approach*, third edition (2007), Sinauer Associates, Inc. ISBN 978–0-87893–407-2.
29. W.Y. Shy, W. He, G.Y. Wen, D.X. Guo, G.Y. Long and Y.G. Liu; Acta Botanica Sinica 31(3) (1989) 209–214.
30. Wealth of India; *A dictionary of Indian raw materials and industrial products*. Raw material series, publication and information Directorate CSIR, New Delhi. vol.2 (1950) 178–179.
31. Y.R. Rao, S.C. Paul and P.K. Dutta; Indian Perfumer 32(1) (1988) 86–89.
32. Y.T. Lin, Y.H. Kuo, T.T. Chen, and S.T. Kao; Perfumer and Flavorist 2(5) (1977) 60.

19
Analysis of Nutrient Content of Underutilized Grain: Chenopodium Album

T. Pachauri[1], A. Lakhani[2] and K. Maharaj Kumari[3]
Department of Chemistry, Faculty of Science, Dayalbagh Educational Institute, Dayalbagh, Agra 282 110.
[3]Email: maharajkumari.k@rediffmail.com

Abstract

The nutritional characteristics of the seeds of C. album and C. quinoa were investigated using standard analytical methods. The results indicate that the seeds of the neglected species of C. album were rich in moisture and ash content, crude proteins, crude lipids and carbohydrates. On comparison, the values of protein and total sugar content in the seeds of C. album and C. quinoa is much lower than that of the other grains. However, the total lipid and minerals like Na, Ca and Mg were found to be much higher in C. quinoa and all varieties of C. album as compared with other grains. Thus, it is recommended for future commercial cultivation.

Introduction

Underutilized plants possess promising nutritional and industrial importance for a variety of purposes for humankind. The underutilized plant species of economic importance are the key to sustainable agriculture in most of the developing countries facing resource constraints as well as rapid depletion of natural resources due to ever-increasing population pressure. Various health and nutritional problems can be improved through wider cultivation and inclusion of underutilized crops in our food habit. Thus, underutilized plants can help to make diets more balanced and hence play an important role in combating silent hunger.

Recently, a resurgence of interest has developed in wild species for their possible medicinal values in diets. Wild plant plays an important role in the diet of inhabitants in different parts of the world. Wild plant species provide minerals, fiber, vitamins and essential fatty acids and enhance taste and color in the diets. In addition, they have anti-bacterial, hepato-protective and anti-carcinogenic properties and are therefore having medicinal values [1, 2].

The family Chenopodiaceae is composed of herbs and shrubs, or rarely small trees that usually grow in alkaline soil. Plants of this family are generally weeds. One of the genera of this family is *Chenopodium*. The genus *Chenopodium* has a worldwide distribution and contains about 250 species of which about eight species are found in India. *Chenopodium quinoa* and *C. album* are such species [3]. The nutritional characteristics of *Chenopodium quinoa* (a food crop of the Andean region of Latin America) have been reported to contain high amounts of proteins, carbohydrates, fat, vitamins, and minerals. The protein content in quinoa (15% dry basis) is much higher than that found in cereals such as wheat, barley, oats, rice and sorghum [4]. Quinoa also contains more carotene, riboflavin, tocopherols and folic acid than wheat, rice, oats and maize, and can supply the daily requirements of certain vitamins and several minerals [5, 6]. Interest in quinoa as a valuable food source has therefore, been renewed in Asia in recent years because of its versatility and its ability to grow under conditions normally inhospitable to other grains [3].

Chenopodium album is a common weed in India. Although eaten as a leafy vegetable in many parts of the country, this plant commonly known as Goosefoot or Pigweed, has been considered only as a source of poultry feed [7]. *C. album* is the most widely distributed and is grown in the Himalayan region. The Himalayan grain Chenopod is comparable to Andean *quinoa* in nutrient composition and is much bet-

M.M. Srivastava, L. D. Khemani, S. Srivastava, *Chemistry of Phytopotentials: Health, Energy and Environmental Perspectives*, DOI:10.1007/978–3-642–23394-4_19, © Springer-Verlag Berlin Heidelberg 2012

ter than wheat, barley, maize and rice. Grain protein quality equals that of milk and contains high lysine (6 g/100 g protein), methionine (2.3 g/100 g protein) and cysteine (1.2 g/100 g protein) [8]. The grains are also used in local alcoholic preparations and out of the four domesticated species, *C. album* is the most widely distributed and is grown in the Himalayan region [9–12]. The crop is also suited to the mixed farming system, particularly the multiple cropping systems. Analysis of foliage of ten species of *Chenopodium* at the National Botanical Research Institute (NBRI), Lucknow, revealed a wide range of variation for protein (26–64 g/kg), carotene (78–190 mg/kg), vitamin C (0.5–2.4 g/kg), nitrate (2.6–5.0 g/kg) and oxalate (9–39 g/kg) [13].

The plant is commonly available in India but day by day its popularity is decreasing due to lack of awareness about its unique medicinal properties. *Chenopodium album* is relatively inexpensive, easily and quickly cooked green leafy vegetable. One does not find much literature on seed of *Chenopodium album* as it is generally discarded from the field as a weed [14]. Since, *Chenopodium album* and *Chenopodium quinoa belong to same genus and thus, it is assumed that the seeds of* Chenopodium album may possess similar nutritional characteristics as are well evident in *C. quinoa.* Data on the nutritional composition of the leaves of this plant is available but no data on nutritional composition of its leaves is available. In this study, an attempt is made to determine the amount of carbohydrates, protein, lipids and to estimate the mineral content in the seeds of this neglected plant so that it may be grown as a crop and provide nutritional support to human society.

Materials and Methods

Collection of Seeds

The seeds of three different varieties of *Chenopodium album*, one cultivated in the botanical garden of Dayalbagh Educational Institute, Dayalbagh (*C. album* B) and two wild varieties were collected from wheat fields (*C. album* A) and from gram fields (*C. album* C) of Dayalbagh. Treated seeds of *C. quinoa* were obtained from USA.

Extraction and Estimation of Nutrients

Proximate Composition

Moisture, ash, protein, carbohydrates (reducing and total sugars) and lipids were analyzed by standard methods. Moisture was determined by drying oven method, by drying 5 g sample in an oven at 105°C for 3 hours. Ash content was determined by incineration of the samples (2 g) in a muffle furnace at 600°C for 6 hours until the ash turned white. The seed samples were extracted in phosphate buffer by centrifugation and the protein content of the seeds was estimated colorimetrically by Lowry method. Reducing sugar was extracted from the seeds with water and determined by Nelson- Somoygi's method. Total sugar was obtained by hydrolyzing the seed samples with conc. HCl and then estimated by potassium ferricyanide method, spectophotometrically. Seeds were extracted with chloroform: methanol mixture and shaken with NaCl to remove non-lipid contaminants and treated for gravimetric estimation of lipids.

Mineral Levels

Mineral elements (Na, K, Mg, Ca and Mn) were determined in homogenized samples after wet digestion. The digestion was done by incineration of seed sample in muffle furnace at 600°C for 8 hours till seeds are converted to white ash. Ash content was weighed and treated with 5 ml conct. HCl in 100 ml volumetric flask and final volume was made up. The digested samples were analyzed by using Dionex ICS 1100 Ion Chromatograph system equipped with guard column (CG12A), analytical column (CS12A), and cation self-regenerating suppressor (CSRS 300) using 20 mM Methane Sulfonic Acid as an eluent.

Results and Discussion

Proximate Composition

The proximate composition of the seed samples (*C. album and C. quinoa*) examined in the study is presented in Table 1.

All samples contained between 11–13 % moisture and 2–3 % ash. The protein content ranged from 2.5 to 3.5 g/100 g with highest content (3.5 g/100 g) in *C. qui-*

noa seeds followed by *C. album* A (3.3 g/100 g). Reducing sugar in all the four varieties ranges between 1.2 to 2.1 %. Among the *C. album* samples, seeds of sample A (sample from wheat fields) contain highest percentage of reducing sugar (1.6 %) which is equal to the reducing sugar content of *C. album C* seeds. The total sugar value in *C. album* and *C. quinoa* ranges between 2.1 to 4.7 %. Among all the album samples, *C. album* A (growing wild in wheat fields) contained highest amount (3.7 %) of total sugar and sample B contained lowest amount (2.1 %) of total sugar. Lipid content as estimated gravimetrically was found to be higher in *C. album* as compared to the seeds of *C. quinoa*. Lipid content of different seed samples varied from 3.5–5.8 %.

Mineral Levels

Mean value for the mineral content of nutritional importance is presented in Table 1. The species analyzed in the study showed remarkably high K content. The K content varied greatly and ranged from 381.4 mg/100 g (*C. album* B, cultivated) to 550.8 mg/100 g (*C. album* from wheat fields). Mn content in all the seeds was nearly equal ranging from 0.8 to 1.8 mg/100 g in three varieties but it is below detection limit in the seeds of *C. album* B. However, the mineral content in *C. quinoa* was found to be high but it is quite comparable with the mineral content of *C. album.*

Discussion

The results of the present study potentially indicate that album species of *Chenopodium* is at par with the quinoa seeds and well endowed with essential nutrients required for human consumption. Table 2 gives a comparison of nutrient composition of *C. album* seeds as compared with other grains.

The findings of moisture and ash content of the seeds analyzed in the study conform to the previously published data for grains. The values of protein in *C. album* seeds in present study are lower than the values of protein in leaves (3.69 %) as reported by [14]. Protein content of *C. quinoa* seeds was estimated to be 3.5 %. Earlier reported literature [3] shows high content of protein in quinoa seeds (12–19 %). The lower values in the present study are probably because the seeds used in the present study were commercially obtained and are debittered seeds in which the protein seed coat was removed. The seeds of *C. album* contain 3.5–5.8 g/100 g of total lipid. The lipid content ranges between the values reported by [3] (6–8 %) whereas in fresh leaves of *C. album* total lipid is reported as 0.6 g/100 g by [4]. Ahamed (1998) reported the value of reducing sugar to be 2–3 % in the seed of *C. quinoa* which is found low with the present study.

On comparison, the values of protein and total sugar content in the seeds of *C. album* and *C. quinoa* is much lower than that of the other grains. However, the total lipid and minerals like Na, Ca and Mg are found to be much higher in *C. quinoa* and all varieties of C. *album* as compared with other grains. The K content was found to be highest in *C. quinoa* (765.2 mg/100 g). On comparing with other grains the K content in C. *album* was much higher than wheat (370 mg/100 g), maize (286 mg/100 g) and rice (150 mg/100 g). However, it is comparable with the K content of barley (560 mg/100 g). The Mn content in *C. quinoa* and C. *album* is much lower than wheat (5 mg/100 g) but it is found to be comparable with the rice (2 mg/100 g) and barley (1.6 mg/100 g).

In all four varieties, the nutritional content of C. *album* A was found to be comparable with that of *C. quinoa*. Thus, the *C. album* growing as a weed in the fields of wheat ad gram can be used as a future crop.

Table 1: Nutritional composition in the seeds of various species of *C. album and C. quinoa* in (g/100 g) while minerals in (mg/100 g)

Seed Sample	Moisture	Ash	Protein	Reducing Sugar	Total Sugar	Lipids	Na	K	Mg	Ca	Mn
C. album A	11	2	3.3 ± 0.5	1.6 ± 0.3	3.1± 0.3	5.8	25.9	550.8	175.8	152.3	1.1
C. album B	12	2	2.5 ± 0.4	1.2 ± 0.2	2.1± 0.1	3.5	14.1	381.4	110.7	58.6	-
C. album C	13	2	3.1 ± 0.6	1.5 ± 0.1	2.5± 0.2	5.1	17.9	471.1	135.3	67.4	0.8
C. quinoa	12	3	3.5 ± 0.5	2.1 ± 0.1	4.7± 0.4	5.6	31.7	765.2	189.6	172.8	1.8

Table 2: Comparative account of nutrient composition of *C. album and C. quinoa* seeds compared with other grains in (%)

NUTRIENTS	Q	A.A	A.B	A.C	M*	W*	R*	B*
Moisture	12	11	12	13	15	13	15	13
Ash	3	2	2	2	2	2	1	3
Protein	3.5	3.3	2.5	3.1	13	14	8	12
Carbohydrate (Total Sugar)	4.7	3.7	2.1	2.5	66	69	78	70
Lipid	5.6	5.8	3.5	5.1	4	2	1	1
Manganese	1.8	1.1	-	0.8	0.5	5	2	1.6
Sodium	31.7	25.9	14.1	17.9	1–16	3	8–9	3
Potassium	765.2	550.8	381.4	471.1	286	370	70–150	560
Calcium	172.8	152.3	58.6	67.4	30–90	29–48	40	10–80
Magnesium	189.6	175.8	110.7	135.3	120–144	128	48–60	120

Q = *C. quinoa*, A.A = *C. album* A (from wheat fields), A.B = *C. album* B (cultivated in Botanical Garden), A.C = *C. album* C (from gram fields), W = Wheat, M = Maize, R = Rice, B = Barley. [*Source: 3].

Conclusions

The nutritional characteristics of the seeds of *C. album* and *C. quinoa* were investigated using standard analytical methods. The proximate analysis shows that the seeds of the neglected species of *C. album* were rich in moisture and ash content, crude proteins, crude lipids and carbohydrates. Elemental analysis in mg/100 g (DW) indicated that the seeds contained Na, K, Ca, Mg and Mn. Comparing the nutrients and chemical constituents with other cereals reveals that the seeds of *C. album* growing as a weed in agricultural fields could be important contributor for improving the nutritional content of rural and urban people. Thus it is recommended for future commercial cultivation.

Acknowledgements

The authors are grateful to University Grants Commission for providing financial support (F. No. 33–280/2007).

References

1. C. Green C; An Overview of production and supply trends in the U. S. specialty vegetable market. *Acta Horticulturae.* 318 (1992) 41–45.
2. W. Bianco, P. Satamaria and A. Elia; Nutritional and nitrate content in edible wild species used in Southern Italy. Proceedings 3rd IS on Diversification of Vegetable Crops. *Acta Horticulturae.* 467 (1998) 71–87.
3. N. T. Ahmed, R. S. Singhal, P. R. Kulkarni, D. D. Kale and M. Pal; A lesser-known grain, *Chenopodium quinoa*: Review of the chemical composition of its edible parts. Food & Nutrition Bulletin. 19 (1) (1998) 61–69.
4. J. Ruales and B. Nair; Nutritional quality of the proteins in quinoa *(Chenopodium quinoa,* Willd), *Plant Foods for Human Nutrition,* 42 (1992) 1–11.
5. J. Ruales and B. Nair; Content of fat, vitamins and minerals in quinoa *(Chenopodium quinoa,* Willd) seeds, *Food Chemistry,* 48 (2) (1993a) 131–136.
6. J. Ruales and B. Nair; Saponins, phytic acid, tannins and protease inhibitors in qulnoa *(Chenopodium qumoa,* Willd) seeds, *Food Chemistry* 48 (2) (1993b) 137–143.
7. Afolayan and Jimoh; Nutritional quality of some wild leafy vegetables in South Africa International Journal of Food Sciences and Nutrition 2009, 60 (5) (2009) 424–431.
8. T. Pratap and P. Kapoor; The Himalayan grain chenopods I. Distribution and ethnobotany. *Agriculture Ecosystems and Environment* 14 (1985) 185–199.
9. M. Tapia, H. Gandarillas, S. Alandia, A. Cardozo, A. Mujica, R. Ortiz, V. Otazu, J. Rea, B. Salas and E. Zanabria; E. *La Quinova Y La Kaniwa: Cultivos Andinos.* CIID and IICA, Bogota, Colombia (1979).
10. T. Pratap; Cultivated grain chenopods of Himachal Pradesh: distribution, variations and ethnobotany. PhD thesis, Dept of Biosciences, Himachal Pradesh University, Shimla, (1982).
11. R. T. Wood. Tale of the food survivor. Quinova. *East-West Journal* (1985)63–67.
12. T. Pratap and P. Kapoor; The Himalayan grain chenopods III. An under-exploited food plant with promising potential. *Agriculture Ecosystems and Environment* 19 (1987) 71–79.
13. V. Joshi, P. L. Gautam, G. D. Bhag Mal, Sharma and S. Kochher; Conservation and Use of Underutilized Crops: an Indian Perspective. Managing Plant Genetic Diversity. IPGRI: (2002) 359–368.
14. E. Yildirim, A. Dursun and M. Turan; Determination of the nutritional contents of the wild plants used as vegetables in upper Coruh Valley. *Turk. J. Bot.* 25 (2001) 367–371.

20

Chemical Analysis of Leaves of Weed *Calotropis Procera* (Ait.) and its Antifungal Potential

R. Verma[1], G. P. Satsangi[2] and J. N. Shrivastava[3]
Department of Botany, Faculty of Science, DEI,
Dayalbagh, Agra-282005, India.
E- mail: ragini26verma@gmail.com

Abstract

Weed Calotropis procera (Ait.) R.Br. of the family Asclepiadaceae is a well known medicinal plant. It showed potent activity against fungal isolates (M. canis, M. fulvum, T. mentagrophytes, A. niger and A. fumigatus). Chemical analysis of ethanol leaf extract was carried out by using GC-MS technique. Spectra showed the presence of fifteen components. The main components with major peaks were 3-Eicosene, (3E)-3-Icosene, Tetratriacontane and 1-Tridecene. The combined bioactivity of these components against selected fungi was estimated in comparison with ethanol crude extract itself. The results indicated that the ethanol crude extract had lower bioactivity as compared to the bioactive fraction.

Introduction

Plants are exploited as medicinal source since ancient age. The traditional and folk medicinal system uses the plant products for the treatment of various infectious diseases. In recent times, plants are being extensively explored for harboring medicinal properties. Studies by various researchers have proved that plants are one of the major sources for drug discovery and development [1–3]. Plants are reported to have antimicrobial, anticancer, anti-inflammatory, antidiabetic, hemolytic, antioxidant, larvicidal properties etc.

Calotropis procera, a member of the Asclepiadaceae, is a woody, broadleaf, evergreen coarse shrub, 3 to 5 m tall, widely growing in the tropics. It is distributed in arid to semiarid regions of the Caribbean, Central America, South America, Africa, India and Israel, mainly appearing on plains and in the uplands. For decades, *C. procera*, also referred to as "The Apple of Sodom" has been used especially in traditional folk medicine because of its pharmacologically active compounds found in its roots, bark, leaves and mainly in its latex which exudates from damaged leaves. Meanwhile, chemical extracts from *C. procera* have been shown to have ascaricidal, schizonticidal, antibacterial, anthelmintic, insecticidal, anti-inflammatory, antidiarrheal, larvicidal, cytotoxic and analgesic effects, thus explaining a growing demand in today's medical research for the different parts of the plant [7, 8, 9 and 14].

Present study attempts to determine the phytochemistry and antidermatophytic effect of the *C. procera* leaves fractions on *Microsporum canis, Microsporum fulvum, Trichophyton mentagrophytes, Aspergillus niger,* and *Aspergillus fumigatus.* The results would enable more rational exploitation of the plant in both traditional and orthodox medicine.

Materials and Methods

Plant Material

C. procera plant was collected from the natural population growing in the wasteland of DEI, Agra, India, during March 2008. The plant has been characterized by Taxonomy Division, Botanical Survey of India (BSI), Allahabad as *Calotropis procera* and the assigned Accession No. is 79385 (BSA).

Processing of the Plant

Plant leaves were collected and washed properly with distilled water. The leaves were shade dried at room

M.M. Srivastava, L. D. Khemani, S. Srivastava, *Chemistry of Phytopotentials: Health, Energy and Environmental Perspectives*, DOI:10.1007/978-3-642-23394-4_20, © Springer-Verlag Berlin Heidelberg 2012

temperature. Dried leaves were uniformly grinded using mechanical grinder. The leaves powder was extracted in different solvents on the basis of their increasing polarity in Soxhlet extraction unit for 48 hours. The extracts were filtered using Whatman filter paper number 1. The filtrates were concentrated using rotary evaporator and dried using lyophilizer. Dried extract were collected in an air tight container and stored at 4 °C.

Test Microorganism

The following five isolates of fungi were used for the study: *M. canis* (MTCC 3270), *M. fulvum* (MTCC 7675), *T. mentagrophytes* (MTCC 7250), *A. fumigatus* (MTCC 8636) and *A. niger* (MTCC 2587). All these cultures were maintained on Sabouraud's agar plates at 4 °C.

Antifungal Sensitivity Test

The extracted powder was dissolved in the respective solvent to make different concentrations and was used to perform antifungal assay in terms of zone diameters using the disc diffusion assay [8]. Two controls were run parallel to the experiment. The positive controls were gresiofulvin and ketoconazole at the same concentration of the crude extract (600 µg/ml). The negative control was of the respective solvent.

Column Chromatography

The ethanol leaf extract was separated on chromatograph over silica gel column. The column was first eluted with petroleum ether and thereafter chloroform was added in order of increasing polarity. Then the column was eluted with different combinations of mobile phase on the basis of increasing polarity of solvents. The fraction with chloroform: methanol (5:1) showed antifungal activity was analyzed by GC-MS.

GC-MS Analysis

The GC–MS analysis of the samples of *Calotropis procera* was performed using a Shimadzu Mass Spectrometer-2010 series system (AIRF, JNU, New Delhi) equipped with a AB inno-wax column (60 m X 0.25 mm id, film thickness 0.25 µm). For GC–MS detection, an electron ionization system with ionization energy of 70 eV was used. Helium gas was used as a carrier gas at a constant flow rate of 1.2 ml min^{-1}. Injector and mass transfer line temperature were set at 270 °C and 280 °C, respectively. The oven temperature was programmed from 50° to 180°C at 3°C min-1 with hold time of min-1 and from 180° to 250°C at 5°C min-1 with hold time 20 min respectively. Diluted samples (prepared in Ethanol) of 0.2 µl were manually injected in the split less mode. Identification of compounds of the sample was based on GC retention time on AB inno-wax column, computer matching of mass spectra with standards (Mainlab, Replib and Tutorial data of GC–MS systems).

Results

Medicinal plants are being proved as an alternate source to get therapeutic compounds based on their medicinal properties. *C. procera* is easily available in most of the agricultural and non agricultural fields and the usage of this plant for medicinal purpose was reported by several researchers.

Table 1: The bioactivity of the eluted fraction compared to the crude extract against five fungal pathogens

Test concentration (µg/ml)	*M. canis*	*M. fulvum*	*T. mentagrophytes*	*A. niger*	*A. fumigatus*
100	12.25±.020	-	09.50±.025	16.00±0.00	10.00±0.00
200	10.50±.025	-	09.75±0.02	07.00±0.00	10.50±.025
300	-	10.50±.025	-	11.00±0.00	11.50±.025
400	13.25±.020	11.80±.056	10.50±.025	-	12.80±.021
500	-	20.30±0.04	13.30±.041	-	16.00±.061
600	18.00±.040	16.80±0.05	19.00±.035	10.50±.025	17.30±.041
Plant extract(600)	17.50±.025	12.50±.025	15.50±.025	11.50±.025	-
Antibiotic (600)	16.25±0.02	15.50±.025	14.50±.025	-	12.50±.025
Control	-	09.25±0.04	-	-	-

Table 2: Components of the chromatographic fraction of the ethanolic leaf extract of *Calotropis procera*

Peak report TIC

Peak #	R. Time	Area	Area%	Name
1	10.743	274843	7.97	1-Tridecene
2	13.238	562444	16.32	3-Eicosene
3	14.298	95392	2.77	8-Pentadecanone
4	15.596	447262	12.98	(3E)-3-Icosene
5	16.548	147087	4.27	(1-Proyloctyl) Cyclohexane
6	17.683	234750	6.81	1-Heptadecene
7	18.717	199263	5.78	1-Nonadecene
8	19.531	205357	5.96	Sulfurous acid
9	20.414	207889	6.03	Di-n-octyl phthalate
10	20.779	217926	6.32	1-Tricosene
11	21.216	202490	5.87	Tetratriacontane
12	21.990	312639	9.07	n-Tetratriacontane
13	22.776	124989	3.63	2-ethylhexyl iso-hexyl ester
14	23.644	169722	4.92	2,6,10,15-Tetramethylheptadecane
15	24.624	44959	1.30	Docosane
		3447012	100.00	

R. Time denotes retention time

Bioactivity of Fraction

The bioactivity of the eluted fraction was estimated against *M. canis, M. fulvum, T. mentagrophytes, A. niger,* and *A. fumigatus* (Table 1). It was found to be positive against these pathogens and showed the zones of 18 mm for *M. canis* at 600 µg/ml, 20.30 mm for *M. fulvum* at 500 µg/ml, 19 mm for *T. mentagrophytes* at 600 µg/ml, 16 mm for *A. niger* at 100 µg/ml and 17.30 mm for *A. fumigatus* at 600 µg/ml. On comparing their bioactivities with that of the crude itself, it was found that the ethanol crude extract had lower activity as compared to bioactive fraction. The inhibition zones were 17.50, 12.50, 15.50, and 11.50 for *M. canis, M. fulvum, T. mentagrophytes, A. niger* and absence of zone against *A. fumigatus* respectively.

Characterization of the Crude Extract

GC-MS Spectra

GC/MS spectra of the collected fraction showed presence of 15 components (Table 2). The major components showed retention times of 13.28, 15.596, 21.99 and 10.743 minutes respectively (Figure 1) named as 3-Eicosene (16.32 %), (3E)-3-Icosene (12.98 %), Tetratriacontane (09.07 %) and 1-Tridecene (07.97 %) accordingly. The chemical properties of the majors GC/MS obtained compounds were represented in Table 3.

Table 3: The chemical properties of four major compounds of the chromatographic fraction of ethanol leaf extract using GC-MS analysis

Library/ID	Retention Time	Mol. Formula	Mol. Weight
1-Tridecene	10.743	$C_{13}H_{26}$	182
3-Eicosene	13.238	$C_{20}H_{40}$	280
(3E)-3-Icosene	15.596	$C_{20}H_{40}$	280
n-tetratriconate	21.990	$C_{34}H_{70}$	478

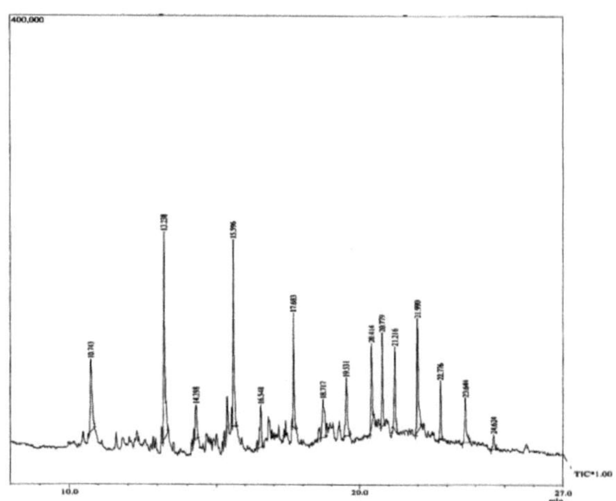

Fig. 1: GCMS Chromatogram (TIC) obtained from 0.2 µl injection of the bioactive fraction of the leaves of *C. procera* shows the appearance of four major compounds at different retention times

Discussion

Earlier studies on the antimicrobial activity of leaves, latex and stem bark of *C. procera* [9–14] revealed its antifungal potential against *A. flavus, A. niger, M. boulardii, C. albicans, M. gypseum, M. canis, T. mentagrophytes* and *T. rubrum*.

We conclude that *C. procera* represents a rich source of valuable medicinal compounds and leaves of *C. procera* contain high antifungal property. The antifungal activity of the bioactive fraction may be due to the combination of different components or may be due single component. However, it is important to note that the activity guided fraction of ethanol extract of *C. procera* leaf need to be further purified through bioactivity guided fractionation to isolate and identify the compound responsible for antifungal activity.

Acknowledgements

We are grateful to Prof. V.G. Das, Director and Prof. D.S. Rao, Head and Dean, Prof. Anil Kumar, Ex-Head and Dean, Dept. of Botany, Faculty of Science, DEI, Agra for their continuous encouragement during the work and all laboratory facilities. We also thank Mr. Ajay Kumar AIRF, JNU, New Delhi for his cooperations.

References

1. De Pasquale A; J. Ethnopharmacol. 11 (1984) 1–16.
2. S.M.K. Rates; Toxicon, 39 (2001) 603–613.
3. M.C. Gordon and J.N. David; Pure Appl. Chem. 77 (2005) 7–24.
4. P. Sharma and J.D. Sharma; J. Ethnopharmacol. 68 (1999) 83–95.
5. S. Dewan, H. Sangraula, V.L. Kumar; J. Ethnopharmacol. 73 (2000) 307–11.
6. S. Kumar, S. Dewan, H. Sangraula, V.L. Kumar; J. Ethnopharmacol .76(2001) 115–8.
7. V.L. Kumar and Y.M. Shivkar; J. Ethnopharmacol. 93 (2004) 377–379.
8. M.J. Pelczar, E.C.S. Chan, G.R. Knieg; Microbiology concepts and applications, McGraw-Hill Inc New York, (1993) 967.
9. F.A. Kuta; Afr. J. Biotech. **7** (2008) 13: 2116–2118.
10. F.A. Kuta; School of Science and Science Education Conference Proceedings, (2006) 27–29.
11. M.K. Rai and S. Upadhyay; Hindustan Antibiot. Bull **30** (1988a) 1–2: 33.
12. S.O. Kareem, J. Akpan, O.P. Oja; Afr. J. Biomed. Res. 11 (2008) 105–110.
13. S.W. Hassan, F.L. Bilbis, M.J. Ladan, R.A. Umar, S.M. Dangoggo, Y. Saidu, M.K. Abubakar and U.K. Faruk; Pak. J. Bio. Sci. **9** (2006) 14: 2624–2629.
14. V. Suvarna and S. Patil; J. Herbal Med. Toxic. 3 (2009) 2: 151–153.

21
Isolation and Characterization of "Flavon-5, 3', 4'-Trihydroxy 7-O-β-D-glucopyranosyl (6"→1"') β-D-glucopyranoside" From Stem Bark of *Quercus Leucotrichophora*

S. C. Sati[1], N. Sati[2] and O. P. Sati[1]
[1]Department of Chemistry,
[2]Department of Pharmaceutical Sciences,
H. N. B. Garhwal University, Srinagar Garhwal, Uttarakhand, India.
Email: sushilsati1983@gmail.com

Abstract

A flavonoidal glycoside, named flavon-5, 3', 4'-trihydroxy 7-O-β-D-glucopyranosyl (6"→1"') β-D-glucopyranoside, has been isolated from the stem bark of Quercus leucotrichophora, together with β-sitosterol, kaempferol, quercetin and 7-methoxy kaempferol. The structure of flavon-5, 3', 4'-trihydroxy 7-O-β-D-glucopyranosyl (6"→1"') β-D-glucopyranoside, by means of rigorous spectroscopic analysis including 2-D NMR measurements.

Introduction

Quercus leucotrichophora vern. Banj belonging to family Fagaceae is an evergreen tree of approximately 40 m height and is commonly found throughout the Himalayan region at altitudes ranging from 800–2000 m (Naithani, 1985). Gum of the *Q. leucotrichophora* is traditionally used for gonorrheal and digestive disorders (Gaur, 1999). The seeds are astringent and diuretic and are used in the treatment of gonorrhea, indigestion, diarrhea and asthma (Chopra et al., 1986). The leaves, seeds and bark are also used in livestoke healthcare (Pande et al., 2006). Previous research showed the isolation of quercetin and its 3-O-disacchride from the leaves (Kalra et al., 1966) and β-sitesterol, 7-methoxy kaemferol and 3-O-[{α-L rhamnopyranosyl-(1"'→4")}{α-L rhamnopyranosyl-(1""→6")}]-β-D-glucopyranosyl quercetin from the stem bark (Sati et al., 2011a). The GC MS analysis of volatile extract of *Q. leucotrichophora* stem bark contained approximately 86.36 % monoterpenoids, 6.53 % sesquiterpenoids and 0.11 % of aliphatic aldehydes. The oxygenated compounds accounted for ≈ 48.71 % of the volatile extract, whereas hydrocarbon compounds were only ≈ 44.29. The major components were 1,8-cineol (40.35 %) followed by γ-terpinene (16.36 %), β-pinene (11.09 %), p-cymene (6.22 %), α-pinene (5.33 %), 4-terpineol (3.70 %), aromadendrene (1.76 %), p-menth-1-en-8-ol (1.60 %) and β-eudesmol (1.05 %) (Sati et al., 2011b). The Ethanolic extract and volatile extract of its stem bark showed potent antimicrobial activity against various micro-organisms (Sati et al., 2011a; Sati et al., 2011b). Herein, we report the isolation and structure elucidation of flavonoidal glycoside, named flavon-5, 3', 4'-trihydroxy 7-O-β-D-glucopyranosyl (6"→1"') β-D-glucopyranoside (compound 1).

Experimental

Collection of Plant Material

The barks of *Q. leucotrichophora* were collected in January 2008 from Nagnath Pokhari, District Chamoli Garhwal, Uttarakhand. The plant was properly identified from Taxonomy Laboratory, Department of Botany, H. N. B. Garhwal University, Srinagar Garhwal, Uttarakhand and the voucher specimen (GUH8835) was kept in the Departmental herbarium.

M.M. Srivastava, L. D. Khemani, S. Srivastava, *Chemistry of Phytopotentials: Health, Energy and Environmental Perspectives*, DOI:10.1007/978–3-642–23394-4_21, © Springer-Verlag Berlin Heidelberg 2012

Extraction and Isolation

The air-dried and chopped bark was defatted with petroleum sprit using Soxhlet. The defatted bark material extracted exhaustively with 85% EtOH at 30–50 °C (for 15 h, 3 times) on a heating mantle and concentrated under reduced pressure. The extract was then fractionated through column chromatography using Chloroform: Methanol as eluting solvent. The polarity of solvent was gradually increased by addition of methanol. The repeated column chromatography afforded Compound 1 along with 7-methoxy kaempferol, β-sitosterol, kaempferol and quercetin.

Compound 1 was isolated as yellow crystalline solid; m.p.: 182–184 °C; molecular formula: $C_{27}H_{30}O_{16}$; molecular weight: 610; UV λ_{max} nm: 245, 282, 340; IR (KBr) γ_{max} cm^{-1}: 3350, 1640, 1610; FABMS (m/z): 610, 593, 576, 448, 286; ^{1}H NMR (400 MHz, DMSO) and ^{13}C NMR (125 MHz, DMSO) δ: Table 1.

Acid Hydrolysis of Compound 1

Compound 1 was refluxed with 8% aqueous HCl (10 ml) for 5h afforded an aglycone and D-glucose identified by co-PC (n- BuOH: H_2O: AcOH :: 4:1:5; Rf value 0.18) with an authentic sample.

Results and Discussion

Compound 1 crystallized as yellow solid having m.p. 184–186 °C. The molecular mass of compound 1 deduced as 610, suggested by its FABMS spectrum, showing the molecular ion peak at m/z 610 $[M]^{+}$. Compound 1 gave characteristic test for flavonoids (green coloration with $FeCl_3$ and positive test with Mg/HCl) and also a Molish's test for carbohydrate, thereby indicating its flavonoidal glycosidic nature. The UV spectrum of compound exhibited the absorption maxima at 245, 282 and 340 nm which also support the flavonoidal nature of compound 1. The IR signals at 3350, 1640 and 1610 cm^{-1} showed presence of hydroxyl group, unsaturated carbonyl group and aromatic moiety in compound 1 respectively.

The ^{1}H NMR spectrum of compound 1 displayed presence of four doublets at δ 6.46, 6.77 (*J*= 1.6 Hz), 7.45 (*J*= 2.4 Hz) and 6.93 (*J*= 8.4 Hz) were assignable to H-6, H-8, H-2' and H-5' respectively and a double doublet at δ 7.43 was ascribed for H-6' suggest presence of aromatic protones. Two doublets at δ 5.05 (*J*= 6.8 Hz) and 4.16 (*J*= 7.2 Hz) showed two β-linked anomeric sugar proton H-1'' and H-1'''. The ^{13}C NMR

Table 1: ^{1}H and ^{13}C NMR data of compound 1 (300,125 MHz, $CDCl_3$)

Position	δ_C	δ_H	HMBC	Position	δ_C	δ_H	HMBC
2	161.24			6'	119.25	7.43 (dd, *J*= 2.4, 8.4 Hz)	2',4'
3	103.18	6.72 s	1', 2, 4, 4a	1''	104.07	5.05 (d, *J*= 6.8 Hz)	7
4	181.95	-	-	2''	76.56	3.08 m	-
4a	105.43	-	-	3''	73.37	2.92 (t, *J*= 8.4 Hz)	-
5	164.61	-	-	4''	75.61	3.61 (t, *J*= 8.4 Hz)	-
6	99.60	6.46 (d, *J*= 1.6 Hz)	4, 5,7,8,4a	5''	69.34	3.25 (ddd, *J*= 3.2, 6.4,9.2 Hz)	-
7	162.94	-	-	6''	68.30	3.91 (dd, *J*= 13.2 Hz)	-
8	94.81	6.77(d, *J*= 1.6 Hz)	4a,8a,6,7	1'''	99.93	4.16 (d, *J*= 7.2 Hz)	6''
8a	157.00	-	-	2'''	76.27	3.25 m	-
1'	121.40	-	-	3'''	73.11	3.35 m	-
2'	113.70	7.45 (d, *J*= 2.4 Hz)	1',3',6'	4'''	75.61	3.61 m	-
3'	145.84	-	-	5'''	69.54	3.25 (ddd, *J*= 2.4, 5.2, 7.8 Hz)	-
4'	150.04	-	-	6'''	65.67	3.67 (dd, *J*= 12.8 Hz)	-
5'	116.14	6.93 (d, *J*= 8.4 Hz)	7,1',2',3',4'				

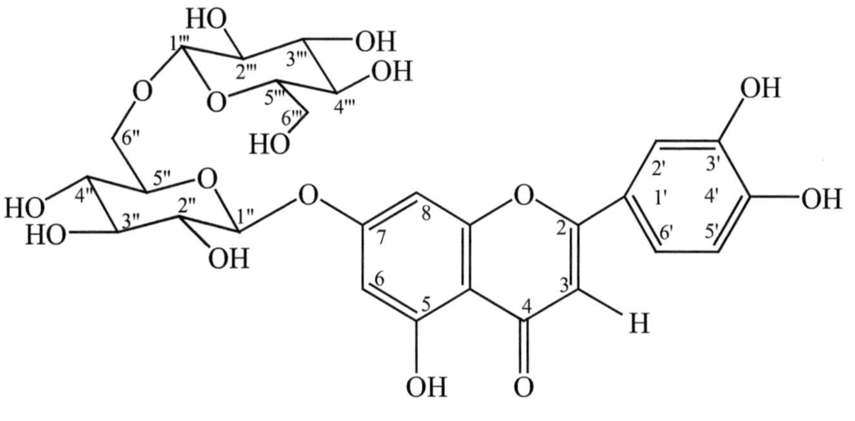

Fig. 1: Structure of compound 1

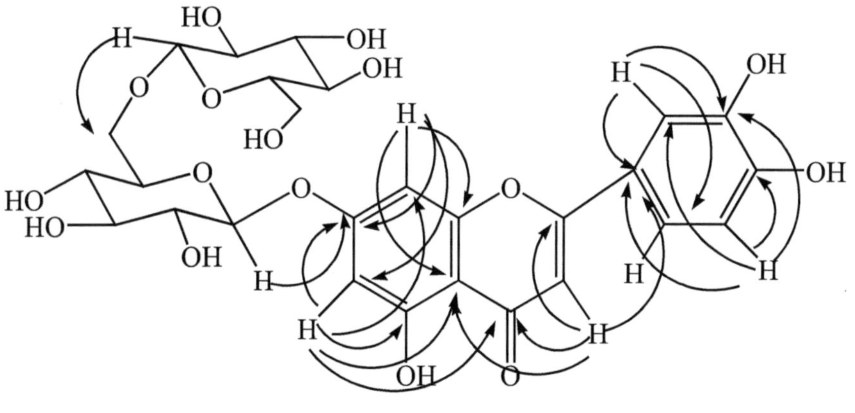

Fig. 2: HMBC correlation of compound 1

of compound 1 displayedsignals for 27 carbon atoms. The downfield signal at δ 181.9 (C-4) was attributed due to carbonyl functional group. Two peaks at δ 104.0 (C-1") and 99.93 (C-1"') were assigned for anomeric carbon of sugar whereas other downfield signals displayed at δ 161.2, 162.9, 150.0 and 145.8 was attributed to four oxygenated carbon atoms. Further the HMBC experiment of compound 1 showed long range coupling of anomeric proton (δ 5.05) with C-7 (δ 162.94), indicated that the attachment of sugar C-7 position. The full NMR data of compound 1 are given in table 1. The acidic hydrolysis of compound Q_9 afforded two molecules of glucose and aglycone as flavonoid (5, 7, 3', 4'-terahydroxy flavonoid). Finally on the basis of above chemical and spectral details compound Q9 was identified as flavon-5, 3', 4'-trihydroxy 7-O-β-D-glucopyranosyl (6"→1"') β-D-glucopyranoside.

Reference

1. B. D. Naithani; Flora of Chamoli. 2nd ed. Botanical survey of India, New Delhi, (1985) 598
2. P. C. Tiwari, L. Pande and H. C. Pande; Aromatic Palnts of Uttaranchal, Bishen Singh Mahendra Pal Singh Dehra Dun-24001, (2006) 238
3. R. D. Gaur; Flora of the District Garhwal North West Himalaya, Trans Media, Media House, Srinagar Garhwal, (1999) 107
4. R. N. Chopra, S. L. Nayar and I. C. Chopra (1986). *Glossary of Indian Medicinal Plants (Including the Supplement).* Council of Scientific and Industrial Research, New Delhi.
5. S. C. Sati, N. Sati, O. P. Sati; Int J Pharm Pharm Sci 3(3) (2011a) 89
6. S. C. Sati, N. Sati, O. P. Sati, D. Biswasc and B. S. Chauhan; Natural Product Research (2011b) Article in press
7. V. K. Kalra, A. S. Kukla and T. R. Seshadri; *Current Science 35* (1966) 204–205

22
Phytochemical Examination of *Anaphalis Busua* Leaves

[1]R. Raturi, [2]S.C. Sati, [1]H. Singh, [2]M.D. Sati and [1]P.P. Badoni
[1]Department of Chemistry, HNB Garhwal Central University Campus Pauri Garhwal, India
[2]Department of Chemistry, HNB Garhwal Central University, Srinagar, Garhwal, India
Email: raaakeshhh@gmail.com

Abstract

Column chromatographic separation of alcoholic extract of the leaves of Anaphalis busua led to isolation of a flavonoidal glycoside 1, together with β- sitosterol and sitigmasterol. The compound 1 identified as tiliroside by spectroscopic and chemical methods.

Introduction

The genus *Anaphalis* (Asteraceae) consists of about 80 species distributed throughout the world and more than 50 species are distributed in china [1] and 31 species reported in India [2]. *Anaphalis busua* (Bugla) is an erect tall herb. Stem usually branched from base, somewhat winged. Leaves sessile, linear lanceolate or oblanceolate most abundantly present in open places of oak and pine forests of submountain and mountain Himalayas. Leaf juice applied on bruises, wounds and cuts [2]. Anaphalisqualenol, anapharenosoic acid and araneosol isolated from the plant previously [3–4]. Isoquercitrin and astragalin, anaphalol, 5-hydroxy 7-O-3'-methyl bute 2'-enylpthalide and 5, 7-dihydroxyphalide, 5-methoxy-7-hydroxy pthalide and β-sitosterol isolated form other species of the plant [5–6]. The present paper deals with the isolation and structure elucidation of a flavonoidal glycoside 1 from alcoholic extract of the leaves of *A. busua.*

Test Plant Material, Extraction and Isolation

The leaves of *Anaphalis busua* was collected from Singoli, Paurikhal, District Tehri Garhwal, Uttarakhand, India and was identified by plant identification laboratory Department of Botany H.N.B. Garhwal University Srinagar Garhwal. A voucher (GUH 2180) specimen was deposited in the Department. The air-dried and coarsely powered leaves of the plant were defatted with light petroleum in a Soxhlet. The defatted mass was exhaustively extracted repeatedly with 90% aqueous EtOH, until the extractive became colorless. All the extracts were mixed and concentrated under reduced pressure using rotatory vacuum evaporator. The concentrated extract was adsorbed on silica gel and fractionated through column chromatography using the solvent system chloroform: methanol (97: 4). The polarity of solvent was gradually increased by addition of methanol. Repeated column chromatography afforded compounds 1 together with β-sitosterol, and stigmasterol.

Compound 1

Pale yellow solid, M.P. 269–271°C(uncorrected), Molecular formula $C_{30}H_{26}O_{13}$, Molecular weight 594 amu, IR(λ_{max} ^{KBr}) 3262(OH), 1654(C=O), 1606(C=C) cm^{-1}, UV (λ_{max} ^{MeoH}) 227, 266, 310 nm, ^{1}H and ^{13}C NMR (DMSO) data are given in Table 1.

Results and Discussion

Compound 1 gave green coloration with $FeCl_3$ and positive Shinoda test (Mg/HCl). Compound showed UV absorption band at 266 and 310 nm. The IR spectrum furnishes a bands at 3262 cm^{-1}(OH), 1654 cm^{-1} (C=C) and 1606 cm^{-1} (C=C). The ^{1}H NMR spectrum showed two signals both integrating for two proton with J=7.7 Hz at δ 7.86(2H, H2'/H6') and at δ 6.71 (4H, H-3'/H-5') clearly indicating the kaemferol derivative [7]. The signal at δ 6.02 and δ 6.18 were assigned to H-6 and H-8 proton respectively. Two other doublets at δ 6.71(J=7.3 Hz) and δ 7.20(J=7.2 Hz)

M.M. Srivastava, L.D. Khemani, S. Srivastava, *Chemistry of Phytopotentials: Health, Energy and Environmental Perspectives*, DOI:10.1007/978-3-642-23394-4_22, © Springer-Verlag Berlin Heidelberg 2012

Table 1

Positions	δ C ppm	δ H ppm	Positions	δ C ppm	δ H ppm
2	158.01		1''	102.72	5.13 (1H, m)
3	133.93		2''	74.50	3.37 (1H, m)
4	178.09		3''	74.44	3.21 (2H, m)
5	161.63		4''	70.42	3.21 (2H, m)
6	98.70	6.02 (1H, d)	5''	76.71	3.38 (1H, m)
7	164.65		6''	63.01	4.19 (2H, m)
8	93.57	6.18 (1H, d)	1'''	125.78	
9	157.07		2'''	129.89	7.20 (2H, d)
10	104.28		3'''	115.48	6.71 (4H, t)
1'	121.40		4'''	159.87	
2'	130.93	7.86 (2H, d)	5'''	115.48	6.71 (4H, t)
3'	114.74	6.71 (4H, t)	6'''	129.89	7.20 (2H, d)
4'	160.21		Cα	113.44	
5'	114.74	6.71 (4H, t)	Cβ	167.53	7.31 (1H)
6'	130.93	7.86 (2H, d)	Cγ	145.26	5.99 (1H, d)

were attributed to H-3'''/H-5''' and H-2'''/H-6''' of the P-Coumaroyl moiety respectively. Two other doublets at δ 5.99 (J=13.2) and 7.31 were assigned to Hγ and Hβ of P-Coumaroyl moiety with trans isomers respectively. The anomeric proton were absorbed at δ 5.13 (J=7 Hz) was assigned for D-glucose moiety. The linkage was found to be β configuration as indicated from coupling constant value (J=7.0 Hz). The ^{13}C NMR spectrum present two intense signals at 130.9 and 114.7, assigned to H-2'/H-6' and H-3'/H-5' respectively. The signals at 121.4 and 160.2 are typical of an unsubstituted β ring of kaemferol like structure. The signals at 63.6 (CH_2) shows that the p-coumaroyl linkage was at C-6 of the glucose unit. In APCIMS the molecular ion peak was observed at 594 amu. The molecular formula of compound 1 was determined to be $C_{30}H_{26}O_{13}$. On the basis of above spectral data compound 1 was identified as **Tiliroside or [Kaemferol-3-O-β-D-6'' (p-coumaroyl) glucopyranoside]** [Fig.1].

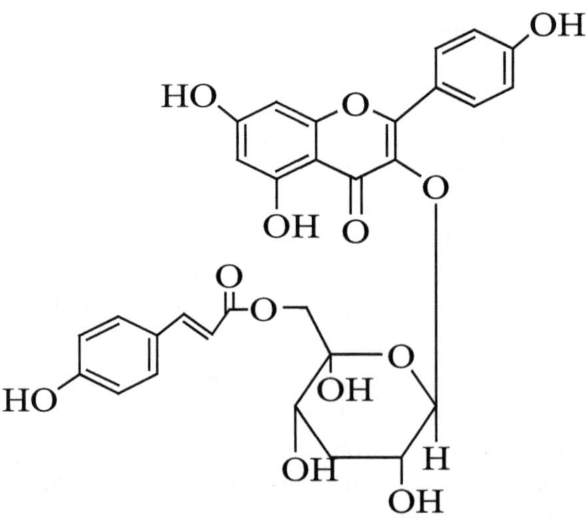

Fig. 1: Structure of compound 1

References

1. Y. Hua and H. Wang; Journal of the Chinese Chemical Society 51(2004) 409.
2. R.D. GaurFlora of the District Garhwal North West Himalayas, vol.1 2nd Edition Transmedia Publisher Srinagar Garhwal India (1999) 556.
3. S.K. Sharma, M. Ali, Journal of Medicinal and Aromatic plant sciences 20 (1998) 352.
4. V.K. Saxena, A. Sahai, G. Samaiya, Indian Perfumes 28(1984) 177.
5. J.H. Lin, Journal of the Chinese chemical society 40 (1993) 93.
6. B. Talapatra, M.K. Roy, S.K. Talapatra. J. Chem. 19 (1980) 927.
7. M.K. Awatef, J. Pharm. Sci. 12 (1998) 101

23
Tannins in *Michelia Champaca* L.

H. Ahmad, A. Mishra, R. Gupta and S. A. Saraf
Faculty of Pharmacy, Babu Banarasi Das National Institute of Technology & Management,
Sector-I, Dr. Akhilesh Das Nagar, Faizabad Road, Lucknow-227105, Uttar Pradesh, India
Email: ahmadhafsa.cog@gmail.com

Abstract

Tannins are a diverse group of poly-phenolic secondary compounds in plants which are redox active, water soluble with sufficiently high molecular weights. Tannins account for a significant proportion of biomass in terrestrial plants and their potential to affect biogeochemical processes is widely appreciated. Michelia champaca L. an Indian medicinal plant rich in its chemistry and subsequent pharmacology was the chosen model for extraction, quantitation and tentative characterization of tannins in this study. An attempt was made to extract crude tannin which was characterized by TLC and UV spectrum. Folin-Denis method was employed to estimate tannin levels in the plant material.

Introduction

Tannins contain sufficient hydroxyl groups and other suitable groups, such as carboxyl's to form effectively strong cross-linked complexes with protein and metal ions such as iron (III) and aluminum (III). On the basis of structures and origin they can be classified as condensed tannins which are composed of flavonol units and are not readily degraded in the gut and hydrolysable tannins which undergo microbial and acid hydrolysis with the release of simpler phenolics. Although tannins are distributed all through the cytoplasm of any vegetal cell, the highest concentration is generally observed in tree barks.[1–4] The role of these compounds in ecosystem ranges from plant defense against herbivores (due to their astringent character) and pathogens (inhibit microbial processes) to a strong influence on degradation of soil organic matter (SOM) and on nitrogen cycling. [1] Tannins enter soil from aboveground living or dead plant material or from roots in complex patterns influenced by interactions between plants and soil. Together with non-tannin phenolic substances, they comprise a substantial soil dynamic flux of C-substrates in the soil that affects numbers, diversity, and functioning of soil biota. [5] Response of soil C and N transformations depends strongly on the type of tannins. [1] Phenolics play a role in pest resistance and are known to affect the activity of various enzymes. Condensed tannins are effective feeding deterrents to many insects and spider mites and also convey effective disease resistance. [4] There is enough scientific evidence to believe that tannins are also useful in improving animal health in general and that of ruminants like cattle, deer, and sheep in particular.

Low to moderate tannin concentration may improve the digestive utilization on feeding, mainly due to reduction in protein degradation in the rumen and subsequent increase in amino-acid flow to small intestine. These effects in nutrition are reflected in better animal health. [3] *Michelia champaca* L. (Magnoliaceae) a tall evergreen tree with yellow fragrant blossoms thrives well in humid conditions. It is reported to have a decent air pollution tolerance potential and recommended for plantation on roadsides and around industrial complexes. It also increases ground water absorption and reduces pollution and also finds mention as a fodder tree especially in the Southern parts of India. [6,7] Since tannin concentration greatly influences organic matter formation, nutrient cycling, nitrogen fixation and microbial processes, it was thought to have a binding on the air purifying effects of *M. champaca* and hence an attempt was made to estimate the tannin levels in its leaves.

M.M. Srivastava, L. D. Khemani, S. Srivastava, *Chemistry of Phytopotentials: Health, Energy and Environmental Perspectives*, DOI:10.1007/978–3-642–23394-4_23,

Methodology

Plant Material

The plant material was collected in and around Lucknow, Uttar Pradesh in the month of August and authenticated by National Botanical Research Institute, Lucknow; also a voucher specimen was submitted for future reference (Ref No. NBRI/CIF/176/2010). The air dried plant material was size reduced to a moderately fine powder (#355/180) and stored in an air-tight container for further studies.

- **Extraction of Tannin:** 5 g of dried ethanolic extract (1) and 5 g dried Soxhlet extracted methanolic fraction previously defatted (2) were digested with boiling water for 30 min. Saturated solution of lead acetate was added to the above to precipitate out tannins. The solution was filtered, filtrate was discarded and the residue containing tannins was collected in water, to which H_2S gas was passed to remove excess of lead acetate as lead sulfide. The solution was again filtered. The residue containing excess of lead acetate was discarded and the filtrate was concentrated to get tannins.
- **Yields of the extracted tannin:** The yield obtained for (1) and (2) was 11.32 % and 2.9 % respectively.
- **Chromatographic characterization of extracted materials:** The extracted materials when subjected to thin layer chromatography afforded single spots (observed as black fluorescence) in UV at 254 nm. (Fig 1)The solvent system employed was n-butanol-acetic acid-water (4:1:5). The respective Rf for (1) and (2) were found to be 0.18 and 0.14.
- **UV Spectrum of extracted materials:** The extracted material in (1) and (2) was dissolved in AR grade methanol and scanned on a range of 200–700 nm in a UV 1700 Pharma Spec Shimadzu UV-Visible Spectrophotometer. A single peak was afforded by (1) at 236 nm with absorbance 1.106 and by (2) at 237.5 nm with absorbance 1.543 as illustrated in Fig 2 and Fig 3 respectively.

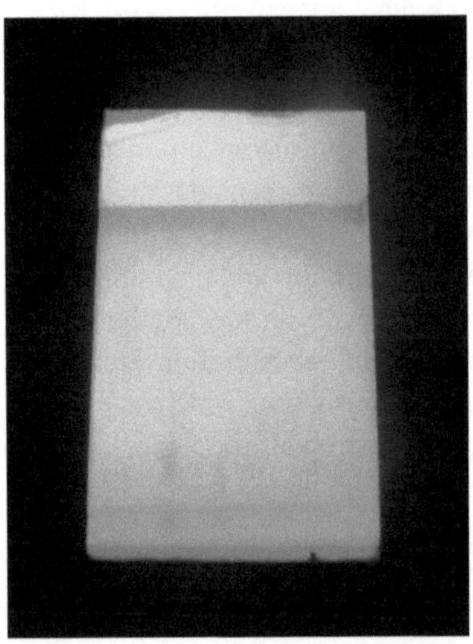

Fig. 1: TLC Chromatogram of extracted material (1) and (2)

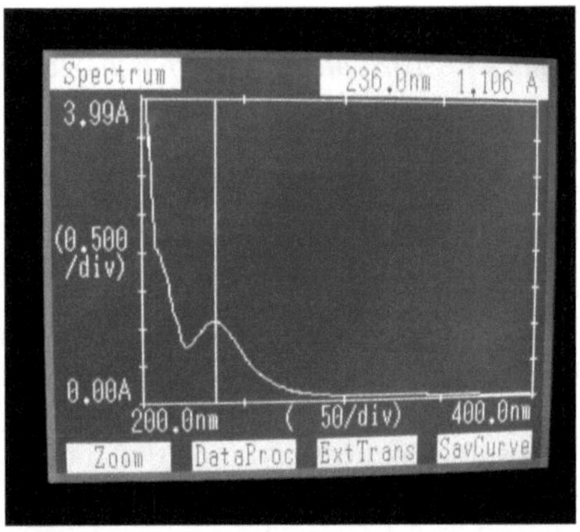

Fig. 2: UV spectrum of extracted material (1); 236 nm, abs 1.106

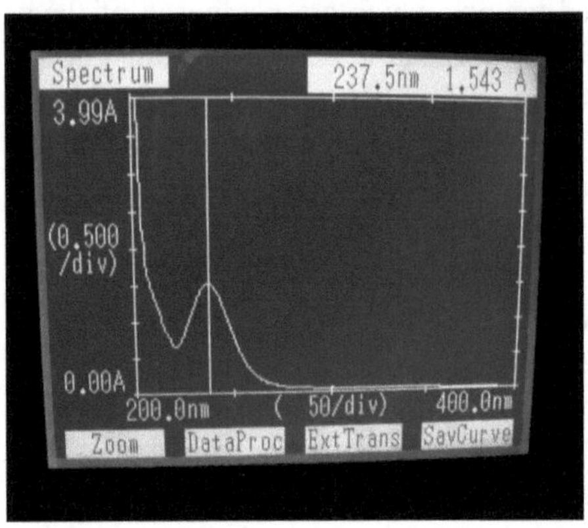

Fig. 3: UV spectrum of extracted material (2); 237.5 nm, abs 1.543

Estimation of Tannins in *Michelia Champaca* Leaves by Folin Denis Method:

Principle: Tannin like compounds reduces phosphotungstomolybdic acid in alkaline solution to produce a highly colored blue solution, the intensity of which is proportional to the amount of tannins. The intensity is measured in a spectrophotometer at 700 nm. [8]

Materials

1) **Folin-Denis Reagent:** 100 g sodium tungstate and 20 g phosphomolybdic acid dissolved in 750 mL distilled water in a suitable flask to which was added 50 mL phosphoric acid. The mixture was refluxed for 2 hours and the volume was made up to 1 L with water. The reagent was protected from exposure to light.
2) **Sodium Carbonate Solution:** 350 g sodium carbonate was dissolved in 1 L of water at 70–80 °C. It was filtered through glass-wool after allowing standing overnight.
3) **Standard Tannic Acid Solution:** 100 mg tannic acid was dissolved in 100 mL distilled water.
4) **Working Standard Solution:** 5 mL of stock solution was diluted to 100 mL with distilled water. (1mL= 5 mg tannic acid)
5) **Sample Solution (Extraction of tannin):** 0.5 g of powdered plant material was transferred to a 250 mL conical flask, to which was added 75 mL of water. The contents of the flask were put to boiling for about 30 minutes. It was then centrifuged at 2000 rpm for 20 minutes in a Refrigerate Centrifuge Sigma 3–18K Sartorius. The supernatant was collected in a 100 mL volumetric flask and the volume was made up with water.

Procedure

1mL of sample extract was transferred to a 100 mL volumetric flask containing 75 mL water. To this was added 5 mL Folin-Denis reagent, 10 mL of sodium carbonate solution and was diluted to 100 mL with distilled water. The contents were shaken well and the absorbance was read at 700 nm in a UV 1700 Pharma Spec Shimadzu UV-Visible Spectrophotometer. For the standard curve the same procedure was utilized taking 0–10 mL of the working standard of tannic acid, and the blank was made with 1 mL of water in place of the plant extract.

Estimation of tannin concentration: The tannin content of samples was calculated as tannic acid equivalents from the standard curve. (Fig 4)

Table 1: Concentration and absorbance values of standard tannic acid and plant extract

S. No.	X (Conc. µg/mL)	Y (Absorbance)
1.	0	0
2.	10	0.007
3.	20	0.046
4.	30	0.077
5.	40	0.1
6.	50	0.132
7.	60	0.154
8.	70	0.19
9.	80	0.202
10.	90	0.22
11.	100	0.249
12.	MC (L)	0.055

MC (L) = *Michelia champaca* leaf

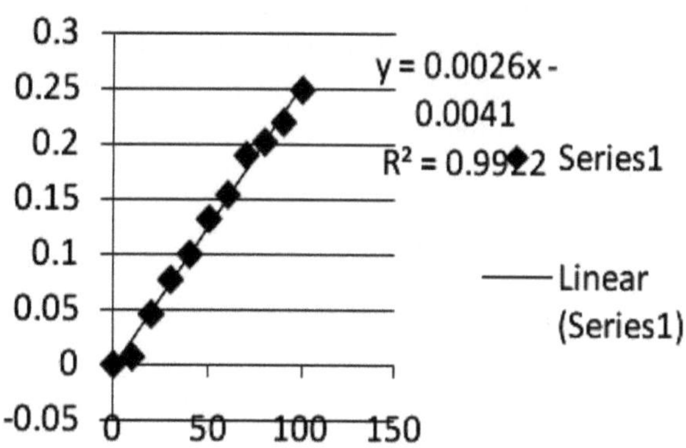

Fig. 4: Tannic Acid Standard Curve
R^2 values represents mean data set of n=3

X axis: Concentration (µg/mL)
Y axis: Absorbance values as read from spectrophotometer
From the regression equation

$$y = 0.002x - 0.004$$
$$\text{Hence } x = 28.5\ \mu g/mL \text{ in MC (L)}$$

i. e. 28.5 µg from 0.5 g drug, hence the tannin concentration in *M. champaca* leaves was found to be 57 µg/g of powdered plant material.

Results and Discussion

The target material was successfully extracted and the yields were found to be 11.32% and 2.9% for (1) and (2) respectively. The extracted material was confirmed as tannin by TLC; affording single spots (black fluorescence at short UV wavelength) with Rf 0.18 and 0.14 for (1) and (2) respectively. The UV spectrum for the same were also recorded which afforded single peaks at 236 nm with absorbance 1.106 for (1) and at 237.5 nm with absorbance 1.543 for (2). Folin Denis method was used to quantify the amount of tannins present in *Michelia champaca* leaves which was found to be 57 μg/g of powdered plant material. Since tannins are invariably involved in numerous bio-geochemical processes and their role in ecosystem and health have been put forward by the authors in the introduction segment; the estimated tannin levels here confirm a positive role in the air-purifying properties of *M. champaca*, also contributing to its fodder quality and thereby improving animal health. The authors are currently generating data on the isolation and HPTLC studies for the same.

Acknowledgements

The authors are thankful to AICTE-MODROBS Grant ,Standardization of selected Indian medicinal plants by High Performance Thin Layer Chromatography. (File No. 8024/RID/BOR-MOD-458), for making the research work possible.

References

1. B. Adamczyk, V. Kitunen and A. Smolander; Soil Biology & Biochemistry, 41 (2009) 2085–2093.
2. P. Schofield, D.M. Mbugua and A.N. Pell; Animal Feed Science and Technology. 91(2001) 21–40.
3. A.M. Elizondo, E.C. Mercado, B.C. Rabinovitz and M.E. Fernandez-Miyakawa; Veterinary Microbiology. 145 (2010) 308–314.
4 P.W Wu and L.S Hwang; Food Research International 35 (2002) 793–800.
5 J.J. Halvorson, J.M. Gonzalez and A.E. Hagerman; Soil Biology & Biochemistry (2011) 1–9 [Article in Press].
6 C. Datt, M. Datt and N.P. Singh; Journal of Forestry Research. 19(3) (2008) 209–214.
7 E.S. Kumar and S. Paulsamy; Environment and Pollution Technology. 5(4) (2006) 591–599.
8 A. Patel, A. Patel, A. Patel and N.M. Patel; International Journal of Pharma Sciences and Research. 1(1) (2010) 66–77.

24
Phytochemical Screening of Some Plants Used in Herbal Based Cosmetic Preparations

N. G. Masih and B. S. Singh
Department of Botany, St. John's College, Agra
Email – masihgneetu@gmail.com

Abstract

With the development of natural product chemistry the potential of chemotaxonomy is new becoming increasingly obvious. The application of chemical data to systematic has received serious attention of large number of biochemists and botanists. Chemically constituents may be therapeutically active or inactive .Several phytochemical surveys have been carried on for detecting diverse groups of naturally occurring phytochemicals. The phytochemical research approach is considered effective in discovering bioactive profile of plants of therapeutic importance. During the present study five medicinal plants i. e. Acacia concina (fruit), Citrus limon (Fruit), Curcuma longa (Rhizome), Rosa indica (Petals) and Spindus mucorosi (Fruit), which were commonly used in various cosmetic preparations were selected 5 gm each of plant samples dissolved in 30 ml ethanol and water separately to obtain plant extracts. The samples were subjected to qualitative phytochemical screening further presence of various phytochemicals present by adopting standard methodology. The results showed the efficacy of ethonolic extracts showing the presence of more phytochemicals in comparison to aqueous extracts. Phytocompounds such as Alkaloids, Tannins, Saponins, Phenols, Flavonoids and Terpenoids are observed to be present in the investigated plants in variable proportions. Ethanolic extracts of C. longa showed presence of all the phytochemicals in comparison to the rest. Phytochemicals play an important role when used in cosmetic preparations as antimicrobial agents as well as antioxidants. Herbal based cosmetics have gained popularity due to technological advances in manufacturing processers. The application of investigated plant species in various cosmetics was based on their phytochemical content and their pharmacological activities.

Introduction

In ancient Greece and Rome, countless ointments and tonics were recommended for the beautification of the hair, skin as well as remedies for the treatment of scalp and skin diseases. Henary and Mondeveille was the first to make a distinction between medicinal therapies intended to treat diseases and cosmetics agents for purpose of beautification [1]. The advancement in the field of cosmetics and knowledge of skin biology and pharmacology has facilitated the formulation of cosmetics [2]. Many infectious diseases have been known to be treated with herbal remedies through the history of mankind [3]. A large number of cosmetics and toiletry formulations have been developed based on Indian herbs recently. Herbal cosmetics, referred as products, are formulated, using various permissible cosmetic ingredients to form the base in which one or more herbal ingredients are used to provide defined cosmetics benefits only, shall be called as "Herbal Cosmetics" [4]. The demand of herbal based cosmetics is increasing rapidly due to lack of their side effects [5]. Bioactive compounds which are extracted from plants are used in many ways including in cosmetic preparations. Several phytochemicals surveyed have been carried on for detecting diverse groups of naturally occurring phytochemicals.

Materials and Methods

Plant Materials

100gms of each of fresh leaves of selected medicinal plants growing under natural habitat were collected early in the morning, they were cut into small pieces,

M.M. Srivastava, L.D. Khemani, S. Srivastava, *Chemistry of Phytopotentials: Health, Energy and Environmental Perspectives*, DOI:10.1007/978-3-642-23394-4_24, © Springer-Verlag Berlin Heidelberg 2012

washed under running tap water than with distilled water. After air drying they were homogenized into fine powder and were stored in air tight bottles till the following analysis were performed.

Phytochemical Analysis

The phytochemical analysis was carried out using the methanol extract with the standard methods of Harbrone J.B., 1998 [6] and Kokate, 2001 [7]. The phytochemicals studied were Alkaloids, Flavonoids, Phenols, Tannins, Saponins and Terpenoids.

Screening Procedure

Test for Alkaloids: Five ml of the extract was added to 2 ml of HCl. To this acidic medium, 1 ml of Mayer's reagent was added. An orange or red precipitate produced immediately indicates the presence of alkaloids.

Test for Flavonoids: One ml of the extract, a few drops of dilute sodium hydroxide was added. An intense yellow color was produced in the plant extract, which become colorless on addition of a few drops of dilute acid indicates the presence of flavonoids.

Test for Phenols: Two ml of the extract was treated with few drops of Neutral ferric chloride solution. Deep green color developed indicated the presence of Phenols.

Test for Saponin: The extract was diluted with 20 ml of distilled water and it was agitated in a graduated cylinder for 15 minutes. The formation of 1 cm layer of foam showed the presence of saponins.

Test for Tannins: Five ml of the extract and a few drops of 1 % lead acetate were added. A yellow precipitate was formed, indicated the presence of tannins.

Test for Terpenoids: Two ml of the plant extract was taken in a test tube and then added few pieces of Tin and 3 drops of Thionyl chloride, violet or purple color developed indicated the presence of terpenoids.

Results and Discussion

In the present study the results showed that maximum six phytocompounds i. e. Alkaloids, Flavonoids, Phenols, Saponins, Tannins and Terpenoids were present in the ethanolic extract of *Curcuma longa* while five phytocompounds, *i. e.,* Alkaloids, Flavonoids, Phenols, Saponins, Tannins were present in aqueous extract. On the other hand minimum phytocompounds, i. e., Alkaloids, Flavonoids and Saponins in the ethanolic extract and Alkaloids, Flavonoids were present in aqueous extract of *spindus mucorossi*. Herbal based cosmetics are in existence from when the men started to use the cosmetic products. So the quality control for efficacy and safety of herbal cosmetic products is of paramount importance. Quality control tests must be carried out for herbal based cosmetics. It is assumed to be safe for longer period of time.

Table 1

S. No	Name of plants	Solvent	A	B	C	D	E	F	G
1	*Acacia conciona*	Ethanol	-	+	+	+	+	-	4
		aqueous	+	-	+	-	-	-	2
2	*Citrus limon*	Ethanol	-	+	+	+	+	-	4
		aqueous	+	+	+	-	+	-	4
3	*Curcuma longa*	Ethanol	+	+	+	+	+	+	6
		aqueous	+	+	+	+	+	-	5
4	*Rosa indica*	Ethanol	-	+	-	+	+	+	4
		aqueous	+	-	+	+	-	+	4
5	Spindus-mucorossi	Ethanol	+	-	+	-	+	-	3
		aqueous	+	-	-	+	-	-	2

A= alkaloids B= tannins C= saponins D= Phenols E= flavonoids F= terpenoids G= No. of phytoconstituents

References

1. R. Martin. Use of at least one extract of the genus chrysanthemum for assisting skin and/ or hair pigmentation. U. S. patent 6726940 (2004).
2. M. J. Teneralli; Neutraceuticals World. (2004) 74–80.
3. R. Rojas, et al., Bustamante B, Bauer J et al., J. Ethanopharmacol. 80 (1998) 199–204.
4. D. K. Chaudhuri, (2006). Sc"F"&H, PCD.
5. A. S. Bouidin et al., Social science medicine, 49 (1999) 279–289.
6. J. B. Harborne. Phytochemical methods: A guide to modern technique of plant analysis (3rd edition). Chapman and Hall co. New York, pp. 1–302 (1998).
7. C. K. Kokate, C. K., Pharmacogonosy. 16th Edn, Nirali Prakashan, Mumbai, India (2001).

25

Cellular Differentiation in the *In Vitro* Raised Zygotic Embryo Callus of *Boerhaavia diffusa* L. to Produce the Flavonoid, Kaempferol

G. Chaudhary[1], D. Rani[1], R. Raj[2], M. M. Srivastava[2] and P. K. Dantu[1]
[1]Department of Botany, [2] Department of Chemistry
Faculty of Science, Dayalbagh Educational Institute (Deemed University),
Dayalbagh, Agra 282110
Email: premdantu@gmail.com

Abstract

Torpedo shaped embryos dissected from young fruits of Boerhaavia diffusa were cultured in semisolid MS basal medium supplemented with 2,4-D and BAP either alone or in various combinations. Callus from the embryos was transferred to fresh medium of the same composition after 20 days. The callus when initiated was creamish, compact to friable and fast growing. With passage of time the callus growth became slow and the callus became brown. Cytological studies revealed that at the end of 30 days in culture most of the callus was composed of elongated cells which were either nucleated or enucleated. Light microscopic studies indicated depositions on the surface of these cells. The brown callus was dried and extracted in 50% ethanol. The extract was re-extracted with pure ethanol. UV scanning of the ethanolic extract gave a twin absorption peak typical of flavonoids. HPTLC of the extract performed against kaempferol as standard revealed the presence of kaempferol at 1.53 µg/mg of callus. The study demonstrated that callus cells growing in vitro are able to differentiate and activate the flavonoid biosynthetic pathway resulting in the production of the important flavonoid, kaempferol.

Introduction

Secondary metabolites render characteristic properties to the plants and their extraction, identification and quantification is a vital step towards any drug development process. Amongst the wide array of secondary metabolites that plant contain flavonoids account for more than half of the naturally occurring phenolic compounds [1]. Flavonoids are a fifteen-carbon skeleton which includes two benzene rings linked by a heterocyclic pyran ring [1]. Flavonoids exhibit significant biochemical and pharmacological properties which includes anti-oxidation, anti-inflammation, anti-platelet, anti-thrombotic action and anti-allergic effects [2–6]. One of the important flavonol, kaempferol, has been reported to possess anti-cancerous properties [7]. The in vivo anti-cancerous activity of kaempferol was successfully tested in BALB/c (nu/nu) mice and was found to inhibit human osteosarcoma cells both in vitro and in vivo [8]. Kaempferol was studied by its scavenging activity of DPPH and was observed to possess significant antioxidant properties [9]. Kaempferol has been found to stimulate osteoblastic activity and thus can be used for the treatment of osteoporosis [10]. The chemopreventive properties of kaempferol inhibit the metabolic carcinogens of cigarette smoke [11]. Kaempferol also has a strong anti-melanin synthesis activity and inhibits COX and has moderate antibacterial activity [12, 13].

Punnarnava (*Boerhaavia diffusa*) is an important medicinal plant used widely in Ayurveda and Unani medicines. Different parts of the plant have been used in several ailments such as dyspepsia, jaundice and spleen enlargement [14–16]. A large number of secondary metabolites such as geranylactone, limonene, indoleresorcinol monoacetate, vanillin, eugenol and kaempferol 3-O-robinobioside from leaves and quercetin 3-O-robinobioside, caffeoyltartaric acid, eupalitin 3-O-galactosyl (1–2)-glucoside, and isomenthone from roots [17] of this plant have been identified.

M.M. Srivastava, L.D. Khemani, S. Srivastava, *Chemistry of Phytopotentials: Health, Energy and Environmental Perspectives*, DOI:10.1007/978–3-642–23394-4_25, © Springer-Verlag Berlin Heidelberg 2012

Plant cell cultures are proving to be effective alternative for producing in vitro secondary metabolites [18, 19]. In this regard a study was undertaken to establish callus cultures of *B. diffusa* for in vitro production of the secondary metabolite, kaempferol.

Materials and Methods

Plant Material

Young immature fruits were collected from the Botanical Garden of the Institute.

Sterilization of Fruits

Fruits were collected and dipped in 2% Savlon (1.5% v/v chlorohexidine Gluconate solution and 30% Cetrimide solution; Johnson & Johnson, UK) for 8 min. The fruits were then washed under running tap water for about 30 min followed by a quick rinse (30 sec) in 90% ethanol and finally surface sterilized with 0.1% (w/v) mercuric chloride for 8 min. Traces of mercuric chloride were removed by five washes in sterile distilled water in a laminar air flow cabinet.

Embryo Isolation

The sterilized young immature fruits were carefully dissected in a laminar air flow cabinet and embryos released by carefully peeling off the seed coat followed by inoculation on appropriate medium.

Preparation of Medium

For all studies MS basal medium [20] was used at normal strength. The medium was supplemented with various growth regulators such as 2,4-D, BAP as described. All media were supplemented with 3% sucrose and gelled with 0.8% agar and pH set to 5.7 with 1N NaOH or HCl. The molten agar containing medium was poured into 25 mm x 150 mm rimless culture tubes and plugged with polypropylene caps. The medium was steam sterilized by autoclaving at 15 psi and 121°C for 15 min.

Callus Establishment and Subsequent Subculture

Young immature embryos were cultured on MS basal medium supplemented with 2,4-D alone at 0, 0.5, 1, 1.5, 2, 2.5 and 3 mg/l.

Callus Maintenance

The resulting callus was subcultured to a medium containing 2,4-D alone or with BAP. Both 2,4-D and BAP were used at 0.5, 1.0, 1.5, 2.0, 2.5 and 3.0 mg/l. Callus was subcultured every 30 days.

Cell Type Composition of the Callus

To determine the type of cells the callus was suspended in 5 ml of 3% sucrose and vortexed to disperse the cells. The suspension was suitably stained with 1% acetocarmine and temporary slides viewed in a Nikon Eclipse E200 with Nikon digital photographic attachment.

Preparation of Callus Extract

Callus obtained from immature embryos was extracted for quantifying their secondary metabolite content. The callus formed was dried in an oven at 40°C, ground into a coarse powder. Powdered callus was extracted thrice in 50% ethanol and the supernatant was filtered. The filtrate was concentrated in Rotary evaporator (Eyela, Japan) and freeze dried in a Lyophilizer (Allied Frost, India). The lyophilized extract was fractionated successively three times through a series of hexane, chloroform, ethanol and water. The ethanolic fraction was further used as the test extract. It was concentrated under reduced pressure and further lyophilized. For further use 1 mg of the extract was dissolved in 1 ml of the ethanol.

Spectrophotometric Conditions

The callus extract was diluted (1 mg/3 ml) with ethanol and scanned at 190–400 nm with ethanol as reference in an UV-Vis Spectrophotometer (Systronics, India).

Preparation of Standard Solution

A 100 ppm stock solution of kaempferol (Sigma-Aldrich, St. Louis) was dissolved in methanol and used as a standard.

HPTLC Conditions

The following chromatographic conditions were used to quantify kaempferol present in the test extract. Stationary phase consisted of Silica gel precoated $60F_{254}$ TLC plates (Merck). Methanol was used as the mobile place. Kaempferol standard was spotted at 25, 20, 15, 10, and 5 µl. Sample was taken at 10 and 5 µl. HPTLC was performed at ambient room temperature. Solvent front i.e. migration distance of the solvent from origin was 8 cm. Standard solution was applied in the form of bands on pre-coated HPTLC silica gel plates (10 × 10 cm) by means of Linomat V automated spray-on band applicator (CAMAG, France). Ascending development of the plates was carried out in Camag HPTLC twin trough chamber saturated with the mobile phase. The optimized chamber saturation time for the mobile phase was 10 min at room temperature. Plates were developed for 20 min up to a distance of 8 cm beyond the origin. After development, the plates were air-dried for 5 min. Densitometric scanning was performed on Camag TLC scanner III in the reflectance-fluorescence mode operated by winCATS TLC software. The source of radiation utilized was deuterium lamp emitting a continuous UV spectrum between 190 and 400 nm. The standard and sample were scanned at 354 nm.

Results

Callus Establishment

In MS basal medium without 2,4-D the embryos neither grew nor survived. 2,4-D at lower concentrations (0.5, 1 and 1.5 mg/l) induced callus formation from the embryos while at the higher concentrations (2, 2.5 and 3 mg/l) callus establishment was not possible. In the higher concentrations embryos turned brown and ultimately died within 15 to 20 days of culture. Best callus initiation and growth from the torpedo embryos was obtained in MS + 0.5 mg/l 2,4-D. The callus was creamish, compact and friable (Fig 1).

Callus Maintenance

The callus from initiation phase could not survive beyond 15 days when transferred to a medium containing either 2,4-D or BAP alone at all concentrations tested. Presence of both 2,4-D and BAP and their concentration was critical for survival of the callus in subsequent subcultures. A combination of 2,4-D at 1 mg/l with BAP at 0.5 mg/l supported the best callus growth.

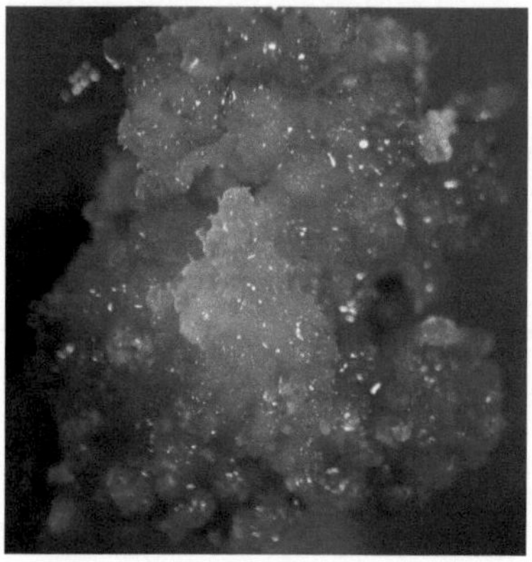

Fig. 1: In vitro raised callus of *B. diffusa* on MS+ 2,4-D + BAP

Types of Cells in the Callus

Microscopic examination indicated that callus from MS + 1 mg/l 2,4-D + 0.5 mg/l BAP at the end of 30 day growth period was composed of three types of cells: (i) small, isodiametric cells with centrally placed nucleus and intensely staining cytoplasm, (ii) elongated cells with sparsely stained cytoplasm with nucleus drifted to one side, and (iii) elongated cells with sparse cytoplasm without traceable nucleus (Fig 2a and 2b). However, as the callus became old the

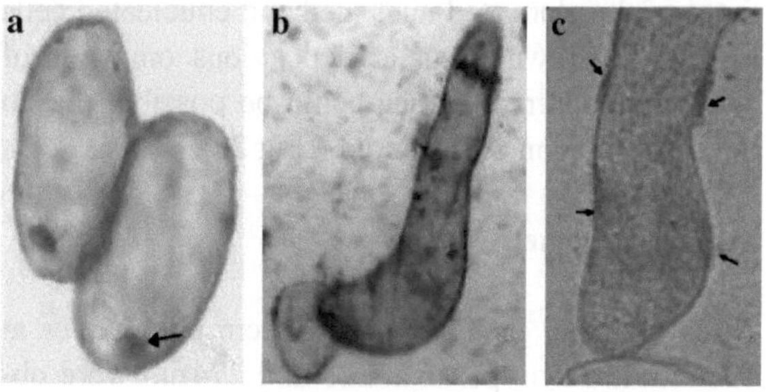

Fig. 2: **(a)** Elongated nucleated cells, **(b)** Elongated enucleated cells, **(c)** Cells showing brown depositions on walls (arrow marked).

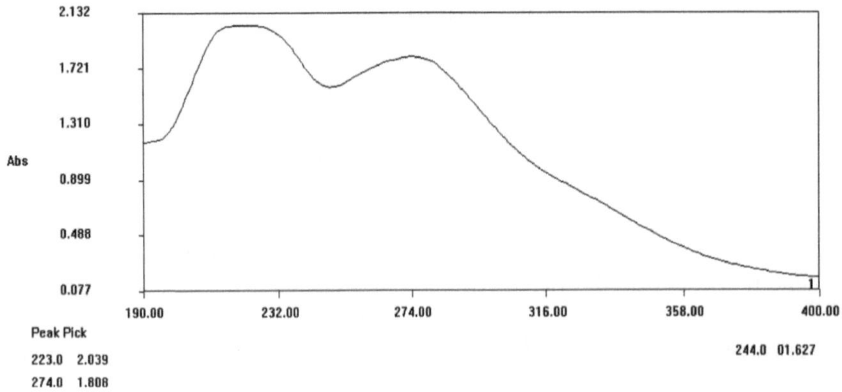

Fig. 3: Scanning of callus extract at 190–400 nm. Note the twin peaks at 223 and 274 nm

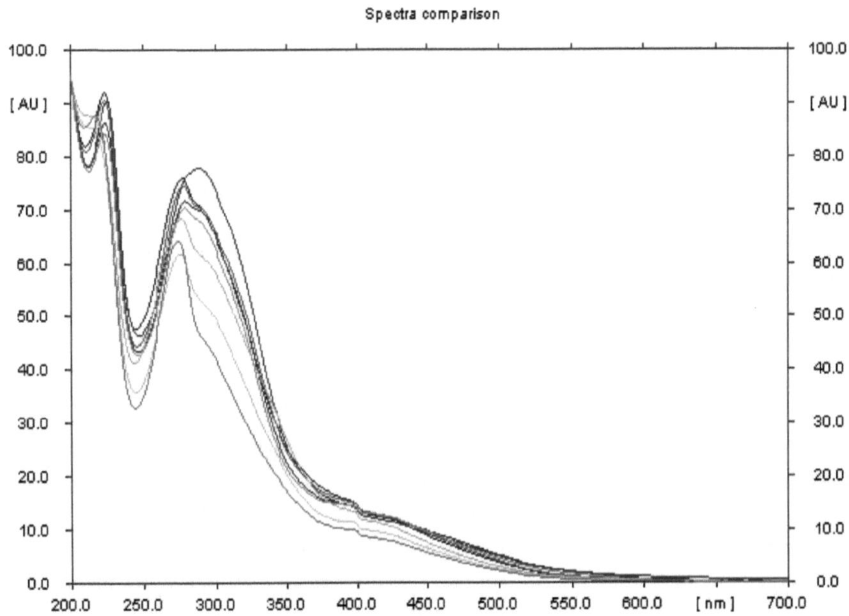

Fig. 4: Overlaying UV spectra of different bands of standard kaempferol and sample

number of small cells decreased dramatically and those of the elongated nucleated and enucleated cells increased. Brown irregular depositions on walls of cells were observed which could be possibly due to the accumulation of flavonoids (Fig 2c).

Spectrophotometric Analysis

After scanning in an UV-Vis Spectrophotometer at 190–400 nm twin peaks at 223 and 274 nm were observed (Fig 3). The twin absorption peaks resembled the characteristic peaks of flavonoids, so further confirmation was done by HPTLC.

HPTLC Fingerprinting

Experimental conditions for carrying out HPTLC, such as mobile phase composition, scan mode, scan speed and wavelength of detection were optimized to provide accurate and precise results. After development with the mobile phase on the silica gel plates, compact and distinct bands were visualized under UV light.

The correlation between amount of standard applied and peak areas obtained showed a linear relation. The overlaying UV spectra showed a similar pattern in the standard and the sample (Fig 4).

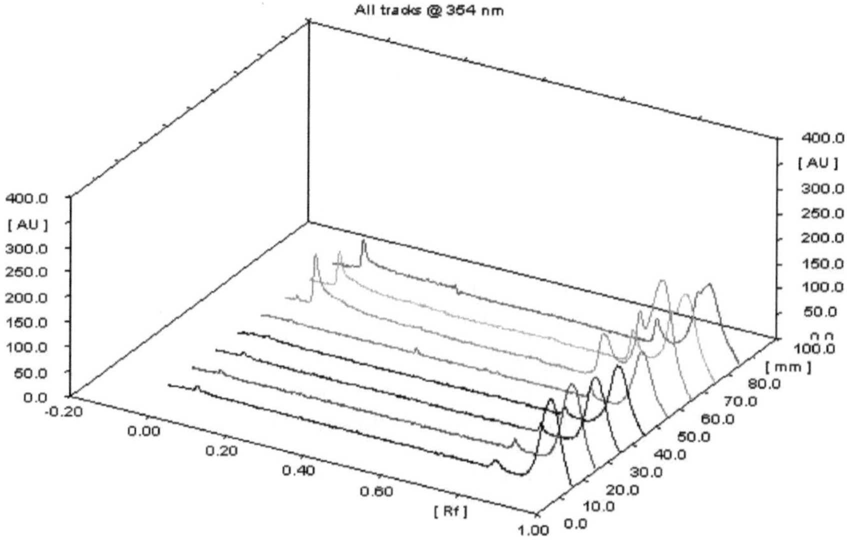

Fig. 5: 3-Dimensional view of spectra of kaempferol standard and sample

The scan densitogram obtained from the test sample gave a selective baseline separation between the standards of flavonoids and the other components in the sample. The amount of kaempferol quantified was 1.532 µg/mg of the brown callus (Table 1) on the basis of calibration curve obtained for standard kaempferol (Fig 6).

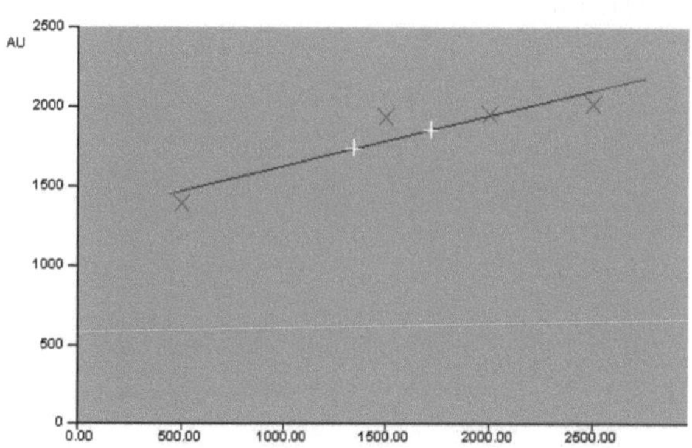

Fig. 6: Calibration curve of kaempferol (cross marks) with sample lying on the curve (+ marks)

Table 1: Validation parameters for quantification kaempferol by HPTLC

Linear Regression	1308.846 + 0.318*X
Standard deviation	7.39
R	0.92
Quantification (µg/mg)	1.532

Discussion

Plant based medicines are gaining popularity as such remedies are often believed to be harmless and can be used for self-medication without supervision [21]. *B. diffusa* is an important medicinal plant in India widely used in Ayurvedic medicine. In vitro production of flavonoid via callus cultures ensures a steady production of flavonoids. Macro and micronutrients have been reported to have considerable influence on growth and biosynthesis of secondary metabolite in cultured plant cells [22]. Increments of nitrate, potassium, ammonium and phosphate support rapid cell growth, whereas the reduction of some of the nutrients leads to growth limitation with a simultaneous enhancement of secondary metabolite production [23]. The type and concentration of auxin and cytokinin, either alone or in combination strongly influence growth of callus as well as secondary metabolite production in tissue cultures. In an earlier report, addition of 2,4-D and kinetin into the media was found to elicit flavonoid production in *Genista tinctoria* [22]. Incorporation of kinetin in combination with 2,4-D enhanced accumulation of valeportiate in *Valeriana gelechomifolia* callus cultures [24].

In the present study we have shown successfully that through appropriate use of growth regulators the callus cells of *B. diffusa* could be made to differentiate and the secondary metabolite pathway gets activated resulting in the in vitro production of the flavonoid, kaempferol. Thus, the current approach of callus mediated biosynthesis of flavonoid could be used to scale up flavonoid production in vitro.

Acknowledgment

The authors wish to thank the Director of the Institute for providing the laboratory facilities for carrying out this work. GC and DR wish to thank the UGC, New Delhi, for the Rajiv Gandhi Fellowship and the Research Fellowship in Sciences for Meritorious Students, respectively.

References

1. J.B. Harborne, H. Baxter and G.P. Moss; In Phytochemical dictionary handbook of bioactive compounds from plants (2nd ed.). London (1999): Taylor and Francis.
2. L.H. Wang and W.H. Li; Pharma.Chem. J. 41 (2007) 46.
3. B. Havsteen; Biochem. Pharmacol. 32 (1983) 1141.
4. R.J. Gryglewski, R. Korbut and J. Robak; J. Sueis. Biochem. Pharmacol. 36 (1987) 317.
5. E.J.R. Middleton and C. Kandaswami; Biochem. Pharmacol. 43 (1992) 1167.
6. N.C. Cooks; S. Samman; J. Nutr. Biochem. 7 (1996) 66.
7. Y. Wang, J. Cao, J.H. Weng and S. Zeng; J. Pharma. Biomed. Anal. 39 (2005) 328.
8. W.W. Huang, Y.J. Chiu, M.J. Fan, H.F. Lu, H.F. Yeh, K.H. Li, P.Y. Chen, J.G. Chung and J.S. Yang; Mol. Nutr. Food Res. 54 (2010) 1585.
9. H.A. Jung, J.J. Woo, M.J. Jung, G.S. Hwang and J.S. Choi; Arch. Pharma. Res. 32 (2009) 1379.
10. C. Prouillet, J.C. Maziere, C. Maziere, A. Wattel, M. Brazier and S. Kamel; Biochem. Pharma. 67 (2004) 1307.
11. D. Puppala, C. G. Gairola and H.I. Swanson; Carcinogenesis. 28 (2007) 639.
12. Y.C. Liang, Y.T. Huang, S.H. Tsai, S.Y. Shiau, C.F. Chen and J.K. Lin; Carcinogenesis 20 (1999) 1945.
13. Y.H. Lim, I.H. Kim, J.J. Seo and J.K. Kim; J. Microbiol. Biotechnol. 16 (2006) 1977.
14. K.R. Kirtikar and B.D. Basu; Indian Medicinal Plants. Vol. III. 2nd Edition. Lalit Mohan Basu, Allahabad, Uttar Pradesh, India. (1956) p. 2045.
15. A.K.S. Rawat, S. Mehrotra, S.K. Tripathi and U. Shama; J. Ethnopharmacol. 56 (1997) 61.
16. B.M. Goyal, P. Bansal, V. Gupta, S. Kumar, R. Singh and M. Maithani; Int. J. Pharm. Sci. Drug Res. 2 (2010) 17.
17. D.M. Pereira, J. Faria, L. Gaspar, P. Valentao and P.B. Andrade; Food Chem. Toxicol. 47 (2009) 2142.
18. S. Roberts and M. Kolewe; Nature Biotech. 28 (2010) 1175.
19. E.K. Lee, Y.W. Jin, J.H. Park, Y.M. Yoo, S.M. Hong, R. Amir, Z. Yan, E. Kwon, A. Elfick, S. Tomlinson, F. Halbritter, T. Waibel, B.W. Yun and G.J. Loake; Nature Biotech. 28 (2010) 1213.
20. T. Murashige and F. Skoog. Physiol. Plant. 15 (1962) 473.
21. Rosidah, M.F. Yam, A. Sadikun, M. Ahmad, G.A. Akowuah and M.Z. Asmavi. J. Ethnopharmacol. 123 (2009) 244.
22. M. Luczkiewics; D. Glod. Plant Sci. 165 (2003) 1101.
23. M.S. Narayan, R. Thimmaraju and N. Bhagyalakshmi; Process Biochem. 40 (2005) 351.
24. N. Maurmann, C.M.B. Decarvalho, A.L. Silva, A.G. Fett-Neto, G.L. Vonposer and S.B. Rech; *In vitro* Cell. Dev. Pl. 42 (2006)5.

26
A Green Thin Layer Chromatographic System for the Analysis of Amino Acids

A. Mohammad and A. Siddiq
Department of Applied Chemistry, Faculty of Engineering and Technology, Aligarh Muslim University, Aligarh
Email: alimohammad08@gmail.com

Abstract

A new green thin layer chromatographic system comprising of silica gel layer impregnated with 1% aqueous urea as stationary phase and double distilled water as mobile phase was identified as the most favorable system for achieving selective separations of lysine and histidine from other amino acids. The detection limits of the amino acids were also determined.

Introduction

Nowadays, the society needs the development of eco-friendly analytical methods where the good selectivity and sensitivity are not sufficient but the methods need to be "Green". The emphasis has been on the use of non-hazardous reagents and minimal generation of chemical waste. Solvents are important components of nature providing one or more liquid phases for chemical reactions and processes [1]. Amino acids are critical to life, and have many functions in metabolism. One particularly important function is to serve as the building blocks of proteins. A number of chromatographic techniques have been used for the analysis of amino acids. Due to several advantageous features such as a) wider choice of stationary and mobile phases, b) open and disposable nature of thin layer chromatographic plates, c) reasonable resolving power, d) minimal sample cleanup, e) reduced need of modern laboratory facilities, thin layer chromatography (TLC) has been most popular for the routine analysis of amino acids and other related substances of pharmaceutical importance [1]. Stationary phases with embedded amide groups were first developed using a two-step modification process where aminopropyl silica was acetylated to form the polar amide groups. Embedded polar group in the bonded silane reduces the hydrophobic properties of the stationary phases and thus alters the overall selectivity. Interestingly, silica gel containing embedded urea groups has shown unique selectivity towards nonpolar and polar compounds. The shielding effect due to the presence of polar urea groups in C18 urea phase has been exploited to separate Neue test mixture at pH value 7.0 [2].

Finding environmentally benign green-solvents is a top priority of the chemists working in the area of organic synthesis, analytical separation, drug analysis and bio-chemical processes [3]. Use of water as a green mobile phase is favorable due to its Non toxicity, non-flammability, easy availability in pure form, low viscosity and pronounced solubility towards hydrophilic compounds. Being single component mobile phase, water can be used repetitively without loss of its chromatographic performance. Thus, the proposed TLC system is novel for achieving important selective separations of amino acids.

Experimental

Acetone, ninhydrin, silica gel G (were from Merck, India), amino acids (leucine, isoleucine, norleucine, phenylalanine, tyrosine, alanine, lysine, proline, serine, glutamic acid, methionine, arginine, histidine and tryptophan) were from CDH, India. All reagents were of Analytical Reagent Grade. 1% w/v aqueous solutions (1%) of all the amino acids were used as analyte. Ninhydrin solution (0.3% w/v) in acetone was used to detect all the amino acids.

M.M. Srivastava, L.D. Khemani, S. Srivastava, *Chemistry of Phytopotentials: Health, Energy and Environmental Perspectives*, DOI:10.1007/978–3-642–23394-4_26, © Springer-Verlag Berlin Heidelberg 2012

Mobile phase:
Double distilled water

Stationary phase:
S1: Silica gel impregnated with 1.0 % aqueous urea
S2: Silica gel impregnated with 5 % aqueous urea

Preparation of TLC Plates

To prepare plates, silica gel (20 g) was homogenized with 60 ml of 1 or 5 % aqueous urea by constant shaking for 5 min. and the resulting slurry was coated immediately onto 20 cm × 3 cm glass plates as 0.25 mm layers by use of a Toshniwal (India) TLC coater. The plates were dried at room temperature and then activated by heating at 100 ± 1 °C for 1h in an electric oven. After activation, the plates were cooled to room temperature and then stored in a closed chamber before use.

Chromatographic Procedure

Test solutions (0.015 μl) were applied to the plates using a Tripette micropipette Germany about 2 cm above the lower edge of the TLC plates. The spots were dried at room temperature (30 ± 5 °C), and the plates were developed in glass jars by ascending technique allowing the ascent up to 12 cm, after presaturation for 10min. After development, the plates were withdrawn from the jars, dried at room temperature, and a glass sprayer was used to apply ninhydrin on the plates to locate the positions of the analyte spots. Plates were then heated for 15–20 min. at 60 °C. All the amino acids except proline (yellow) appeared as violet spots. The amino acids were identified on the basis of their R_F values, which were calculated from R_L (R_F of leading front) and R_T (R_F of tailing front) for each spot, where as

$$R_F = (R_L + R_T)/2$$

For the separation, mixture containing lysine or histidine in combination of other amino acids as reported in Table 1 was applied to TLC plate (S_1). The plate was then chromatographed as described above.

Limit of Detection

The limits of detection of all amino acids were determined by spotting successively the decreasing amounts of these amino acids on S_1 plate until no spot was detected. The minimum detectable amounts of these amino acids were taken as their limits of detection.

Results and Discussion

Results of this study have been summarized in Tables 1–3. The results are discussed below. When silica gel impregnated with 1.0 % aqueous urea with water as mobile phase were used for chromatography of amino acids, compact spots with differential mobility of amino acids were realized| (Table 1). In case of silica gel impregnated with 5 % aqueous urea as stationary phase and water as mobile phase no improved separations were observed. From, this Table it is clear that combination of silica gel impregnated with 1.0 % aqueous urea as stationary phase and double distilled water as mobile phase gives the most satisfactory results of varying retention pattern of amino acids. From these results it can be inferred that polarity of stationary phase has some influence on migration behavior

Table 1: Mobility in terms of R_f values of amino acids on different stationary phase and water as mobile phase

Amino acids	S_1	S_2
Leucine	0.61	0.71
Isoleucine	0.77	0.82
Norleucine	0.75	0.75
Phenylalanine	0.72	0.78
Tyrosine	0.95	0.95
Alanine	0.94	0.95
Lysine	0.42	0.66
Serine	0.96	0.97
Glutamic acid	0.97	0.99
Methionine	0.81	0.84
Histidine	0.37	0.59
Arginine	0.63	0.70
Tryptophan	0.82	0.88
Proline	0.60	0.69

of amino acids. It appears that silica impregnated with polar urea may have a water layer tightly bonded near the underlying silica surface due to hydrogen bonding- ability and thus, provides unique separation opportunity of amino acids. Lysine and histidine have been successfully resolved with other amino acids (Table 2 and Table 3). The limits of detection of all amino acids under study obtained on S1 with water as eluent fall in the range 0.05–0.24 ug per spot.

Table 2: Selective separation of lysine from other amino acids

1.	Lysine-Leucine
2.	Lysine-Isoleucine
3.	Lysine-Norleucine
4.	Lysine-Phenylalanine
5.	Lysine-Tyrosine
6.	Lysine-Alanine
7.	Lysine-Serine
8.	Lysine-Glutamic acid
9.	Lysine-Methionine
10.	Lysine-Arginine
11.	Lysine-Tryptophan

Table 3: Selective separation of histidine from other amino acids

1.	Histidine-Leucine
2.	Histidine-Isoleucine
3.	Histidine-Norleucine
4.	Histidine-Phenylalanine
5.	Histidine-Tyrosine
6.	Histidine-Alanine
7.	Histidine-Serine
8.	Histidine-Glutamic acid
9.	Histidine-Methionine
10.	Histidine-Arginine
11.	Histidine-Tryptophan

References

1. A. Mohammad, N. Haq, and A. Siddiq; J. Sep. Sci. 33 (2010) 3619.
2. C.R. Silva, I.C.S.F.Jardim, C. Airoldi; J.Chromatogr. Sci. 987 (2003) 139.
3. S.K. Sharma, A. Mudhoo, *Green Chemistry for Environmental Sustainability,* CRC Press, (2010), p 450.

27

High Performance Thin Layer Chromatographic Method for the Estimation of Cholesterol in Edible Oils

S. Medhe, R. Rani, K. R. Raj and M. M. Srivastava
Department of chemistry, Dayalbagh Educational Institute, Dayalbagh, Agra
Email: dei.smohanm@gmail.com

Abstract

Cholesterol was detected in six edible market available oil brands using high performance thin layer chromatography. Standard conditions have been optimized based on simulation in R_f values under experimental conditions of polarity of mobile phase and saturation time of solvent chamber. The peanut oil contains highest (0.71 %) while coconut oil contains lowest (0.15 %) cholesterol level. Among the oils studied, no oil was found cholesterol free.

Introduction

Edible oils are directly linked with the human health. Reports highlight that approximately 75 % of the World's production of oil and fats come from plant sources [1]. The oilseed plants commonly used worldwide include; coconut, soybean, cotton, palm, rape, sunflower, mustard, groundnut etc [2]. Many vegetable oils are consumed directly or used as ingredients in food [3]. Cholesterol has been found in vegetable oils as major component, where it could make up to 5 % of the total sterols [4]. Cholesterol is produced by the liver and is found in all body tissues where it helps to organize cell membranes and control their permeability [5]. It is a health-promoting substance and critical component of cell membranes as well as the precursor to all steroid hormones.

Due to increasing awareness of the health implications of high cholesterol in our diets, most people now prefer to purchase cholesterol free vegetable oils. The development of chemical and instrumental methods for the identification and quantification of individual components in food and beverages has become extremely important for establishing the oil quality and their genuineness. Commonly used techniques for the analysis of constituents of edible oils are GC and HPLC [6]. High Performance Thin Layer Chromatography (HPTLC) is recently introduced technique for the analysis of food products without chemical treatment of the sample and has the advantages of simplicity, speed, reproducibility and cost effectiveness [7]. It is an offline technique: the subsequent steps are relatively independent, allowing parallel treatment of multiple samples during chromatography, derivatization and detection. Unlike other methods, HPTLC produces visible chromatograms in which the complex information about the entire sample is available at a glance.

The present work reports standardization of HPTLC method for the estimation of cholesterol in market purchased six popular edible (Coconut, Soybean, Peanut, Mustard), less popular (Taramira) and cholesterol free (Sunflower) oils. Simulation in R_f values as a function of polarity of mobile phase and time of saturation of solvent chamber has been carried out. The proposed HPTLC method has been validated according to ICH guidelines [8] based on selectivity, linearity, accuracy in terms of recovery %, limit of detection and quantification and precision.

Materials and Method

1. Chemicals and Reagents

Pure cholesterol was obtained from E. Merck (Darmstadt, Germany). Six popular edible (Coconut, Soybean, Peanut, Mustard), less popular (Taramira) and cholesterol free (Sunflower) oils available in the market were considered for the study. HPTLC plates (sil-

M.M. Srivastava, L.D. Khemani, S. Srivastava, *Chemistry of Phytopotentials: Health, Energy and Environmental Perspectives*, DOI:10.1007/978–3-642–23394-4_27, © Springer-Verlag Berlin Heidelberg 2012

ica gel 60 F254, 20×10 cm) purchased from E. Merck (Darmstadt, Germany) were used for analysis. Plates were developed in a chromatographic chamber using optimized solvent system comprising of n-Hexane-Diethyl ether- MeOH. The solvent was allowed to migrate up to a height of 80 mm from the lower edge of the plate and then dried it.

2. Sample Preparation and Analysis

Standard solution of cholesterol was freshly prepared by dissolving cholesterol (0.05 mg/mL) in toluene. Edible oils were mixed with toluene and sonicated for 30 minutes for proper mixing and then injected on the HPTLC plates for the analysis. HPTLC system (Camag, Muttanz, Switzerland) consisted of a TLC scanner which is connected to a PC running WinCATS; an auto sampler Linomat V using 100 μL and 500 μL syringes, connected to a nitrogen cylinder; a UV scanner. Each HPTLC plate contains different tracks of samples and standards under following conditions: band width 6 mm; distance between bands 3 mm; application volume of standard cholesterol 2–14 μL; gas flow rate10 s/μL. UV scanner was set for the maximum light optimization with the following settings: slit dimension, 4.00 mm × 0.30 mm, micro; scanning speed, 20 mm/s; data resolution, 100 μm/step. Remaining parameters were left as default settings. Regression analysis and statistical data were automatically generated by the WinCATS software.

Table 1A: Effect of polarity of solvent system on the R_f value of cholesterol using silica gel

Solvent system (v/v/v)	**Saturation**	**R_f**
n-Hexane- Diethyl ether- MeOH (5:2:0.1)	No	**0.24**
n-Hexane- Diethyl ether- MeOH (5:2:0.3)	No	**0.27**
n-Hexane- Diethyl ether- MeOH (5:2:0.5)	No	**0.37**
n-Hexane- Diethyl ether- MeOH (5:2:1.5)	No	**0.66**
n-Hexane- Diethyl ether- MeOH (5:2:2.5)	No	**0.71**

Table 1B: Effect of saturation time on the R_f value of cholesterol using silica gel

Solvent system (v/v/v)	**Saturation**	**R_f**
n-Hexane- Diethyl ether- MeOH (5:2:1.5)	No	**0.66**
n-Hexane- Diethyl ether- MeOH (5:2:1.5)	15 minutes	**0.71**
n-Hexane- Diethyl ether- MeOH (5:2:1.5)	30 minutes	**0.75**

Results and Discussion

Preliminary tests on silica gel, alumina and cellulose coated HPTLC plates indicated that silica gel layer gave the best resolution of the cholesterol. Therefore, all subsequent analyses were done on silica gel layers. Optimization of solvent system has been achieved based on simulation in R_f values obtained in differently designed solvent system as a function polarity (Table 1A) and saturation time (Table 1B).

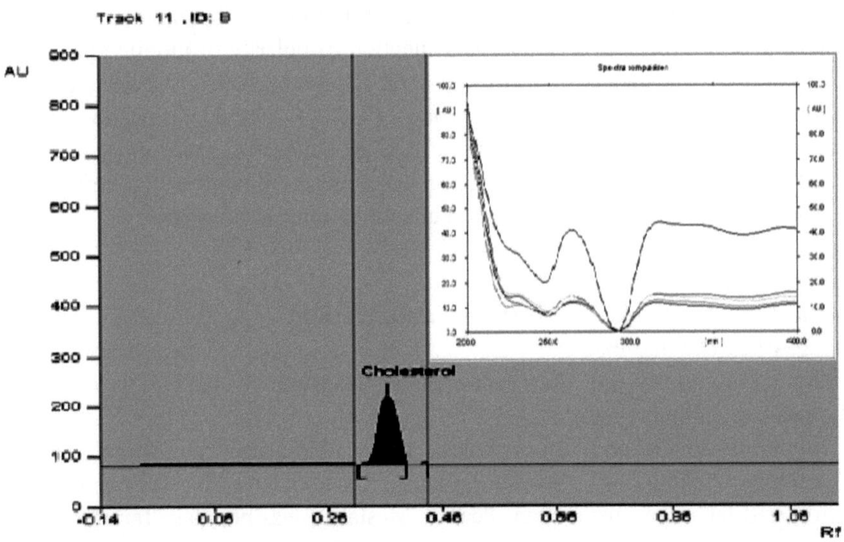

Figure 1: Chromatogram and UV spectra of cholesterol

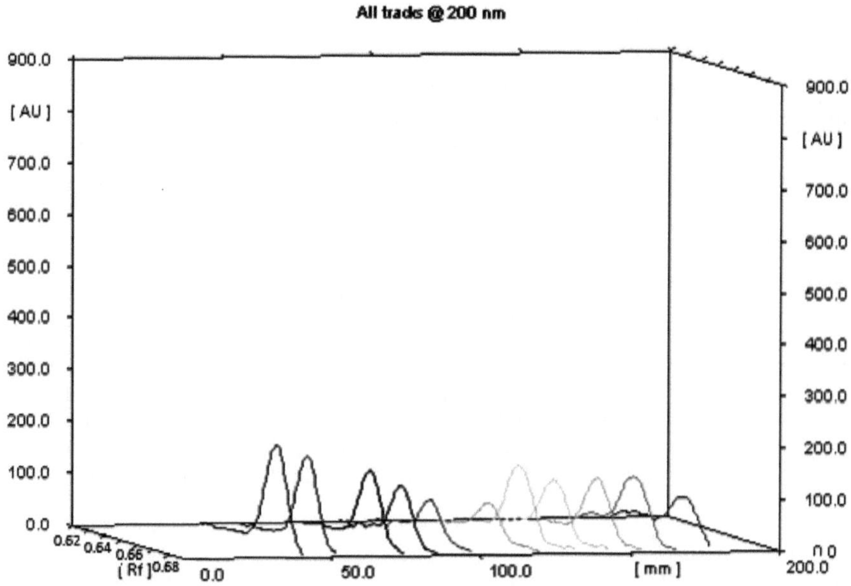

Figure 2: 3D display of cholesterol peaks

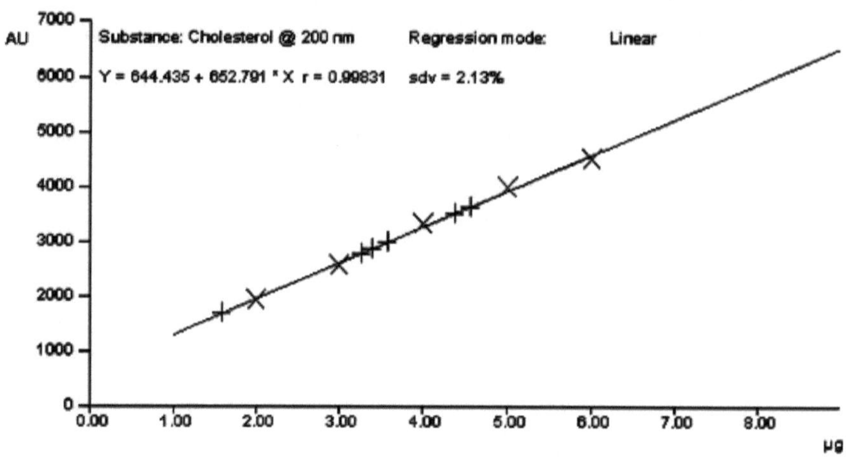

Figure 3: Calibration curve of cholesterol

The chromatographic profile of cholesterol in the samples was simple, showing cholesterol as the main component (Figure 1).

Presence of cholesterol was confirmed by recording the spectra of standard cholesterol and edible oil samples in the range of 200–400 nm.

Peak of cholesterol was identified using optimum solvent system: n-Hexane- Diethyl ether- MeOH (5 : 2 : 1.5, v/v/v) with the R_f value of 0.66 ± 0.01 and there was no overlap with any other analyte of the sample at 200 nm (Figure 2).

The developed chromatographic method was validated according to ICH guidelines. Validation parameters include selectivity, linearity, accuracy in terms of recovery %, limit of detection and quantification and precision [9]. Selection of wavelength (200 nm) is specific for the detection of cholesterol and enabled its detection at R_f value of 0.66 ± 0.01. The linearity of the proposed method was confirmed in the range of 100 – 700 ng of standard cholesterol. This range was suitable for the determination of cholesterol content in edible oils. A linear regression of the data points for standard cholesterol is resulted in a calibration curve with the equation $Y = 644.436 + 652.7908x$ [regression coefficient (r^2) = 0.99831, standard deviation (S. D.) = 2.13 %] (Figure 3).

The correlation coefficient was found to be greater than 0.998 which manifests a linear relationship between concentration and the peak area. Cholesterol content in edible oils was found to be in the range of 150 – 710 ng selected for study (Table 2).

Table 2: Cholesterol levels in edible oils

Edible oils	Cholesterol content [%]
Coconut	0.15
Mustard	0.28
Taramira	0.32
Soybean	0.42
Sunflower	0.56
Peanut	0.71

The peanut oil contains highest (0.71 %) while coconut oil contains lowest (0.15 %) cholesterol level. Among the oils studied, no oil was found cholesterol free. The calibration curve was accurate within the specified concentration range with a mean recovery [10] of 90.32 ± 1.04 % (Table 3).

Table 3: Recovery % of cholesterol

Standard added (ng)	Found (ng)	Recovery (%)
500	449.20	89.84
600	537.60	89.60
700	640.64	91.52
Mean ± S.D = 90.32 ± 1.04		

The limit of detection (LOD) and quantification (LOQ) was found to be 10 and 32 ng respectively. Precision (repeatability) was determined by running a minimum of four analyses and the coefficient of variability was found to be 1.662 %.

Conclusion

Proposed method is simple, sensitive, rapid, specific and could be applied for quality and quantity monitoring of cholesterol content in different edible oils. Among the oils studied, no oil was found cholesterol free.

Acknowledgements

The authors gratefully acknowledge Prof. V. G. Das, Director, Prof. L. D. Khemani, Head, Department of Chemistry, Dayalbagh Educational Institute, Dayalbagh, Agra. The authors are also thankful to Ministry of Human Resource Development (MHRD), Govt. of India, for rendering financial assistance.

References

1. H. P. Raven and B. G. Johnson; Biology. pp. 1231. Mc Graw-Hill, New York. 1999.
2. E. J. Behrman and G. Venkat; J. Chem. Edu. 82 (2005) 1791–1793.
3. K. M. Anderson, W. P. Castelli and D. Levy; J. Am. Med. Assoc. 257 (1987) 2176–2180.
4. V. K. S. Shukla, P. C. Dutta and W. E. Artz; J. Am. Oil Chem. Soc. 79 (2002) 965–969.
5. E. O. Aluyor, C. E. Ozigagu, O. I. Oboh and P. Aluyor; Scientific Res. Essay. 4 (2009) 191–197.
6. A. Cert, W. Moreda and C. Perez; J. Chromat. A. 881 (2000) 131–148.
7. G. Bazylak, H. Brózik and W. Sabanty; Polish J. Environ. Stud. 9 (2000) 113–123.
8. ICH Guidelines. http://www.ich.org/products/guidelines.html
9. P. K. Tiwari and P. Sathe; Adv. Biosci. Biotechnol. 1 (2010) 131–135.
10. E. A. Abourashed and J. S. Mossa; J. Pharm. Biomed. Anal. 36 (2004) 617–620.

28
Vegetable Seed Oil Based Waterborne Polyesteramide: A "Green" Material

F. Zafar*, H. Zafar[2], M. Yaseen Shah[1], E. Sharmin[1] and S. Ahmad[1]
[1]Materials Research Lab., Dept. of Chemistry, Jamia Millia Islamia, New Delhi-110025, India
[2]Inorganic Chemistry Lab., Department of Chemistry, Aligarh Muslim University, Aligarh, India
*corresponding author; Emails: fahmzafar@gmail.com

Abstract

In present work we have taken an attempt to develop vegetable seed oil [VSO] based microwave assisted (green route) waterborne polyesteramide (zero toxicity), a green material. VSO based waterborne polyesteramide [WBPEA] was synthesized by simple route through microwave irradiation within 4–5 min by amidation and condensation of oil. The structure of the material was confirmed with the spectral techniques. The thermal degradation analysis of WBPEA by TGA showed that the polymer was stable up to 220°C, suitable for use as degradable "green polymeric material" for a variety of applications.

Introduction

Nowadays, biomass such as rosin, chitosan, starch, cellulose, fats, proteins, lignin, jute, wood flour, cashew nut shell liquid, bagasse, seeds and others are used as alternatives of petrobased resources [1]. Among synthesized bio-based products from agricultural resources, natural oils are useful raw materials in the synthesis of polymers to replace traditional petrochemicals based polymers. Natural oils are inexpensive, non toxic, biodegradable, found in abundance and most importantly renewable in nature. Natural oils such as linseed and tung oil have long found various uses in coatings, printing inks and varnish industries. Soybean oil, safflower oil, sunflower oil, canola oil and karani oil have also been used in industrial applications including: plastics, lubricants, adhesives, inks, paints, coatings, fuels/biodiesel, solvents, surfactants [2–3]. Due to the presence of sufficient unsaturation, unique molecular structure and diverse functionality of vegetable oils, they play an important role towards the development of polymers such as alkyds, polyesters, polyurethanes, epoxies, polyols, polyurethaneamides, polyesteramides, for different applications and compete over petro-based materials[4]. Some drawbacks are associated with these polymers during synthesis and applications, which are elevated reaction temperatures, longer reaction times, cumbersome multi-step processes, low yield and use/emission of harmful organic solvents (VOCs) [3]. In order for its easy applications as eco-friendly material-green materials, the foregoing undesirable characteristics of these resins can be overcome by the use of alternative route that is "Green Chemistry".

Green Chemistry means using green source (biomass) through green route with zero negative impact on environment (zero toxicity), and meeting green end (biodegradable) that is from source to end everything is green. Green route involves development of green polymers via solvent free, enzymatic and microwave irradiation methods [5–6]. Features of microwave irradiation, i.e., solvent-free reactions, low waste, energy efficiency, high yield, short reaction time, and possible use of alternative solvents can also play an important role in the development of Green Chemistry methods [7].

Waterborne polymers- green materials have gained much popularity for the most obvious reason and perhaps the most important benefit of reduced emissions of volatile organic compounds [VOCs]. They are classified as: water-soluble/water-reducible (solutions), water-dispersible/ colloidal (dispersions) and emulsions (latex) paints. The physical properties and performances of each type mentioned above depend upon the choice of the resin. The resins generally used are vinyls, two-component acrylics, epoxies, polyes-

M.M. Srivastava, L.D. Khemani, S. Srivastava, *Chemistry of Phytopotentials: Health, Energy and Environmental Perspectives*, DOI:10.1007/978–3-642–23394-4_28, © Springer-Verlag Berlin Heidelberg 2012

ters, styrene-butadiene, amine-solubilized, carboxyl-terminated alkyd and polyurethanes. Interest in water borne resins arises due to their non-polluting, easy to handle, quick drying, economic and environmentally friendly nature [8]. Waterborne polymer formulations exhibit a wide range of uses and properties that can frequently match and even exceed the conventional solventborne polymers. They have been commercially utilized in a wide variety of applications for civil engineering, adhesives, structural end-uses, fiber sizing, glass, synthetic and carbon fibers, tie-coats, primers and laminates, textiles, sealants, insulating materials, electronic photo imaging, and coatings-floor, corrosion and microbial resistant [9–16]. Synthesis of oil based water borne polymer is a challenging task due to hydrophobic nature of oil triglyceride chains. Aigbodion et al have synthesized waterborne alkyds based on rubber seed oil for coating applications by treating the glycerolysis product of molinate or fumarized oils with phthalic anhydride and neutralizing the final product with amine [17]. Several other reports on oil waterborne resins are also available in literature [8, 18–22].

Therefore, in present work, we have taken an attempt to synthesize VSO based waterborne polyesteramides [WBPEA] by microwave irradiation (green route) as a small step for development of green polymer for clean environment. VSO based polyesteramide is a polyester containing both ester (-COOR) and amide (-CONRR') linkages in one polymer chain. Several polyesteramides have been synthesized from different VSO and have found application as protective materials [23]. However, their overall synthesis strategy by conventional heating methods is time consuming, multi-step process [24–26]. Literature survey reveals that no work is reported yet on the synthesis and characterization of such new material.

Experimental

Materials

VSO procured from local market. Diethanolamine, phthalic anhydride (S.D Fine Chemicals, India), sodium metal, methanol and xylene (Merck, India), were of analytical grade.

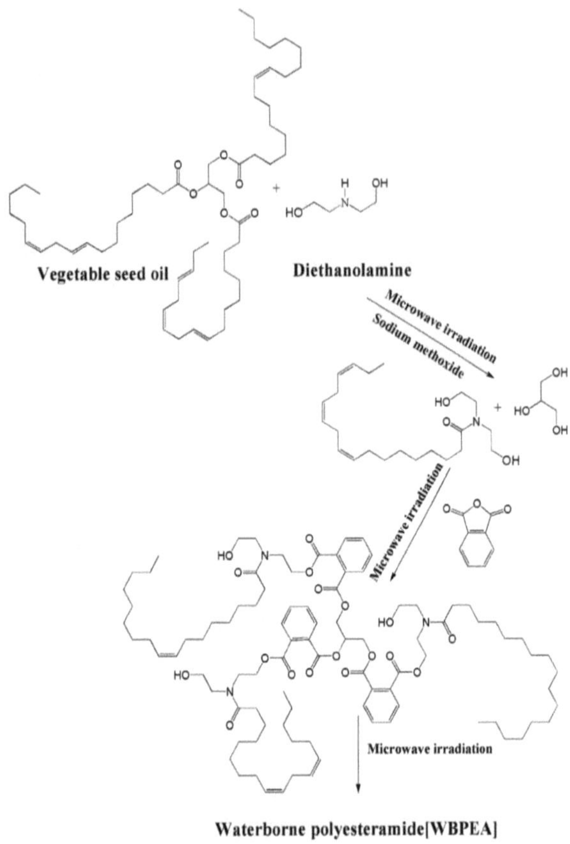

Fig. 1: Synthesis of WPEA

Synthesis of WBPEA

The synthesis was carried out in domestic microwave oven model LG MS 1927C operating at 230V- 50 Hz frequency. The synthesis of WBPEA was carried out by reacting calculated amount of VSO, diethanolamine, and freshly prepared sodium methoxide as a catalyst in an Erlenmeyer flask for 3–4 min under microwave irradiation. Thin layer chromatography [TLC] and FTIR spectral technique were used to monitor the progress of the reaction. After completion of the reaction, calculated amount of fine powder of phthalic anhydride was added in same flask containing reaction mixture under microwave irradiation further for 1–2 min. Acid value was used to monitor the progress of the reaction. The reaction was carried out till the product attained the desired acid value. The reaction was stopped at this point; WBPEA was cooled and preceded for characterizations.

Instruments and Test Methods

Domestic microwave oven of microwave frequency range 230V-50 Hz (LG, Model No.LGMS 1927C) was used for synthesis reactions. FT IR spectra of the resin was taken on Perkin-Elmer 1750 FTIR spectrophotometer (Perkin Elmer Instruments, Norwalk, CT) using a NaCl cell. ^{1}H-NMR and ^{13}C-NMR spectra were recorded on JEOL GSX 300MHZ FX-1000 spectrometer using deuterated chloroform as solvent and tetramethylsilane (TMS) as an internal standard. The thermogravimetric analysis (TGA) was done with thermogravimetric analyzer, TA-51 (T.A. Instruments U. S. A.) at 20 °C/min in nitrogen atmosphere. Solubility of the resin was tested in various polar and nonpolar solvents by taking 25 mg of each resin in 5 ml of different solvents in a closed test tube and set aside for a day.

Results and Discussions

WBPEA was synthesized by simple route through microwave irradiation within 4–5 min. Fig. 1 shows the synthesis of polymer, which reveals that the synthesis was carried out in situ by two step reaction. In the first step amidation of VSO with diethanol amine in the presence of sodium methoxide results in the formation of diol fatty amide and glycerol in 2–3 min followed by second step, that is condensation polymerization with phthalic anhydride in 1–2 min under microwave irradiation. The condensation polymerization of the reacting mixture of the first step with phthalic anhydride results in WBPEA. In the synthesis, glycerol plays a vital role for branching; it acts as a core molecule.

The presence of characteristic peaks of polyesteramide in FTIR, ^{1}H-NMR and ^{13}C-NMR spectra of WBPEA confirmed the structure [25]. These are as follows:

FTIR (cm^{-1}): 3374 (-OH), 1724.2 (υ, > C = O ester), 1622 (> C = O, amide), 1278 (υ, C – C (=O) – O –C); ^{1}H-NMR (δ, ppm): 7.1–7.3 (phthalic ring proton), 4.1–4.3 {CH_2– O – C (=O) –Ar/ CH– O – C (=O) –Ar }, 3.1–4.0 (CH2 – N< / CH_2OH) , 5.3 (-OH), 0.88 (-CH3); ^{13}C-NMR: 171, 173 (amide carbonyl), 165, 167 (carbonyl of ester and acid), 61–62.5 [CH_2 OC(=O)Ar/ CH OC(=O)Ar], 44–48 (CH_2N<), 28–29 (chain CH_2), 14 (-CH_3).

Solubility was checked in polar and non polar solvents that showed the resin was soluble in non polar as well as polar solvents. Solubility of the polymer in polar solvents especially in water is due to the presence of hydroxyl functionality and substituted amide linkages on branched structure collectively responsible for excellent water solubility.

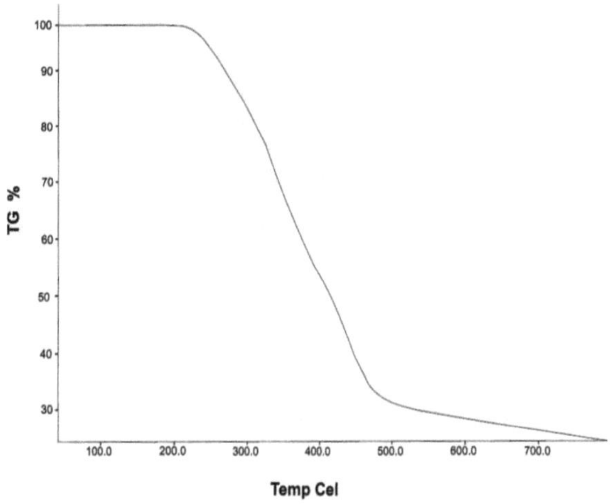

Fig. 2: TGA thermogram of WBPEA

The thermal degradation of WBPEA studied by TGA (Fig. 2) showed that the polymer was stable upto 220°C but exhibited weight loss of 5 %, 10 %, 20 %, 50 % and 70 % at 240–250°C, 275–280 °C, 350–355 °C, 410–420 °C and 545–550 °C, respectively. The 5 % weight loss can be correlated to the entrapped moisture. 10 – 70 % weight losses can be correlated to the degradation of ester and amide linkages followed by aromatic and hydrocarbon moieties. The thermal degradation trend of the polymer can be attributed to the presence of ester and amide linkages along with aromatic ring in same polymeric chain in addition to its branched structure.

Conclusion

VSO was successfully utilized for the synthesis of waterborne polyesteramide [WBPEA] by simple route through microwave irradiation within 4–5 min in situ compared to the conventional heating method, which involves 6–8 hours and is a multi-step reaction. The spectral studies confirmed the formation of polyesteramide. The synthesized oil (biomass) based microwave assisted (green route) waterborne (zero toxicity) polymer can be used as thermally stable, novel, "green" polymeric material for variety of applications such as adhesives, biodegradable coatings and antimicrobial agents. The biodegradable studies of WBPEA are under process.

Acknowledgements

Dr Fahmina Zafar (Pool Officer) and Dr Eram Sharmin (Pool Officer) are thankful to CSIR, New Delhi, India for Senior Research Associateship against grant nos. 13(8385-A)/Pool/2010 and 13(8464-A)/2011-Pool, respectively. Dr Fahmina Zafar (Pool Officer) and Dr Eram Sharmin (Pool Officer) are thankful to the Head, Dept. of Chemistry, Jamia Millia Islamia, New Delhi, for providing facilities to carry out the research work.

Reference

1. A. Corma, S. Iborra, and A. Velty. Chem. Rev. 107(2007) 2411
2. L. Zengshe, M. D. Kenneth, and R.A. Holser, Green Chem. 11 (2009) 1774
3. V. Sharm, and P. P. Kundu, Prog. Polym. Sci. 33 (2008) 1199
4. E. Sharmin, S. M. Alam, R. K. Philips, and S. Ahmad. Prog. Org. Coat. (2010) inpress
5. W.A. Hermann, A. M. J. Rost, E. Tosh, H. Riepl, and F. E. Kiihn, Green Chem. 10 (2008) 442
6. S. Miao, S. Zhang, Z. Su, and P. Wang. J. Polym. Sci. Part A: Polym. Chem. 47 (2009) 295
7. R. Hoogenboom, and U. S. Schubert, Macromol. Rapid Commun. 28 (2007) 368
8. H. Mutlu, and M.A. R. Meier, Eur. J. Lipid Sci. Technol. 112 (2010) 10
9. C. G.Sanchez, M. Lobert, R. Hoogenboom, and U. S. Schubert, Macromol. Rapid Commun., 28 (2007) 456
10. C. L. Bao, L. S. Wang, A. Zhang, and J. Taiwan Ins. Chemi. Eng. 40 (2009) 174
11. A. I. Aigbodiona, and C. K. S. Pillai, Prog. Org. Coat. 38 (2000) 187
12. C. Quintero, S. K. Mendon, O. W. Smith, and S. F. Thames, Prog. Org. Coat. 57 (2006) 195
13. G. L. Gyorik, T. Pajkossy, and B. Lengyel, Prog. Org. Coat. 6 (2006)
14. V. Kesari, A. Das, and L. Rangan, Biomass Bioenergy 34 (2010) 407
15. B. Ni. L. Yang, C. Wang, L. Wang, and D. E. Finlow. J. Thermal Analy. Calorim. 100 (2010) 239
16. E. Hablot, D. Zheng, M. Bouquey, and L. Avérous, Macromol. Mater. Eng. 293 (2008) **922**
17. E. U. Ikhuoria, M. Maliki, F. E.Okiemen, A. I.Aigbodion, E. O.Obaze, and I. O.Bakare. Prog. Org. Coat. 59 (2007) 134
18. E. U.Ikhuoria, A. I.Aigbodion, and F. E.Okeimen; Prog. Org. Coat. 52 (2005) 238
19. Y.Lu, and R. C.Larock; Biomacromol. 8 (10) (2007) 3108
20. D. Colombini, M. J. Deriss, O. J.Karlsson, and F. H. J. Maurer. Macromol. 37 (2004) 2596
21. S. T.Wang, F. J.Schork, G. W.Poehlein, and J. W. Gooch, J. Appl. Polym. Sci. 60 (12) (1996) 2069
22. Y. Lu, L.Tighzert, F.Berzin, and S.Rondot, Carb. Polym. 6 (2005) 174
23. F. Zafar, S. M. Ashraf and S. Ahmad. J. App. Polym. Sci. 104 (2007) 1143
24. F. Zafar, M. H. Mir, M. Kashif, Eram Sharmin, and S. Ahmad; J. Inorg. Organomet. Polym. 21 (2011) 61
25. U. Riaz, and S. Ahmad, J Appl. Polym. Sci.121(2011) 2317
26. F. Zafar, E. Sharmin, M. Kashif, O. Rehman, S. Pathan, M. S. Khan, and S. Ahmad; Bhartiya Vaigyanik Evam Audyogik Anusandhan Patrika 1 (2010) 38
27. F. Zafar, S. M. Ashraf, and S. Ahmad; Chem. Che. Technol. 2(4) (2008) 285

29
QSAR Analysis of Anti-Toxoplasma Agents

R. Mishra, A. Agarwal and S. Paliwal
Department of Pharmacy, Banasthali University, Banasthali,
Rajasthan- 304022, India.
E-mail: ruchimishra_0209@rediffmail.com

Abstract

The QSAR models were developed using multiple linear regression (MLR) and partial least square (PLS) for 2, 4-diamino-8-deazafolate analogues generated comparable models with good predictive ability and all other statistical values, r, r^2, r^2_{cv}, r^2 (test set) F and s values were 0.94, 0.88, 0.86, 0.84, 58.70 and 0.13 respectively for MLR and r^2_{cv}, r^2 (test set) and statistical significance values were 0.87, 0.84 and 0.99 respectively for PLS, were satisfactory. The analysis indicate the role of first atom E-state, number of H-bond doners and VAMP mean polarizability in determining the activity of 2, 4-diamino-8-deazafolate analogues as potent antitoxoplasmal agents.

Introduction

Toxoplasma gondii, the causal agent of toxoplasmosis, is an important water and food borne protozoan parasite ubiquitous throughout the world [1]. Toxoplasma gondii is a highly prevalent obligate intracellular parasite that has no host specificity and infects all warm-blooded vertebrates including mammals and birds. There are three principal routes of transmission of the parasite: ingestion of infective oocysts shed by the cat, carnivory (consumption of raw or undercooked meat) and congenital transmission.

The metabolism of folate plays an important role in the biosynthesis of nucleic acid precursors. During the synthesis of purines and thymidylate, the cofactor tetrahydrofolate is oxidized to 7,8-dihydrofolate and subsequently converted back to tetrahydrofolate by the enzyme dihydrofolate reductase (DHFR). The inhibition of DHFR causes the depletion of tetrahydrofolate and disrupts DNA synthesis, leading to cell death. For this reason, DHFR inhibitors such as methotrexate (MTX) have been used as antitumor, antibacterial and antiprotozoal agents. Because MTX and other classical antifolates require an active transport mechanism for their uptake, they are not effective for the treatment of infections caused by Toxoplasma gondii (tg) that lack these mechanisms [2–3].

T. gondii infections are the principal cause of death in patients with AIDS, and also affect patients with other immune disorders [4]. Trimetrexate (TMQ) and piritrexim (PTX) are potent lipophilic inhibitors of pcDHFR and tgDHFR taken up by passive diffusion, but inhibit mammalian DHFR to a greater extent [5–6]. This results in toxicity to mammalian tissue and requires that PTX or TMQ be co-administered with leucovorin (5-formyl-5,6,7,8 tetrahydrofolate), a reduced folate which is taken up by active transport and protects the host tissue [7]. Treatment with leucovorin is costly and subject to serious side effects that may require interruption of treatment. As such, there is great interest in developing potent and selective inhibitors of T. gondii DHFR. In the present paper we describe the QSAR studies of inhibition of DHFR in Toxoplasma gondii by 2, 4-diamino-8-deazafolate analogues. The study was aimed to identify the physicochemical parameters associated with DHFR inhibitor which are more selective for T. gondii than for mammalian DHFR.

M.M. Srivastava, L.D. Khemani, S. Srivastava, *Chemistry of Phytopotentials: Health, Energy and Environmental Perspectives*, DOI:10.1007/978-3-642-23394-4_29, © Springer-Verlag Berlin Heidelberg 2012

Material and Methods

The structures of 44 2, 4-diamino-8-deazafolate analogues [8] (Table 1) were sketched using Chem Draw software and were imported on TSAR (Version 3.3; Accelrys Inc, oxford, England) software. The generated 3D models of all derivatives created were cleaned up and subjected to charge calculation and energy minimization.

More than 300 molecular descriptors were calculated for all the compounds under consideration. TSAR affords the calculation of the following descriptors: atomic attributes (like molecular properties, dipole moment and verloop steric parameters), atomic indices (like shape, connectivity and topological indices) and Vamp electrostatic properties (total energy, HOMO, LUMO, heat of formation, etc) [9]. To reduce data redundancy, pairwise correlation analysis was carried out [10]. Among the highly intercorrelated descriptors the one that had high correlation with biological activity was kept and other was discarded. This process was repeated number of times and finally three descriptors were retrieved that were highly correlated with biological activity and were not having intercorrelation among each other.

To develop QSAR models, stepwise MLR analysis with leave out groups of rows cross-validation was applied to the training set. The molecules of the series were divided randomly into training set (31 molecules) and test set (13 molecules). Training set was used to build linear models so that an accurate relationship could be found between structure and biological activity [11]. The test set of thirteen molecules was not used to develop the regression model but served to check the predictive power of the developed model. In addition to MLR, partial least squares (PLS) [12] analysis was also performed to check the predictive ability and robustness of the developed model.

Results and Discussion

Multiple linear regression (MLR) and partial least squares (PLS) were used to derive the QSAR equations. The statistically significant model was constructed from the training set by using 3 parameters. In order to improve the predictivity of the model, 5 potential outliers namely 5, 8, 25, 37 and 44 (which exhibited high residual value and were to far away from regression line) were identified and deleted.

The final regression equation obtained from MLR analysis (final 26 molecules in training set) after deleting the outliers is represented as equation 1

$$Y = 0.24*X1 - 0.82*X2 + 0.056*X3 - 1.11 \quad \textbf{[1]}$$

$r = 0.94$ $r^2 = 0.88$, $r^2_{cv} = 0.86$, $F = 58.70$, $S = 0.13$, predictive r^2 for test set = 0.84

This best model was selected on the basis of various statistical parameters such as coefficient of determination (r^2), predictive power of model (r^2_{cv})standard deviation (SD), sequential Fisher test (F) and test for statistical significance (t). The value of r^2 should always be greater than 0.6 (a good model should have an $r^2 > 0.9$) and the value of r^2_{cv} could fall into three categories [13]:

- $r^2_{cv} > 0.6$: The model is fairly good.
- $0.4 < r^2_{cv} < 0.6$: The model is questionable.
- $r^2_{cv} < 0.4$: The model is poor.

PLS analysis was also performed on the same data set to check the soundness of the MLR model. The resulted r^2_{cv} value of 0.87 clearly demonstrates the high predictive ability of the developed PLS model equation 2.

$$Y = 0.24*X1 - 0.82*X2 + 0.056*X3 - 1.11 \quad \textbf{[2]}$$

Statistical significance = 0.99, $r^2_{cv} = 0.87$, $r = 0.88$, Predictive r^2 for test set = 0.84

In equation [1] and [2], X1 is first atom E-state (subst. 1), X2 is Number of H-bond Doners (subst. 1), X3 is VAMP Mean Polarizability (whole molecule).

Since for a well defined problem, both MLR and PLS should generate comparable results [14], the r^2_{cv} values of MLR and the PLS models were evaluated and it was found that both the models have comparable r^2_{cv} value of 0.86 and 0.87 for MLR and PLS respectively. The predictive ability of the model was also validated using the external test set of 13 compounds in context of minimum difference between the actual and predicted biological activity values of MLR and PLS analysis for training and test which is shown in table 2 and their respective plots are depicted in figure 1 and 2.

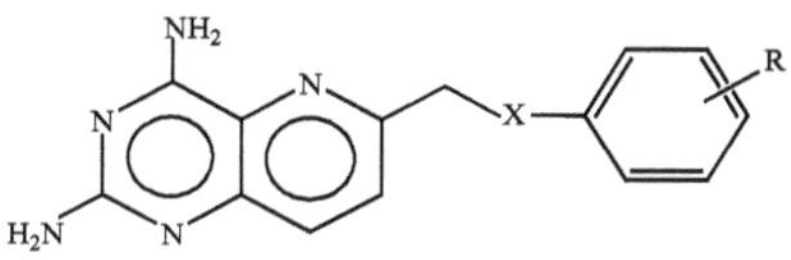

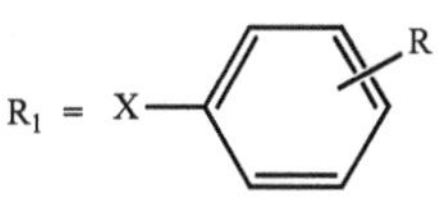

Table 1: Series of 2, 4-diamino-8-deazafolate analogues

Compounds name	R	X
3	S	H
4	NH	H
5	NMe	H
6	NH	2'-OMe
7	NMe	2'-OMe
8	NH	3'-OMe
9	NMe	3'-OMe
10	NH	4'-OMe
11	NMe	4'-OMe
12	NH	2'-Cl
13	NMe	2'-Cl
14	NH	3'-Cl
14a	NMe	3'-Cl
15	NH	2', 4'-$(OMe)_2$
16	NMe	2', 4'-$(OMe)_2$
17	NH	2', 5'-$(OMe)_2$
18	NMe	2', 5'-$(OMe)_2$
19	NH	3', 4'-(OMe)2
20	NMe	3', 4'-(OMe)2
21	NH	2',4'-Cl_2
22	NMe	2',4'-Cl_2
23	NH	2',5'-Cl_2
24	NMe	2',5'-Cl_2
25	NH	2',6'-Cl_2
26	NMe	2',6'-Cl_2
27	NH	3',4'-Cl_2
28	NMe	3',4'-Cl_2
29	NH	3',4',5'-$(OMe)_3$
29a	NMe	3',4',5'-$(OMe)_3$
30	NH	3',4',5'-Cl_3
31	NMe	3',4',5'-Cl_3
32	NH	3',4',6'-Cl_3
33	NMe	3',4',6'-Cl_3
34	S	2',3'-C_4H_4
35	S	3',4'-C_4H_4
36	NH	2',3'-C_4H_4
37	NH	3',4'-C_4H_4
38	NMe	2',3'-C_4H_4
39	NMe	3',4'-C_4H_4
40	NH	4'-$COCH_3$
41	NMe	4'-$COCH_3$
42	N-propargyl	4'-$COCH_3$
43	NMe	4'-$COCF_3$
44	N-propargyl	4'-$COCF_3$

Table 2: Series of 2, 4-diamino-8-deazafolate analogues along with their actual & predicted biological values

Compounds name	Actual (Log 1/c)	Predicted (MLR)	Predicted (PLS)
3	0.88	0.83	0.83
4	1.07	0.97	0.97
5**	2.07	--	--
6	0.92	1.12	1.12
7*	1.58	1.81	1.81
8**	1	--	--
9	1.82	1.80	1.80
10	1.26	1.12	1.12
11	1.79	1.79	1.79
12	0.95	1.02	1.02
13	1.82	1.74	1.74
14	0.88	1.02	1.02
14a	1.69	1.69	1.69
15*	0.85	1.26	1.26
16	1.85	1.95	1.95

Compounds name	Actual (Log 1/c)	Predicted (MLR)	Predicted (PLS)
17*	0.92	1.26	1.26
18*	1.60	1.94	1.94
19	1.04	1.26	1.26
20	2.00	1.93	1.93
21	1.30	1.15	1.15
22*	1.30	1.77	1.77
23	1.04	1.06	1.06
24	1.60	1.77	1.77
25**	1.55	--	--
26*	1.52	1.78	1.78
27	1.24	1.06	1.06
28	1.56	1.73	1.73
29	1.39	1.40	1.40
29a	2.32	2.06	2.06
30	1.06	1.10	1.10

Compounds name	Actual (Log 1/c)	Predicted (MLR)	Predicted (PLS)
31*	1.42	1.77	1.77
32	1.33	1.14	1.14
33*	1.35	1.78	1.78
34	1.30	1.34	1.34
35	1.31	1.31	1.31
36	1.58	1.52	1.52
37**	0.79	--	--
38*	1.74	2.22	2.22
39*	1.79	2.17	2.17
40*	1.56	1.82	1.82
41*	1.82	2.48	2.48
42*	1.69	2.44	2.44
43	1.33	1.44	1.44
44**	1.26	--	--

* Compounds included in Test set.

** Outliers (not included in the final regression model).

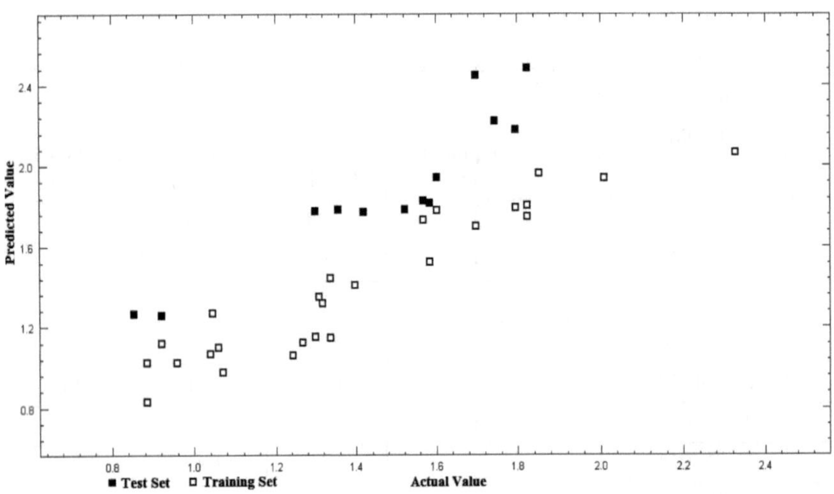

Fig. 1: Actual vs. predicted activity for the training and test set of compounds derived from MLR analysis

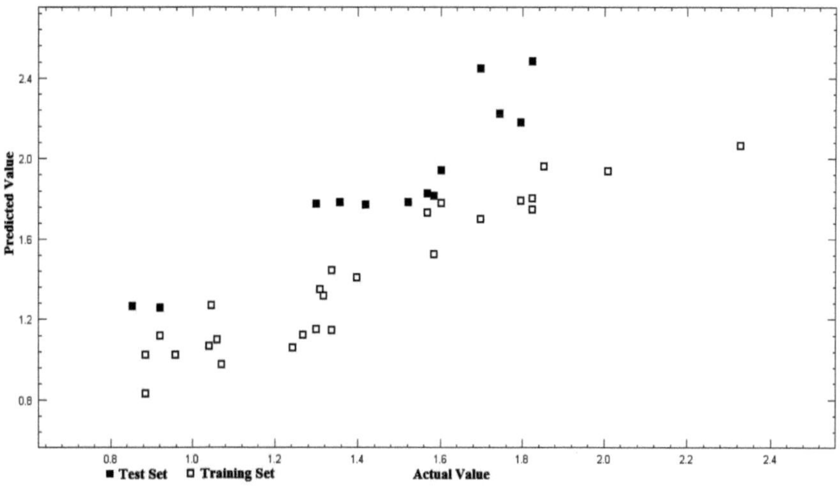

Fig. 2: Actual vs. predicted activity for the training and test set of compounds derived from PLS analysis

Conclusion

On the basis of present study, it can be concluded that the physicochemical descriptors have sufficient reliability to relate the biological activity of 2, 4-diamino-8-deazafolate analogues synthesized with their structural features. A highly predictive QSAR model has been obtained using the MLR. It was validated using external test set of thirteen compounds and PLS to all molecules. Based on the thorough analysis of the experimental data of the target property, this computational approach will help to add up new molecules as antitoxoplasma agents and to combat with existing problems of this class. The findings of present study will certainly aid in the design of more potent antitoxoplasma agents with improved activity and reduced mechanism based side effects of traditional antitoxoplasma agents.

Acknowledgement

Computational resources were provided by Banasthali University, and the authors thank the Vice Chancellor, for extending all the necessary facilities.

References

1. M.E. Grigg and Y. Suzuki; 5 (2003) 685.
2. E. Warren, S. George and J. You; Pharmacotherapy. 17 (1997) 900.
3. R. Behbahani, M. Moshfeghi and J.D. Baxter; Ann. Pharmacother. 29 (1995)760.
4. J.A. Kovacs, J.W. Hiemenz and A.M. Macher; Ann. Intern. Med. 100 (1984) 663.
5. C.J. Allegra, J.A. Kovacs and J.C. Drake; J. Clin. Invest. 79 (1987) 478.
6. J.A. Kovacs, C.J. Allegra and J.C. Swan; Antimicrob. Agents Chemother. 32 (1988) 430.
7. J.A. Kovacs, C.J. Allegra, J. Beaver, and D. Boarman; J. Infect. Dis. 160 (1989) 312.
8. A. Gangjee, Y. Zhu, S.F. Queener and P. Francom; J. Med. Chem. 39 (1996) 1836.
9. S. Paliwal, A. Narayan and S. Paliwal; QSAR Comb Sci. 28 (2009) 1367.
10. S.K. Paliwal, M. Pal and A.A. Siddiqui; Med. Chem. Res 19 (2010) 475.
11. A. Golbraikh and M. Shen, Z. Xiao; Journal of Computer-Aided Molecular Design. 17 (2003) 241.
12. Y. Yu, R. Su, L. Wang; Med. Chem. Res. (2009).
13. R.Y. Prasad, R.P. Kumar and J.D. Smiles; Arkivoc. 11 (2008) 266.
14. R.D. Cramer; Drug Discov Des. 1 (1993) 269.

30
A QSAR Study Investigating the Potential Anti-Leishmanial Activity of Cationic 2-Phenylbenzofurans

A. Agarwal[1], R. Mishra[1] and S. Paliwal[1]
[1]Department of Pharmacy, Banasthali University, Banasthali, Rajasthan-304022, India.
Email: aankita72@yahoo.com

Abstract

The QSAR models were developed using multiple linear regression (MLR) and partial least squares (PLS) for 41 2-phenylbenzofurans derivatives against leishmania donovani. The MLR and PLS generated comparable models with good predictive ability and all other statistical values, r, r^2, r^2_{cv}, r^2 (test set) and F and S values, were 0.86, 0.74, 0.72, 0.70 and 38.56, 0.28, respectively, for MLR and r^2_{cv}, r^2 (test set) and statistical significance value were 0.71, 0.70 and 0.99, respectively, for PLS, were satisfactory. The results indicate the importance of dipole moment and Kier Chi in determining the activity of 2-phenylbenzofurans as potent antiprotozoal agents.

Introduction

Visceral leishmaniasis (VL) is caused by the protozoan parasite Leishmania donovani [1] and transmitted by the bite of around 30 species of phlebotamine sandflies [2]. Research over the past decade has identified a number of drugs and formulations that offer improved treatment for this disease [3]. The drugs for leishmaniasis treatment are sodium stibogluconate (pentostam) and meglumine antimonate (glucantime), but they exhibit renal and cardiac toxicity. Alternative drugs, such as pentamidine, amphotericin B, and some azo-derivatives are also very toxic and exhibits serious side effects [4]. Although newer treatments exist, they are not optimal due to problems of toxicity, high price or difficulty in administration. Over the last two decades, immunotherapy (immune based therapies), either alone or combined with chemotherapy, has been developed as an additional approach in the treatment of leishmaniasis, but presents severe gastrointestinal problems [5]. Pentavalent antimony, the most widely prescribed drug to treat leishmaniasis patients, has serious side effects, requires a prolonged course of treatment and is losing its efficacy in some regions due to increasing parasite resistance. Although newer treatments exist, they are not optimal due to problems of toxicity, high price or difficulty in administration [6]. Given the problems of toxicity, need for hospitalization, associated with the currently available drugs for leishmaniasis, it is clear that patients urgently need new, improved, treatments to replace or complement these drugs [7]. So it will be beneficial to optimize existing anti-leishmanial agents using QSAR modeling techniques to identify the important molecular properties required for the effective inhibition of parasite. In view of this we decided to develop models from classical QSAR descriptors using multiple linear regressions and partial least square.

Material and Methods

A QSAR study was performed on the series of 49 2-phenylbenzofurans, compound number 4, 5, 7, 13, 16, 31, 40 and 48 were not having exact IC50 value, therefore these eight compounds were not included for analysis. Hence, we developed model with total 41 compounds in data set [8] (Table 1). The structures of 41 cationically substituted 2-phenylbenzofurans were sketched using Chem Draw software and were imported on TSAR (Version 3.3; Accelrys Inc, oxford, England) software. The generated 3D models of all derivatives created were cleaned up and subjected to charge calculation and energy minimization.

More than 300 molecular descriptors were calculated for all the compounds under consideration.

M.M. Srivastava, L. D. Khemani, S. Srivastava, *Chemistry of Phytopotentials: Health, Energy and Environmental Perspectives*, DOI:10.1007/978–3-642–23394-4_30, © Springer-Verlag Berlin Heidelberg 2012

TSAR affords the calculation of the following descriptors: atomic attributes (like molecular properties, dipole moment and verloop steric parameters), atomic indices (like shape, connectivity and topological indices) and Vamp electrostatic properties (total energy, HOMO, LUMO, heat of formation, etc). To reduce data redundancy, pairwise correlation analysis was carried out [9]. Among the highly intercorrelated descriptors the one that had high correlation with biological activity was kept and other was discarded. This process was repeated number of times and finally two descriptors were retrieved that were highly correlated with biological activity and were not having intercorrelation among each other.

To develop QSAR models, stepwise MLR analysis with leave out groups of rows cross-validation was applied to the training set. The molecules of the series were divided randomly into training set (35 molecules) and test set (6 molecules). Training set was used to build linear models so that an accurate relationship could be found between structure and biological activity [10]. The test set of six molecules was not used to develop the regression model but served to check the predictive power of the developed model. In addition to MLR, partial least squares (PLS) [11] analysis was also performed to check the predictive ability and robustness of the developed model.

Results and Discussion

Multiple linear regression (MLR) and partial least squares (PLS) were used to derive the QSAR equations. The statistically significant model was constructed from the training set by using 2 parameters. In order to improve the predictivity of the model, 5 potential outliers namely 8, 27, 35, 36 and 22 (which exhibited high residual value and were too far away from regression line) were identified and deleted. The final regression equation obtained from MLR analysis (final 30 molecules in training set) after deleting the outliers is represented as equation 1

$$Y = -0.16 * X1 - 0.36 * X2 + 2.07 \quad [1]$$

$r = 0.86$, $r^2 = 0.74$, $r^2_{cv} = 0.72$, $F = 38.56$, $S = 0.28$
predictive r^2 for test set = 0.70.

This best model was selected on the basis of various statistical parameters such as coefficient of determination (r^2), predictive power of model (r^2_{cv}), standard deviation (SD), sequential Fisher test (F) and test for statistical significance (t). The value of r^2 should always be greater than 0.6 (a good model should have an $r^2 > 0.9$) and the value of r^2_{cv} could fall into three categories [12]:

- $r^2_{cv} > 0.6$: The model is fairly good.
- $0.4 < r^2_{cv} < 0.6$: The model is questionable.
- $r^2_{cv} < 0.4$: The model is poor.

PLS analysis was also performed on the same data set to check the soundness of the MLR model. The resulted r^2_{cv} value of 0.71 clearly demonstrates the high predictive ability of the developed PLS model (Equation 2).

$$Y = -0.16 * X1 - 0.36 * X2 + 2.07 \quad [2]$$

Statistical significance = 0.99, $r^2 = 0.74$, $r^2_{cv} = 0.71$, $r = 0.77$, Predictive r^2 for test set = 0.70

In equation [1] and equation [2], X1 is Dipole Moment Y component (whole molecule) and X2 is Kier Chi4 (Path) index (whole molecule).

Since for a well defined problem, both MLR and PLS should generate comparable results [13], the r^2_{cv} values of MLR and the PLS models were evaluated and it was found that both the models have comparable r^2_{cv} value of 0.72 and 0.71 for MLR and PLS respectively. The predictive ability of the model was also validated using the external test set of 6 compounds in context of minimum difference between the actual and predicted biological activity values of MLR and PLS analysis for training and test which is shown in Table 2 and their respective plots are depicted in Figure 1 and 2.

Table 1: Series of 2-Phenylbenzofurans used in QSAR analysis

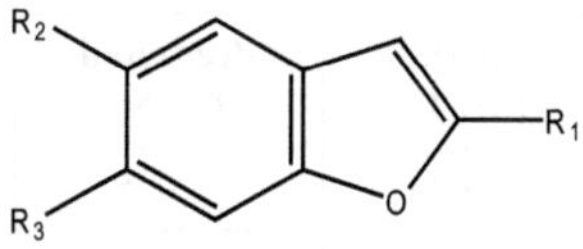

Am

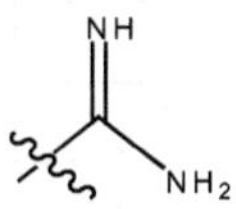

i-PrAm

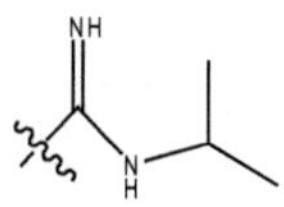

Im

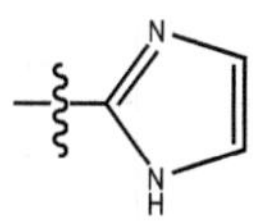

BzIM

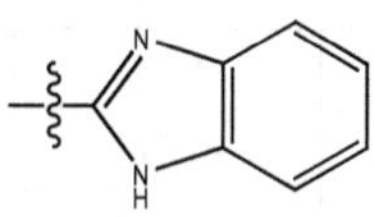

Compounds Name	R1	R2	R3	R4
1	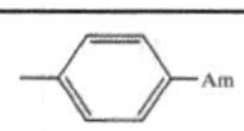	Am	H	H
2	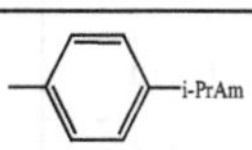	i-PrAm	H	H
3	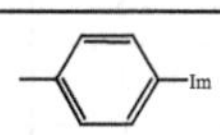	Im	H	H
6	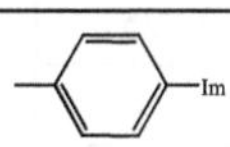	BzIM	H	H
8	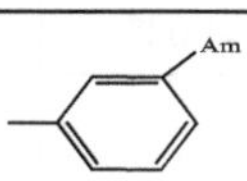	Am	H	H
9	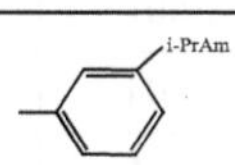	i-PrAm	H	H
10	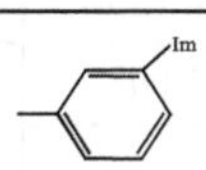	Im	H	H
11	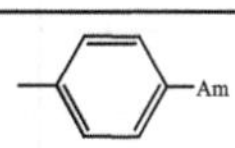	H	Am	H
12	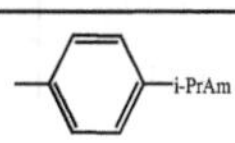	H	i-PrAm	H
14	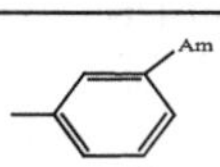	H	Am	H
15	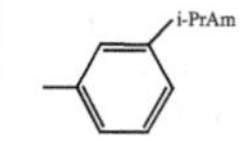	H	i-PrAm	H
17	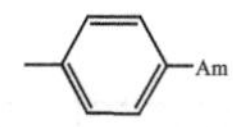	Am	H	OMe
18	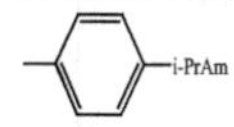	i-PrAm	H	OMe
19	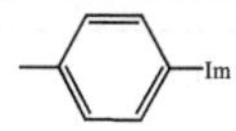	Im	H	OMe
20	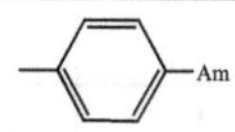	Am	H	OH
21	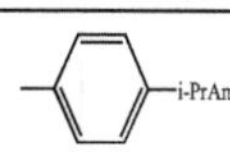	i-PrAm	H	OH
22	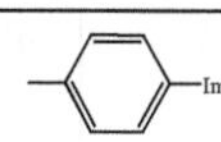	Im	H	OH
23	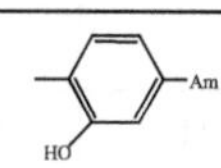	Am	H	OH
24	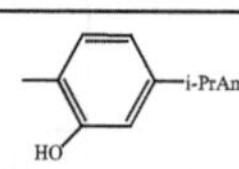	i-PrAm	H	OH
25	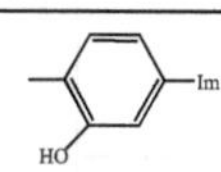	Im	H	Oh
26	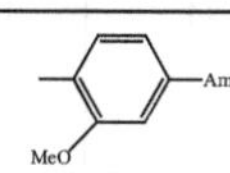	Am	H	H
27	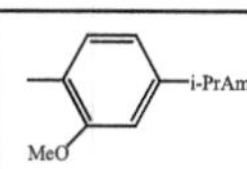	i-PrAm	H	H
28	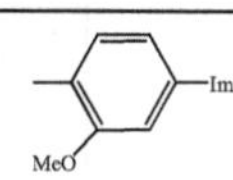	Im	H	H
29	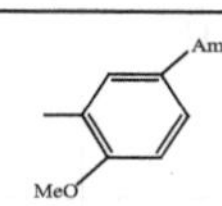	Am	H	H
30	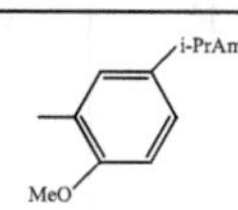	i-PrAm	H	H
32	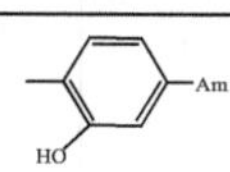	Am	H	H
33	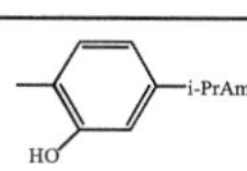	i-PrAm	H	H

34	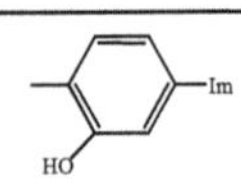	Im	H	H
35	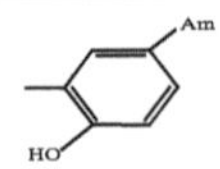	Am	H	H
36	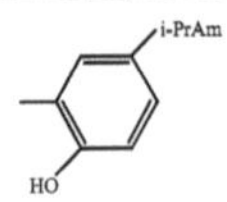	i-PrAm	H	H
37	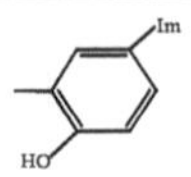	Im	H	H
38	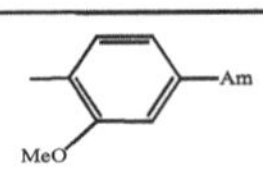	H	Am	H
39	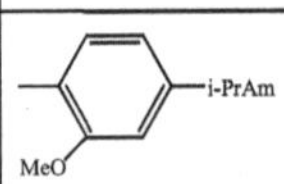	H	i-PrAm	H
41	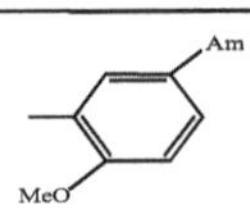	H	Am	H
42	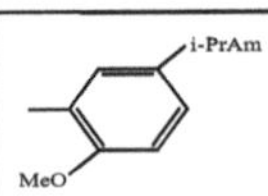	H	i-PrAm	H
43	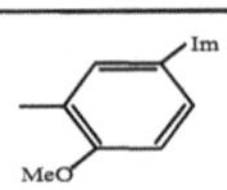	H	Im	H
44	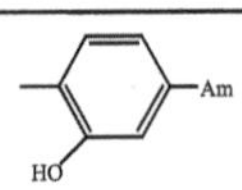	H	Am	H
45	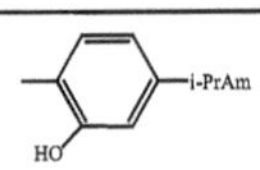	H	i-PrAm	H
46	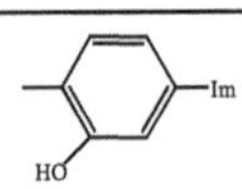	H	Im	H
47	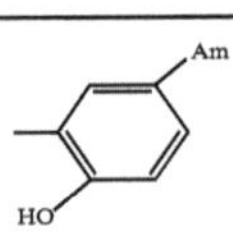	H	Am	H
49	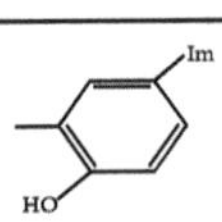	H	Im	H

Table 2: Series of 2-Phenylbenzofurans along with their actual and predicted biological values

Compounds Name	Actual (Log 1/c)	Predicted (MLR)	Predicted (PLS)
1	0.004	-0.024	-0.025
2	-0.755	-1.018	-1.016
3	-1.113	-1.070	-1.070
6	-0.491	-0.585	-0.586
8**	-1.763	-	-
9	-1.079	-1.225	-1.225
10	-1.380	-1.643	-1.644
11	-1.591	-1.593	-1.593
12	-1.113	-1.247	-1.247
14	-1.612	-1.286	-1.288
15	-1.176	-0.718	-0.717
17	-0.204	-0.614	-0.614
18	-1.380	-1.396	-1.396
19	-1.662	-1.283	-1.283
20*	-0.720	-0.536	-0.536
21	-0.255	-0.448	-0.448
22**	-0.869	-	-
23*	-1.146	-1.514	-1.514
24	-1.204	-1.516	-1.516
25	-0.568	-0.804	-0.805
26	-0.556	-0.925	-0.926
27**	-0.176	-	-
28	-1.397	-1.087	-1.086
29*	-0.755	-2.284	-2.284
30	-1.146	-1.001	-1.002
32	-1.322	-0.904	-0.903
33	-0.477	-0.125	-0.125
34	-0.653	-1.007	-1.006
35**	-1.301	-	-
36**	-1.602	-	-
37	-1.301	-1.033	-1.033
38	-0.301	-0.350	-0.350
39*	-1.505	-0.446	-0.446
41	-1.778	-1.289	-1.289
42	-0.255	-0.602	-0.602
43	-0.763	-0.671	-0.670

Compounds Name	Actual (Log 1/c)	Predicted (MLR)	Predicted (PLS)
44*	-1.462	-0.015	-0.015
45	-1.755	-1.744	-1.744
46	-1.732	-1.655	-1.655
47	-1.531	-1.677	-1.677
49*	-1.544	-1.005	-1.005

* Compounds included in Test set.
** Outliers (not included in the final regression model).

Conclusion

On the basis of present study, it can be concluded that the physicochemical descriptors have sufficient reliability to relate the biological activity of 2-phenylbenzofurans series synthesized with their structural features. A highly predictive QSAR model has been obtained using the MLR. It was validated using external test set of six compounds and PLS to all molecules.

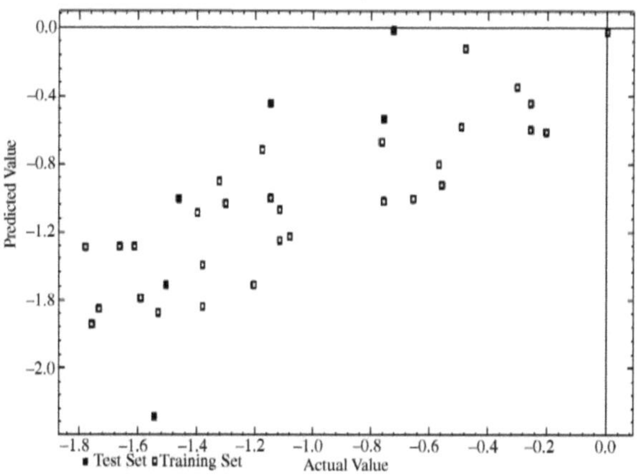

Fig. 1: Actual vs. predicted activity for the training and test set of compounds derived from MLR analysis

Based on the thorough analysis of the experimental data of the target property, this computational approach will help to add up new molecules as anti leishmanial agents and to combat with existing problems of this class. The findings of present study will certainly aid in the design of more potent anti leishmanial agents with improved activity and reduced mechanism based side effects of traditional anti leishmanial agents.

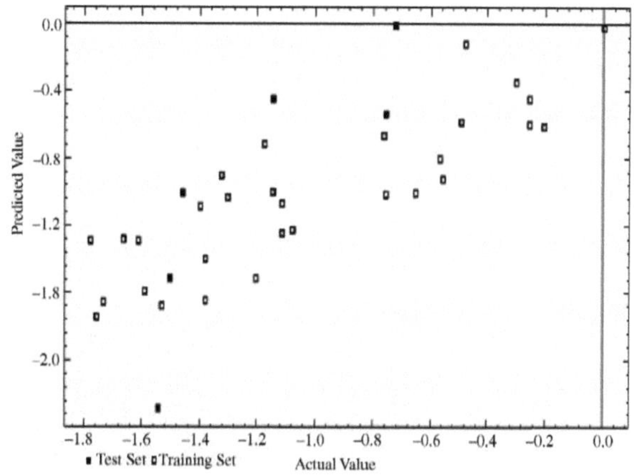

Fig. 2: Actual vs. predicted activity for the training and test set of compounds derived from PLS analysis

Acknowledgements

Computational resources were provided by Banasthali University, and the authors thank the Vice Chancellor, for extending all the necessary facilities.

References

1. S.K. Bhattacharya, S. Dipika and K. Juntra; Indian J. Med. Res. 123 (2006) 353.
2. R.S. Barnetson, R.S. Ridley and H.W. Wheate; Trans R. Soc. Trop. Med. Hyg. 72 (1978) 516.
3. H.W. Murray; Int. J. Infect. Dis. 4 (2004) 158.
4. Mcgregor; Lancet. 351 (1998) 575.
5. H. Sangraula, K.K. Sharma, S. Rijal and S. Dwivedi; J. Assoc. Physicians India. 51 (2003) 686.
6. N. Singh; Indian J. Med. Res. 123 (2006) 411.
7. P.B. Carvalho, M.A.G. Arribas and E.I. Ferreira; Braz. J. Pharm. Sci. 36 (2000) 69.
8. S.A. Bakunov, S.M. Bakunova, T. Wenzler and T. Barszcz; J. Med. Chem. 51 (2008) 6927.
9. S.K. Paliwal, M. Pal and A.A. Siddiqui; Med. Chem. Res. 19 (2010) 475.
10. Golbraikh, M. Shen, Z. Xiao, and Y.D. Xiao; J. Computer-Aided Molecular Design. 17 (2003) 241.
11. Y. Yu, R. Su, L. Wang and W. Qi; Med. Chem. Res. DOI 10.1007/s00044–009-9266–9 (2009)
12. R.Y. Prasad, R.P. Kumar, J.D. Smiles and A.P. Babu; Arkivoc. 11 (2008) 266.
13. R.D. Cramer; Drug Discov Des. 1 (1993) 269.

31
2D QSAR Study of Some TIBO Derivatives as an Anti HIV Agent

L. K. Ojha[1], M. Thakur[2], A. M. Chaturvedi[1], A. Bhardwaj[1], A. Thakur[3]
[1]Department of Chemistry, Govt. Madhav Science College, Ujjain 456001.
[2]Department of Chemistry, Softvision College, Indore,452001
[3]Department of Chemistry, Indore Institute of Science & Technology(II), Indore, Research & Development Division, DJ Laboratories Ltd., Indore,453331
Email: ojha_lokendra@rediffmail.com

Abstract

A set of sixteen Tetraimidazolebenzodiazepine -1- one (TIBO) derivatives with inhibitory concentration (pIC50) activity was subjected to the two dimensional quantitative structure activity relationships studies using computational drug design. Drug Designing module contain various combinations of physiochemical, electronic, topological and indicator parameter. TIBO taken as the lead molecule and QSAR model developed using multiple regression approach. For each set of descriptors, the best multilinear QSAR equations were obtained by the stepwise variable selection method using leave-one out cross-validation as selection criterion. Value of pIC50 was taken as dependent variable and physiochemical and topological parameter was taken as independent variable. The best QSAR model (r^2 = 0.9672, Fisher test value F=38.706, Se = 0.2628) has acceptable statistical quality and predictive potential. From the build model it seems to be clear that indicator parameter (presence of halogen atom at X position) and Balaban index along with Molecular Refractivity (MR) of the molecule is very much responsible in the binding affinity of anti HIV drug. Thus this validated model brings important structural insight to aid the design of novel anti HIV agents.

Introduction

HIV-1, the etiological agent of AIDS, has been identified in the beginning of the 1980s [1, 2]. There is currently no cure for AIDS. Highly active antiretroviral therapy (HAART) is a powerful HIV treatment that was introduced in the mid-90s. The therapy consists of three or more different drugs that are used in combination to suppress the virus. The combination drugs consist of nucleoside reverse transcriptase inhibitors (NRTIs), non-nucleoside RT inhibitors (NNRTIs), and protease inhibitors (PIs). NRTIs and NNRTIs get incorporated into newly synthesized DNA strands by HIV-1 reverse transcriptase resulting in chain termination and inhibition of genomic DNA synthesis. PIs specifically inhibit the virus- associated protease.

Unlike NRTIs, NNRTIs do not require cellular activation to inhibit HIV-1 RT. They are not incorporated into nascent viral DNA, are noncompetitive inhibitors, and bind into a hydrophobic "pocket" in the p66 subunit of HIV-1 RT located close to (but distinct from) the NRTI binding site [3,4].

NNRTI binding distorts the nearby RT polymerase active site, thus affecting the chemical step of polymerization [5,6]. The NNRTI binding pocket does not exist in the un liganded RT and is formed upon binding by the side chains of aromatic (including Y181 and Y188) and hydrophobic amino-acid residues [7,8]. NNRTIs are highly specific for HIV-1 and do not inhibit HIV-2 or any other retrovirus. NNRTI resistance mutations affect the binding of the inhibitors to their binding pocket. These mutations alter the size, shape, or polarity of the NNRTI binding pocket or affect the access of NNRTIs to this site [7].

A quantitative structure activity relationship (QSAR) is an approach to build an activity model based on the structural and physicochemical properties of the known active compounds through multivariate analysis. Traditional QSAR does not require receptor information and is a ligand-based computer-

M.M. Srivastava, L.D. Khemani, S. Srivastava, *Chemistry of Phytopotentials: Health, Energy and Environmental Perspectives*, DOI:10.1007/978–3–642–23394–4_31, © Springer-Verlag Berlin Heidelberg 2012 aided drug design (CADD) approach. Some important QSAR models have been established in the study of the HIV non-nucleoside inhibitor. The QSAR models have provided critical help to identification, screening, and design of new potent HIV inhibitors. There is no much QSAR has been reported for NRTIs.

Observations related to the relationships between activity/property and compounds structure has been actually reported before the apparition of the QSAR/QSPR concepts. In 1868, Crum-Brown and Fraser stipulate the idea that the activity of a compound is a function of its chemical composition and structure [9]. In 1893, Richet and Seancs shown for a set of organic molecules that the cytotoxicity was inverse related with water solubility [10]. Mayer suggests in 1899 that the narcotic action of a group of organic compounds is related with solubility into olive oil [11]. Ferguson introduced in 1939 a thermodynamic generalization to the correlation of depressant action with the relative saturation of volatile compounds in the vehicle in which they were administered [12].

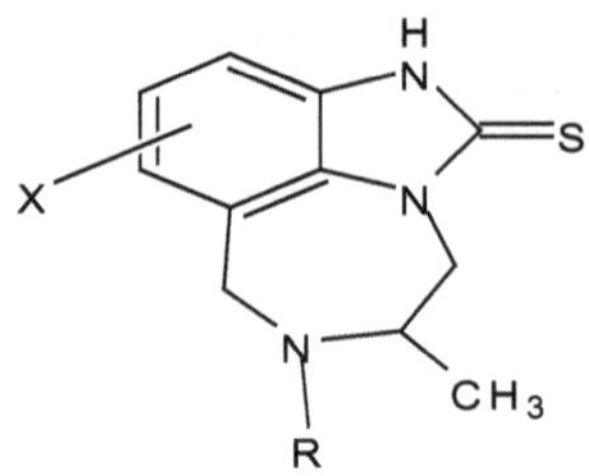

Fig. 1: Parent Structure of TIBO Compound

Table 1: Structures of the compound used in present study

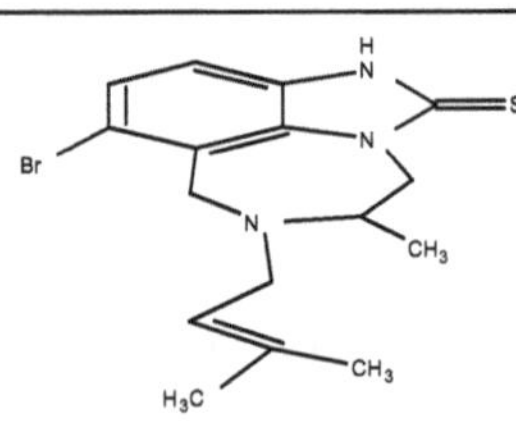 Comp1	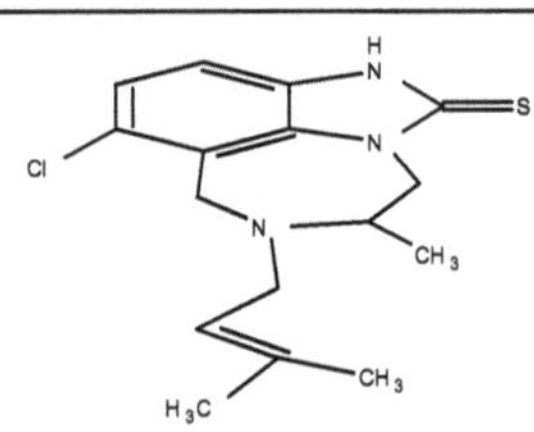 Comp2	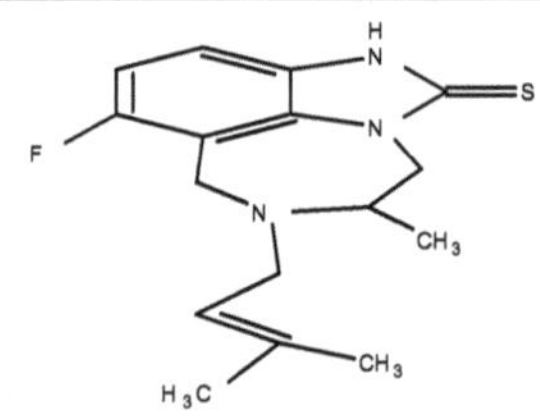 Comp3	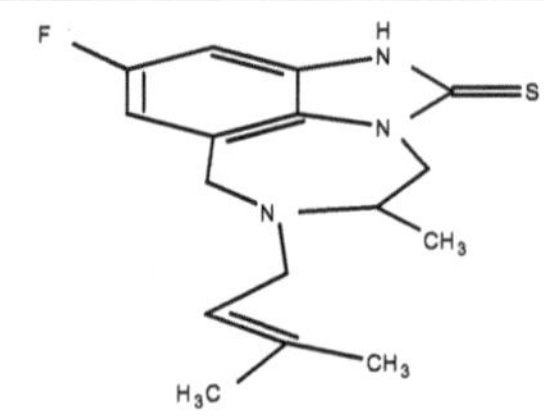 Comp4
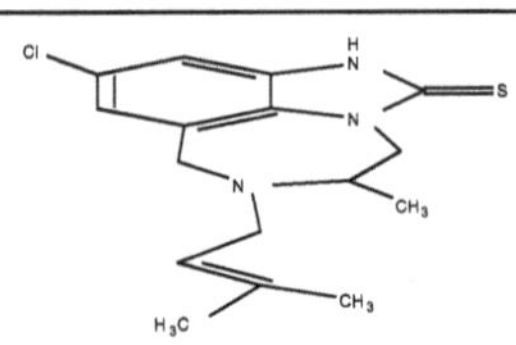 Comp5	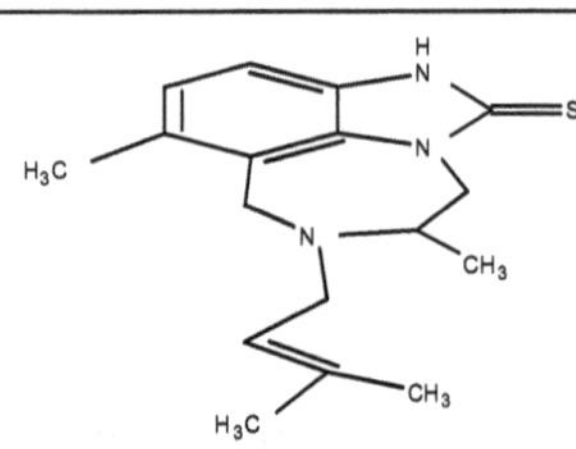 Comp6	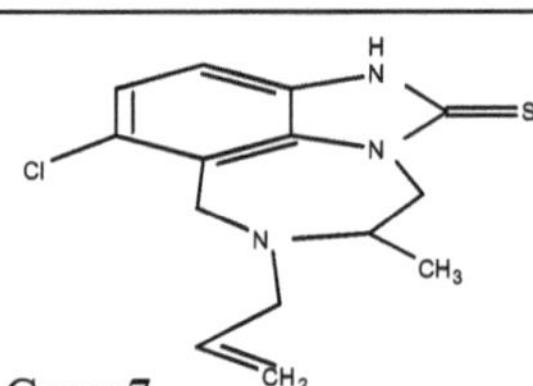 Comp7	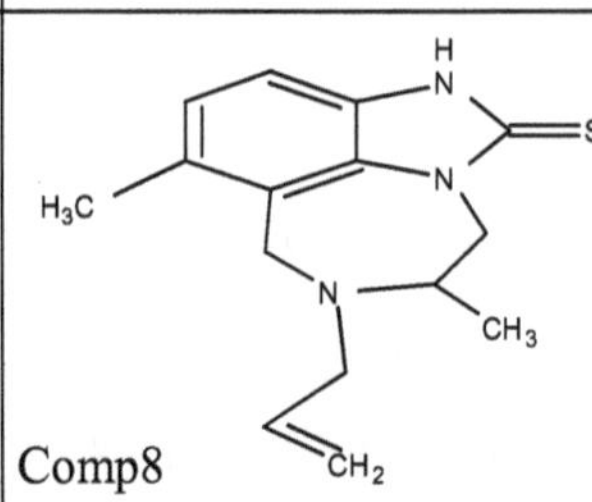 Comp8
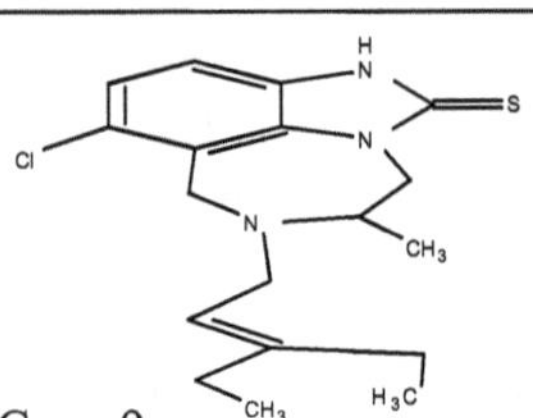 Comp9	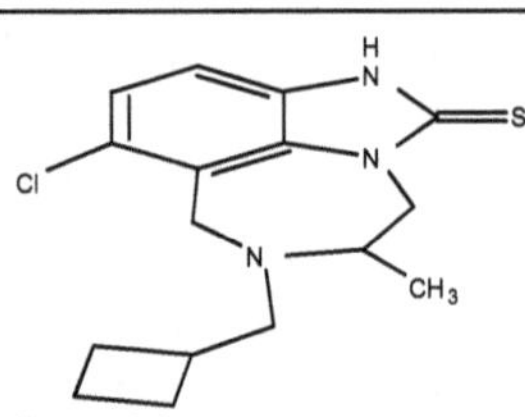 Comp10	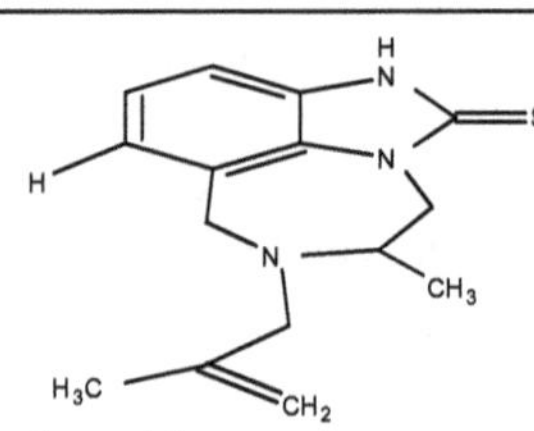 Comp11	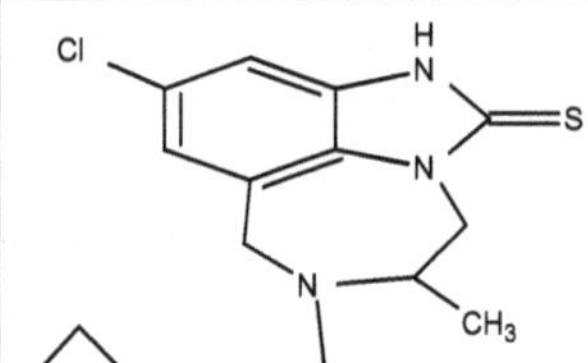 Comp12
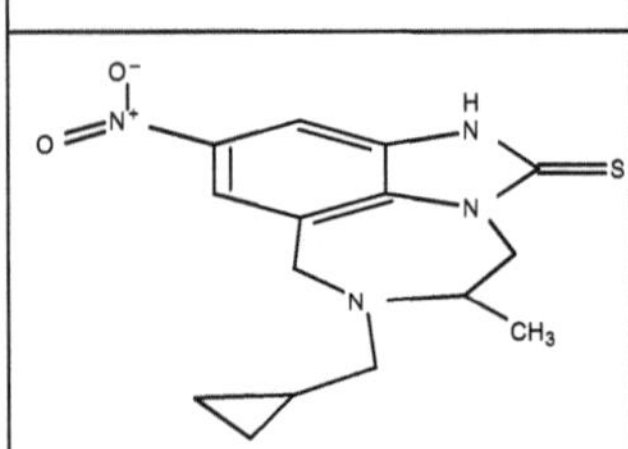 Comp13	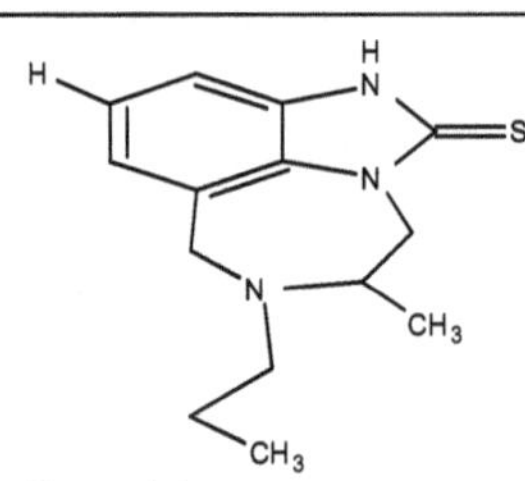 Comp14	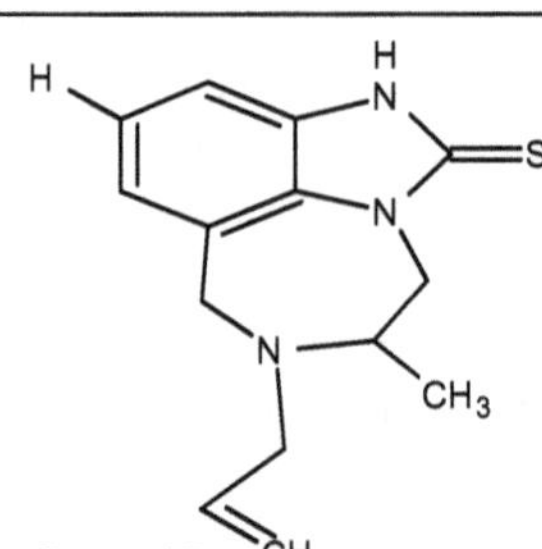 Comp15	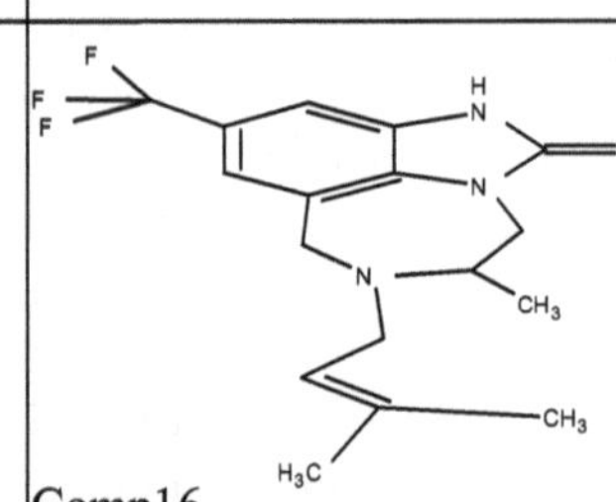 Comp16

Material and Method

Data Set

As a part of ongoing efforts to design novel molecules with potent anti-HIV activity, a QSAR analysis was performed to relate HIV non-nucleoside reverse transcriptase inhibitors activity of TIBO derivatives to its 3D properties. The selected series consists of total 16 numbers of compounds. The biological activities were expressed in terms of inhibitory concentration (pIC50). The parent structure of the TIBO is given in the Fig. 1. Table 1 shows the all 16 derivatives of TIBO used in the present study. Experimental anti-HIV activity (pIC50) complied from references [13–21]. Table (2) shows the value of the parameter which gives the better result in the QSAR model. Table (3) describes the observed and calculated pIC50 value.

Regression Analysis

The relationship between biological activities and various descriptors (Physiochemical and Topological) were established by sequential multiple regression analysis (MLR) using in house made software "ANALYSIS" in order to obtain best QSAR models and to know the structural requirement of the drug in particular set of compound.

Calculation of the Parameter

Some physiochemical parameter like Molecular Refractivity (MR), Molecular Volume (MV), Parachor (Pc), Index of Refraction (IR), Surface Tension (ST), Density (D), Polarizability (Pol), were calculated using chemsketch and Different topological [8, 9] and constitutional descriptors for each molecule is calculated Dragon freeware software in this study.

Results and Discussion

The calculated descriptors were gathered in a data matrix. First, the descriptors were checked for constant or near constant values and those detected were discarded from the original data matrix. Then, the descriptors were correlated with each other and with the activity data. Among the collinear descriptors detected, the one most highly correlated with activity was retained and the rest were omitted. Finally, MLR with stepwise selection and elimination of variables was applied to the development of the model.

Table 2: Values of Descriptors used

Comp No.	J[a]	MR[b]	IX[c]
1	1.78	94.86	1
2	1.78	91.97	1
3	1.78	87.25	1
4	1.759	87.25	1
5	1.759	91.97	1
6	1.78	91.76	0
7	1.86	82.79	1
8	1.86	82.59	0
9	1.72	101.23	1
10	1.557	90.22	1
11	1.799	82.37	0
12	1.509	90.22	1
13	1.564	86.8	0
14	1.817	78.2	0
15	1.817	77.97	0
16	1.8	92.13	0

[a]= Balaban Index, [b]= Molecular Refractivity, [c]= Indicator parameter in which presence of halogen at X position of TIBO is taken as 1 otherwise 0.

For each set of descriptors, the best multi-linear regression equations were obtained by the stepwise selection methods of MLR subroutine of the software. The squared correlation coefficient (r2), standard error of estimate (Se) and Fisher's value (F) which represents the F-ratio between the variance of actual and predicted activity, were employed to judge the validity of regression equation. In order to validate the generated QSAR models leave one out (LOO) method was used. Only the indicator parameter (IX) shows the greater (>0.5) correlation with the binding affinity of the drug. Best monovariate combination from the correlation matrix are given below

$$pIC50 = 1.5034(\pm .5336)\, IX + 6.4886$$

$N = 16$, $r = .6015$, $Se = 1.0588$, $F = 7.938$ **Eq(1)**

Table 3: Values of observed and calculated pIC50 for given sets of compound

Com-pound no	Observed pIC50	Calculated pIC50	Residual
1	8.52	8.579	-0.0591
2	8.34	8.369	-0.0294
3	8.24	8.027	0.2132
4	7.838	7.918	-0.0801
5	8.48	8.261	0.2194
6	7.85	7.029	0.8208
7	8.33	8.118	0.2124
8	7.85	6.778	1.0719
9	7.92	8.731	-0.8106
10	7.88	7.087	0.7928
11	7.85	6.446	1.4039
12	6.38	6.839	-0.4586
13	5.61	5.55	0.0597
14	5.78	6.237	-0.4567
15	4.17	6.22	-2.05
16	6.31	7.16	-0.8496

The model above shows the dominance role of indicator parameter over other parameter. In order to get best model we were tasted various biparametric combination and the best models are given below-

pIC50= 1.7369(± .5155) IX + 4.3191(± 2.4631) J -1.1853
N= 16, r= .6956, Se= .9881, F=6.095 **Eq(2)**

In the above model Balaban index and indicator parameter both are very important to enhance the binding affinity of the drug but there is lower value of F is not suggest that this one is the best model and hence from here we tested several triparametric combination and got the best model as follows-

pIC50= 1.3250(± .5572) IX + 5.1800 (± 2.4050) J + 0726(± .468) MR -8.8505
N= 16, r= .7550, Se= .9388, F=5.303 **Eq(3)**

Now, this model is giving much better relationship with the drug binding affinity. But the higher value of standard error (Se) and lower value of F is not support our findings. So, from here we outlier several compound (misfit or higher residual) from the date set.

pIC50= 1.2075(± .4018) IX + 5.3713 (± 1.7291) J + .0314(± .356) MR -5.3188
N= 15, r= .8002, Se= .6746, F=6.526 **Eq(4)**

pIC50= 1.1091(± .3390) IX + 5.5388 (± 1.4496) J -0.0322(± .0327) MR -2.6207
N= 14, r= .8356, Se= .5649, F=7.714 **Eq(5)**

pIC50= .7477(± .2964) IX + 6.0880 (± 1.1510) J + .0251 (± .0272) MR -5.5278
N= 13, r= .8920, Se= .4415, F=11.678 **Eq(6)**

pIC50= .5898(± .1805) IX + 7.6002 (± .7751) J + .0360(± .0164) MR -9.1079
N= 12, r= .9672, Se= .2628, F=38.706 **Eq(7)**

From equation (4–7) shows the gradual improvement in the value of correlation coefficient (.8002-.9672) and at the same time standard error of estimation become low (.6746-.2628). The greater value of F stat also suggests the deleted compound is not best fitted in the set.

Equation (6) is the result of 3 outlier compound, which could be the best model, because of the thumb rule for the deletion of the compound we could not perform the operation from equation (6), and hence equation (7) is dishonor of the thumb rule. But tremendous improvement in the value of r and F and decrease in the value of Se, we outlier the fourth compound.

So, from the above mathematical model present, it can be said that the role of indicator parameter is very essential. It will definitely increase the binding affinity of the drug receptor interaction. Not only the indicator parameter, but molecular refractivity is also important. It is the property of the drug which is related to the polarity of the drug. More polar drug can very easily penetrate the blood brain membrane.

All compound from the data for MLR resulted in the generated model with improved statistical significance and predictive ability, this generated model can be developed for the series. All these models were screened on the basis of $q^2 > 0.7$ and the intercept to best fit line. The plots of calculated activity and the observed activity (pIC50) are represented in Fig. (2). Low standard error of estimate of this model (<0.2) demonstrates the accuracy of the model.

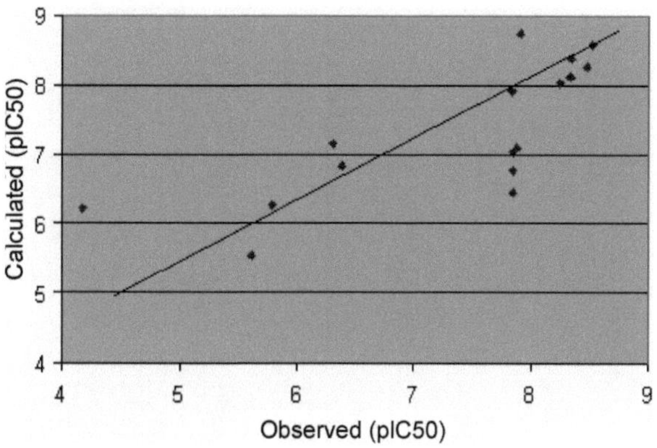

Fig. 2: Plot of observed and calculated pIC50

Conclusion

In conclusion, above QSAR study reveals that the indicator parameter (presence of halogen on benzene ring) along with Balaben (J) and MR is vital in this set of TIBO compound. This suggests that, by increasing in Balaben index J, will be helpful for designing of more potent anti HIV agents. This strategy may, therefore, be followed for designing higher potency compounds for future synthesis. These guidelines may, therefore, provide a basis for rationalizing substituent selection in the future designing of selective TIBO receptor ligands. The study may also help in proposing the possible mode of action of TIBO.

Acknowledgements

Author (Lokendra Kumar Ojha) is thankful to the Dr. Usha Shrivastava (Principal, Govt. Madhav Science College, Ujjain) and Dr. U. C. Maiwal (Professor & Head, Department of Chemistry, Govt. Madhav Science College, Ujjain) and other faculty members of the chemistry department for providing the necessary facilities to complete this work.

References

1 F. Barre-Sinoussi, J.C. Chermann, F. Rey, M.T. Nugeyre, S. Chamaret, J. Gruest, C. Dauguet, C. Axler- Blin, F. Vezinet-Brun, C. Rouzioux, W. Rozenbaum, L. Montagnier, Science 220, 868 (1983).

2 R.C. Gallo, S.Z. Salahuddin, M. Popovic, G. .M. Shearer, M. Kaplan, B.F. Haynes, T.J. Palker, R. Redfield, J. Oleske, B. Safai, Science 224, 500 (1984).

3 L.A. Kohlstaedt, J. Wang, J.M. Friedman, P.A. Rice, T.A. Steitz, Science 256, 1783 (1992).

4 J.D. Pata, W.G. Stirtan, S.W. Goldstein, T.A. Steitz, Proc. Natl. Acad Sci. USA, 101, 10548 (2004).

5 E.D. Clercq, *Antiviral Res.* 38, 153 (1998).

6 R.A. Spence, W.M. Kati, K.S. Anderson, K.A. Johnson, Science 267, 988 (1995).

7 K. Das, P.J. Lewi, H. Hughes, E. Arnold, *Prog. Biophys. Mol. Biol.* 88, 209 (2005).

8 D.W. Rodgers, S.J. Gamblin, B.A. Harris, S. Ray, J.S. Culp, B. Hellmig, D.J. Woolf, C.Debouck, S.C. Harrison, Proc. Natl. Acad. Sci. USA, 92, 1222 (1995).

9 Crum-Brown, A.; Fraser, T.R. On the Connectionbetween Chemical Constitution and Physiological Ac-tion. Part I. On the Physiological Action of the Salts ofthe Ammonium Bases, derived from Strychnia, Brucia,Thebaia, Codeia, Morphia, and Nicotia. *Philosophical Transactions of the Royal Society of London* 1868, *25*, 151–2003.

10 Richet, C.R. *Comptes Rendus des Seances de la Societe de Biologie et de ses Filiales*, 1893, *9*, 775. 4. Meyer, H. Zur Theorie der Alkoholnarkose Erste Mittheilung. Welche Eigenschaft der Anästhetica bedingt ihre narkotische Wirkung? *Naunyn-Schmiedeberg's Archives of Pharmacology* 1899, *42*(2- 4), 109–118.

11 Ferguson, J. *Proceedings of the Royal Society B: Biological Sciences* 1939, *127*, 387.

12 Hammett, L.P. Some Relations between Reaction Rates and Equilibrium Constants. *Chemical Reviews* 1935, *17*, 125–136.

13 M. Mahmoudian, J. Mol. Graphics Modell. 15 (1997) 149.

14 Z. Zhou, M. Madrid, J.D. Madura, Proteins. 49 (2002) 529.

15 M.B.K. Smith, M.L. Lamb, J. Tirado-Rives, W. Jorgensen, C.J. Michejda, S.K. Ruby, R. Smith Jr., Protein Eng. 13 (2000) 413.

16 S.J. Titmuss, P.A. Keller, R. Griffith, Bioorg. Med. Chem. 7 (1999) 1163.

14 R. Silvestri, M. Artico, G. De Martino, R. Ragno, S. Massa, R. Loddo, C. Murgioni, A.G. Loi, P.L. Colla, A. Pani, J. Med. Chem. 45 (2002) 1567.

18 Y.Z. Chen, X.L. Gu, Z.W. Cao, J. Mol. Graphics Modell. 19 (2001) 560.

19 M.L. Barreca, A. Carotti, A. Carrieri, A. Chimirri, A.M. Monforte, M.P. Calace, A. Rao, Bioorg. Med. Chem. 7 (1999) 2283.

20 M.A.L. Eriksson, J. Pitera, P.A. Kollman, J. Med. Chem. 42 (1999) 868.

21 R.H. Smith Jr., W. Jorgensen, J. Tirado-Rives, M.L.Lamb, P.A.J. Janssen, C.J. Michejda, M.B.K Smith, J. Med. Chem. 41 (1998) 5272.

32
Indole Derivatives as DNA Minor Groove Binders

S. P. Gupta,[1] P.Pandya,[1] G. S. Kumar[2] and S. Kumar[1*]
[1]Department of Chemistry, Dayalbagh Educational Institute, Dayalbagh, Agra-282110
[2]Biophysical Chemistry Laboratory, Indian Institute of Chemical Biology, Kolkata-700032
Email: kumar.surat@gmail.com

Abstract

DNA binding drugs form an important class of compounds which can be utilized as anticancer and antibacterial agents. We have investigated the DNA binding ability of small indole derivatives using fluorescence and molecular docking techniques. The results clearly demonstrate that small indole derivatives bind with DNA double helix showing binding affinity of 10^4–10^5per mole. All the indole derivatives were found to be binding in the minor groove of DNA. The binding was non-specific nature and dominated by van der Waals and hydrophobic forces. The experimental and computational results were found to be in good agreement as evident by the high correlation constant values.

Introduction

Indole derivatives are extensively used in several industrial processes as substrate for chemical reactions and used to synthesize complex synthetic molecules. It is one of the most important groups of compounds used in the pharmaceutical industry. Several synthetic and natural indole derivatives have shown significant DNA binding ability as well as DNA sequence specificity. Compounds like ellipticine, Hoechst-33258, cryptolepine, etc are known to interact with double stranded DNA. They possess indole nucleus in their structure. Ellipticine is a well known DNA topoisomerase-II inhibitor also having significant DNA intercalative binding [1, 2]. On the other hand Hoechst-33258 is a well studied DNA minor groove binder [3,4]. Cryptolepine is an antimalarial and cytotoxic drug known to intercalate between cytosine-cytosine site in the DNA duplex [5]. It is, therefore, clear that indole ring present in the molecule plays an important role in the DNA binding ability of such molecules. We have investigated a series to small indole derivatives for their DNA binding ability using fluorescence quenching experiments along with molecular docking technique.

Materials and Methods

Eight indole derivatives (I-VIII) (Figure 1) were selected based on the variation of position and length of the side chain attached to the main indole nucleus. Four DNA decamer sequences (Figure 2) were used based on their central GC/AT specific core.

All the chemicals were purchased from Sigma-Aldrich chemicals Pvt. Ltd. New Delhi. The fluorescence titrations were conducted in 20 mM sodium phosphate buffer of pH 7.4 using Hitachi F-4010 spectrofluorimeter (Hitachi Ltd, Tokyo, Japan). Emission spectra of free and bound alkaloid were measured according to the reported method by Le Pecq and Paoletti [6]. The excitation and emission bandpass were fixed at 5 nm while scan speed was 240 nm/min. Double reciprocal method was used to calculate the binding constants (K_{exp}) of the drug-DNA complex titrations where the binding affinity was small [7]. K_{exp} values were used to calculate ΔG_{exp} using equation 1.

Molecular docking experiments were conducted using DNADock program based on the minor groove binding protocol [8]. Before docking, all the compounds were energy minimized to eliminate bad geometries and steric clashes. The docked structures obtained were then subjected to theoretical energy calculations using PreDDICTA program [9]. PreDDICTA produced binding free energy values

M.M. Srivastava, L. D. Khemani, S. Srivastava, *Chemistry of Phytopotentials: Health, Energy and Environmental Perspectives*, DOI:10.1007/978–3-642–23394-4_32, © Springer-Verlag Berlin Heidelberg 2012

(ΔG_{PreDD}) which were used to calculate K_{PreDD} values using equation 1. The structural analyses of the docked conformations were carried out using Discovery Studio visualizer from Accelrys Inc. USA.

$$\Delta G = -RT \ln K \quad \text{Eq (1)}$$

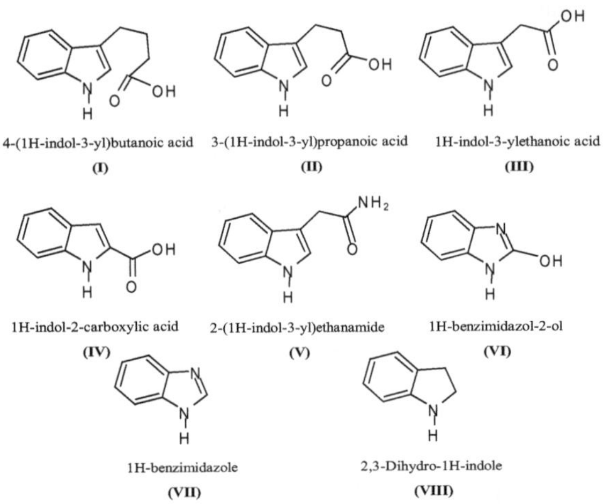

Fig. 1: Sketch of the Indole derivatives used in this study

Name	Sequence	% GC content	Molar extinction coefficient ε	Wavelength (nm)
DNA-1	5'-d-(GATGGCCATC)$_2$	60%	95000	260
DNA-2	5'-d-(GATCCGGATC)$_2$	60%	96600	260
DNA-3	5'-d-(GGCAATTGCC)$_2$	60%	92600	260
DNA-4	5'-d-(GGCTTAAGCC)$_2$	60%	93200	260

Fig. 2: Four decamer DNA sequences with their GC content and molar extinction coefficient values

Results and Discussion

Fluorescence quenching was observed (Figure 3) in the case of all indole derivatives upon DNA addition clearly suggested the interaction between small molecules and the DNA duplexes. The binding constant values obtained from double reciprocal method (K_{exp}) (Figure 4) are listed in Table 1. It is evident from the binding constant values that side chain length affects the binding affinity of indole derivatives. Compound (I) showed highest overall binding affinity of the order of 10^5 per mole while (VII) and (VIII) showed lowest binding of the order of 10^4 per mole. The effect of side chain position on the binding affinity is not very significant since compound (III) and (V) showed different binding affinities in spite of having similar position and length of the side chain attached to the indole nucleus. Compound (III) has a strongly electronegative OH group attached at the side chain terminal while the (V) has a slightly weaker electronegative NH_2 group. Therefore, the binding of (III) has a greater electrostatic contribution than (V). Molecular docking method furnished docked structures which were used in the analysis of binding affinity, specificity and forces responsible for binding.

The docked structures (Figure 5) obtained clearly suggest that electrostatic and van-der-Waals interactions were the main driving force in the complexation of selected indole derivatives with DNA oligomers. Inter-molecular H-bonding was observed in several complexes indicating strong attraction between the small molecules and the DNA. Since the binding affinity profile obtained from experimental and computational methods are in good agreement with each other (correlation coefficient between ΔG_{PreDD} and ΔG_{cal} was 0.83 and between K_{exp} and K_{PreDD} was 0.83), it is suggested that the docking protocol produces the experimental conformations. The present study has revealed that these molecular modeling approaches were adequately capable of assessing the results from solution studies like fluorescence, UV-absorbance and other similar techniques. However, the details about the precise inter-atomic distances can only be further secured by NMR or X-ray methods.

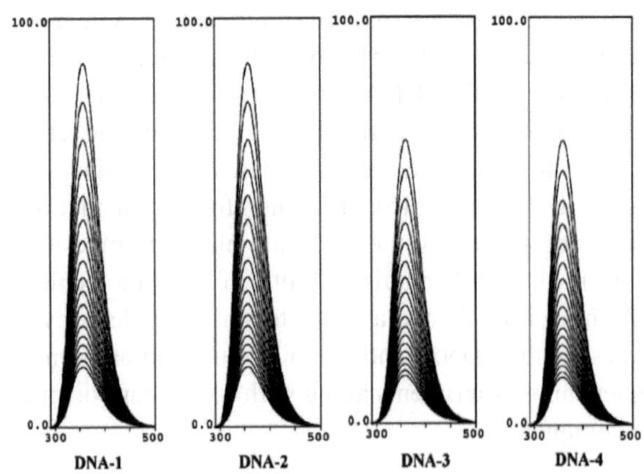

Fig. 3: Fluorescence quenching spectra of indole-3-acetamide (V) with four DNA decamer sequences

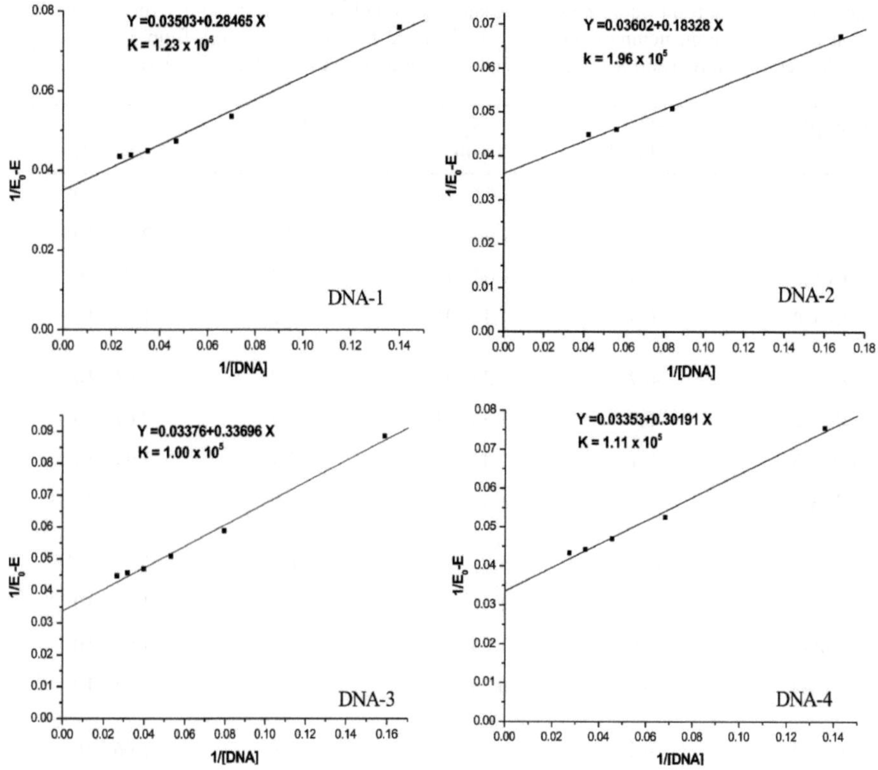

Fig. 4: Double reciprocal plots showing binding constant K values

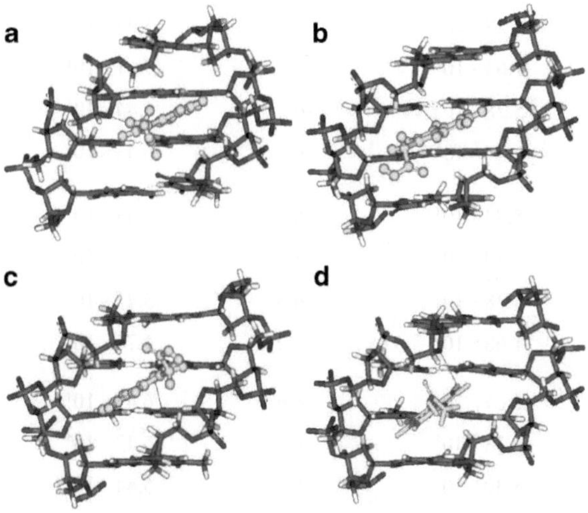

Fig. 5: Representative docked structure of Indole-3-acetic acid (III) with four DNA decamer sequences. (a) docked structure with 5'-d(GATGGCCATC)$_2$ (DNA-1), (b) with 5'-d(GATCCGGATC)$_2$ (DNA-2), (c) with 5'-d(GGCAATTGCC)$_2$ (DNA-3) and (d) with 5'-d(GGCTTAAGCC)$_2$ (DNA-4)

Table 1: Binding constant and free energy values from fluorescence quenching data and molecular docking experiments. K_{exp} (experimental binding constant from fluorescence quenching data), ΔG_{cal} (calculated binding free energy from K_{exp}), ΔG_{PreDD} (predicted binding free energy from docked structures) and K_{PreDD} (predicted binding constant from docked structure)

Drug	Complex with	ΔG_{cal} (Kcal/mole)	K_{exp} (mole^{-1})	ΔG_{PreDD} (Kcal/mole)	K_{PreDD} (mole^{-1})
Indol-3-butanoic acid (I)	DNA-1	-6.9	1.23×105	-7.0	1.26×10^5
	DNA-2	-6.9	1.96×105	-7.3	2.10×10^5
	DNA-3	-6.8	1.00×105	-7.2	1.76×10^5
	DNA-4	-6.9	1.11×10^5	-7.2	1.76×10^5
Indol-3-propionic acid (II)	DNA-1	-6.7	7.33×10^4	-6.7	7.61×10^4
	DNA-2	-6.7	7.30×10^4	-6.7	7.61×10^4
	DNA-3	-7.1	1.41×10^5	-7.1	1.49×10^5
	DNA-4	-7.1	1.38×10^5	-6.9	1.06×10^5
Indol-3-ethanoic acid (III)	DNA-1	-6.5	5.81×10^4	-6.5	5.44×10^4
	DNA-2	-6.9	1.11×10^5	-6.9	1.06×10^5
	DNA-3	-6.9	1.04×10^5	-6.9	1.06×10^5
	DNA-4	-6.6	6.39×10^4	-6.6	6.43×10^4
Indol-2-carboxylic acid (IV)	DNA-1	-6.9	1.00×10^5	-6.9	1.06×10^5
	DNA-2	-7.0	1.19×10^5	-7.0	1.26×10^5
	DNA-3	-7.1	1.48×10^5	-7.2	1.76×10^5
	DNA-4	-6.8	9.44×10^4	-7.1	1.49×10^5
Indol-3-ethanamide (V)	DNA-1	-6.5	5.28×10^4	-6.5	5.44×10^4
	DNA-2	-6.7	7.64×10^4	-6.7	7.61×10^4
	DNA-3	-6.5	5.28×10^4	-6.5	5.44×10^4
	DNA-4	-6.6	6.47×10^4	-6.6	6.43×10^4
Benzimidazol-2-ol (VI)	DNA-1	-6.6	5.93×10^4	-6.6	6.43×10^4
	DNA-2	-6.7	7.49×10^4	-6.7	7.61×10^4
	DNA-3	-6.5	5.78×10^4	-6.5	5.44×10^4
	DNA-4	-7.2	1.68×10^5	-7.2	1.76×10^5
Benzimidazole (VII)	DNA-1	-6.6	6.27×10^4	-6.6	6.43×10^4
	DNA-2	-6.6	6.33×10^4	-6.6	6.43×10^4
	DNA-3	-6.5	5.30×10^4	-6.5	5.44×10^4
	DNA-4	-6.7	7.67×10^4	-6.7	7.61×10^4
2,3-Dihydro-1H-indole (VIII)	DNA-1	-6.5	5.27×10^4	-6.8	8.99×10^4
	DNA-2	-6.9	1.05×10^5	-6.9	1.06×10^5
	DNA-3	-6.7	8.23×10^4	-6.8	8.99×10^4
	DNA-4	-6.9	1.02×10^5	-6.9	1.06×10^5

References

1. A. Canals, M. Purciolas, J. Aymam'i and M. Coll; Acta Crystallographica Section D 61 (2005) 1009.
2. N. C. Garbett and D. E. Graves; Current Medicinal Chemistry—Anti-Cancer Agents 4 (2004) 149.
3. P. E. Pjura, K. Grzeskowiak, R. E. Dickerson; J. Mol. Biol. 197 (1987) 257.
4. M. K. Teng, N. Usman, C. A. Frederick, A. H. J. Wang; Nucleic Acids Res. 16 (1988) 2671.
5. J. N. Lisgartenl, M. Coll, J. Portugal, C. W. Wright & J. Aymami; Nature Structural Biology 9 (2001) 57.
6. J. B. Le Pecq, C. Paoletti; J.Mol.Biol. 27 (1967) 87.
7. H. Benesi, J. Hildebrand; J. Am. Chem. Soc. 71 (1949) 2703.
8. A. Gupta, A. Gandhimathi, P. Sharma, B. Jayaram; Protein and peptide letters 14 (2007) 632.
9. S. A. Shaikh, B. Jayaram; J. Med. Chem. 50 (2007) 2240

33
Structure Determination of DNA Duplexes by NMR

K. Pandav[1], P. Pandya[1], R. Barthwal[2] and S. Kumar[1*]
[1]Department of Chemistry, Dayalbagh Educational Institute, Dayalbagh, Agra-282110
[2]Department of Biotechnology, Indian Institute of Technology, Roorkee-247667
Email: kumar.surat@gmail.com

Abstract

Identification of the structure of DNA is an important step towards understanding its function in various biochemical processes. X-ray crystallography and NMR are the two most important tools for the evaluation of the structure of the DNA double helix. The present work employs NMR method to identify the structural details of the DNA oligomer sequences. 2D ^1H NOESY experiments were used to find out the chemical shift values of each cross peak. Complete structural assignment was furnished for DNA oligomers using 2D-NOESY technique. The cross peak volumes obtained from most of the NOESY peaks clearly indicated that most of the nucleotides in the DNA oligomer were in the B-form with a few residues in the A-form conformation. All the nucleotides were shown to have anti-conformation of deoxyribose sugar with respect to the nuclear base.

Introduction

DNA is a genetic material vital for several living processes. Various types of DNA duplexes exist in nature, viz., A-DNA, B-DNA, Z-DNA, etc. The structure of various forms of DNA differs significantly from each other. Several methods exist for the determination of the structural features of DNA duplexes. NMR spectroscopy and X-ray crystallography are the two main techniques used for this purpose. We have investigated the structural parameters of two DNA duplexes viz., Dickerson dodecamer (DNA-I) and a tailor made decamer sequence (DNA-II) using multidimensional NMR methods and compared them with the predicted model.

Materials and Methods

DNA-I [5'-d(CGCGAATTCGCG)$_2$] and DNA-II [5'-d(GGCAATTGCC)$_2$] were purchased from Sigma Aldrich as normal desalted bases. The duplex of each DNA sequence was generated with the help of annealing process. The annealing was carried out in Na-Phosphate buffer of 20 mM and 1 mM EDTA. The DNA bases in buffer were heated till 60°C at the rate of 1 °C /min. The DNA solutions were then allowed to cool slowly to the room temperature.

NMR experiments: All experiments were conducted in 20 mM Na-Phosphate in D_2O. NMR experiments on DNA-I using following parameters [1–3]: Temperature = 297K, Solvent = D_2O, Mixing Time = 300 ms, Number of Scans = 32, Total number of experiments in first dimension= 256, FID Resolution = 2.934 Hz/point, Sweep width = 6009 Hz and Pulse program = noesyesgpph.

2D-NOESY ^1H-NMR experiments were conducted on DNA-II using the following parameters: Temperature = 297 K, solvent = D_2O, Mixing time = 300 ms, Initial delay = 1.5 seconds, Number of scans = 48, Total number of experiments in first dimension = 256, FID resolution = 2.93 Hz/point, Sweep width = 6009 Hz and Pulse program = noesyphpr. The analysis of the NMR spectrum was accomplished on SPARKY software developed at UCSF [4].

Results and Discussion

Structural assignment of DNA base pairs and their connectivities (Figure 1) were accomplished using 2D NMR NOESY experiments. Complete assignment of NOE cross peaks were accomplished with DNA-I and DNA-II.

Tables 1 and 2 include details of the assignments of non-exchangeable proton resonances secured from

M.M. Srivastava, L.D. Khemani, S. Srivastava, *Chemistry of Phytopotentials: Health, Energy and Environmental Perspectives*, DOI:10.1007/978–3-642–23394-4_33, © Springer-Verlag Berlin Heidelberg 2012

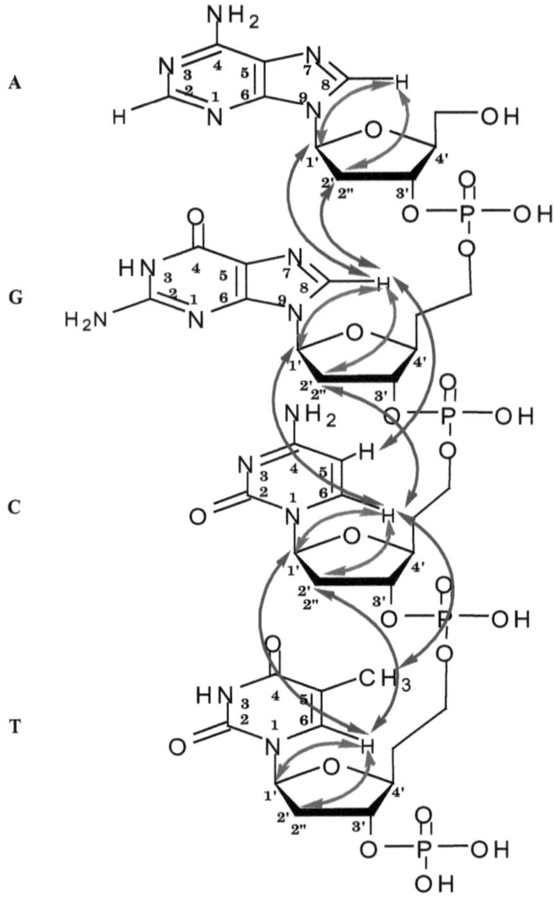

Fig. 1: Intra-nucleotide and Inter-nucleotide connectivities between various protons. These connectivities are observed as NOE cross-peaks in the NMR spectra of DNA. Blue color indicates inter-nucleotide connectivity while red color indicates intra-residue connectivity. A = adenosine, G = guanosine, C = cytidine and T = thymidine

2D-NOESY experiments for DNA-I and DNA-II respectively. The corresponding spectra for 'fingerprint' region of each of the DNA sequences are shown (Figures 2 and 3). Base proton to sugar H1' proton region (ω1 = 5.3 – 6.3 ppm vs ω2 = 7.1 – 8.4 ppm) is known as the fingerprint region of the spectrum and contains the H8/H6 protons of the nitrogenous bases and H1' protons of the deoxyribose sugar residues. This region contains two types of peaks, viz., the inter-nucleotide peaks and the intra-nucleotide peaks.

In case of DNA-I, strong intensity of cross peaks for base protons and H1' protons were assigned for all the residues. Cross peaks for base protons to H1', H2', H2" protons, Cytidine H6 to H5 protons and Thymidine H6 protons to methyl protons for all the residues were also unambiguously assigned. Significantly, assignments of H3' and H4' protons were also furnished for 12 base pairs.

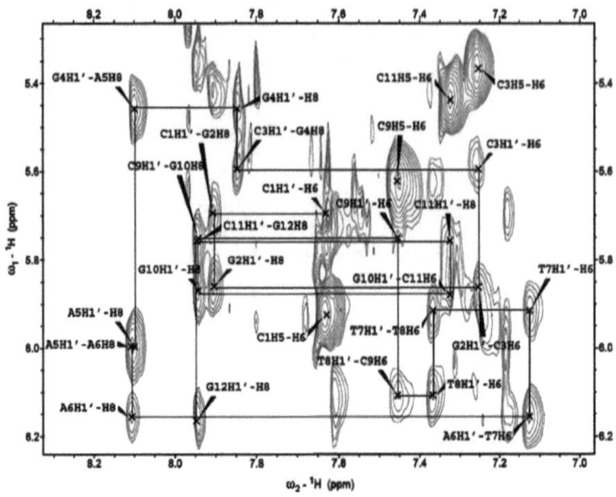

Fig. 2: NOESY spectrum of DNA-I showing sequential NOE connectivities between base protons and sugar H1' protons

Similarly, in case of DNA-II (Table 2), base protons for residue G1 and G2 were found to be overlapped and the stronger intensity of cross peaks for H8-H1' proton of guanosine residues. This resulted in the same value of 7.72 ppm for both the H8 protons of guanosine residues. Cross peaks for base protons to H1', H2', H2" protons, cytidine H6 to H5 protons and thymidine H6 protons to methyl protons for all the residues were also unambiguously assigned. Assignments of H3' and H4' protons were also furnished for 9 base pairs with a few overlapping peaks.

The assignment of majority of H1' protons out of 10 nucleotide bases were accomplished first from the base protons to H1' protons region. However, the peaks were identified in H1' to H2' and H2" region and other regions (Figure 3) where intra-nucleotide cross peaks were observed. This way, all the non-exchangeable protons were accomplished securing their assignment in several regions of the NOESY spectrum.

Cross peak volumes were also calculated for base protons to H2'/H2" cross peaks and base protons to H3' cross peaks in NOESY spectrum in order to identify the DNA duplex types (viz., A-form or B-form). In case of B-form DNA, the sugar pucker is C2'-endo conformation while the A-form DNA shows C3'-endo conformation (Figure 4).

It was observed in case of DNA-I that certain residues were in the A-form configuration while other residues were in B-form configuration. Residues G2 and G10 were found to be in A-form configuration as evident from the larger peak volume of base proton to H3' cross peak as compared to base proton to H2' cross peak (Table 1). The larger volume of the cross peak indicates smaller distance between base proton and H3' proton of G2 and G10 residues.

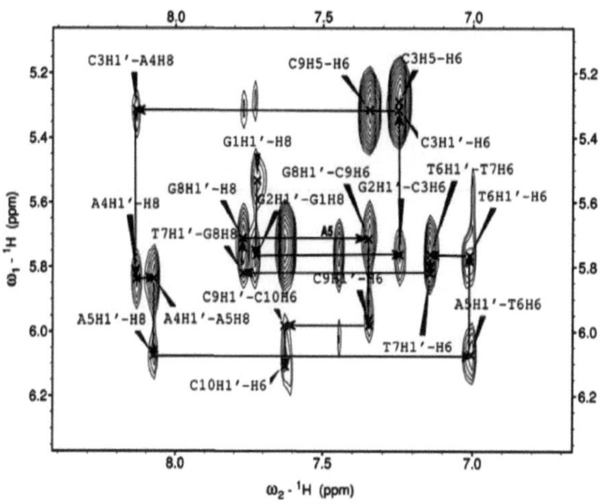

Fig. 3: NOESY spectrum of DNA-II showing sequential NOE connectivities between base protons and sugar H1' protons.

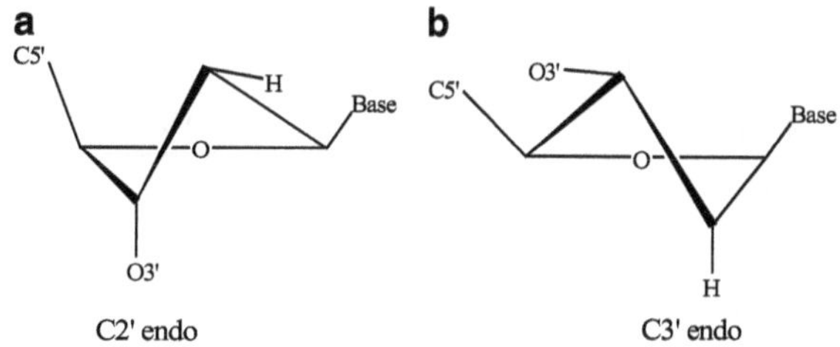

Fig. 4: Sugar conformations: (a) B-form DNA with C2'-endo sugar pucker. (b) A-form DNA with C3'-endo sugar pucker

The cross peak volumes of DNA-II were also compared. Based on the cross peak volume data, it was concluded that DNA-II also has B-form structure with C2'-endo sugar pucker.

The present work underscores the strength of 2D-NMR methods in the structure determination of small DNA oligomer sequences in solution state. NMR is also helpful in identifying small structural variation in the structure of DNA sequences due to dynamic nature of different nucleotide residues in the solution.

Table 1: Assignment table showing chemical shift values in ppm for non-exchangeable proton resonances DNA-I [5'-d(CGCGAATTCGCG)$_2$] along with the cross peak volume between base proton and H2'/H3' protons

Residue	H8/H6	H5	Me	H1'	H2'	H2"	H3'	Base-H2' Cross Peak Vol.	Base-H3' Cross Peak Vol.
C1	7.631	5.926	-	5.694	1.975	2.408	4.702	8.69×10^7	1.27×10^7
G2	7.907	-	-	5.862	2.065	2.656	4.980	2.29×10^7	2.40×10^8
C3	7.254	5.367	-	5.595	1.834	2.268	4.796	4.92×10^8	3.67×10^8
G4	7.848	-	-	5.458	2.264	2.659	4.986	2.97×10^7	1.35×10^7
A5	8.100	-	-	5.996	2.683	2.700	5.049	3.93×10^8	2.97×10^7
A6	8.107	-	-	6.156	2.572	2.947	5.012	3.53×10^8	3.05×10^7
T7	7.125	-	1.272	5.915	2.016	2.631	4.842	8.46×10^7	1.38×10^7
T8	7.368	-	1.535	6.107	2.181	2.568	4.946	1.21×10^8	3.75×10^8
C9	7.455	5.622	-	5.751	2.067	2.428	4.842	1.34×10^8	2.61×10^8
G10	7.942	-	-	5.880	2.369	2.649	4.967	9.17×10^7	1.17×10^8
C11	7.325	5.438	-	5.757	1.896	2.694	4.820	1.36×10^8	2.78×10^8
G12	7.947	-	-	6.165	1.950	2.384	4.688	4.27×10^7	2.98×10^7

Table 2: Assignment table showing chemical shift values in ppm for non-exchangeable proton resonances of DNA-II [5'-d(GGCAATTGCC)$_2$]

Residue	H8/H6	H5	Me	H1'	H2'	H2"	H3'	H4'
G1	7.72	-	-	5.53	2.31	2.50	NA	NA
G2	7.72	-	-	5.80	2.49	2.62	4.85	4.00
C3	7.25	5.3	-	5.77	1.88	2.21	4.71	4.01
A4	8.14	-	-	5.84	2.66	2.82	4.95	4.00
A5	8.07	-	-	6.08	2.49	2.82	4.92	4.14
T6	7.01	-	1.18	5.77	1.88	2.39	4.71	4.05
T7	7.14	-	1.46	5.82	1.98	2.30	4.81	4.02
G8	7.77	-	-	5.72	2.49	2.61	4.85	4.00
C9	7.35	5.32	-	5.92	2.08	2.35	4.68	4.07
C10	7.62	5.73	-	6.11	2.16	2.16	4.41	4.01

References

1. J. Chen, and J. Stubbe; Biochemistry 43 (2004) 5278
2. Y. Zheng and T. L. Sheppard; Chem. Res. Toxicol 17 (2004) 197
3. I. Goljer, S. Kumar and P. H. Bolton; The Journal of Biological Chemistry 270 (1995) 22980
4. T.D. Goddard, D.G. Kneller; (2006). SPARKY- 3 Software. San Francisco: University of California, San Francisco.

34
Pharmacotechnical Assessment of Processed Watermelon Flesh as Novel Tablet Disintegrant

S. Pushkar[1], Nikhil K. Sachan[2] and S. K. Ghosh[3]
[1,2]University Institute of Pharmacy, C. S. J. M. University, Kanpur – 208024
[3]Department of Pharmaceutical Sciences, Dibrugarh University, Dibrugarh – 786004
Email: seemapushkar@gmail.com

Abstract

The biomaterial from hydro-extracted aqua soluble fraction of Citrullus lanatus was characterized and evaluated for its possible use as novel disintegrating agent in the formulation of solid dosage forms. The product was valuated as excipient for repose angle, bulk density, tapped density, water activity, particle size distribution, sorption behavior, compressibility, color, pH of the solution and excipient qualification parameters as per the guidelines of 'International Pharmaceutical Excipient Council'. This investigation revealed the hydro-extracted watermelon flesh powder as a promising new excipient material for pharmaceutical usage.

Introduction

The Pharmaceutical excipients are having key role to play in formulation of drug delivery systems. They can affect the overall efficiency and cost effectiveness of the dosage form [1]. There has always been a search for the better pharmaceutical excipients in the field of tabletting technology because the tablets have been the ruling dosage form since years, and are one of the most challenging in all the pharmaceutical products to design and manufacture [2]. Substantial amount of research have been poured in the field of formulation excipients, exploring these supposedly inert materials for their better compatibility with active ingredients, improved stability, cost effectiveness and associated desirable features demanded by the product for handling, aesthetic acceptance, and aid in the delivery of drug [1]. Presently, in highly competitive and increasingly global pharmaceutical market there is a mounting pressure on the research and development (R&D) to shorten development time and get products to the market faster with a cost effective formulation. In this connection, the pharmaceutical utility of hydro-extracted water soluble fraction of watermelon fruit pulp as potential tablet disintegrant which being a food stuff can be classified as 'GRAS' (Generally regarded as safe) - a necessary status for any new excipient to be used in the food or pharmaceuticals for regulatory purpose [3]. Getting 'GRAS' status for a new substance is a high cost and time consuming affair, and because of this there are only a handful of substances from numerous researched materials which could be scaled up for industrial use. Therefore, the proposed novel excipient from fruit pulp of *Citrullus lanatus* holds potential for saving both the time and money for such regulatory approval as well as it is cheaper, readily available, capable of multitude of chemical modifications, and is biocompatible because of its natural origin. In addition, this can be a potential marketing tool on account of its good flavor and in the 'herbal boom worldwide'. Today consumers opt for the natural ingredients in the food, drug and cosmetics as they, believe that anything natural will be safer and devoid of side effects as compared to their synthetic counterparts [1].

In all solid dosage forms, disintegrants may be added to a drug formulation as functional fillers. These excipients assist in the disintegration of the capsule or tablet when it is introduced to moisture. Disintegrants are agents added to tablet and some encapsulated formulations to promote the breakup of tablets and capsule form "slugs' into smaller fragments in an aqueous environment there by increasing the available surface area and promoting a more rapid release of the drug substance. They promote moisture penetration and dispersion of the tablet matrix.

M.M. Srivastava, L.D. Khemani, S. Srivastava, *Chemistry of Phytopotentials: Health, Energy and Environmental Perspectives*, DOI:10.1007/978–3-642–23394-4_34, © Springer-Verlag Berlin Heidelberg 2012

Ideally, it should cause the tablet to disrupt, not only into the granules from which it was compressed, but also into powder particles from which the granulation may prepare. Combinations of swelling and/or wicking and/or deformation are the mechanisms of disintegrant action [4]. The overall objective of this investigation holds a three tier approach for exploring the pharmaceutical potential of dried hydro-extracted fruit pulp of *Citrullus lanatus* as natural disintegrating agent in solid dosage forms. This involves collection, procession, characterization and subsequent formulation aspects.

Fig. 1: Watermelon (*Citrullus lanatus*) Fruits

Chemicals and Instruments

Poly, N-1–2-Vinyl Pyrollidone (Merck chemicals, Mumbai, India); Calcium Carbonate, Magnesium Stearate, Talc (Central Drug House, New Delhi, India), Maltodextrin (Hi- Media, Mumbai, India), Paracetamol was supplied by Merck Chemicals, Mumbai, India.

Hand Blender (Koryo, KHB 5011), Homogenizer (Remi India), Vertex Mixer (SPINIX), Bulk density Apparatus (Hicon), Rotary Vacuum Evaporator (Evator Rotatory Evaporator, Medica Instruments, Ltd, India), Spray Dryer (Custom made, Biotech Park, LKO), Refrigerator (Samsung), Vacuum Oven (Hicon, India), Grinder (Philips), Sieve # 10 (B.P. Standard), Projection Microscope (Olympous), Camera (Cannon), Lab. Lyophilizer (STARTEK), U.V visible spectrophoto-meter (Shimadzu 1700), High Sensitivity Electronic Balance (KM-2, Deva Inc), USP – Tapped density apparatus (Electrolab, India), pH meter (LI 127, Elico Ltd, India), Mechanical Stirrer (Remi, India) and Desiccator *etc.*

Collection and Authentication of Fruit

Watermelon (*Citrullus lanatus*) fruits were purchased from fruit and vegetable market Chakarpur and the same were confirmed with University Department of Life Sciences. It being a well known fruit did not call for a voucher specimen to keep. Same lot of the watermelon collected was used in all type of drying and subsequent evaluation for study.

Processing and Extraction of Flesh

The watermelon flesh taken off and blended (Koryo blender KHB 5011) to liquidize the pulp. It was then passed through sieve #10 and the strain collected was homogenized (Remi, India) and shaken in a vertex shaker (Spinix, India) so as to obtain the desired hydro-extracted biomaterial in strain and to remove undesired fruit parts. This blend filtered through a muslin cloth to have water soluble fraction which was dried to powder using different drying methods *viz.* direct sun drying, vacuum drying, spray drying and lyophilization. For sun drying, isolated liquid was kept in direct sunlight under contemporary day climate, between 10.30 am and 5.30 pm daily till strain attained constant weight, due care was taken to avoid any exposure to rainfall and drying material were covered with thick polyethylene bags after removing from open area [5]. Parameters like humidity, air velocity, temperature etc. were not studied. The vacuum drying was performed in a Rotary Vacuum Evaporator (Evator Rotatory Evaporator, Medica Instruments, Ltd, India) followed by vacuum oven (Hycon, India). Two hundred grams of processed juice was put into the instrument flask under 200 mbar vacuum pressure, heated constantly by immersing it in a 55 °C water bath warmed with a 1500 Watt electrical power heater [6].

Operation optimized with constant monitoring and after the time when sufficient solidification is achieved and evaporation did not remain effective, it was transferred to vacuum oven and dried under vacuum at a temperature <40 °C. For obtaining spray dried powder of test material, a high efficiency spray drier (custom made Bio. Tech. Park, Lucknow) was employed for the spray drying process. Drying operation was optimized at the aspirator rate of 60 %, flow rate of 600 L/h, pressure of 4.5 mbar, and feed temperature of 20 °C after conducting several trial runs. Four inlet air temperatures were investigated, *i. e.* 145 °C, 155 °C, 165 °C and 175 °C. Maltodextrin (3 %) was added according to the weight of the watermelon juice. The dryer was washed with water at the desired parameter setting for 10 min before and after the spray drying process. The spray-dried powder was collected in clean container with known weight. The powders produced were weighed, sealed in bottle and stored in desiccators till further use [7].

Another drying technique employed was freeze drying performed in a lyophilizer (Startek Inc) with glycol chamber. Required quantity of prepared strain was carefully filled in beaker and was frozen using deep freezer (-20 °C). The frozen unit was placed in the freeze dryer and dried at (-50 °C). The dried powder so obtained was stored in desiccators till further use [9]. All the dried powder samples were taken for further qualitative physicochemical and pharmacotechnical characterization to assess their suitability for pharmaceutical usage as excipient in solid dosage forms, and to study effect of drying techniques as process parameter.

Bulk and Tapped Density

Both loose bulk density (LBD) and tapped bulk density (TBD) were determined. A quantity of 20gm of powder from each method, previously lightly shaken to break any agglomerates formed, was introduced into a 50 ml graduated cylinder. After the initial volume was observed, the cylinder allowed to fall under its own weight onto a tapping pad from the height of 0.5–1.5 cm at 5 sec intervals. The tapping has continued until no further change in volume was noted. LBD and TBD were calculated using formulae given below [7, 8].

LBD = weight of the powder/volume of packing
TBD = weight of the powder/tapped volume of packing

Compressibility Index

The compressibility and flow property of the granules were determined by Carr's compressibility index and Hausner's ratio calculated using the formula (1) and (2) respectively as below [7, 8]:

(1) Carr's Index (%) = [(TBD – LBD) ×100/TBD]
(2) Hausner's ratio = [TBD / LBD × 100]

Particle Size

The particle size of the dried-powder was determined by the microscopic method using a calibrated ocular micrometer (n=500), and average particle size was calculated, the study was performed for each sample obtained through different drying techniques [7].

Angle of Repose

The fixed funnel and free-standing cone method was utilized which employ a funnel that is secured with its tip at given height, H, which was kept 2 cm, above graph paper that is placed on a flat horizontal surface. With r, being the radius of base of conical pile and h is the height of pile [9]:

$$\tan \theta = h / r$$

pH of Solution

The pH of the solution obtained by dispersion of dried powder of watermelon flesh and was measured using pH meter (LI 127, Elico Ltd, India) [7].

Color Measurement

The color analysis was performed by colorimetric measurements using spectrophotometer (Shimadzu, Japan). Duplicate samples were analyzed and the mean was recorded [7].

Sorption Behavior

Sorption study was carried out in a 100 ml stoppered graduated cylinder. The initial bulk volume of 1 gm of dried powder was recorded. Water was then added in sufficient quantity to yield 100 ml of a uniform dispersion. The sediment volume of the swollen mass was measured after 24 hour, stored at room temperature. The swelling ratio was calculated by taking the ratio of the swollen volume to the initial bulk volume [10].

Water Activity

The water activity of sample was determined using a reported chilled mirror dew-point method [11, 12].

Moisture Content

The moisture content was determined based on Association of Analytical Communities (AOAC) method. Triplicate samples of watermelon powder (20 mg each) were weighed and then dried in a vacuum oven at 70 °C for 24 h. The samples were removed from the oven, cooled in a desiccator and weighed. The drying and weighing processes were repeated until constant Weight were obtained [7].

Results and Discussion

Compressibility Index which is the ratio of tapped density to bulk density gives an insight to the degree of densification of powders which could occur during tabletting. The lower the ratio the less the propensity of the powder to densify. This phenomenon may cause tablets which lack uniformity of weight and active content to be produced.

The pH of solution on dispersion was 5.79±0.56, to evaluate physicochemical properties of the watermelon juice used for drying. It can be seen that the watermelon juice had a high pH value (5.79) which implied that it is susceptible to microbial grow.

From the observations, there was hardly any powder accumulated in the collector if maltodextrin was not added in the feed. The particles produced were very sticky and mainly deposited onto the wall of drying chamber and cyclone and could not be recovered. Therefore, maltodextrin of 3 % of total feed solution was added to the juice prior to spray drying to investigate its effect on the spray drying product.

The moisture sorption capacity was found to be 26.84 % for vacuum dried powder; 24.42 % for sun dried powder and 30.16 % for lyophilized powder under contemporary experimental condition for the sample taken. This indicated good swelling which would result in quick rapture of tablet and disintegration property. The spray dried powder undergo dissolution into the water medium and hence sorption results could not be obtained. Water activity (*a*w) is an important index for dried powder because it can greatly affect the shelf life of the powder produced. Water activity is different from moisture content as it measures the availability of free water in a food system that is responsible for any biochemical reactions, whereas the moisture content represents the water composition in a food system. High water activity indicates more free water available for biochemical reactions and hence, shorter shelf life. Generally, food with $aw < 0.6$ is considered as microbiologically stable and if there is any spoilage occur, it is induced by chemical reactions rather than by microorganisms. From the results (Table 1), the water activities of the powders were in the range of 0.24–0.91. This meant that the spray-dried powders produced were relatively stable microbiologically. However, the storage conditions also played an important role in this matter. As the spray-dried powder contain high amount of sugar, they were highly hygroscopic and should be stored properly in air-tight container and kept in cool dry place.

Table 1: Comparative Results of Micromeritic and Derived Properties of Powder

Parameters	Drying Methods			
	Sun Drying	Vacuum Drying	Spray Drying	Lyophilization
Bulk Density (g/ml)	0.51±0.007	0.52±0.005	0.48±0.003	0.50±0.004
Tapped Density (g/ml)	0.76±0.006	0.76±0.003	0.73±0.002	0.72±0.003
Compressibility Index (%)	32.9±0.12	31.6±0.11	34±0.004	30±0.007
Particle Size (μm)	88.95±0.003	86.06±0.002	95.06±0.004	97.46±0.005
Angle of Repose (°)	35.67±0.01	35.96±0.03	36±0.04	33±0.05
Sorption Behavior (%)	24.42±0.04	26.84±0.07	--	30.16±0.03
Water activity	0.91±0.02	0.59±0.06	0.24±0.11	0.25±0.08
Moisture Content (%)	9.26±0.26	6.24±0.41	1.62±0.20	4.78±0.21

Moisture content of the powder was lower at the higher inlet drying as shown in Table 1. This is because at higher inlet drying temperature, the rate of heat transfer to the particle is greater, favoring the moisture evaporation [13]. Besides the effect of drying temperature, addition of maltodextrin also affected moisture content. A study on the dried watermelon powder using a spray dryer showed that an increased amount of maltodextrin resulted in a lower moisture content of the final powder Besides showed that an increased amount of maltodextrin the effect of drying temperature, addition of maltodextrin also affected moisture content. A study on the dried watermelon powder using a spray dryer resulted in a lower moisture content of the final powder [7]. In a spray drying system, the water content of the feed had an effect on the final moisture content of the product [13].

With higher initial water content in the feed, the final product also had higher moisture content. This phenomenon did not relate to the temperature [14].

Color is the most obvious change that occurs in many fruits. Colormetric analyses showed that the L^*, a^*, b^* values changed with the inlet temperatures as well as during storage at room temperature. The value of L* in ranged from 30.0 to 40.1. The redness or a* value ranges from +25.8 to +29.4, while the yellowness b* value range from +16.8 to +21.2. During storage at 21.1 °C, the flesh of watermelon can change their intensive to red color [15–16].

Conclusions

In conclusion, these studies depict that collection of particles of hydro-extracted watermelon flesh have acceptable pharmacotechnical parameters for further compression into tablets. Water activity of the product indicates that the spray dried powder is relatively stable microbiologically, when compared to the powders from other drying methods. In view of the favourable biochemical stability, it should stored in airtight container at cool place till used. Further pharmacotechnical properties of the powder glow a clue towards its potential use as a tablet disintegration agent.

References

1. N.K. Sachan, and D. Singh; J. As Sc Soc. 46(2) (2006) 20–22.
2. M.C. Gohel, *et al.* Tablet: the ruling dosage form since years. Pharmapedia, (2005).
3. C. Rados, GRAS: Time-Tested, and Trusted, Food Ingredients. FDA Consumer Magazine (2004).
4. P.S. Mohanachandran, P.G. Sindhumol, and T.S. Kiran; Int. J. Pharma. Sci. Review and Res., 6 (1) (2011) 105–109.
5. S.A. Bankole, A.Osho, A.O. Joda, and O.A. Enikuomehin; Afr. J. Biotechnol.4(8)(2005) 799–803.
6. R. Assawarachan and A. Noomhorm; Int. J. Agric. Biol. Eng. 3(1)(2010)74–84.
7. S.Y. Quek, N.K. Chok, and P. Swedlund; Chemi. Eng. Processing 46 (2007) 386–392.
8. N.K. Sachan and S. Pushkar; Int. J. Pharm. Sci. Tech. 2(1) (2009) 14–21.
9. K. Rajitha, Y. Shravan Kumar, D. Adukondalu, R. Gannu, and Y. Madhusudanrao; Int. J. Pharma. Sci. Nanotech. 1(4) (2009) 327–334.
10. K.K. Mehta, H.H. Patel, N.D. Patel, C.N. Vora, and N.J. Patel; Int. J. Pharmacy Pharm. Sci. 2(3) (2010) 102–108.
11. G.D. Anagnostopoulos; J. Gen. Microbiology 77 (1973) 233–235.
12. A.J. Fontana; Current Protocols Food Anal. Chem. (2001) A2.2.1-A2.2.10.
13. D. Halliday and J. Walker; Drying Technique & Process. Brisbane: John Wiley and Sons, Inc, 2001.
14. S. Thankitsunthorn, C. Thawornphiphatdit, N. Laohaprasit and G. Srzednicki; Int. Food Res. J. 16 (2009) 355–361.
15. J.M. Lutz, R.E. Hardenburg and A.K. Thompson, eds., Blackwell Publishing, (1968).
16. E.W. Yau, S. Rosnah, M. Noraziah, N.L. Chin and H. Osman; Int. Food Res. J. 17 (2010) 327–334.

35

Evaluation of Assam Bora Rice as a Natural Mucoadhesive Matrixing Agent for Controlled Drug Delivery

Nikhil K. Sachan[1], S. Pushkar[1] and S. K. Ghosh[2]
[1]University Institute of Pharmacy, C. S. J. M. University, Kanpur – 208 024 India
[2]Department of Pharmaceutical Sciences, Dibrugarh University, Dibrugarh – 786 004 India
Email: nikhilsachan@gmail.com, FAX: 91–5122570173

Abstract

The Present study envisages pharmaceutical utility of Assam Bora rice as biopolymeric excipient in drug delivery. The sustained release potential from the proposed matrix backbone was investigated through in vitro dissolution studies of microbeads prepared by an industrially feasible conventional ionotropic gelation method using the blends of pregelatinized Bora rice along with sodium alginate, as per SUPAC-MR guidelines. The prepared beads were characterized for surface morphology, drug-polymer compatibility, mucoadhesion and other pharmacotechnical parameters. The outcomes of studies have demonstrated that Assam Bora rice holds the promise of being used as a drug release modulator in the drug carrier systems.

Introduction

Polymers have become an indispensable part of the drug delivery systems, be it be conventional drug delivery or novel drug delivery. They have drastically changed the mode of drug delivery by introducing lot of flexibility. Polymers have taken a long stride from oral controlled release dosage forms to polymeric stents, implants, microchips *etc* [1]. Recent advances pertain to drug delivery systems incorporate different type of polymers within the matrix of drug delivery systems to protect the active ingredient and to induce slow release characteristics [2]. The use of natural and modified natural polymers in the drug delivery continues to be an area of intensive research despite the advent of several new synthetic polymers for a number of reasons, as they are economical, readily available, capable of chemical modifications and they are potentially biodegradable and biocompatible due to their natural origin [3]. The naturally occurring bioadhesives especially the polysaccharides have received extensive attention due to their biocompatibility, low-cost and because of a "getting out of oil" policy [4]. The present investigation proposed and examined the pharmaceutical utility of *Bora rice* in drug delivery, which is a variety of glutinous rice [5], contains very low amylose generally less than 3 % by weight [6], cultivated mainly in Assam region. The approach to use *Bora rice* flour as release controlling polymer in drug delivery is potentially very interesting because the rice is a common foodstuff and is biocompatible, readily available, and may be classified as 'Generally regarded as safe' (GRAS). Numerous new smart polymers have been created and used to develop environment sensitive drug delivery systems, but none of them have been used in the commercial formulations to date because, for first time use, these polymers required to be labeled as 'GRAS' which is a high cost affair [7]. The *Assam Bora rice* has been used in food industry in bakery as binder but there has been no report of its use in pharmaceuticals. This variety of rice is also known as sticky or waxy rice is having high proportion of amylopectin, composed of highly branched structure with a degree of polymerization ranging from 10 to 60 glucose units connected to each other by the alpha-1, 6 linkages. Amylose most likely accounts for the disintegration properties of starch. On the other hand, amylopectin is a good binder. It also retards the disintegration of matrix and dissolution of the drug from matrix. The present study intended to examine the mucoadhesive property, potential of modulating drug release from the matrix, compatibility with the matrixed chemical species and other biopharmaceutical and pharmacotechnical pa-

M.M. Srivastava, L.D. Khemani, S. Srivastava, *Chemistry of Phytopotentials: Health, Energy and Environmental Perspectives*, DOI:10.1007/978–3-642–23394-4_35, © Springer-Verlag Berlin Heidelberg 2012

rameters of *Assam Bora rice* and fabricated micro-devices to investigate its utility in pharmaceutical drug delivery system. The *Assam Bora rice* starch as such does not swell/dissolve in cold water for that it was taken pregelatinized that can take up water from the medium and is miscible with the hydrogel system of other natural polymers.

Experimental

Materials

Bora rice was procured from the village near to Dibrugarh University and was confirmed so by local people. Sodium alginate (Loba Chemi Pvt. Ltd. Mumbai), Calcium chloride (Qualigens Mumbai, India), Glacial acetic acid (Qualigens Mumbai, India), were procured from the commercial sources. Metformin hydrochloride was kindly gifted by Ms Ranbaxy Pharmaceuticals, India. All other reagents were of analytical grade laboratory reagents and were procured from commercial sources.

Methods

Preparation of Matrix Micro Devices

The drug loaded micro devices were prepared by an industrially feasible micro-orifice ionic gelation method [8] using the blends of pregelatinized *Assam Bora rice* and sodium alginate in varying proportions. At later stage the method was modified to dissolve the $CaCl_2$ in acetic acid and adding co-solvent to the curing solvent in order to improve drug entrapment efficiency.

Optimized beads with *Bora rice* backbone were coated with HPMC using solvent evaporation method [9]. Beads were dispersed in 10% strength of HPMC in acetone and the solvent was evaporated in a rotary evaporator by applying vacuum 300 mmHg and rotation rate was 50 rpm, then vacuum dried in desiccators. All the batches were evaluated on the basis of release study.

Determination of the Drug Content

A spectrophotometric determination was adopted from the assay method of 'Metformin tablets-BP' to estimate the content of Metformin hydrochloride in microbeads and its release as a function of time. The drug entrapment efficiency of prepared microbeads was calculated as below:

$$\text{Entrapment efficiency} = \frac{\text{Estimated percentage drug loading}}{\text{Theoretical percentage drug loading}} (100)$$

$$\text{Percentage drug loading} = \frac{\text{Amount of drug in microspheres}}{\text{Amount of microspheres}} (100)$$

Particle Size Measurements

The particle size and size distribution of prepared microbeads was studied by the optical microscopy and using phase contrast microscope (LaboMed XLR II). The mean diameter was calculated. The effect of drug concentration, concentration of cross-linking agent, and ratio of the two polymers on the average particle size was studied.

$$\text{Mean particle size} = \sum n.d / \sum n$$

Swelling Studies

The swelling studies were performed gravimetrically in aqueous swelling media with 0.1N HCl, buffer pH 7.4 and water alone at 37.5 ± 0.5 °C. The swelling ratio, S_{wt}, calculated from the following expression [10]:

$$S_{wt} = [(W_t - W_0)/ W_0] \text{ ' } 100$$

Where, W_t and W_0 are the weight of sample swollen at time t and the weight of the original sample, respectively.

Evaluation of Mucoadhesive Property

The mucoadhesive property of the fabricated drug delivery systems was evaluated by *in vitro* was-off test as reported by Lehr *et al.* (1992) [8]. The retention time of the microbeads to the excised mucosa was compared with a non bioadhesive material in different release environments.

Water Vapor Uptake Study

The prepared beads were kept at controlled humidity environments (32%, 52%, 75% and 92%), using saturated salt solutions and equilibrium moisture content was determined by weight difference attained due to the moisture uptake by the beads [11].

In Vitro Drug Release Studies:

The release of Metformin hydrochloride from the microbeads was studied in aqueous medium using USP XXVI Dissolution test apparatus basket type (Campbell Electronics, Mumbai) at 37± 0.2 0 C with a rotating speed of 50 rpm [12, 13]. The *in vitro* dissolution profile studies for characterization of release kinetics were carried as per the SUPAC-MR guidelines [14].

Scanning Electron Microscopy

The surface morphologies of the blank microspheres, drug loaded microspheres and microspheres collected after dissolution study, were investigated by using HITACHI, S – 3600 N, Scanning Electron Microscope at 15kv.

Drug-polymer Compatibility Study

The proposed polymer backbone was subjected to drug-polymer compatibility testing, using various techniques as below:

Ultraviolet Spectroscopy (UV)

The UV scan was taken for the drug solution and of the extracted medicament from the formulation to find out any new peak observed that can provide a clue about the chemical interaction between the drug and excipients.

FTIR Spectral Studies

The FTIR spectra of the drug, *bora rice*, sodium alginate, blank and drug-loaded, were obtained in KBr pellets using a Perkin-Elmer model 883 spectrophotometer and, at mid IR region (wavelength 25 m to 2.5 m, wave-number from 400 cm^{-1} to 4000 cm^{-1}), in order to identify the possibility of any interaction between the drug and polymer materials.

Thermal Analysis (DSC)

The DSC thermograms of the pure drug, drug-loaded microbeads, blank microbeads and the two polymers (bora rice and sodium alginate) were obtained using the METLORs DSC system, to identify any interaction between the components of formulation.

X-ray Powder Diffraction (XRPD) Study

The X-Ray diffractograms of the drug, drug loaded microbeads, blank microbeads and the two polymers were obtained using a X-Ray diffractometer (Philips, PW117/PW118) to study any change in the crystallinity of the drug in formulation and to presume about any drug polymer interaction during the process of encapsulation.

Results and Discussion

Industrially feasible modified ionic gelation method was adopted successfully to prepare the drug loaded microbeads. Several preformulation trials were undertaken to optimize the matrix micro-carriers. The use of acetic acid and co-solvents was sought because of the low entrapment efficiency in trial batches due to high water solubility of drug. Prepared micro-carries were of smooth surface with sufficient mechanical strength those become rough on drying. The drying conditions had shown some impact on the drug release property as the oven dried microbeads were getting cracks on surface and produced the irregular release pattern, the rotary vacuum dryer (Tempo, India) was used to dry the prepared beads at a temperature below 40 °C. Stirring speed of 150 rpm was found suitable producing almost spherical beads as the higher speed of stirring reduced the entrapment efficiency and lower stirring produced elongated beads. An improvement in the drug entrapment and prolonged release time was observed through alteration in order of blending the two polymeric gels, and addition drug to pre-gelatinized *Bora rice*, subsequently ultrasonicated and allowed to stand for some time, before it was finally blended for preparation of beads. The drug entrapment efficiency of the beads was in range of 19–53 % depending upon the drug polymer ratio, stirring speed, curing time, gel strength and medium of microencapsulation.

The physical characterization using ordinary and phase contrast microscope showed the almost spherical shape of prepared microbeads in a size range of about 0.726 ± 0.008 mm to 1.16 ± 0.009 mm. Particle size distribution within a batch was in a narrow range while in different batches it was found to be affected by the drug-polymer ratio, stirring speed, gel strength and size of precision device used. The microbeads exhibited a variable swelling under different ionic environments. It was 30 % in acidic medium, 56 % in DM water and 84 % in a 7.4pH buffer solution. The low swelling in acidic medium is conceptually favorable as it will retard the drug release rate in the gastric environment while as the system sink down in gastrointestinal (GI) tract the beads will swell up more for maximum drug exhaust. The microbeads prepared with *Assam Bora rice* backbone have shown good mucoadhesive potential when compared with non bioadhesive material (Eudragit) and exhibited a retention time up to 8 hours in an *in vitro* wash off test. Water vapor uptake studies demonstrated the hygroscopic nature of microbeads (Table 1).

Table 1: Wash off profile of microbeads

S. No.	Time (h)	% of beads adhering
1.	01	96
2.	02	84
3.	03	82
4.	04	74
5.	05	64
6.	06	40
7.	07	38
8.	08	28

The *in vitro* drug release studies demonstrated the capability of proposed matrix backbone in modulating the drug release behavior (Fig.1). The release of medicament from the beads was sustained and continued up to 9–11 hours depending upon the drug load, release environment, cross-linking agent, particle size and drying conditions. The *Bora rice*/Sodium alginate ratio had not shown any significant difference if the gel strength maintained constant. The coated beads with HPMC exhibited better release profiles with the mean t_{90} values up to up to 11 hours in a buffer at pH 7.4.

The scanning electron micrographs taken for different batches prepared with varying formulation and process variables revealed the surface morphology of matrix micro devices, confirmed the coating and supported the presumed mechanism of drug release from the microcarrier systems (Fig. 2). The beads showed rough surfaces those are uncoated, become smooth on coating. The SEM pointed out the cracks produced in the oven dried beads then those were corrected by adopting rotary vacuum drying. There are holes on to the surface of microbeads after dissolution supporting the diffusional drug release. The uncoated microbeads became eroded and fluffy after dissolution in alkaline medium showing that drug released through a combined mechanism of diffusion and erosion.

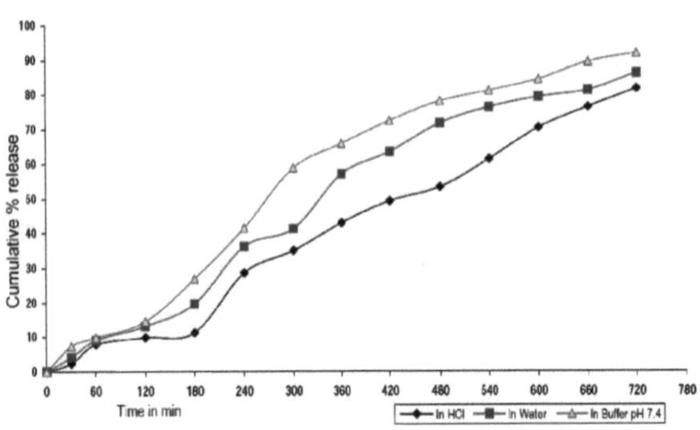

Fig. 1: Drug release profile of optimized coated beads

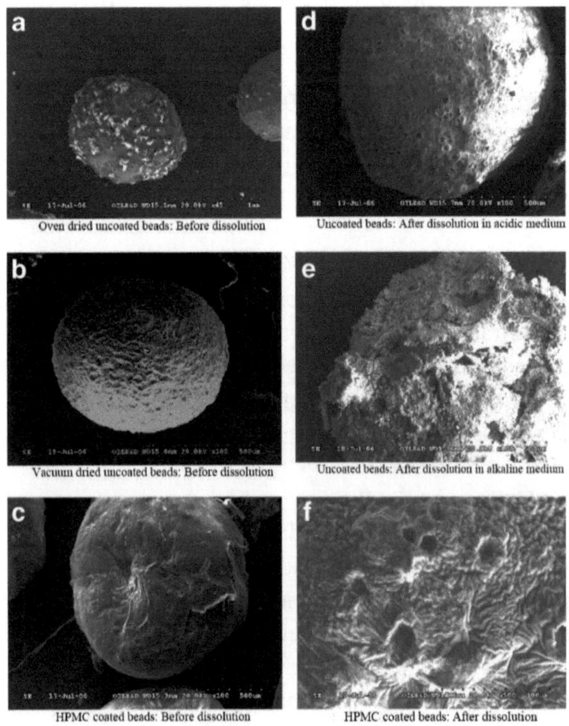

Fig. 2: Scanning electron micrographs of microcarriers

The characteristic N–H stretching band of drug remained unchanged in the FTIR spectra in drug-loaded microbeads. Also the peaks corresponding to C – N vibrations, C = N stretching, N = N stretching, and C – H (alkane) stretching were retained in the FTIR spectrum of the drug loaded microbeads reflecting the identity features of the Metformin hydrochloride (Fig. 3). This finding indicated no chemical interaction between the drug and polymer backbone that is further supported by DSC. A sharp endotherm at 224°C was observed in DSC thermograms of drug and drug-loaded beads corresponding to the melting point of metformin hydrochloride. Another broad band was observed at 60–70°C that is comparable to the transition temperature of *Bora rice* starch and first endothermic peak of the sodium alginate. The DSC thermograms, of the drug and drug loaded beads when compared along with the thermograms of the individual components and of the blank beads, reflected that the drug-loaded beads had comparable endothermic peaks to that of drug and the blank beads. There was neither formation of any new peak nor the shifting of any peak significantly in the DSC thermograms indicating that the drug had not undergone any chemical interaction with the polymer backbone during microfabrication [15].

X-ray powder diffraction is a useful method in studying crystalline phases [16]. The samples are fixed and sealed between two mica windows and placed in the beam of monochromatic X-rays. The scattered radiation satisfying Bragg's law: $n \varsigma = 2d \sin \theta$, where n is the order of the diffraction pattern, ς is the radiation wavelength, d is the distance between planes in the crystal and θ is the scattering angle, gives a diffraction pattern [17].

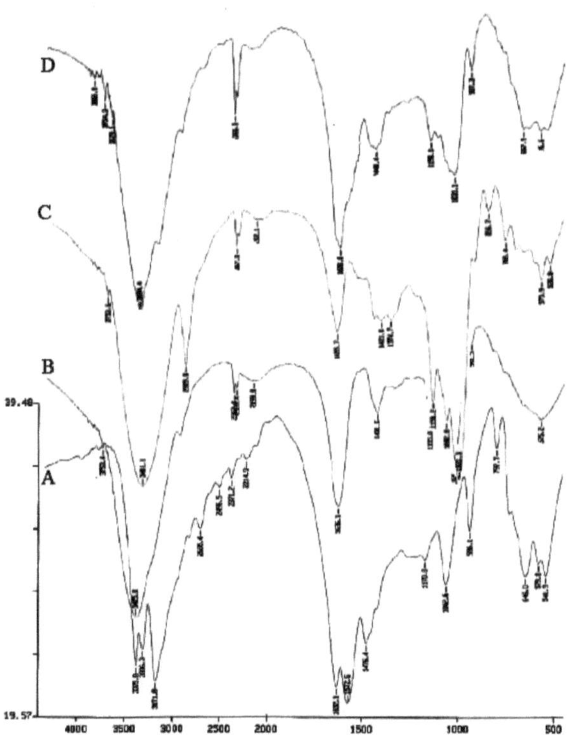

Fig. 3: FTIR Spectra: **(A)** drug sample **(B)** placebo **(C)** bora rice polysaccharide **(D)** drug loaded micro carriers

This technique has been extensively used in the literature for investigating the structure of crystalline phases in polymer matrices [18–20]. In X-ray diffractograms the Metformin hydrochloride shown the characteristic significant peaks at the 2q of 17, 22, 29 and 31°, the same diffraction behavior was observed in case of the drug-loaded beads, exhibiting the similar peaks but with notable decrease in intensity of the signal indicating towards no chemical interaction between the drug and other components of formulation. The decrease in intensity of diffraction signal is probably because of the dilution effect and a decrease in the intensity of the drug due to encapsulation into the polymer matrix.

Conclusions

The present investigation explored the utility of *Bora rice* (glutinous rice) starch in the pharmaceutical dosage forms as a natural mucoadhesive polymer for drug delivery. The carrier backbone so prepared exhibited a good mucoadhesive potential in an in vitro wash off test, and was able to modulate the drug release from the matrices. In conclusion, the proposed biopolymer, with proper research efforts, holds promise for scale up and further exploitation for utilization in pharmaceutical drug delivery systems.

Acknowledgements

The authors are thankful to R&D center Oil India Limited, Duliajan for providing the instrumental facilities for SEM and X-ray diffraction studies and to the Indian Institute of Technology, Guwahati for conducting DSC study without any charges.

References

1. N. Kashyap, S. Modi, J.P. Jain, I. Bala, S. Hariharan, R. Bhardwaj, D. Singh, R. Mahajan, N. Kumar, M.N.V. Ravi Kumar; CRIPS 5, 3 (2004) 7–12.
2. J.K. Vasir, K. Tambwekar and S. Garg; Int. J. Pharm. 255 (2003) 13–32.
3. M.P. Satturwar, V.S. Fulzele and A.K. Dorle; AAPS Pharm SciTech, 4, 4 (2003) Article 55.
4. S. Suzuki, et al.; Biomacromolecules, 6 (2005) 3238–3242.
5. P.K. Pathak, T. Ahmed, K.K. Sharma and A.K. Pathak; 32 (1995) 48–50.
6. B.O. Juliano, In Proceedings of a Workshop on Chemical Aspects of Rice Grain Quality. International Rice Research Institute, Los Banos, Laguna, Philippines (1979) 251–256.

7. C. Rados; FDA Consumer Magazine. 2004 (March-April).
8. C.M. Lehr, J.A. Bouwstra, E.H. Schacht and H.E. Junginger; Int. J. Pharm. 78 (1992) 43–48.
9. A. Paharia, A.K. Yadav, G. Rai, S.K. Jain, S.S. Pancholi and G.P. Agrawal; AAPS PharmSciTech; 8 (2007) 12.
10. K.C. Gupta and M.N.V. Ravi Kumar; J. Mater. Sci.: Mater. Med. 12 (2001) 753- 759.
11. APhA: Hand Book of Pharmaceutical Excipient, 10th Ed. American Pharmaceutical Association. Pharmaceutical Society of Great Britain. England (1983) 364.
12. M. Asarafi, J.A. Chowdhary and S.M. Reza; J. Pharm Sci. 4, 1 (2005).
13. R.P. Patel, et al.; Drug Delivery, 5, 1 (2005).
14. FDA Guidance for Industry, Center for Drug Evaluation and Research (CDER), Sept 1997.
15. P.T. Tayade and R.D. Kale; AAPS PharmSciTech 6, 1 (2004).
16. M. Scherlund, PhD Thesis, Faculty of Pharmacy, Uppasala University, Sweden; 2000.
17. G.H. Stout, L.H. Jensen; X-ray Structure Determination. A Practical Guide. John Wiley and Sons, Inc. New York (1989).
18. V. Luzzati; Xray diffraction studies of lipidwater systems. In: D Chapman (Ed.) Biological Membranes. Academic Press, New York (1968) 71–123.
19. K. Fontell; Adv. Colloid Interface Sci.; 41 (1992) 127–47.
20. K. Larsson: Lipids-Molecular Organization, Physical Functions and Technical Applications. Vol. III. Scotland: The Oily Press Ltd, West Ferry, Dundee (1994).

36

Utilization of Some Botanicals for the Management of Root-Knot Nematode and Plant Growth Parameters of Tomato (*Lycopersicon Esculentum* L.)

S.A. Tiyagi, I. Mahmood and Z. Khan
Plant Pathology and Nematology Lab, Department of Botany, Aligarh Muslim University, Aligarh-202002
Email: Zehra.khan08@gmail.com

Abstract

Efficacy of some botanicals such as Calotropis procera, Calotropis gigantia, Thevatia peruviana, Euphorbia neriifolia, Argemone mexicana, Solanum xanthocarpum, Eichhornia echinulata, Nerium indicum and Pedilanthus tithymoides were evaluated against root-knot nematode, Meloidogyne incognita under field conditions. Significant reduction was observed in the number of root galls caused by Meloidogyne incognita in those plants treated with these botanicals. A much improvement was observed in the growth parameters like plant weight, per cent pollen fertility, chlorophyll content and ascorbic acid content.

Introduction

Tomato (*Lycopersicon esculentum* L.) is one of the important vegetable crops grown and processed in almost all the countries in the world. They have high degree of lycopene and ascorbic acid content [1]. There are several factors responsible for low production and diseases could be one of them. Plant-parasitic nematodes become a potential threat to the production of vegetables including tomato. Yield loss of tomato due to root-knot nematodes ranges from 39.71–46.0 % in India. It has been estimated for decades that chemical nematicides are the main effective means to control the nematodes. This indiscriminate use to manage these phytopathogens causes environmental degradation and leaves harmful effects on human and cattle population and subsequently shows direct toxicity to predators, pollinators, beneficial organism etc. There is now tremendous pressure on farmers to use eco-friendly methods of disease control which do not only to pollute or degrade these botanicals as an organic matters seems to be the most effective method to manage the root-knot nematode *M. incognita*. The main aim is to investigate the potential of these botanicals against *M. incognita* in field condition.

Materials and Methods

A field experiment was conducted at the University Agriculture Research farm of Aligarh Muslim University during the last two winter seasons of 2008–10 to investigate the efficacy of some botanicals such as *C. procera*, *C. gigantia*, *T. peruviana*, *E. neriifolia*, *Argemone mexicana*, *Solanum xanthocarpum*, *N. indicum* and *P. tithymoides* against root-knot nematode *M. incognita* and on the growth parameters of tomato (*lycopersicon esculentum* L.) Cv. "Pusa ruby". The field was protected with barbed wire and thoroughly ploughed. The small beds of 6 m^2 were prepared leaving 0.5 m buffer zone between them. The beds were treated separately with these botanicals at 110 kg N/ha. Untreated beds served as control. The treatments were randomized with five replication and after 10 days, seedling of tomato were transplanted in each beds. During the four months growing period, watering and weeding were done whenever required.

Plant growth parameters like plant weight, per cent pollen fertility, chlorophyll content and ascorbic acid content were measured at time of termination of experiment. Pollen fertility (percentage) was estimated by the method of [2] using stain ability of pollen grains in 1 % acetocarmine solution. Chlorophyll content of leaf was determined by the method of [3]. The

M.M. Srivastava, L.D. Khemani, S. Srivastava, *Chemistry of Phytopotentials: Health, Energy and Environmental Perspectives*, DOI:10.1007/978-3-642-23394-4_36, © Springer-Verlag Berlin Heidelberg 2012

Table 1: Effect of some botanicals on different growth parameters of tomato (*Lycopersicon esculentum*) Cv. Pusa Ruby and root-knot development caused by *Meloidogyne incognita*

Treatments	Plant shoot weight (g)	Plant root weight (g)	Total weight	Pollen fertility (%)	Chlorophyll content (mg/g)	Ascorbic acid content (mg/100 g)	No of root galls/plant
Untreated	50.33	20.56	70.89	45.60	2.023	17.8	327.5
Calotropis procera	130.49	46.63	177.12	96.72	4.145	35.3	60.2
Thevatia peruviana	124.72	44.28	169.00	78.00	3.636	32.7	82.3
Euphorbia neriifolia	122.58	43.30	165.88	76.50	3.569	32.2	87.9
Argemone mexicana	112.29	4.064	152.93	66.87	3.407	30.5	103.3
Solanum xanthocarpum	120.94	43.59	164.53	82.40	3.560	32.0	92.2
Eichhornia echinulata	117.80	42.00	159.80	73.16	3.473	30.9	73.5
Calotropis gigantia	127.77	45.84	173.61	95.60	3.934	33.7	66.2
Pedilanthus tithymaloides	126.19	44.73	170.90	94.75	3.738	33.0	69.0
Nerium indicum	105.50	37.45	142.95	66.52	3.301	26.5	93.6
C.D (P=0.05)			7.13	5.59	0.144	2.80	8.53

ascorbic acid content of tomato fruit tissues was determined by the method of based on the reduction of 2,6-diculorophenol endophenol dye [4]. The mean value of two years have been pooled and analyzed statistically. The number of root-galls per plant was counted.

Results and Discussion

The data presented in table1 clearly revealed that soil application of some botanicals as organics matter significantly improved the plant-growth, percent pollen fertility, chlorophyll content and ascorbic acid content of tomato in all the treatments as compared to untreated control. The highest improvement was noted in growth in the beds treated with *Calotropis procera* and the least in the plants grown in the beds treated with *N. indicum.* The root-knot development was also found reduced in all the treated beds but the highest being in treated with *C. procera* and the least in *N. indicum.* The application of organic matter to soil is known to have beneficial effects on soil nutrients, soil physical condition, soil biological activities and improvement of crop. The improvement of tomato achieved by the application of botanicals was attributed directly to increase in the nutrient status of the soil. This helps the plat to tolerate nematode attack [7]. The decomposed organic matter leading to the control of *M. incognita* may be due to the nematostatic substances [6] present in the botanicals.

Various research workers reported that organically produced vegetables are considered valuable and it is also believed that post-harvested losses of such vegetables are usually lower as compared to conventionally growth vegetables [5]. Such type of produced vegetables is free from only any health hazard and also safe for environment.

References

1. C. Kaur and H.C. Kapoor; Antioxidant activity in tomato Science publisher U.S.A (2008).
2. G.T. Brown, pollen slide studies. C.C. Thomas springerfield, Illinois, U.S.A. (1949).
3. J.D. Hiscox and G.F. Israelstam; Can. J. Bot. 57 (1979) 1332–1334.
4. J.H. Roe; Method of biochemical analysis 1 (1954) 139.
5. K.P. Prasanna and S. Rajan; S. Indian Hortic. (2001) 49.
6. M.W. Khan, A.M. Khan and S.K. Saxena; Acta Bot. Ind. 120–2 (1974) 128.
7. R. Rodriguez-Kabana, G. Morgan-Jones and I. Clift; Plant Soil 100 (1987) 237–247.

37
Statistical Media Optimization for Enhanced Biomass and Artemisinin Production in *Artemisia Annua* Hairy Roots

N. Patra, S. Sharma and A. K. Srivastava
Department of Biochemical Engineering & Biotechnology,
Indian Institute of Technology, Hauz Khas, New Delhi
Email: nivedita.patra@dbeb.iitd.ac.in

Abstract

Artemisinin is an effective anti-malarial drug which is widely used to cure multi-drug resistant malaria caused by Plasmodium falciparum. Production via hairy root cultivation of plant cell lines features an attractive alternative for in-vitro mass production of artemisinin. Identification of appropriate medium recipe for growth and product formation is extremely important for increased biomass and/or Artemisinin production. Statistical media design was used to establish optimal concentration of the media components in the cultivation medium. Using optimized media it was possible to enhance the biomass concentration in the shake flask hairy root cultivation to 5.70 g/L and Artemisinin content to 1.9 mg/g in 15d. These values are significantly greater than the corresponding values from the un-optimized medium.

Introduction

Artemisinin is a pharmaceutical compound of immense importance. It is commonly used to treat multi-drug resistant Malaria (Cerebral malaria) caused by *Plasmodium falciparum*. Invariably the supply of this drug is much lower than its demand as it is obtained from the annual herb *Artemisia annua* primarily because field grown plants are severely affected by seasonal and commercial limitations. Chemical synthesis of Artemisinin is tedium and yields lower concentrations and are thereby uneconomical. Development of *in-vitro* production technologies has been attempted by different researchers. One alternate production protocol could be its production via hairy root culture system [1]. Hairy roots are obtained by genetic transformation of *Artemisia annua* through *Agrobacterium rhizogenes* infection and cultivation of infected plant part in the defined medium after its appropriate co-cultivation in bacteria. The hairy root cell lines generated has been reported to contain Artemisinin content up-to 0.4 % mg/g DW [2]. The success of this newer technology is highly dependent on identification mass scale propagation media and cultivation conditions of hairy roots in shake flasks [2, 3] and different bioreactor configurations.

Media composition has major effect on the production of Artemisinin from hairy roots. A basal medium containing major salts, minor salts and vitamins is generally suitable for growth and maintenance of hairy roots. MS (and its dilutions) and B5 are the commonly used culture media for hairy root cultivations. The type of carbon source present in the media significantly affects secondary metabolite production. Sucrose is reported to be better source for growth of biomass in *Artemisia annua* hairy roots as compared to glucose or fructose [4]. Nitrate and ammonium ions supply the nitrogen requirements of the growing hairy roots. It has been observed that the ratio of Ammonium to Nitrate features better buffering capacity and therefore leads to enhanced growth and product formation. Low ammonium to nitrate ratio in the culture medium favors growth in hairy roots. Supplementation of KNO_3 at a later stage is also known to promote biomass growth in hairy roots. Four factor analysis of growth and terpenoid production in *Artemisia annua* hairy roots established definitive conditions for optimizing growth but not terpenoid production [5]. The method of media sterilization like autoclaving or filter sterilization has been reported to affect the variability of Artemisinin content. It has been reported that about 9–25 % of sucrose is hydrolyzed in autoclaved media

M.M. Srivastava, L.D. Khemani, S. Srivastava, *Chemistry of Phytopotentials: Health, Energy and Environmental Perspectives*, DOI:10.1007/978-3-642-23394-4_37,

Table 1: Actual concentration range of components in Plackett Burman experiments

Component	High conc.	MS (Basal medium)	Low conc.
KNO_3	30 mM	19mM	5.1 mM
NH_4NO_3	5.15 mM	20.6mM	5.15 mM
NO_3/NH_3 Ratio	5.82	-	1.0
KH_2PO_4	1.24 mM	1.24 mM	0.31 mM
GA_3	0.01 mg/l	-	0.005 mg/l
Sucrose	50 gm/l	30 gm/l	30 gm/l

prior to inoculation [4]. Modified MS medium has been used for optimal Artemisinin production in which major components like sucrose concentration, nitrate, phosphate and pH has been optimized by studying one factor at a time [3]. Similarly, modified B5 medium containing half strength calcium chloride (dehydrate) and 30 g/L sucrose was used for maintenance and subculture of hairy roots of *Artemisia annua* [6,7]. In the present study, the cumulative effect of different media components sucrose, nitrate/ammonia, phosphate and the blooming agent GA3 in the design of medium recipe was investigated and established to improve the biomass growth and product formation.

Methodology

The medium recipe consisted of Basal MS/4 medium and addition of major effectors like NO_3/NH_4 ratio, KH_2PO_4, GA_3, Sucrose etc was tested for its effect on biomass and/or product formation. The range of the concentrations of different effectors were derived from the literature reported studies for *A. annua* /or similar plant culture system.

NO_3/NH_3 molar ratios were tested in the range of: 6:1 (High) and 1:1 (Low). Low concentration of Ammonium nitrate, Potassium phosphate was same as that of MS/4. Sucrose concentration was kept as 50 g/L as the upper (higher) value and 30 g/L was selected as the lower sucrose concentration in the MS/4 medium.

Statistical optimization software Stat Ease ver 5.09 (Stat Ease Corporation USA) was used to design eight experiments. These experiments were then experimentally conducted and their responses with respect to biomass and Artemisnin were analyzed as shown in Table2. Detailed analysis of these responses was done to calculate "t" values as indicated in Table 3.

After the identification of the effective nutrients from the Plackett-Burman protocol, Response Surface Analysis was used to determine the optimum concentrations of the selected nutrients. A 2^4 full factorial Central Composite Design (CCD) for four independent variables each was employed to fit a second order polynomial model.

Experimental Results

Table 3: The value of regression coefficient (t-value) from Plackett Burman experiments

S. No.	Factor	t-value for Biomass	t-value for Artemisinin
1.	NO_3/NH_4 ratio (N)	1.89[1]	1.35[1]
2.	KH_2PO_4 (P)	1.53[2]	0.44[2]
3.	Sucrose (S)	-0.33	0.046[3]
4.	GA_3 (GA)	-1.05	-0.54[4]

Table 2: Result of Plackett Burman experiment

	Factor NO_3/NH_4 Ratio	Factor Phosphate mM	Factor Sucrose g/l	Factor GA3 µg/L	Response Biomass gm/L	Response Artemisinin content mg/g
Experiment 1	1.00	1.50	30.00	10.00	2.68	0.384
Experiment 2	6.00	0.31	50.00	5.00	3.18	1.39
Experiment 3	6.00	0.31	30.00	10.00	2.72	0.525
Experiment 4	6.00	1.50	30.00	5.00	3.79	3.13
Experiment 5	1.00	0.31	30.00	5.00	1.71	1.61
Experiment 6	6.00	1.50	50.00	10.00	2.6	2.48
Experiment 7	1.00	1.50	50.00	5.00	2.98	0.39
Experiment 8	1.00	0.31	50.00	10.00	1.37	1.53

Results of Central Composite Design

Model Equation:

$$\text{Biomass} = +5.25 + 0.24 * A - 0.79 * B - 0.19 * C + 0.26 * D - 0.42 * A^2 - 0.57 * B^2 - 0.56 * C^2 + 0.012 * D^2 + 0.13 * A * B - 0.19 * A * C - 0.16 * A * D - 0.092 * B * C + 0.15 * B * D - 0.17 * C * D \quad \text{Eqn (1)}$$

Model Equation:

$$\text{Artemisinin} = -0.67 + 0.42 * A - 0.99 * B + 0.18 * C + 0.53 * D + 1.06 * A2 + 1.05 * B2 + 0.93 * C2 + 0.91 * D2 + 0.19 * A * B + 1.04 * A * C + 0.78 * A * D + 0.21 * B * C - 0.67 * B * D + 0.89 * C * D \quad \text{Eqn (2)}$$

Where A= Nitrate/Ammonia, B=Phosphate, C= Sucrose, D= GA_3 in Equation 1 and 2.

Discussion

Plackett Burman experiments showed that nitrate and Potassium phosphate were the major factors affecting biomass and Artemisinin production in hairy roots as indicated by their high t-value for both biomass and Artemisinin (Table 3). Sucrose and GA3 were the least affecting factors for obtaining maximum biomass and/or Artemisinin. The negative t-value of

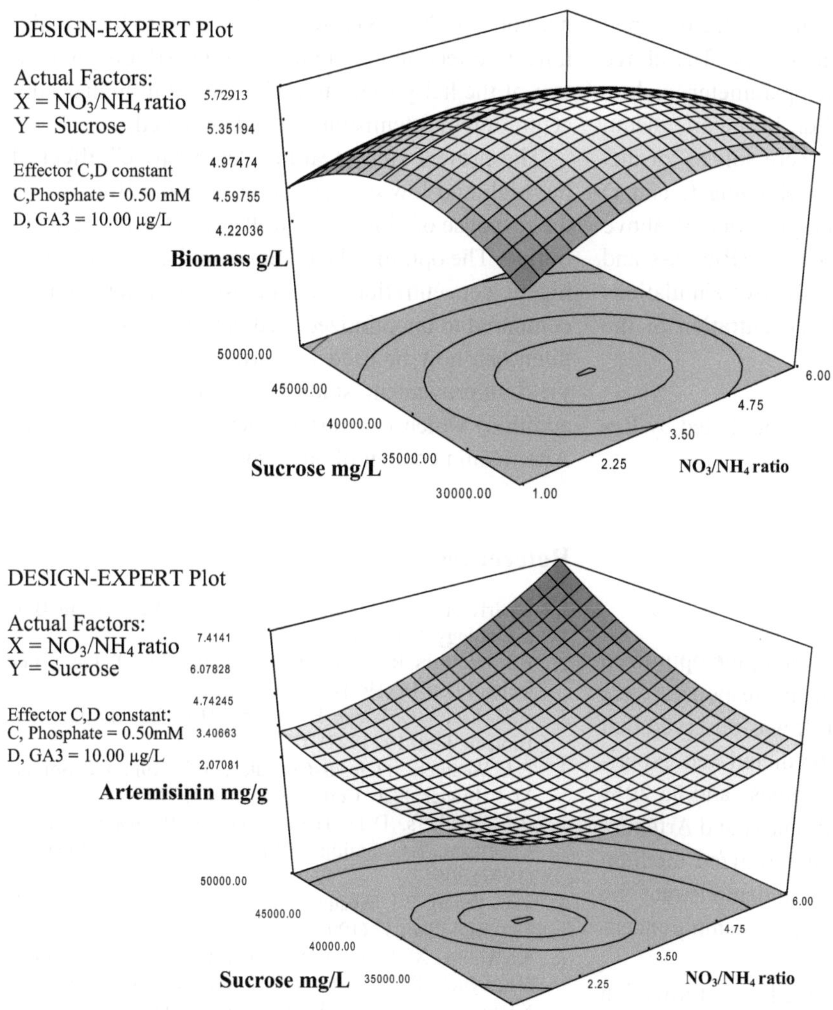

Figure 1: 3-D Contour plots to show the combined effect of Sucrose and Nitrate source on Artemisinin accumulation in the hairy roots of *Artemisia annua*.

sucrose shows that it sparsely affects biomass production. However it had positive t-value for Artemisinin and therefore a major factor for further optimization. Similarly high negative t-value for GA_3 for Biomass and Artemisnin indicated it is least affecting factors for the two responses. However based on literature reports, it has been reported that GA_3 has significant effect on artemisinin production in both shoot and hairy root culture [6]. Therefore in-spite of low "t" value by Plackett Burman experiments (Table 3) it was decided to include GA_3 in the list of major effectors for the study of responses. The final list of effectors had interaction between Nitrate to Ammonium ratio, Phosphate, Sucrose and GA_3 The interaction between low and high value of concentrations were studied in detail using the designed experimental recipe of RSM. Thirty experiments were designed using above four effectors and the results of responses were obtained in terms of biomass Artemisinin content. The above experiments formed the basis for parameter evaluation of the models for biomass (Eqn 1) and artemisnin (Eqn 2). These model equations were solved for different values of effectors A & B Keeping C and D constant. The contour plots emerging out of above analysis are described in the Figure 1 (Biomass and Artemisnin). From the analysis of above simulations and contour plots, the optimal concentration of the medium emerged as shown below:-

Optimum concentration of nutrients predicted by Response Surface Methodology:
NO_3/NH_3 ratio: 3.5
KH_2PO_4: 0.5 mM
Sucrose: 37.134 g/l
GA_3: 10 µg/l in MS/4 medium.

Shake flask cultivation was done using the optimized medium and the actual experimental responses were determined as shown in Table 4. It was observed that the correlation between model predicted and experimental values was 99.65% for biomass and 86.29% for Artemisinin. The values of Biomass and Artemisinin content in optimized and unoptimized medium are given in Table 4 which clearly demonstrate the improvement in the biomass and artemisnin concentrations.

In literature, conventional media optimization studies to optimize the effect of above four factors have been reported but by using one factor at a time [3] which does not ensure the complex intriguing interactions between the effectors. Similar interactive studies were also reported to study the combined effect of Sucrose, phosphate, nitrate and inoculum age [5] It is important to note here that above studies did not consider NO_3/NH_3 which is known to have a significant effect on the biomass and/or Artemisnin content of the hairy roots and with the result higher concentration of artemisnin was not achieved.

Table 4: The experimental and model predicted values of biomass and Artemisinin obtained in optimized medium given by RSM.

	Experimental value		Model predicted
	Unoptimized media	**Optimized media**	
Biomass	3.92 g/L	**5.70 g/L**	5.72 g/L
Artemisinin	0.829 mg/g	**1.94 mg/g**	2.248 mg/g

The present study features "cumulative" effect of the high and low concentrations of four effectors on the response of biomass as well as artemisnin concentration. The optimized medium exhibited significantly higher concentrations of biomass and artemisnin as compared to unoptimized medium (Table 4). The medium can now be used for further concentration and yield improvement studies (e.g. elicitor, precursor addition) which can further increase biomass and/or Artemisinin content of hairy roots.

References

1. S. Srivastava, A.K. Srivastava; Critical Reviews in Biotechnology 27 (2007) 1
2. P.J. Weathers, R.D. Cheetham, E. Follansbee, T. Teoh; Biotechnol Lett 16 (1994) 1281
3. C.Z. Liu, Y.C. Wang, F. Ouyang, H.C. Ye, and G.F. Li; Biotech. Lett. 19 (1997) 927
4. P.J. Weathers, DeJesus-Gonzalez, Y.J. Kim, F.F Souret, M.J. Towler; Plant Cell Rep. 23 (2004) 414
5. P.J. Weathers, D.D. Hemmavanh, D.B. Walcerz, R.D. Cheetham, T.C. Smith; In Vitro Cell Dev Biol Plant 33 (1997) 306
6. T.C. Smith, P.J. Weathers, R.D. Cheetham; In Vitro Cell Dev Biol Plant 33 (1997) 75
7. K. Wobbe, X. Zhang, P. Weathers, Radical biology:Flores HE, Lynch JP, Eissenstat D, Eds. Rockville, MD: American Society of Plant Physiologists. (1998) p 432.
8. G. Prakash and A.K. Srivastava; Proc Biochem 40 (2005) 3795.

38

Formation and Characterization of Hydroxyapatite/Chitosan Composite: Effect of Composite Hydroxyapatite Coating and its Application on Biomedical Materials

S. Mulijani[1] and G. Sulistyso[2]
[1]Department of Chemistry, Faculty of Mathematics and Natural Sciences, Bogor Agricultural University, Bogor Indonesia.
[2]National Atomic Energy Agency (BATAN), Serpong Indonesia.
Email: smulijani@hotmail.com

Abstract

Natural bone is actually an inorganic/organic composite mainly made up of nano-structure hydroxyapatite ($Ca_{10}(PO_4)_6(OH)_2$, HAp) and collagen fibers. It is of most importance to synthesize nano-composites of inorganic/organic in order to have good biocompatibility, high bioactivity and great bonding properties. In this study, HAp nano-particle and HAp/chitosan (CTS) nanocomposite with a homogeneous microstructure were prepared and characterized. It is proposed that the nano-structure of hydroxyapatite/chitosan composite will have the best biomedical properties in the biomaterials applications. The mechanism of formation of the composite and the effects of inorganic component (n-HA) on the porous morphologies were investigated by using of FTIR, XRD, and SEM. Cathodic electrophoretic deposition has been utilized for the fabrication of composite hydroxyapatite-chitosan coatings on 316L stainless steel substrates. The addition of chitosan to the hydroxyapatite suspensions promoted the electrophoretic deposition of the hydroxyapatite nanoparticles and resulted in the formation of composite coatings. The obtained coatings provided the corrosion protection for the 316L stainless steel substrates.

Introduction

Recently, hydroxyapatite (Hap) have attracted the attention as they provide specific advantages over conventional component of bone. It has been used extensively for biomedical implant applications and bone regeneration due to its bioactive, biodegradable and osteoconductive properties [1–4]. Synthetic HAp is a biocompatible prosthetic material, bonding strongly to the bone and promoting the formation of bone tissue on its surface. In addition, HAp can accelerate the formation of bonelike apatite on the surface of the implant [5].

Owing to its brittleness, the instability of the HAp particulate is often encountered when the particles are mixed with saline or patient's blood, making it unable to be used as bone regenerating template. To solve the problem, incorporation of HAp into polymer matrix has been carried out to increase osteoconductivity and biodegradability with significant enhancement of mechanical strength. Since the natural bone is a composite mainly consisted of nano-sized, needle-like HAp crystals and collagen fibers, many efforts have been made to modify HAp by polymers. Some researchers modified HAp by polylactic acid [6,7]. Others tried to improve performance by employing different kind of polymers such as collagen [8], chitosan (Ch) [9], polyethylene [10,11].

There are many methods to synthesize HAp/Ch composite. Chen *et al.* was synthesize HAp/chitosan in aqueous solutions, Jingxiao *et al.* prepared HAp/Ch composite in the simulated body fluids (SBF) containing appropriate amount of chitosan *via in situ* co-precipitation method [12] and Muzzarelli *et al.* deposited of chitosan on plasma-sprayed HAp coatings on Ti–6Al–4V alloys. In another study, composite coatings were obtained by electrolytic deposition of brushite ($CaHPO4·2H2O$) and electrochemical depo-

M.M. Srivastava, L.D. Khemani, S. Srivastava, *Chemistry of Phytopotentials: Health, Energy and Environmental Perspectives*, DOI:10.1007/978–3-642–23394-4_38, © Springer-Verlag Berlin Heidelberg 2012

sition of chitosan. The brushite/chitosan composites were converted to HAp/Ch composites by heating the coatings in aqueous solutions of sodium hydroxide.

The present study however attempts to enhance the integration of implants and to promote the regeneration of bone, chitosan also has been utilized in combination with HAp to meet these requirements. We studied the possibility of deposition of composite coatings using a combined method, based on the electrophoretic deposition (EPD) of HA and sol gel of chitosan. Chitosan is one of the most widely-used natural polymers in tissue engineering research that can be obtained by partially deacetylation of chitin which can be extracted from shells of crustacean (prawns, crab, squids, etc).

Experimental

Synthesis of HAp

Raw eggshells were calcinated at 950◦C. After a short time (30 min), the color of eggshells turned to black, and after 6 h they became white. The color change suggested that most of the organic materials were burnt out. As obtained powders were crushed in an agate mortar and milled with P_2O_5 powder for 6 h. The composition of P_2O_5 and CaO powder was 46,43 % and 50,52 %. Sintering of samples was performed in air at 900 or 1000 ◦C for 2 h.[14]. For phase identification, X-ray diffraction (XRD) analysis was carried out using Cu Kα radiation (k = 1.5406 A). Fourier transform infrared analysis (FT-IR, Vortex 70, Bruker) of starting HAp powder was carried out in the range of 400– 4000 cm_1. Detailed SEM analysis was performed using a JEOL 2000 FXII microscope operated at 200 kV.

Sample Preparation

Type 316L SS and 316 SS were used as the metal substrate. Type 316L SS alloy in as received mill annealed condition was cut into coin formed (r= 1.5 cm) pieces. Prior to the study, the samples were polished using silicon carbide papers of 120, 220, 320, 400 and 600 grit. Final polishing was done using coarse (1 μm) diamond pastes in order to produce scratch-free mirror-finish surface. The polished specimens were submerged into H2SO4 : HCl : H2O (1:1:1) for 1 h at 60^{0}C and thoroughly submerged with NaOH 10M for 24 h and finally the sample was rinsed in deionized water, dried and used for further studies.

Electrophoretic Deposition

Electrodeposition (EPD) was performed from chitosan solutions (3 % in acetic acid) containing HAp nanoparticles in a mixed ethanol–water solvent (17 vol.% water). The composition of chitosan solution such as 0; 0,1; 0,5; 1,0; dan 1,5 ml Before the electrodeposition, the solutions were ultrasonicated for 1 h to achieve a homogeneous dispersion of the HA nanoparticles.. The electrochemical cell for the deposition included a cathodic substrate (metal of 316L SS and 316 SS) and a anodic was carbon electrode. The electrodeposition was performed at a constant current density of 0.1 mA/cm^2. Cathodic deposits were obtained 316L stainless steel and 316 stainless steel The deposition conditions were done at 120V for 2 minutes. The coated specimen was then dried in air oven at 90 °C for 1h, cooled and corrosion studies were made. X Ray Diffraction was performed for investigate the deposition of HAp/Chitosan.

Results and Discussion

X-Ray Diffraction Studies

A typical XRD profile of HAp powder synthesized by calcinations' method has been shown in Fig. 1. The XRD phase analysis has been performed referred to Standard Reference Material 2910. The samples heated at 950 °C show marked fractions of crystalline phases, which increases with heating temperature.

The HAp powder before calcinated consist $CaCO_3$, $Ca_3(PO_4)_2$ and $MgCO_3$. After calcinations process at 950 ◦C for 6 h the main lines of $Ca(OH)_2$ appeared, as can be observed in Fig. 1a. Presumably, the CaO phase was the calcination product, but as the powder sample was in contact with ambient atmosphere after firing the product turned to Ca(OH)2 (as in Fig. 1). Some small diffraction lines from CaO (JCPDS-PDF 82–1691) and MgO, JCPDS-PDF 78–0430 can be also observed. After a successive and intensive milling process, and by treating the powder with P_2O_5, next to the main lines of $Ca(OH)_2$, new phases appeared ($CaHPO_4$, JCPDS-PDF 01–0653) that are characteristic can be seen in Fig. 2. The difractogram of HAp synthetic compared with HAp commercial showed that HAp synthetic was exhibited strong crystalline phase.

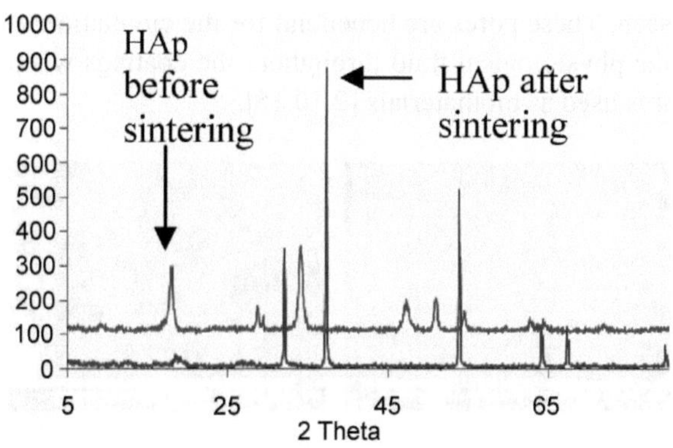

Figure 1: X-ray diffraction patterns of HAp powder

A common phenomenon reported during sintering of HAp based composites is the decomposition of HAp due to dehydroxylation (removal of OH^- ion). Such decomposition results in a combination of HAp and dissociated tricalcium phosphate (TCP) phase, which is termed biphasic calcium phosphate (BCP). In recent years, BCP-based ceramics have received attention as an ideal bone substitute due to their controlled biodegradation. Literature reports indicate that it is possible to alter the HAp:TCP ratio to form BCP offering a combination of biodegradability and stability [16].

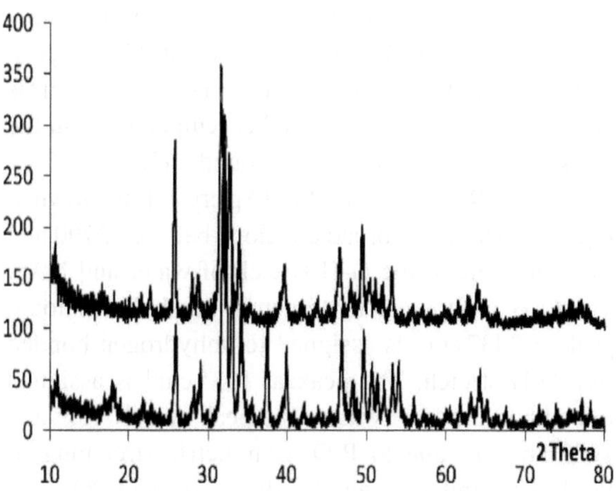

Figure 2: Difractogram of HAp synthetic and HAp Commercial

The morphological evolution of raw eggshell, calcinated at 950°C for 6 h (Fig. 2), followed by a mechanical activation. A lamellar nanostructure can be observed (consisting of 10–20 nm columns) in the case Ca(OH)2 as it was obtained after calcination (Fig. 2). After mechanical activation for 6 h, this microstructure have been transformed to 1–20 µm grains which are consisting of Ca(OH)2 and CaHPO4 nanograins as resulted from XRD measurements. A macroscopic view about calcium phosphate foams and testing bars realized by ball milling of eggshell and P_2O_5 at different mixing ratio and sintered at 950°C for 6 h is shown in Fig. 3. As can be seen after sintering sample was keeping the initial shape of a bar and transformed to hydroxyapatite $(Ca_{10}(PO4)_6(OH)_2)$, JCPDS-PDF 74–0565, Fig. 2).

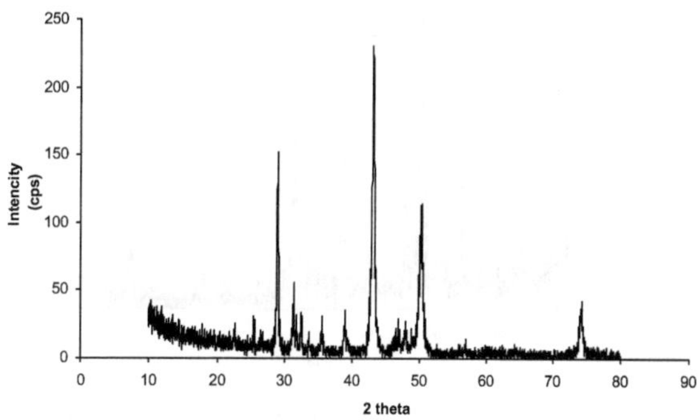

Figure 3: Difractogram of HAp/Chitosan composite

The XRD patterns of the HAp /chitosan composites are shown in Figure 3. The peak appeared approximately at 32° was assigned to chitosan. The sharp diffraction characteristic peaks appeared at around 42° is the nano-HAp/chitosan composites correspond to the peaks of HA powder. These results confirm the existence of HAp compound in the chitosan matrix [2].

However, XRD results of metal (Fig.4a-4c) indicate the crystallinity of the chitosan prepared by electrodeposition from the mixed ethanol–water solvent. Pang *et al.*[2] reported that thermogravimetry analysis showed thermal degradation of the chitosan deposit starts at 280–300°C. This result is in agreement with the literature data, according to which the thermal degradation temperature of chitosan is about 280–300°C. Chitosan is soluble in water only when protonated in acidic solutions.

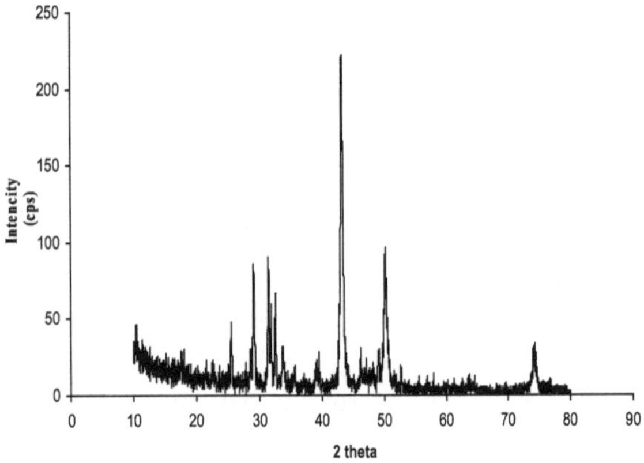

Figure 4a: Difractogram of 316L SS metal HAp/chitosan electrodepositon (0.2:0)

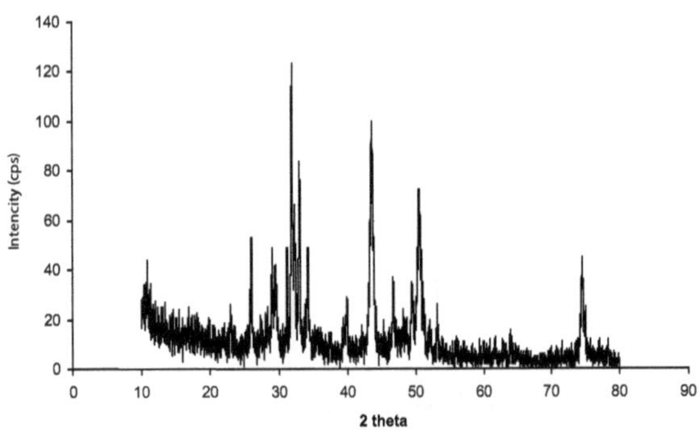

Figure 4b: Difractogram of 316L SS metal HAp/chitosan electrodepositon (0.2:0.1)

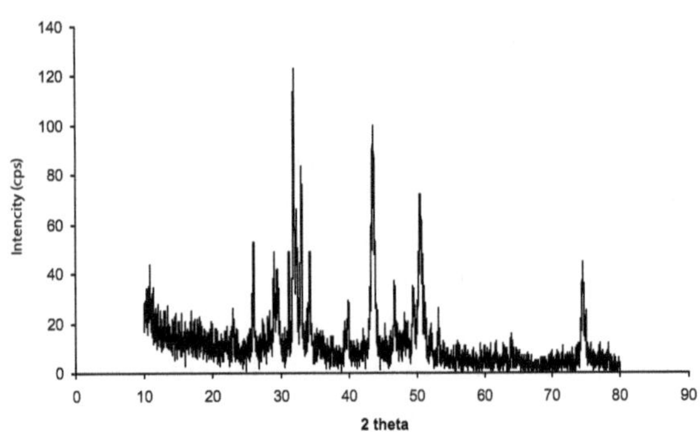

Figure 4c: Difractogram of 316L SS metal HAp/chitosan electrodepositon (0.2:0.5)

Scanning Electron Microscopy (SEM) Analysis

Figure 5a and b shows the SEM micrograph of HAp synthetic and HAp commercial respectively. There are many spherical agglomerations and few crystallites of 1–10 micrometers in size with pores between them were observed. At higher temperature (950°C), the material agglomerate in bigger size due to the growth of crystalline phases and the pores were also seen. These pores are beneficial for the circulation of the physiological fluid throughout the coatings when it is used as biomaterials [2,10,15].

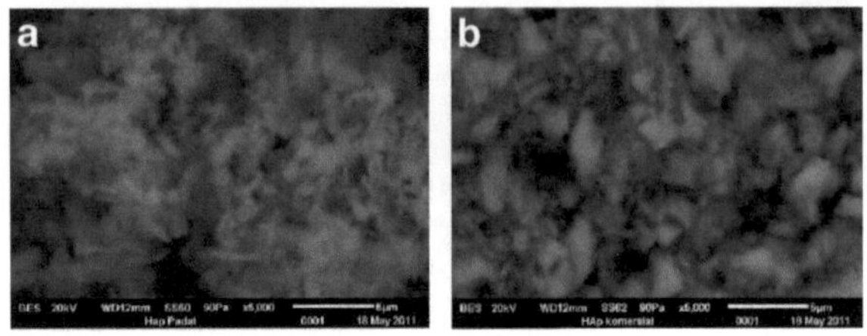

Figure 5: SEM micograph of HAp: (a) HAp synthetic and (b) HAp commercial

There are many spherical agglomerations and few crystallites of 1–10 micrometers in size with pores between them were observed. At higher temperature, the material agglomerate in bigger size due to the growth of crystalline phases and the pores were also seen. These pores are beneficial for the circulation of the physiological fluid throughout the coatings when it is used as biomaterials [16].

FT-IR Spectral Studies

Figure 6 shows FT-IR spectra of hydroxyapatite heated at 900 °C which suggest that HAp powder unheated are with carbonate substitution. Broad bands appearing at wave number values of around 875, 1420 and 1480 cm^{-1} are indicative of the carbonate ion substitution which might have come from the source (eggshell) and for the sample heated at 900 °C, peak at 875 $cm^{-1}$and 1480 cm^{-1} disappears. This confirms the elimination of CO_3^{2-} at higher temperature but the weak CO_3^{2-} band still appear around 1420 cm^{-1}.

The FT-IR spectrum of the HAp crystals is shown in Figure 6. There is a broad envelope between 3700 and 2700 cm^{-1} due to the O-H stretch of water and HAp. The O-H groups are hydrogen bonded. The smooth peak at 3437 cm^{-1} is assigned to unhydrogen bonded free O-H stretch. The peak at 1635 cm^{-1} is assigned to bending mode of water. The peak at 1110 cm^{-1} and 1029 cm^{-1} are due to P-O asymmetric stretching of PO_4^{3-}. The stretching and bending modes of PO_4^{3-} appeared at 602 and 560 cm^{-1} [12].

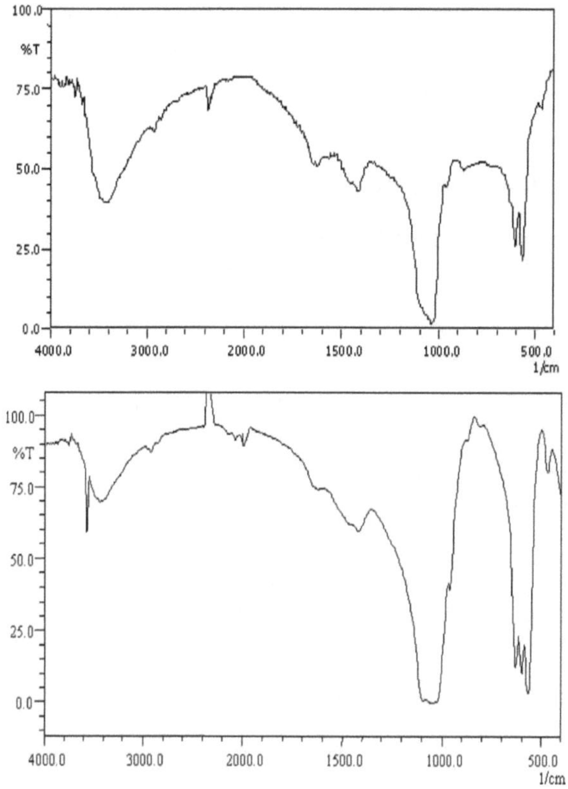

Figure 6: FT-IR spectra of HAp synthetic

Electrophoretic Deposition Analysis

According to Zianwei (2008) electrophoretic deposition is achieved by the motion of charged particles toward an electrode under an applied electric field. As such, coatings produced using EPD is governed by electrical parameters as well as the properties of the suspension. During EPD, the electrophoretic velocity (v) of the charged particles can be described by the following equation:

$$v = QE/4\pi r\eta \quad (1)$$

where Q, r, η, and E represent the charge, particle radius, viscosity of the suspension, and the potential difference applied to the suspension respectively.

In suspensions with low solid concentration, η is often considered constant. Under this condition, the electrophoretic velocity is mainly a function of the electric field and the particle size. When E is constant, the suspension used for EPD usually has a distribution of particle sizes, and particles with different Q/r ratios have different electrophoretic mobility, thereby resulting in segregation during the EPD process. In addition, in a suspension of particles with different radii, preferential deposition of fine particles is expected because of their increased mobility compared to that of larger particles.

Mobility of particles can also be improved by increasing the applied potential. This condition enables larger particles to be deposited besides finer particles. These theories help to explain the phenomenon observed in this study, substantiating the production of a porous and roughened coating at a higher electric field and the production of a dense coating of fine particles at a lower electric field.

The HAp powder used in this study was synthesized from eggshell and contained many agglomerated particles. It was difficult to control the particle size to a narrow size range, which meant that use of these powder particles would affect coating integrity. By applying dynamic voltage during EPD, it was observed that coatings with continuous gradient could be produced. The coating layer closest to the substrate was observed to be dense, whereas the outer coating layers porous. The production of dense coatings with initial voltage of 120 V suggested that only fine particles had enough speed to arrive at the cathodic 316L SS or 316 SS substrate.

It is important to note that electrophoresis results in the accumulation of charged particles at the electrode surface. The deposit formation is achieved via the particle coagulation, which is influenced by the electrode reactions, solvents, additives, and other factors [15]. In this work, EPD of HAp nanoparticles was achieved from pure HAp suspensions in a mixed ethanol–water solvent. The addition of HAp to the chitosan solutions

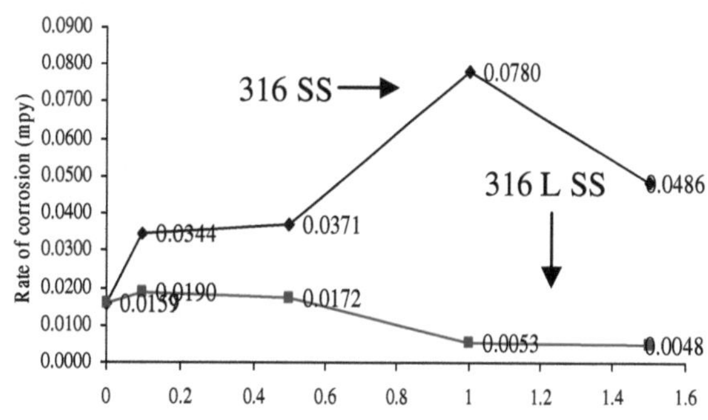

Figure 7: Corrosion rate curves of HAp/Chitosan coated 316L SS and 316 SS

enabled the co-deposition of HAp and chitosan. The results indicate that chitosan promotes the EPD of HAp. The formation of composite is achieved via the heterocoagulation of HAp nanoparticles and chitosan macromolecules at the electrode surface.

The corrosion rate performed to evaluate the potential of HAp/chitosan as a biomaterial coated metals. Figure 7 illustrated of corrosion rate of stainless steel, these decreased as increasing concentration of chitosan addition. Metal 316 SS exhibited more corrosives compare with 316L SS.

The experimental results presented above indicate the possibility of the fabrication of composite containing different amounts of chitosan and HAp. The composition can be varied by the variation of HAp concentration in the chitosan solution. The deposition yield can be changed by the variation of the deposition time at a constant current density.

Conclusion

A novel method has been developed for the fabrication of HAp/chitosan composite material. Method of synthesis of HAp powder is very simple and cost-effective, since it has been synthesized from the egg-shell, which is considered to be a waste material after egg's usage. Electrophoretic deposition of 316L SS was used to achieve uniform distribution of HAp/chitosan deposits. The electrochemical assessment indicates that the breakdown potential of acetic acid treated and HAp/chitosan coated 316L SS has been shifted towards the nobler direction compared with the 316 SS. Thus, the presence of HAp/chitosan coating over acetic acid treated 316L SS place a dual role in preventing the release of metal ions (rendering it more corrosion resistance) and in making the metal surface more bioactive.

Acknowledgements

The authors are grateful to National Atomic Energy Agency Indonesia for rendering financial support to carry out the work.

References

1. F.Chen, Z.C. Wang and C.J. Lin; J. Materials Letters 57 (2002) 858–861.
2. X. Pang and I. Zhitomirsky; J. Materials Chemistry and Physics 94 (2005) 245–251.
3. J.M. Gomez-Vega and E. Saiz.; Biomaterials 21(2000) 105.
4. R.H. Doremus; J. Mater. Sci. 27 (1992) 285.
5. A. Sabokbar, R. Pandey, J. Diaz, J.M.W Quinn and D.W. Murray; J. Mater Sci: Mater Med 12 (2001) 659–664.
6. T. Furukawa and Y. Matsusue; Biomaterials 21 (2000) 889.
7. T. Kasuga and Y. Ota; Biomaterials 22 (2001) 19.
8. M.C. Chang and T. Ikoma; J. Mater. Sci. Lett. 20 (2001) 1199.
9. S. Viala, M. Freche and J.L. Lacout; Ann. Chim. Sci. Mater. 23 (1998) 69.
10. M. Wang and W. Bonfield; Biomaterials 22 (2001) 1311.
11. M. Wang, S. Deb and W. Bonfield; Mater. Lett. 44 (2000) 119.
12. L. Jingxiao, Fei Shi, Ling Yu, Liting Niu and Shanshan Gao; J. Mater. Sci. Technol. 25 (2009) 4.
13. R.A.A. Muzzarelli, G. Biagini, A. DeBenedittis, P. Mengucci, G.Majni and G. Tosi, Carbohydr. Polym. 45 (2001) 35.
14. C. Balazsi, F. W'eber, K. Zsuzsanna, E.Horv'ath and C. Nemeth; Journal of the European Ceramic Society 27 (2007) 1601–1606.
15. M. Xianwei, K. Tae-Yub and K. Kyo-Han; Dent Mater J. 27 (5) (2008) 666–671.
16. S. Nath, K.. Biswas and B. Basu; Scripta Materialia 58 (2008) 1054–1057.

39
A Wonder Plant; Cactus Pear: Emerging Nutraceutical and Functional Food

R. C. Gupta,
Nagaland University
Medziphema- 797106, India.
Email:ramesh_gupta101@hotmail.com

Abstract

Since ages, cactus bears colorful flowers and fruits yet because of its thorny nature it was always looked with suspension. In recent time an increased interest in its health improving capacity has been registered. Its protective spectrum include; anticancer, antiviral, anti-inflammatory and anti-diabetic, neurological and cardiovascular actions. On functional food and nutraceutical front, it is good source of fiber, minerals rich juice and edible pulp. While to look it's "inside" actions, it is interesting to observe high concentration of taurine which has protective effect in organs dysfunctions and is involve in "host defense". It seems that taurine is major contributor to beneficial properties of cactus pear.

Introduction

Since centuries cactus is associated with a caution, "handle with care" because of this label it was not explore for a long and this has hampered to understand its biological and chemical nature resulting restricted utility. Infect cactus pear has been ignored by the scientific community till late 70's. It is only in 1980's when reports on its biological function start to accumulate with more linkage of its chemical constituents to biological function. Its *opuntia* spp has find prime place and have attracted attention as food, feed, and pharmaceutical plant. [1]

On food and feed front, its pad is used for vegetable, juice is rich in vitamins, minerals, sugar and it is utilized to make specialized jam, jelly, and icecream. It is believed that pad as vegetable taste like green beans and slimy like okra. It is also part of salads, omelets, and salsa. This plant is also loved by live stock and it is rich source of water for animals under dry conditions [2] Biological properties of any substance is the cumulate index of its physic-chemicals properties in return the molecules presents as chemical their arrangement, interaction, association, provides synergic effect, cactus pear contains a large number of phenolic compounds in form of flavonoids, which are known to have properties of scavenging of free radicals with anti-oxidant action resulting beneficial effects in number of diseases and disorders It is very interesting to note that cactus pear bears a significant amount of sulfur containing amino acid, taurine as free amino acid. Taurine chemically 2-amino ethane sulfonic acid is a conditional amino acid. Taurine beneficial action has a board spectrum, and many of its protective action are now well established. It is surprising to know that, cactus pear has almost all such beneficial properties as taurine has, hence it is logical to conclude that it might possible 'that in ;: inside story of action mechanism of cactus pear beneficial activities; taurine may be a major contributing agent of cactus pear biological action profile [3–4].

M.M. Srivastava, L.D. Khemani, S. Srivastava, *Chemistry of Phytopotentials: Health, Energy and Environmental Perspectives*, DOI:10.1007/978-3-642-23394-4_39, © Springer-Verlag Berlin Heidelberg 2012

Cactus Constituents and Chemistry

Cladodes (stem) and fruits contain carbohydrate, fiber, lipids, protein with high calcium, A mean chemical composition of de-spine *opunita* spp; cladodes is as below;

Water	Dry matter basis g /100 g	Fresh weight basis g /100 g
water		88.95
carbohydrate	46–71	3–7
Ash	19–23	1–2
Fiber	18	1–2
protein	4–10	5–1
lipids	1–4	2

Main Technological Parameters: Chemical and Mineral Composition of Cactus-pear Pulp.

Technological Parameters	Range
Pulp (%)	43–57
Seeds (%)	2–10
Peel (%)	33–55
pH	5.3–7.1
Acidity (% of citric acid)	0.05–0.18
Brix	12–17
Total solids	10–16.20
Chemical Composition of the pulp	**Range**
Moisture (%)	84–90
Protein (%)	0.2–1.6
Fat (%)	0.09–0.7
Fiber (%)	0.02–3.1
Ash (%)	0.3–1
Total sugars (%)	10–17
Vitamin C (mg 100 g)	1–41
Minerals	**Range**
Ca(mg 100 g)	12.8–59
Mg(mg 100 g)	16.1–98.4
Fe(mg 100 g)	0.4–1.5
Na (mg 100 g)	0.6–1.1
K (mg 100 g)	90–217
P as Po_4 (mg 100 g)	15–32.8

Amino acids	Maximum content
Proline	1768.7
Glutamine	574.6
Taurine	572.1
Serine	217.5
Alanine	96.6
Glutamic acid	83.0
Methionine	76.9
Lysine	53.3

Cactus Pear is Fast Becoming Active Constituent of Functional Foods and Nutraceuticals

Functional food: Cactus pear is now part of several functional food sold in the market. Fruit juice of *opuntia* spp is used to make alcoholic beverage as well as bear. Being low caloric fruit higher in vitamin C, calcium and potassium with little sodium and no cholesterol or saturated fat make it fit for part of several health tonics. Prickly pear is used to make jelly, candy, ice cream and margaritas

Nutraceuticals: are health promoting compounds or products that have been isolated or purified from food source and they are generally sold in a medicinal (usually pill) form.

Cosmaceuticals: Juice from cladodes is now part of shampooing conditioners, lotions, soaps and sun protectors, cladodes juice is also known to improve hair growth. Pectin extracted from fruits contains enough alacturonic acid to use as cosmetic additive. Having high water holding capacity they serve as thickening or emulsifying agent and from viscous or gelatinous colloids.

Fuel production: It is sure that fossil fuel is not forever, hence search for alternative combustibles are on and in one such alternative cladodes when subjected to fermentation reasonable yields were achieved after acid and enzymatic hydrolysis of cladodes releasing mono and disaccharides for further fermentation to produce ethanol.

Cactus as animal feed: Cactus pear used for animal feeding are abundant, easy and cheap to grow 'palatable and can with stand prolonged droughts such characteristic makes it a potential important feed supplement for livestock, particularly during periods of drought and seasons of low feed availability.

Cactus as Content of Medicine/ Drug/ Pharmaceuticals

Since centuries cactus spp are used to serve as therapeutic agents. In folk medicine it has been used for treatment of gastritis, fatigue and liver injuries. Cactus pear was known as "village pharmacy", now with discovery of its new uses, cactus plant is regarded as "plant for cure". It is believed that presence of poly phenol components and some known vitamins are the core of its therapeutic potential. It seems that cactus pear exhibits beneficial action through its antioxidant activities. The flavonoids present in cactus pear are well recognized agent for anti-oxidation. However it is interesting to note that cactus pear contains significantly very large amount of free amino acid, It is observed that highly reactive molecules such as free radical scavengers may be damaged during processing of food such as juice preparation but on analysis, it is found that concentration of cystine and vitamin C is five and ten fold lower, where as beta carotene and GSH were totally lost mainly due to thermal degradation. On the contrast, vitamin E is less susceptible but taurine appeared to be preserved, betalains were not lost during processing, but polyphenolics decreases after 6 days. All these observation suggest that thermally taurine is most stable molecule this supports the view that taurine is major contributor in beneficial action of cactus pear. [5]. Taurine is known for its beneficial action is verities of diseases, dysfunctions of organs and its now proved beyond that taurine such action are mediated through its antioxidant nature, management of Ca^{+2} ions, stabilization of cell membrane, osmo-regulation. Hence it is logical to think beneficial action of cactus pear to presence of taurine in such high concentration.

Antioxidant Action

Beside taurine several substituted flavonoids are responsible for its antioxidant action; efficient radical scavenger activity against neuronal cell damage cause by H_2O_2 and xanthin/ Xanthin oxides has been recorded. Cactus pear fruit positively affect the body redox balance, decreases the oxidative damage to lipid and improve the antioxidant status in healthy humans.

Antiproliferative Effect

Betanin isolated from plant has shown anticancer effects on human chronic myeloid leukemia cell line K 562 and betanin induced cell death was also elucidated. It also suppresses tumor related genes. Hence cactus pear is a natural cancer chemo- preventive agent. It inhibits the proliferation of cervical ovarian and bladder cancer.

Antialcohol

Prickly pear is often used to relieve the symptoms associated with excess intake of alcohol including dry mouth and nausea. Prickly pear juice is also part of several hangover prevention formulas. It is reported that extract of prickly pear when used 2 hours before consuming alcohol renders the uses hangover free up to three days. It is also named, "amazing hangover cure prevent your hangover before it happens".

Protector from Liver Injury

In animal models carbon tetra chloride induces liver injury was protected by use of prickly pear juice, hence constituents of juice has protective and curative effect against the hepatic damage. This could be due to the presence of vitamin C, betalains and taurine.

Antidiabetic Effect

The hypoglycemic activity of purified extract from prickly pear cactus was evaluated on STZ induced diabetic rats. Blood glucose and glycated hemoglobin levels were recorded to normal value by combined treatment of insulin and opuntia extract but even alone prickly pear extract maintained normal glycemic state in diabetic rats.

Antiinflammatory

Antiinflammatory actions were demonstrated by opuntia in an induced rat paw edema model. Fresh stems of prickly pear have potent antiinflammatory activity in an adjuvant induced chromic inflammation model in mice. Reduction of acute inflammation by alcoholic extract of stem was ascribed to a lower leucocytes migration.

Hyperlipidemia

The effect of prickly pear in serum and lipid parameter in rats shows an increase in high density lipoprotein cholesterol and reduction in serum cholesterol. Raw plant has also beneficial effects on hyper cholesterolemia in rat.

Effect on Bile Acids and Bile Salt

Bile salt is requiring for the emulsification of dietary fats and pancreatic lipase is one of the most vital of these salts. Besides emulsification process, bile salts also play a role in gastric emptying and transit time. Bile salts are derivations of cholesterol and their emulsification capacities are further enhanced by taurine and glycine.

Neuro Protection

It is found that flavonoids extract of cactus pear protect against neural injuring induced in primary cultured rat cortical cells and their antioxidant activities by using different cell free bioassay. Methanol extract from dried fruit, protected against global ischemic injuries. Fruit extract may be help in alleviating the toxic neural damage induce by global ischemia.

Antiaging effect: (Immunosuppressive effect), it is found that prickly pear polyphenolic compounds show an increase Ca^{2+} in human junket T-_cell lines. It is believe that polyphenolic compounds of pear induce Ca^{2+} via ER pool; and opening of CRAC channels and exert immunosuppressive effects in Junket T- Ca^{2+} in human junket T-_cell lines. It is believe that polyphenolic compounds of pear induce Ca^{2+} via ER pool; and opening of CRAC channels and exert immunosuppressive effects in Junket T-Cells.

Improvement of platelet function and cardiovascular action: Prickly pear consumption significantly reduced the platelet proteins and (Platelet factor 4 and beta thrombogluobulin, ADP induced platelet aggregation) and improved platelet sensitively (against PG12 and PGE1).

Improvement of stomachache: Pectin component in prickly pear does not dissolve in water and becomes a thick syrupy liquid, which is thought to coat and protect the stomach and gastrointestinal tract, thereby helping to reliance digestive problem release stomachache and irritable bowed syndrome.

Wound healing: Evidences are there which show the topical application of polysaccharide extracts from prickly pear cladodes enhanced repair and healing of large full thickness wounds in rat model. It also has protective effect in treating joint diseases.

Antivirus effect: Cladode extract from, was reported to exhibit antiviral properties towards DNA viruses, such as herpes, and RNA viruses, such as influenza type A and human immunodeficiency virus (HIV)-1. The active principle was located in the outer non-cuticular tissue and ascribed to a protein with unknown mechanisms of action. Both the replication of DNA and RNA viruses was inhibited while the extract from the parenchyma acted both preventively and post infection.

Other effect: Use to treat eye inflammation, rabies pimples, dysentery and diarrhea. It is believed that pectin contents bind bacteria and cause them to be eliminated from the body and is therefore helpful in treating diarrhea and other conditions. Prickly pear was also used to treat gonorrhea.

Prickly pear joints are split open and applied as a healing pad for rheumatic and as thematic symptoms. Applied to the skin in a slave it is also used to provide topical relief for sun/wind burn/wounds/rash/minor burns, hemorrhoids, insect bites and abrasions.

Conclusion

Cactus pear for centuries was regarded as irritant plant and always looked with danger but today this plant has attain a new height and almost every part of cactus pear has its our glorified story, the useless spine was also used as gramophone needle. The whole plant justify as food, feed, fodder and pharmaceuticals and follow the concept of 'let food by your medicine'. So called nutraceuticals and the functional food make this old tent a new reality. Today this plant is now integral part of human and biosphere. Man and animal both use it as a component of food. Its application in large number of diseases and disorder and dysfunction of organs are increasing. To name few; diabetes to smart functioning of heart are best examples. As longevity is increasing so the desire of people to look better is also enhancing resulting discovery of antiageing and cosmeceuticals, cactus pear is part of several anti-ageing and cosmeceuticals. Currently we are much more concern with toxicological safety and we always love for nature organic material/product. Cactus pear has a spectrum of coloring substances which can be core of natural dyes. Hence to accelerate all such finding, interdisciplinary efforts on molecular basis are required to integrate nutrition related health and diseases research. Phytochemicals are future biomolecules for improving human health and preventing diseased states. The in depth study of techno-functional properties of various constituents of this plant is urgently needed and if it is done in proper from then this plant can be efficiently exploited for many fold for food ‘cosmetic and medicine application. I am sure with such happening this plant can be analogies to the multiple uses of Aloe Vera or even more in coming decades.

References

1. A. Piga. J. PACD (2004) 9.
2. F.C. Stintzing and R. Carie; Mol. Nutr. Food Res.49 (2005) 175.
3. R.C. Gupta and S.J. Kim; Critical Care and Shock, 6 (2003) 171.
4. R.C. Gupta, T. Win and S. Bittner; Current Med. Chem. 12 (2005) 2021.
5. L. Tesoriere, M. Fazzari, M. Allejra, M.A. Livrea; J. Agric. Food Chem. 5 (2005) 7851.

Section B Energy Perspectives

40

A Clean and Green Hydrogen Energy Production Using Nanostructured ZnO and Fe-ZnO via Photoelectrochemical Splitting of Water

P. Kumar[1], N. Singh[1], A. Solanki[1], S. Upadhyay[1], S. Chaudhary[1], V. R Satsangi[2], S. Dass[1] and R. Shrivastav[1]

[1]Dept. of Chemistry, Dayalbagh Educational Institute, Dayalbagh, Agra

[2]Dept. of Physics and Compt. Science, Dayalbagh Educational Institute, Dayalbagh, Agra

Email: rohitshrivastav_dei@yahoo.co.in

Abstract

To meet tomorrow's demand of cheap, environment friendly and renewable hydrogen, the industries must find new ways of producing it. One solution, with a high potential, is extracting hydrogen by direct splitting of water using solar energy. This study deals with photoelectrochemical splitting of water by using nanostructured ZnO and Fe-ZnO thin films as a working photoelectrode. ZnO and Fe-Zno films were synthesized by sol-gel spin coating method and were characterized for formed crystalline phase by XRD analysis, band gap energy by spectrometric measurement and surface topography by AFM analysis. Subsequently, these were used as working electrode in PEC cell in conjunction with a platinum counter electrode, saturated calomel reference electrode and 150 W Xenon Arc Light Source for illumination, and PEC current densities were recorded, using aqueous solution of NaOH and KOH.

Introduction

To meet tomorrow's demand of cheap, environment friendly and renewable hydrogen, the industries must find new ways of producing it. One solution, with a high potential, is extracting hydrogen by direct splitting of water using solar energy. The investigation on photoelectrochemical (PEC) cells, converting solar energy to electrical or chemical energy was initiated in 1972 by Honda and Fujishma [1]. The development of solar-hydrogen production technology, based on photoelectrochemical device, requires new photosensitive semiconductor materials serving as photoelectrodes. Photo-electrodes are likely to be made of inexpensive polycrystalline materials rather than expensive single crystals. It is important to realize that the photo-sensitivity of polycrystalline materials is strongly influenced, if not determined, by the progress, in the science and engineering of materials interfaces.

In photoelectrochemical splitting of water, the light, illuminating a photoelectrode (semiconductor), is used to split water and generate Hydrogen, a valuable fuel [2]. The semiconductor electrolyte junction, forming depletion layer, is the main functional unit in such systems, where light is absorbed and electron hole pairs are generated. During the photoelectrolysis, the electrons and holes migrate from electrodes to electrolyte to participate in redox process, in which electrolytic conditions also play a major role.

In recent years the possibility of using nanostructured ZnO for PEC splitting of water to produce hydrogen has been explored by some researchers [3,4,5]. The energy levels and band gap energy of ZnO are quite similar to TiO_2, which is the most extensively investigated material for PEC splitting of water, so far. The conduction and valence band edges of ZnO straddle H_2O/H_2 and OH^-/O_2 redox levels and, thus, satisfy a mandatory requirement for spontaneous photosplitting of water. Further, electron mobility in ZnO is much higher compared TiO_2 and it can be an advantage for fabricating current generating devices such as PEC/solar cells [4]. Moreover, defects induced low intensity absorption in visible region, so common in ZnO, can be exploited to convert it into a good solar light absorber [6]. Doping by suitable dopant is seen

M.M. Srivastava, L.D. Khemani, S. Srivastava, *Chemistry of Phytopotentials: Health, Energy and Environmental Perspectives*, DOI:10.1007/978–3–642–23394–4_40, © Springer-Verlag Berlin Heidelberg 2012 as one of the key strategies by different researchers not only to extend the absorption of light to the visible region by ZnO but also to modify its structural and electrical properties [7,8]. This motivated us to investigate the effect of Fe doping on structural, electrical and optical properties of ZnO in the light of its subsequent use in PEC splitting of water.

Experiment

Zinc acetate dihydrate (ZAD), $Zn(COOCH_3)_2.2H_2O$, was used as the starting compound to prepare ZnO thin films. To a solution of ZAD in 2-methoxyethanol (0.9 mol/L), monoethanolamine (MEA), used as stabilizer, was added at room temperature. The molar ratio of MEA to ZAD was kept to 2.0. The mixture was stirred for 1 h at 50–60 °C till a clear and transparent solution was obtained which was spin coated on TCO glass plates (having SnO_2: F overlayer on one side) at 3000 rpm for 20 s.). Coated films were dried on hot plate at 60 °C for 10 min., preheated at 250 °C for 30 min, and then finally annealed at 600 °C for 1h. For iron doping $Fe(NO_3)_2$ was used as a dopant source.

Results and Discussion

As synthesized films were subjected to XRD, UV-Visible, AFM analysis. Following which Fe-ZnO films were used as working electrode (WE) in PEC cell in association with platinum foil counter electrode and a saturated calomel reference electrode (SCE) using NaOH solution of different pH as electrolyte.

In the preparation method used MEA acts as complexing agent and also increases the pH, which promotes the formation of ZnO. In the ensuing sequence of reactions, three nucleophilic species (MEA, HO^- and CH_3COO^-) compete for the Zn^{2+} Lewis acid center. The attack of an HO^- group leads to the formation of small zinc-oxo-acetate oligomers, which are probably evolved from gradual complexation during aging. The progressive condensation of the hydrolyzed moieties then gives rise to colloids or precipitates.

The films were free from pinholes and adhered well to the substrate. Moreover, these were stable in contact with electrolyte under illumination and did not peel off or dissolve. Measured density of films was found to be in the range of 3.90–3.98 g/cm³ which is ≈ 70 % of theoretical density (5.60 g/cm³) of ZnO. This suggests that the films are porous and can provide a larger contact area with electrolyte when used in PEC cell. The measured thickness was around 2 µm.

Observed XRD pattern of undoped and 1 % at. Fe doped ZnO thin films are shown in Fig. 1. The peaks identified correspond to (100), (002), (101) and (200) plane of wurtzite ZnO. Thermodynamically ZnO is more stable with the wurtzite phase due to its ionicity that resides exactly at the borderline between the covalent and the ionic materials [4, 8]. It is interesting to note that no phase other than hexagonal ZnO wurtzite phase evolved in this case. Among these diffraction peaks, the (002) diffraction peak located at 34.6 is the most intensive suggesting c-axis oriented preferred growth. The mechanism of the c-axis oriented growth can be attributed to the low value of surface free energy on this plane. On doping, the intensity of peaks increased slightly indicating improved crystallinity. The particle size was calculated by Debye Scherrer equation and was found to be 45 nm for undoped and 48 nm for Fe doped ZnO. In case of Fe doped ZnO, the (002) peak shifted slightly towards lower angle which suggests the expansion of unit cell. The ionic radii of Fe^{2+}, Fe^{3+} and Zn^{2+} are 0.078, 0.068 and 0.074 nm, respectively. Ionic radius of Fe^{2+} is relatively larger than Zn^{2+} therefore; it appears that to accommodate Fe^{2+} ion, unit cell of ZnO expands. XRD data was also utilized to calculate different other parameters, given in Table 1. Absorption curve of undoped and Fe doped ZnO indicated that on iron doping the absorption maxima shifted towards visible portion of light. The band gap was determined by a plot of $(\alpha h\nu)^2$ vs. $h\nu$

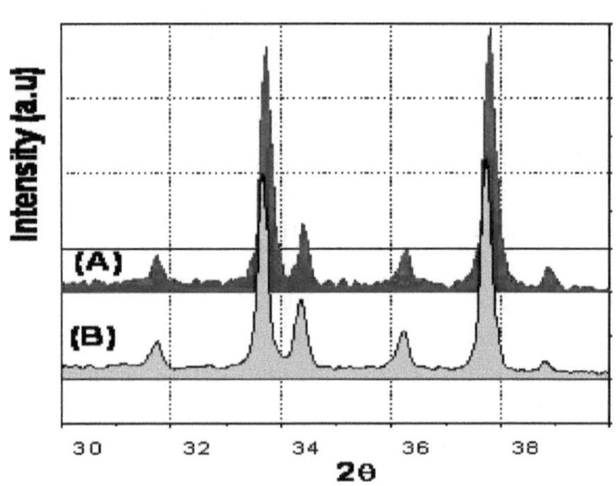

Fig. 1: A: X-ray Diffraction pattern. (A) Undoped ZnO, (B) 1at% Fe doped ZnO

Table 1: Microstructural and Surface Properties of Films

Calculated Parameters	**Undoped ZnO**	**1 % at. Fe-ZnO**
Particle Size	45	48
Lattice Constt. (a)	3.192 Å	3.196 Å
Lattice Constt. (c)	5.213 Å	5.218 Å
Bond Lenght	1.95 Å	1.95 Å
Strain (along c axis)	0.11 %	0.21 %
Dislocation Density (10^{-14})	4.93	4.34
Microstrain (10^{-3})	0.53	2.96
Average Roughness	24.56 nm	27.84 nm

(Fig. 2). The computed band gap for undoped film was 3.18 eV and was reduced to 3.15 eV after iron doping. AFM images are shown in Fig. 3. It is clear that the films are homogeneous, free from any major structural defects, compact, uniform and void free. The computed roughness parameters for the film are depicted in Table 1.

To evaluate the photoelectrochemical properties of ZnO and Fe doped ZnO films, the current voltage (I-V) curves were recorded under illumination using 150 W Xenon arc lamp. The films were converted into electrode and employed as working electrode in PEC cell. The observed (1–5) curves of ZnO films demonstrated a typical feature of n-type semiconductor electrodes. To see the effect of material properties on PEC response, the I-V measurements were made in two different electrolytes, viz., KOH & NaOH (Fig 4 and 5).

As recorded 1–5 curves are shown in Fig. 4 and Fig. 5 and high photocurrent was observed with Fe

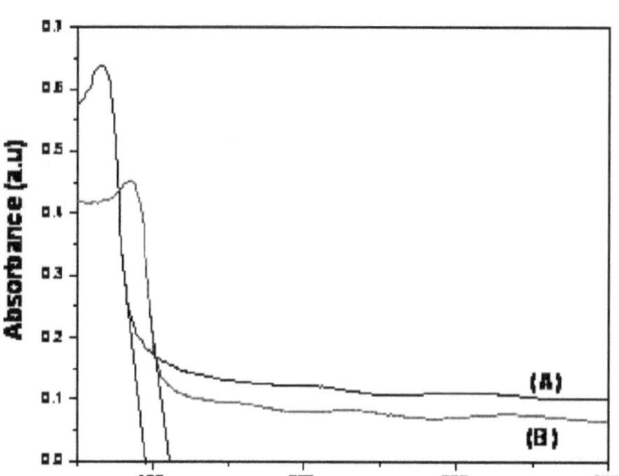

Fig. 2: Plot of $(\alpha h\nu)^2$ vs. hν. (A) Undoped ZnO, (B) 1at% Fe doped ZnO

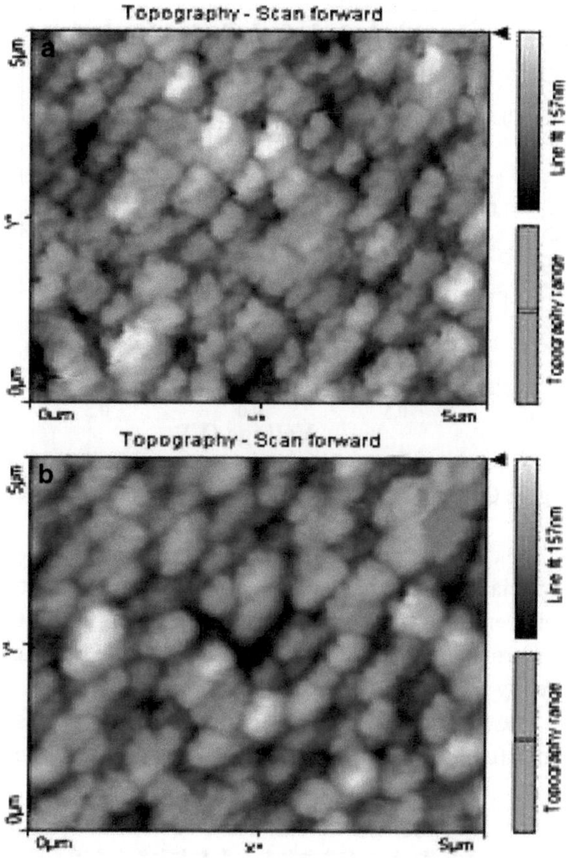

Fig. 3: AFMimajes, **(a)** UndopedZnO, **(b)** 1% at. Fe doped ZnO

doped ZnO. From the 1–5 curves it can be elucidate that the recorded current density under illumination is higher for Fe doped ZnO than undoped ZnO. As depicted from XRD analysis that on iron doping, the crystallinity of the film and the orientation about c-axis increased which may lead in the reduction of defects in the crystal. It is known that defects typically act as recombination centers that can kill photon-generated electron–hole pairs before they can reach surfaces and drive reactions. The electron–hole pairs are generated by the absorption of photons with energies larger than the band gap and separated by the electric field of the depletion region. In the case of n-type semiconductors, the excited electrons move through bulk region to the counter electrode where water reduction occurs. The generated holes move towards semiconductor/electrolyte interface where water oxidation takes place. The aligned nanocrystals along the *c*-axis provide high surface area and superior carrier transport along the *c*-axis, leading to the increased in-

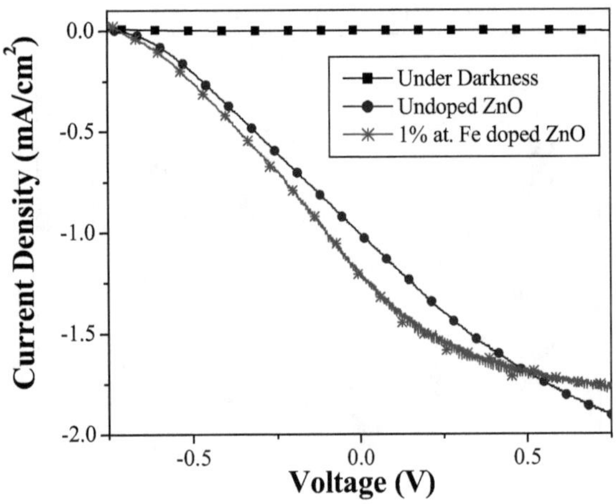

Fig. 4: Current – Voltage Curves using 0.1M NaOH as electrolyte

terfacial reaction sites and reduced recombination rate between the electrons and holes. As a result, the PEC performance of the iron doped ZnO thin film was greatly enhanced.

The values of short circuit current and series resistance (resistance near 0V) are given in Table 2. A high photocurrent about 1 mA was recorded with undoped ZnO while about 1.2 mA current was recorded with Fe doped ZnO. Series resistance which reflects the electron – hole separation, also decreased on doping. This suggests the reduction in electron hole recombination resulted in high photocurrent.

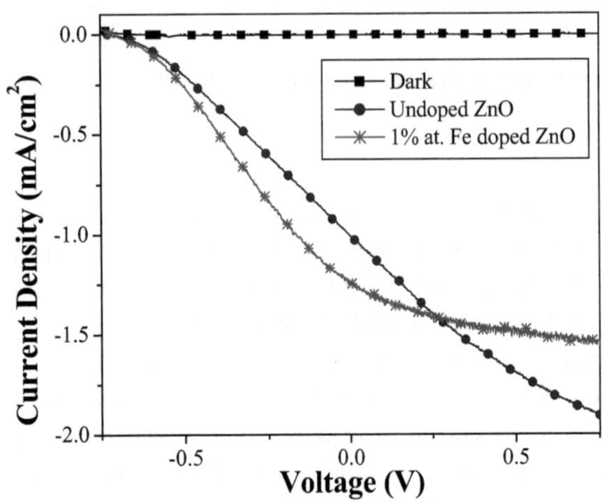

Fig. 5: Current – Voltage Curves using 0.1M KOH as electrolyte

Table 2

	Undoped ZnO		**1 % at. Fe doped ZnO**	
	NaOH	KOH	NaOH	KOH
Short Circuit Current (mA)	1.016	1.012	1.200	1.239
Series Resistance (kΩ)	0.61	0.59	0.46	0.46

Since, there is no additional redox couple in the electrolyte, the significant value of photocurrent density can be attributed to electrochemical splitting of water, which was well indicated by evolution of gases in the form of bubbles on the electrode surface.

References

1. A. Fujishima, K. Honda, Nature; 238 (1972), 37–38.
2. T. Bak, J. Nowotny, M. Rekas and C. C. Sorrel; Int. J. Hydrogen Energy, 27 (2002), 991.
3. K. Soon Ahn, S. Shet, T. Deutsch, C. S. Jiang, Y. Yan, M. Al-Jassim and J Turner; Journal of Power Sources 176 (2008) 387–392.
4. M. Gupta, V. Sharma, J. Shrivastava, A. Solanki, A. P. Singh, V. R. Satsangi, S Dass and R Shrivastav; Bull. Mater. Sci., 32 (2009), 23–30.
5. M. Gupta, V. Sharma, J. Shrivastava, A. Solanki, A. P. Singh, V. R. Satsangi, S Dass and R Shrivastav; Advanced Mater. Research, 67 (2009), 95.
6. Q. Zhang, C. S. Dandeneau, X. Zhou and G Cao; Adv. Mater. 21 (2009), 4087.
7. Y. Li and J. Z. Zhang; Laser and Photon. Rev., (2009) 1–12.
8. V. Sharma, P. Kumar, J Shrivastava, A Solanki, V. R. Satsangi, S. Dass and R Shrivastav; J Mater Sci, (2011), DOI 10.1007/s10853–011–5293–2.

41
One Pot and Solvent-Free Energy Efficient Synthesis of Metallophthalocyanines: A Green Chemistry Approach to Synthesize Metal Complexes

R. K. Sharma, S. Gulati* and S. Sachdeva
Green Chemistry Network Centre, Department of Chemistry, University of Delhi, Delhi-110007
Email: rksharmagreenchem@hotmail.com

Abstract

A novel one pot and solvent-free synthesis of metal(II)phthalocyanines in microwave is described. This method in comparison to conventional route offers high yields, eliminates the use of toxic and expensive organic solvents, and reduces multistep synthesis to one-step, reaction temperature and reaction time from several hours to few minutes which leads to the development of clean and environmentally benign procedure.

Introduction

Phthalocyanines are one of the first man-made macrocycles that were synthesized and used as model compounds to mimic the biologically important porphyrins. Metallophthalocyanine complexes [1] have attracted considerable attention due to their impressive and useful chemical and physical properties. These complexes have wide applications in dyes and printing inks, laser printers, optical devices, photosensitizers in treatment of cancer, photovoltaic materials in solar cells, materials for fabrication of light emitting diodes (LED), non-linear optical materials, photoconductors in xerography, differential staining agents, gas diffusion electrodes, liquid crystals, photosensitizers in treatment of cancer, active compounds in gas sensing systems, as well as in diverse catalytic systems [2].

We have developed a novel one pot and solvent-free method for the synthesis of metalloporphyrins [3] and metallophthalocyanines in microwave. This method in comparison to conventional method offers high yields, eliminates the use of toxic organic solvents, and reduces multistep synthesis to one-step, reaction temperature and reaction time from several hours to few minutes which leads to the development of clean and environmentally benign procedure. The use of microwave technology as a non-polluting mode of activation accelerates the reaction and reduces the reaction time and energy input compared to conventional heating method. On the other hand the use of solvent free system in which microwaves interact directly with the reagents, can more efficiently drive the chemical reaction towards completion. The wastage on account of the ligand synthesis is also minimized since the process was carried out in one pot i. e. there

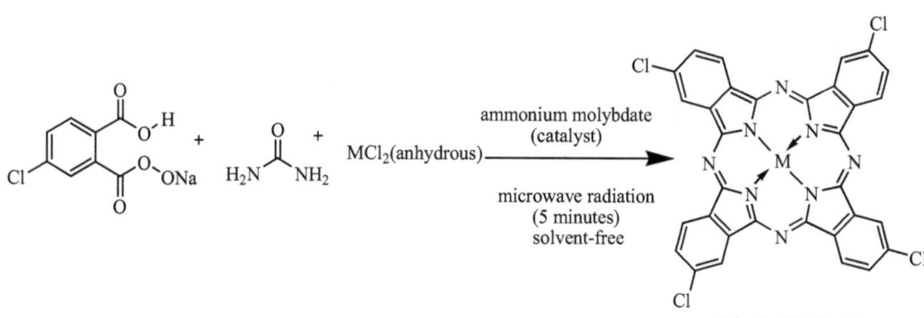

Scheme 1:. One pot and solvent-free synthesis of Metal(II)phthalocyanines

M.M. Srivastava, L.D. Khemani, S. Srivastava, *Chemistry of Phytopotentials: Health, Energy and Environmental Perspectives*, DOI:10.1007/978–3-642–23394-4_41,

is an in-situ generation of the ligand from the starting material followed by the insertion of metal to form a complex. Thus these advanced reaction conditions reduces the cost of production of commercially important phthalocyanine complexes.

Discussion

The solvent free method adopted for the direct synthesis of metal(II)phthalocyanines in one pot using microwave involves cyclotetramerization of derivatives of phthalic acid in presence of urea as the source of nitrogen, in which template effect is afforded by a suitable metal cation (Scheme 1). The use of ammonium molybdate $[Mo_7O_{24}(NH_4)_6.4H_2O]$ as a catalyst results in maximum utilization of starting materials and amount of energy used is also minimized. The microwave irradiation accelerates the cyclotetramerization process and reduces the reaction time and energy input compared to conventional heating method. The complexes were characterized by elemental analysis, IR, TGA and UV/Vis and AAS techniques.

Conclusion

An efficient, economical, rapid and green protocol for the direct synthesis of metal(II)phthalocyanines using microwave under solvent-free conditions is developed. Short reaction time, high yields, milder reaction conditions, no use of toxic organic solvents are the advantages of present protocol which leads to the development of clean and environmentally benign procedure.

References

1. R.K. Sharma, C. Sharma and I.T. Sidhwani, J. Chemical Ed. (Green Chemistry Lab-Expt) 88 (2011) 86–87.
2. R.K. Sharma and C. Sharma; Tetrahedron Letters, 51 (2010) 4415–4418.
3. R.K. Sharma, G. Ahuja and I.T. Sidhwani, Green Chem. Lett. Rev. *2* (2009) 101–105.
4. R.K. Sharma, S. Gulati and S. Sachdeva, Green Chemistry Letters and Reviews, 2011, DOI:10.1080/17518253.2011.581701

42
Photoelectrochemical Hydrogen Generation Using Al Doped Nanostructured Hematite Thin Films

P. Kumar[1], P. Sharma[1], R. Shrivastav[2], S. Dass[2] and V. R. Satsangi[1]
[1]Department of Physics & Computer Science, Dayalbagh Educational Institute, Dayalbagh, Agra
[2]Department of Chemistry, Dayalbagh Educational Institute, Dayalbagh, Agra
Email: vibhasatsangi@gmail.com

Abstract

Photoelectrochemical (PEC) water splitting using solar energy is the most intriguing rout of hydrogen generation in the emerging green economies. Role of Al doing in nanostructured hematite thin films for hydrogen generation via solar splitting of water in PEC cell have been investigated in detail and presented. PEC data have been analyzed with the help of structural, optical, and Mott-schottky studies. Highest photocurrent density of 2.81 mA/cm² at 0.7 V/SCE was exhibited by 1 at% Al doped hematite. The rate of hydrogen generation for same sample was 19.2 lh^{-1}/m^2.

Introduction

Researchers are continuously searching for green, clean and sustainable source of energy which can meet the growing demand of energy in human life. Hydrogen has great potential for its use as energy carrier, due to its high heat of combustion and high energy storage capacity. Most important is that use of hydrogen is environmentally benign, as combustion of hydrogen with oxygen produces water [1].

Hydrogen generation from photoelectrochemical water splitting using solar energy stand for a holy grail in energy science and technology, as water and solar energy are most abundantly available [2]. Also, hydrogen once produced from solar energy, provides storage to the world's most abundant but intermittent source of energy.

Hematite (α-Fe_2O_3) is an attractive material for its use as photoelectrode in photoelectrochemical (PEC) cell for solar generation of hydrogen due to its favorable optical band gap (2.2eV), chemical stability in oxidative environment, abundance and low cost [3]. However, reported efficiency for α-Fe_2O_3 is much below the level of commercialization due to many factors, such as relatively poor absorptivity, very short excited-state lifetime, rapid electron-hole recombination in the hematite resulting in short diffusion lengths of charge carriers [4]. Many methods, including doping have been investigated to improve the performance of hematite in PEC cell [5–12].

Recently, advent of nanotechnology has opened new avenues to modify properties of semiconductors in the development of efficient and economically viable PEC system for hydrogen production using solar energy. Nanostructured semiconductors material hold potential advantages in PEC application as compared to bulk material due to their large surface area and size-dependent properties, such as increased absorption coefficient, increased band-gap energy, and reduced carrier-scattering rate.

Among the various method of synthesis of nanostructures of α-Fe_2O_3, electrodeposition has many advantages including low cost, electrical contact, precise control of nanostructure and the potential for co-depositing dopant elements [8].

In present piece of work, we have prepared Al doped nanostructured hematite thin films by the electrochemical route. Further the effect of Al doping on the structural, morphological, optical and photoelectrochemical properties of hematite thin films has been investigated.

M.M. Srivastava, L.D. Khemani, S. Srivastava, *Chemistry of Phytopotentials: Health, Energy and Environmental Perspectives*, DOI:10.1007/978–3-642–23394-4_42, © Springer-Verlag Berlin Heidelberg 2012

Experiment

The thin films of undoped and Al doped α-Fe_2O_3 were prepared on fluorine-doped tin oxide (FTO) conductive glass substrates by electrodeposition [12]. For Al doping, aluminum nitrate nonahydrate at different atomic percentages from 0 to 15.0 at. % was added to the electrodeposition solution. Obtained films were sintered in muffle furnace at 450°C for 1 h to obtain the desired crystalline phase.

To determine the phase and crystallinity of the prepared thin films XRD was carried out with X-ray diffractometer (Bruker AXS, D8 Advanced) using Cu Ka radiation (λ=1.5418 °A).The surface morphology of the samples was studied using a field emission scanning electron microscope (FE-SEM), (Carl Zeiss SUPRA 40VP). Absorption band edge and bandgap energy was determined with absorption curves obtained by using UV-Visible spectrophotometer (Shimadzu, UV-2450). Thin films prepared were used as working electrode in PEC cell and current voltage characteristics were recorded under darkness and visible light illumination with a 150 W Xenon lamp (Bentham) by using a potentiostat (PAR,Versa Stat II). PEC cell consist of three electrodes: working electrode, Pt counter electrode and SCE reference electrode immersed in 1 M NaOH. To obtain flat band potential and donor density, capacitance at semiconductor/electrolyte junction was measured using an LCR meter (Agilent Technology, 4263 B) at varying electrode potential, under dark conditions in the same three-electrode configuration.

Results and Discussion

Structural Study

The XRD pattern obtained for all the doped/undoped iron oxide thin films have been shown in Fig. 1. The peaks, namely, (012), (104), (110), (113), (024) and (116) provides evidence for the existence of α-Fe_2O_3 (hematite) phase of iron oxide. The peaks indexed by the * symbol correspond to the SnO_2 in the FTO substrate. The crystallite sizes as calculated from Scherer's equation on the basis of (104) reflection were ~38 nm for undoped, ~47 nm for 1 % Al, and a reduction in particle size to ~22 nm for 15 % Al doping.

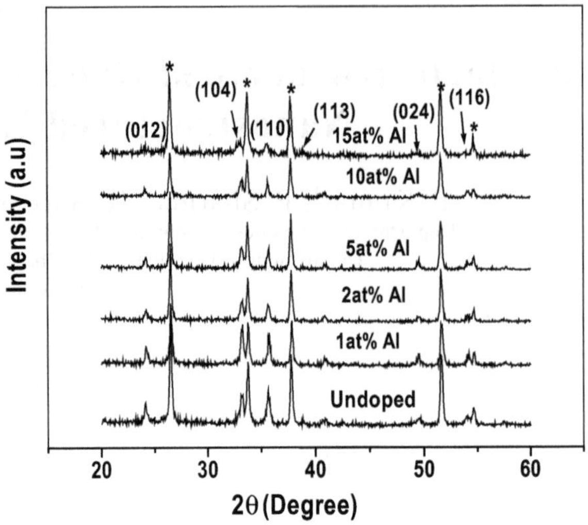

Fig. 1: XRD for undoped and Al doped samples.

Optical Study

To determine optical properties such as absorption coefficient and optical band edge energy of undoped and Al-doped hematite film, absorption spectrum was recorded in the wavelength range of 200–800 nm as shown in Fig. 2. The absorption band edge for all the samples were approximately at ~2.1 eV. No significant change in absorption band edge was observed after Al doping.

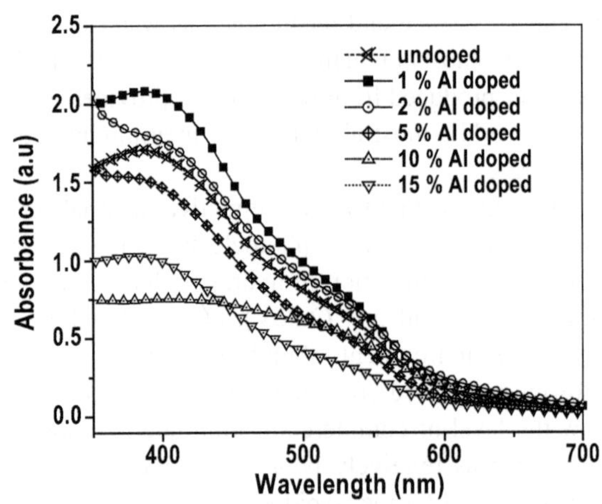

Fig. 2: Absorption spectra for undoped and Al doped samples.

Morphological Studies

Scanning electron micrographs (SEM) showed that all the films were porous in nature as shown in Fig.3. Grain size increased at lower doping concentration while fragmentation of grain was observed at higher doping level. These results are in good agreement with XRD.

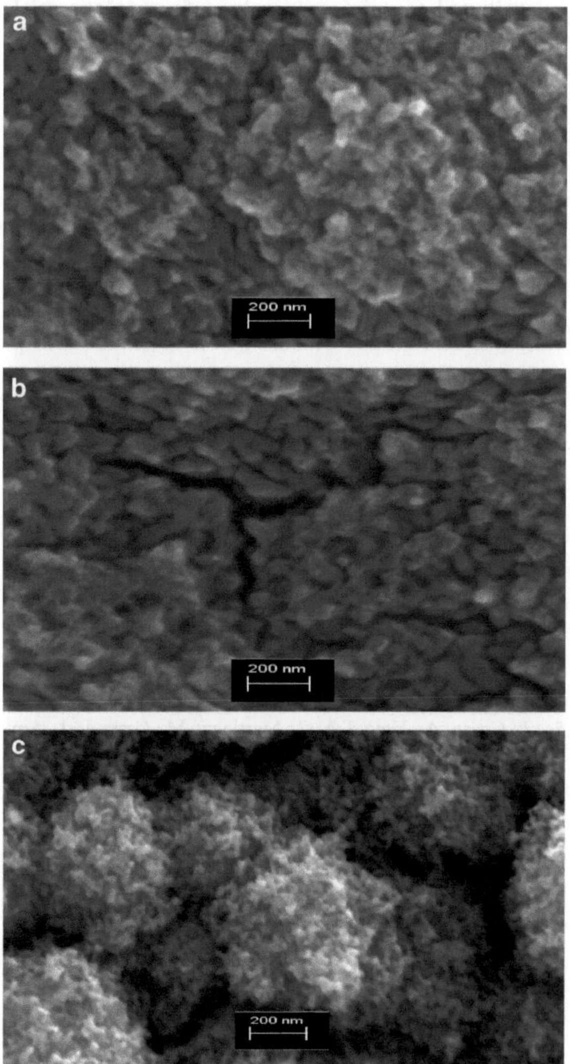

Fig. 3: SEM micrographs for **(a)** Undoped, **(b)** 1at % Al doped and **(c)** 15at % Al doped

Photoelectrochemical Studies

All the samples were used as photoelectrode in PEC cell and current–voltage (I–V) curves were obtained in the dark and under 150 W Xenon lamp illumination. Photocurrent density versus potential curves for undoped and Al doped α-Fe_2O_3 thin films have been presented in Fig. 4. Al doping showed a significant improvement in the photocurrent as compared to undoped sample. This improvement is especially pronounced at lower doping concentration. Highest photocurrent density of 2.81mA/cm^2 at 0.7V/SCE was exhibited by 1at % Al doped sample. At higher doping level the photocurrent density was observed to decrease. A significant negative shift in the photocurrent onset was also observed from ~500 mV for undoped sample to ~200 mV for 1.0 at % Al doped sample. The flat band potential and donor density were calculated from Mott-Schottky curves shown in Fig. 5. Flat band potential was observed to increase from –0.5V/SCE to –0.83V/SCE for 1at % Al doped sample. The more negative flatband potential the better is the ability of the semiconductor film to facilitate the charge separation in PEC application [12]. Donor density of $8.9\times10^{20}\,cm^{-3}$ was maximum for 1at % Al doped hematite as compared to $1.7\times10^{20}\,cm^{-3}$ for undoped sample. Hence maximum value of flatband potential and donor density strongly support to the best photoelectrochemical response of 1 at % Al doped hematite . The decrease in the photocurrent at high Al doping content could be related to the segregation of alumina to the surface of hematite thin films by creating barrier for inter particle electron transport [9].

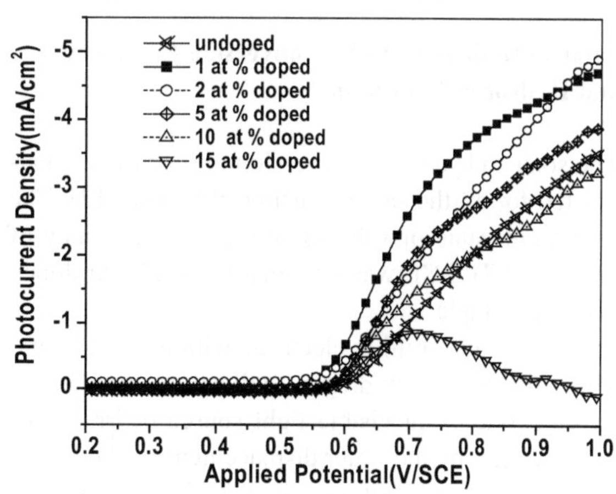

Fig. 4: Photocurrent density versus applied potential curves under visible light illumination for Al doped and undoped hematite samples

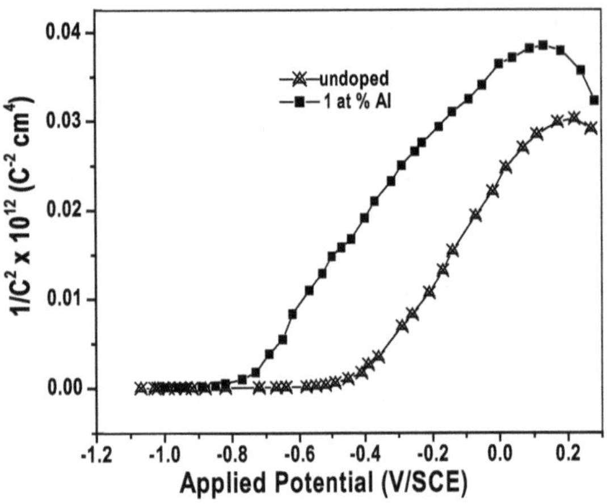

Fig. 5: Mott-Schottky plots ($1/C^2$ versus V) for undoped and 1at% Al doped samples

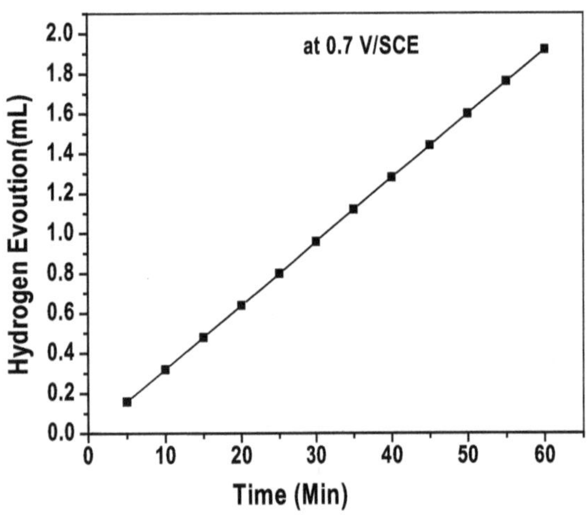

Fig. 6: Hydrogen evolution rate versus time curve under visible light illumination for best performing electrode

Solar to Hydrogen (STH) Conversion Efficiency and Hydrogen Collection

The solar to hydrogen conversion efficiency was calculated for all the samples using PEC data [12]. The maximum value of solar to conversion efficiency of 1.7% at 0.7V/SCE was obtained for 1 at% Al doped hematite sample.

The best performing electrode with area ~1×1 cm^2 was utilized for hydrogen collection over counter electrode using visible light source of irradiance 150 mW/cm^2 by the water displacement method. Collected volume of hydrogen at 0.7 V/SCE has been plotted with respect to time as shown in Fig. 6. The measured rate of hydrogen production was 1.92 mlh^{-1} at 0.7 V/SCE, which is equivalent to 19.2 lh^{-1}/m^2. The photoelectrode was observed to be stable even after multiple scans (15 times) and no change in the physical appearance of the sample was observed. This confirms the stability of the given photoelectrode for water splitting reactions.

Conclusion

We have investigated the effect of aluminum doping on the performance of hematite thin films in photoelectrochemical cell for hydrogen generation. 1 at% Al doped hematite thin film exhibited best photoelectrochemical density of 2.81 mA/cm^2 at 0.7V/SCE, which is attributed to minimum value of onset potential, maximum donor density and more negative flatband potential. Hydrogen evolution rate was 19.2 lh^{-1}/m^2 for best performing sample of doped hematite.

Acknowledgements

The financial support provided by University Grant Commission, New Delhi, India vide project no: 37–128/2009 for this work is gratefully acknowledged.

References

1. J.M. Ogden; Annu. Rev. Energy Environ. 24 (1999) 227.
2. Y. Li and J.Z. Zhang, Wiley-Vch Verlag GmbH & Co. KGaA, Weinheim (2009).
3. V.R. Satsangi, S. Dass, R. Srivastav; Solar Hydrogen and Nanotechnology: Nanostructured α-Fe_2O_3 in PEC Generation of Hydrogen, John Wiley & Sons (Asia), Pvt Ltd, (2010)349.
4. A.B. Murphy, P.R.F. Barnes, L.K. Randeniya, I.C. Plumb, I.E. Grey, M.D. Horne, J.A. Glasscock; Int. J. Hydrogen Energy. 31 (2006) 1999.
5. I. Cesar, A. Kay, J.A.G. Martinez, M. Gratzel; J. Am. Chem. Soc. 128 (2006) 4582.
6. S. Kumari, A.P. Singh, Sonal, D. Deva, R. Shrivastav, S. Dass, V.R. Satsangi; Int. J. Hydrogen Energy. 35 (2010) 3985.
7. I. Cesar, K. Sivula, A. Kay, R. Zboril and M. Grätzel; J. Phys. Chem. C. 113 (2009) 772.
8. Y.S. Hu, A.K. Shwarsctein, A.J. Forman, D. Hazen, J.N. Park, E.W. McFarland; 20 (2008) 3803.
9. A.K. Shwarsctein, M.N. Huda, A. Walsh, Y. Yan, G.D. Stucky, Y.S. Hu, M.M. Al-Jassim, E.W. McFarland; *22* *(***2010)** 510.
10. A.K. Shwarsctein, Y.S. Hu, A.J. Forman, G.D. Stucky and E.W. McFarland; J. Phys. Chem. C. 112 (2008) 15900.
11. A. Kay, I. Cesar, M. Gratzel; J. Am. Chem. Soc. 128 (2006) 15714.
12. P. Kumar, P. Sharma, R. Shrivastav, S. Dass, V.R. Satsangi; Int. J. Hydrogen Energy. 36 (2011) 2777.

43

Proton Conducting Membrane from Hybrid Inorganic Organic Porous Materials for Direct Methanol Fuel Cell

N. K. Mal[1] and K. Hinokuma[2]

[1]Green Chemistry and Catalysis, Tata Chemical Limited Innovation Centre, Ghotavde Phata, Urawde Road, Pirangut Industrial Area, Pune – 412 108.

[2]Solid State Ionics Research Laboratory, ML, SONY Corporation, 4–16-1, Okada, Atsugi-Shi, Japan

*Email: Nawalkishor.mal@tatachemicals.com

Abstract

5 to 20 mol% hybrid inorganic organic porous silica (SO_3H-Si) and polymer polytetrafluoroethane (PTFE) was mixed together to produce 10 to 300 µm thick membrane ($S0_3H$-Si-PTFE). Proton conductivity of resultant hybrid silica before and after mixing with PTFE determined in dry nitrogen and 100% relative humidity (RH) using impedance spectroscopy. 20% $S0_3H$-Si-PTFE shows proton conductivity up to 10^{-2} S/cm, similar to Nafion117 at room temperature at 100% relative humidity. Membrane electrode assembly (MEA) was developed and used in direct methanol fuel cell technology. Power density of the MEA was 55 mW/cm2 at 40 °C. Materials were characterized using various instrumental techniques, such as ICP, XRD, TG-DTA, XPS, N_2 sorption, ^{29}Si and ^{13}C MAS NMR, FT-IR, particle size analyzer and TDS.

Introduction

Fuel cell technology is one of the alternative arrangement of energy is strongly desirable due to environmental problem and limited stock of conventional energy. Fuel-Cell (FC) technology is green technology, which will provide energy for stationary and mobile application in near future, free from pollution or greatly reduced the emission of CO_2. Nafion is an electrolyte used in fuel cell, but its higher cost is still problem to commercialize the technology. We have recently developed hybrid inorganic organic porous silica ($S0_3H$-Si) as alternative to nafion, which cost is minimum one order less than nafion and utilized as proton conductor (electrolyte) in direct methanol fuel cell (DMFC).

20 mol% HS functionalized mesoporous silica has been reported without developing membrane [1,2]. Inorganic-hybrid SO_3H-Si can not be used as electrolyte, unless membrane was developed by mixing with suitable organic polymer. Moreover, we developed 20 mol% hybrid porous silica (SO_3H-Si) in one step using dodecylamine ($C_{12}NH_2$) as surfactant, where as references [1,2] prepared it in two steps (post synthesis). We report for the first time synthesis of 20 mol% hybrid porous SO_3H-Si in one step followed by proton conduction membrane was developed using hybrid inorganic organic porous silica ($S0_3H$-Si) and PTFE ($S0_3H$-Si-PTFE) and thereafter membrane electrode assembly (MEA) was developed and used in direct methanol fuel cell technology.

Experimental

Synthesis of 20 mol% hybrid porous SO_3H-Si was carried out using following molar composition: 0.80TMOS : 0.20MPTS : 0.25 $C_{12}NH_2$: 10 EtOH : 8 H_2O_2 : 100 H_2O, where as TMOS = tetramethoxysilane, MPTS = 3 mercaptopropyltrimethoxy silane, $C_{12}NH_2$ = dodecylamine, EtOH = ethanol. In a typical procedure, 6 mmol of dodecylamine was dissolved in 9 g of ethanol at 70 °C and then diluting the surfactant by adding 36 g of water preheated at 70 °C. 16 mmol of TMOS and 4 mmol of MPTS added to surfactant together at 70 °C for 30 min. Finally, 0.16 mole of aqueous H_2O_2 was added to the gel. Gel was sealed in propylene bottle and aged at 70 °C for 3 days. Solid was then filtered off, washed, dried. 5 and 10 mol% hybrid porous SO_3H-Si was also prepared using above

M.M. Srivastava, L.D. Khemani, S. Srivastava, *Chemistry of Phytopotentials: Health, Energy and Environmental Perspectives*, DOI:10.1007/978–3-642–23394-4_43,

method for comparison. Surfactant was extracted using an acidic ethanol solution by taking 1 g of solid, 1 mL of HCl (1 Molar) and 100 mL of ethanol at 50 °C for 16 hrs [3].

Hybrid Porous SO_3H-Si– PTFE Membrane

0.54 g of *Hybrid Porous SO_3H-Si* in 2 g of dimethylformamide (DMF) was ball milled using 4 ø zirconia balls (150 g) for 12 hrs with 300 rpm. It was kept at 40–50 °C in open air to become pest type. 0.100 g of Polytetrafluroethylene (PTFE) (60 % solution dispersed in H_2O) was then added and mixed for 15 min. It was placed on glass plate (15 cm x 15 cm) and pressed to get 150 µm thickness membrane with 10 % content PTFE in SO_3H-MCM-41-PTFE membrane, followed by drying at 60 °C under vacuum for 1 h. Similarly, different ratio of SO_3H-MCM-41 to PTFE was prepared.

Anode Catalyst

0.355 g of Pt/Carbon black (TEC66E50, E-Tek Co.) (Pt-Ru 50 wt%, Pt : Ru = 1:1 atomic ratio) was mixed with 0.75 g of nafion (20 % aqueous) and 2 g of isopropanol. It was then hybrid mixture (HM) for 30 min and ultrasonic for 30 min. Gel was transferred into Petridis and dried at room temperature for overnight. Known amount of catalyst (0.03 g) was placed in 10 mm diameter die and pressed with 1.5 ton for 2 second at ambient temperature.

Cathode Catalyst

0.355 g of Pt/Carbon black (TEC10E60E, E-Tek Co.) (60 wt% Pt) was mixed with 0.75 g of nafion (20 % aqueous) and 3 g of isopropanol. It was then hybrid mixture (HM) for 30 min and ultrasonic for 30 min. Gel was transferred into Petridis and dried at room temperature for overnight. Known amount of catalyst (0.03 g) was placed in 10 mm diameter die and pressed with 0.50 ton for 2 second at ambient temperature.

Membrane Electrode Assembly (MEA)

5 cm x 5 cm dimension nafion 112 membrane was sandwich between anode (5 mm diameter) and cathode catalyst pellet (5 mm diameter) and hot pressed at 135 °C at 0.20kN for 15 min. MEA was placed in single cell hardware. The performance of single cell direct methanol fuel cell (DMFC) was tested at 25 to 80 °C with input flow rate of methanol at the anode was 3 mL min^{-1} and the cathode input O_2 flow rate was 200 mL min^{-1}.

Characterization

Elemental analyses were done using ICP (Shimadzu, ICPV-1017) for silica and CE instruments EA1110 for CHNS. XRD patterns were recorded with MAC Science, MO3X with Ni filtered Cu-Kα radiation ($\lambda = 0.15406$ nm). Transmission electron micrographs (TEM) were taken on a 2000 JEOL electron microscope operating at 200 kV. FT-IR spectra of template extracted samples were obtained with a JASCO FT/IR-230 using KBr pellets (3 mass% catalyst). BET specific surface area and pore size were obtained from nitrogen adsorption isotherms measured at −196 °C using Bellsorp 28 instrument (BEL JAPAN, Inc.). Thermogravimetric analyses (TGA) and differential thermal analyses (DTA) were performed on a Seiko SSC/5200 apparatus. Solid-state ^{29}Si MAS-NMR spectra were collected on a JEOL, model CMX400 (9.4 Tesla) at 79.42. Polydimethylsilane (PDMS) were used as external chemical shift standard. ^{13}C CP MAS NMR spectrum measurements were performed using VXR S-400WB instrument. X-ray photoelectron spectra (XPS) were obtained using a Fisons Escalab 200R. To evaluate the amount of acid sites (SO_3H) in the sulfonic acid containing samples, 57 mg dried sample was mixed with 50 mg of H_2O and stirred for 20 h at room temperature. Finally, it was titrated against 0.01 M NaOH using Automatic Potentiometric Titratior AT-510 (Kyoto Electronic MFG Co. Ltd.).

Proton conductivity of silica derived proton conductor were measured using Solartron impedance analyzer type 1260. Proton conductivity of solid electrolyte membrane were measured using Solartron impedance analyzer type 1260 at different RH% and various temperature ranging from 25 to 130 °C. The performance of single cells DMFC was measured using a Solartron SI 1287 (Electrochemical Interface Chem.).

Results and Discussion

Aqueous H_2O_2 was added in the gel during the synthesis, which oxidized Mercapto- or thiol- (SH) of mercaptoproyl silane (HS-CH_2-CH_2-CH_2-Si) into sulfonic acid. This sulfonic acid group work as proton conductor in all samples. Chemical analyses results of 10 and 20 mol% Inorganic-hybrid porous SO_3H-Si are shown in Table 1. ICP and CHNS analyses indicate that SH-$SiO_{1.5}$/(SH-$SiO_{1.5}$+SiO_2) ratio in the gel and product is nearly the same. XRD profiles of both the surfactant extracted Inorganic-hybrid SO_3H-Si samples, with different SH-$SiO_{1.5}$/(SH-$SiO_{1.5}$ + SiO_2) mol% are shown in Fig. 1. One intense peak was observed at lower two theta angle for both samples, which are characteristics of wormhole structure.[4] Intensity of peak decreases with increasing the loading of mercapto (HO_3S/HS-) and d_{001} peak shifted to further lower order angle. d_{001} spacing for Samples 1 and 2 are 3.2 and 3.7 nm, respectively (Table 2). TEM image of Sample 2 confirms the wormhole structure shown in Figure 2, as observed by Yutaka Mori et al [4]. Physico-chemical properties of surfactant extracted Samples 1 and 2 are shown in Table 2. N_2 sorption isotherms show a feature of a type IV isotherm with sharp capillary condensation at p/p_0 ca. 0.2 and 0.12 for Samples 1 and 2, respectively. Pore size distributions calculated from adsorption isotherm shows peak pore diameter at diameter measured from adsorption branch of isotherms.

Table 1: Chemical composition of the surfactant extracted hybrid porous HO_3S-Si

Sample[a]	**(HS-$SiO_{1.5}$)/ (HS-$SiO_{1.5}$+ SiO_2)**[b]		**Wt%**		
	in Gel	**in product**	C	H	S
1	0.10	0.10	5.28	1.03	4.70
2	0.20	0.21	9.64	1.88	8.57

[a]Molar composition of Samples 1 and 2 are $(HS\text{-}R\text{-}SiO_{1.5})_{0.07}$ $(HO_3S\text{-}R\text{-}SiO_{1.5})_{0.03}$ $(SiO_2)_{0.90}$ and $(HS\text{-}R\text{-}\mathbf{SiO}_{1.5})_{0.12}$ $(HO_3S\text{-}R\text{-}SiO_{1.5})_{0.09}$ $(SiO_2)_{0.79}$, respectively, where R = -CH_2-CH_2-CH_2-, and molar composition evaluated based on Si and CHNS analysis.
[b]HS- refers to both mercapto (SH) and sulfonic acid (SO_3H).

Table 2: Physico-chemical properties of hybrid porous HO_3S-Si samples[a]

Sample	**d_{001} (nm)**	**S_{BET} ($m^2 g^{-1}$)**	**V_p ($cm^3 g^{-1}$)**	**Peak pore diameter (nm)**
1	3.2	740	0.60	2.0
2	3.7	602	0.45	1.2

[a]S_{BET} = BET surface area; V_p = Pore volume; Peak pore

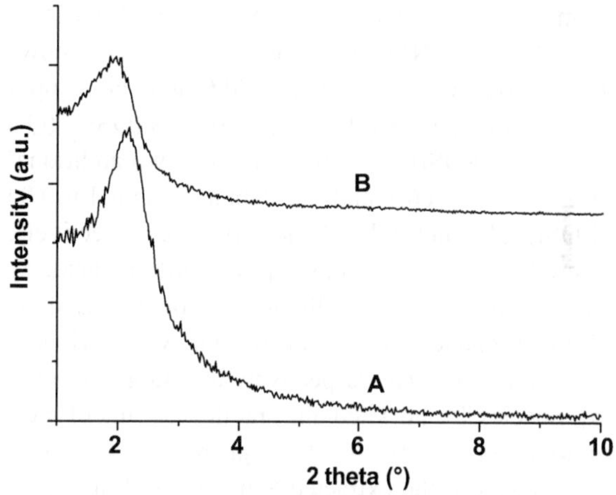

Figure 1: XRD profiles of 10 (A, Sample 1) and 20 mol% (B, Sample 2) of hybrid porous silica (HO_3S-Si)

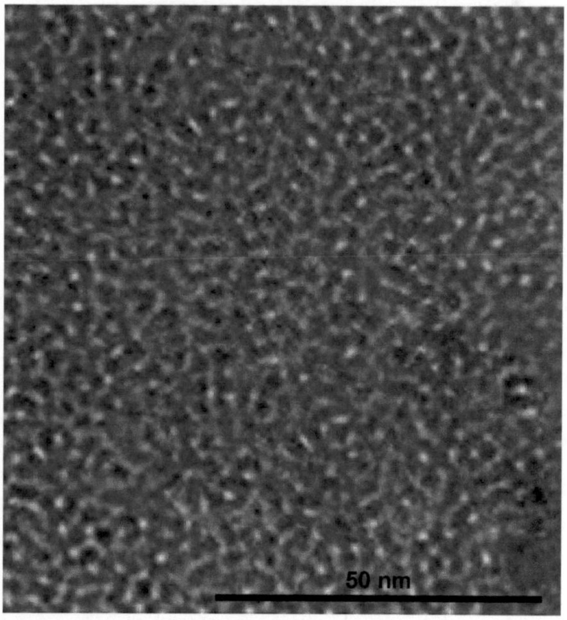

Figure 2: TEM image of 20 mol% hybrid porous silica ($S0_3H$-Si) (Sample 2)

2 and 1.2 nm, respectively. Surface area of Samples 1 and 2 are 740 and 602 $m^2 g^{-1}$, respectively (Table 2). Pore volume of Samples 1 and 2 are 0.60 and 0.45 $cm^3 g^{-1}$, Sample 2 possesses 1.21 meq.acid/g-sample based on titration against 0.01 M NaOH. Based on molar composition derived using CHNS analysis as presented in Table 1, 1.15 meq.-acid/g was expected for Sample 2, which is nearly close to the titration value. ^{13}C MAS NMR further indicates that 75 mol% of SH oxidized to SO_3H in 20 mol% hybrid SO_3H-Si (Sample 2) as discussed later in ^{13}C MAS NMR section. ^{29}Si MAS NMR spectrum for Sample 2 shows three resonances at –110.5, –100.6 and –91.2 ppm corresponding to **Si**$(OSi)_4$ (Q^4), (OH)**Si**$(OSi)_3$ (Q^3) and $(OH)_2$**Si**$(OSi)_2$ (Q^2), respectively. Two additional resonances at –66.3 and –56.3 ppm assigned to (R) **Si**$(OSi)_3$ (T^3) and (OH)(R)**Si**$(OSi)_2$ (T^2), respectively, where R is mercapto propyl / propylsulfonic acid. As reported, in the case of sulfonic acid containing SBA-15, the resonances at –57 and –66 ppm were attributed to T^2 and T^3 species, respectively by Margolese *et al* [5]. Species T^3 and T^2 confirming the grafting of mercaptopropyl moiety in silica sphere. ^{13}C CP MAS NMR result of the extracted Sample 2 and their proposed assignments for the observed resonance signals have been collected in Table 3. Intense resonance peaks at 11.3, 18.4 and 56.4 are assigned to C^3, C^2 and C^1 carbons of propyl sulfonic acid (C^1 is adjacent to –SH functionality). A weak signal was observed at 30.7 attributed to the central carbon atom (C^2) of propyl group of mercaptopropyl moiety, while the carbon atom adjacent to the SH moiety (C^1) produces a peak at 33.1 ppm. In addition to that an intense peak can be distinguished at 41.3 ppm, which could be assigned to the C^1 atom adjacent to S atoms in dipropyldisulfide species. The C^3 resonance of mercaptopropyl would appear close to C^3 of propyl sulfonic acid, contributing in this way to the broad signal observed at ca. 11.3 ppm [5,6]. ^{13}C MAS NMR results confirms that 75 % of the SH (thiol) group oxidized to SO_3H in Sample 2. External surface of the Sample 2 has been characterized by XPS. XPS spectrum of the Sample 2 shows two S 2s peaks at 226.7 and 232.7 ev assigned to thiol (-SH) and sulfonic acid (-SO_3H) [7]. The % area ratio of HS-R-Si : Si-R-SO_3H is 0.75 : 1. This result confirms that 75 % of the surface thiol group is oxidized to sulfonic acid. This result is good agreement with ^{13}C MAS NMR where 70 % of bulk thiol is oxidized to sulfonic acid. FTIR spectrum of the Sample 2 shows band at 2511 cm^{-1} assigned to thiol (SH). The peak at 1066 and 1214 cm^{-1} was observed that correspond to the symmetric and asymmetric band of SO_3H, respectively [8]. The hump band at 2874 and 2981 cm^{-1} are attributed to -CH_2 group in mercaptopropylsulfonic acid. The TG analysis of samples 1 and 2 show four distinct weight losses in TG diagram. Below 120 °C corresponds to desorption of physisorbed water and ethanol, between 120 and 700 °C corresponds to breakage, decomposition and combustion of mercaptopropyl/propylsulfonic acid, and water losses resulting from dehydroxylation reaction. The weight loss for Samples 1 and 2 found to be 12.2 and 24 %, respectively. The expected wt loss of Samples 1 and 2 are 11.95 and 23.46 %, respectively, based on molar composition (Table 1).

Table 3: Assignment of the ^{13}C CP MAS NMR spectra

δ (ppm)[a]	Atom[b]	Species
11.3	C^3	≡S-CH_2-CH_2-CH_2-SO_3H
18.4	C^2	≡S-CH_2-CH_2-CH_2-SO_3H
30.7	C^2	≡Si-CH_2-CH_2-CH_2-SH
33.1	C^1	≡Si-CH_2-CH_2-CH_2-SH
41.3	C^1	≡Si-$(CH_2)_3$-S-S(O_2)-$(CH_2)_3$-Si≡
56.4	C^1	≡S-CH_2-CH_2-CH_2-SO_3H

[a]Chemical shift from polydimethylsilane (PDMS).
[b]Cabon atoms are numbered from the one closest to S atom.

Proton Conductivity and Direct Methanol Fuel Cell

Proton conductivity of 10 and 20 mol% hybrid SO_3H-Si and Nafion117 is shown in Figure 3. Proton conductivity of Sample 2 is 1×10^{-2} S/cm, which is slightly lower than nafion117 (1.8×10^{-2} S/cm) at 100 % RH. If the proton conductivity of the membrane is more than 10^{-3} S/cm, then it is good enough to use an electrolyte in fuel cell technology. In Figure 4, polarization curve of 10 (A) and 25 wt% (B) PTFE in 20 mol% hybrid SO_3H-Si-PTFE polymer electrolyte membrane in direct methanol fuel cell is presented. 10 % PTFE shows better performance than 25 % PTFE containing 20 mol% hybrid SO_3H-Si-PTFE membranes in direct methanol fuel cell, because of higher proton conductivity. The performance of single cell in direct methanol fuel cell (DMFC) was tested at 40°C with input flow rate of methanol at the anode was 5 mL min^{-1} and the cathode input O_2 flow rate was 175 mL min^{-1}. Maximum power density observed was 55 mW/cm^2 at 0.40 V and 137.5 mA/cm^2 by using 10 wt% PTFE in 29 mol% SO_3H-MCM-41-PTFE membrane.

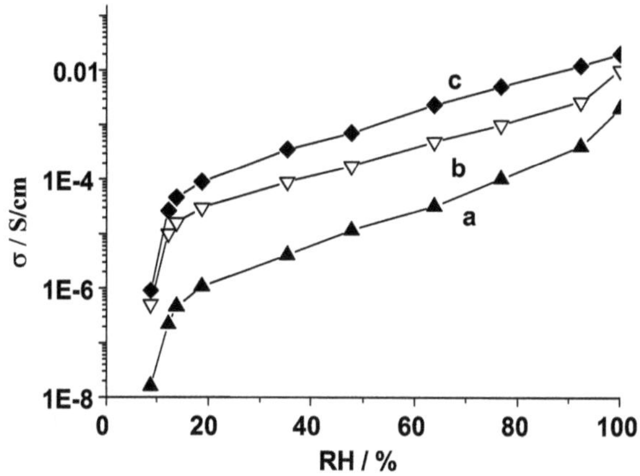

Figure 3: Influence of relative humidity (RH%) on the 10 (Sample 1) (a), 20 mol% (Sample 2) (b) of hybrid porous HO_3S-Si and nafion117 (c)

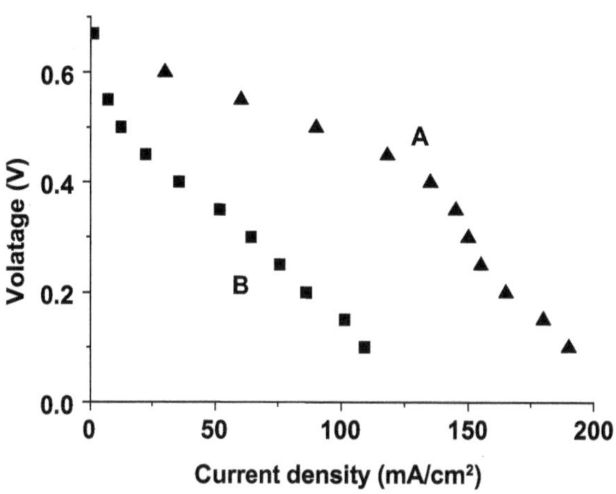

Figure 4: Polarization curve of 10 (A) and 25 wt% (B) PTFE in hybrid porous HO3S-Si-PTFE membrane

Conclusion

Power density of 10 to 25 wt% PTFE in 20 mol% hybrid SO_3H-Si-PTFE polymer electrolyte membranes in direct methanol fuel cell was measured. Maximum power density observed was 55 mW/cm^2 at 0.40 V and 137.5 mA/cm^2. In this report where organic polymeric membranes such as Nafion, polybenzimidazole (PBI), Polyvenylalcohol (PVA), etc. were replaced by much cheaper and higher stable inorganic hybrid porous silica membrane (SO_3H-Si-PTFE). This work may lead an option to use an inorganic electrolyte in future.

References

[1] S. Hamoudi, S. Royer, and S. Kaliaguine; J. Micro. Meso. Mater. 71 (2004) 17.

[2] S. Mikhailenko, D.D. Giscard, C. Danumah, and S. Laliaguine; J. Micro. Meso. Mater. 52 (2002) 29.

[3] I. Diaz, C.M. Alvarez, F. Mohino, J.P. Pariente, and E. Sastre ; J. Micro. Meso. Mater. 44 (2001) 295.

[4] Y. Mori, and T.J. Pinnavaia. Chem. Mater. 13 (2001) 2173.

[5] D. Margolese, J.A. Melero, S.C. Christiansen, B.F. Chmelka, and G.D. Stucky; Chem. Mater. 12 (2000) 2448.

[6] D. Isabel, M.A. Carlos, M. Federic, P.P. Joaquin and S. Enrique; J. Catal. 193 (2000) 283.

[7] C.D. Wagner, W.M. Riggs, L.E. Devis, J.F. Moulder and G.E. Muilenberg; X-Ray photoelectron Spectroscopy. Perkin-Elmer, Eden Prairie, MN.), 1979.

[8] M.H. Zong, Z.Q. Duan, W.Y. Lou, T.J. Smith and H. Wu; Green Chemistry. 9 (2007) 434.

44
Environmental Friendly Technology for Degradation of Dye Polluted Effluent of Textile Industries Using Newly Developed Photo Catalyst

R. B. Pachwarya
Motilal Nehru College (University of Delhi)
Benito Juarez Marg New Delhi-110021
Email: **pachwarya@gmail.com**

Abstract

The treatment of azo dye polluted water is challenging for researchers and environmentalists. Azo dye polluted water is hazardous for living world. Degradation of these azo dye by recently developed environmental friendly photo catalyst Methylene Blue immobilized Resin Dowex-11 is cheaper and very good alternative to replace costly traditional treatment and less efficient technologies for industrial application. This catalyst has potential to degrade all dye molecules and we recover 99.99 % clean and transparent water from dark color textile effluent wastewater within few hour treatments by Methylene Blue immobilized Resin Dowex-11, photo catalyst. We use Sudan IV common name is Scarlet red as model dye for experimental purpose; this is Azo dye and used in textile industries. The size of catalyst particle is 20–50 mesh so it can filter easily. Activity of catalyst remains un-effected on continuous use.

Introduction

Treatment of wastewater of textile industries, paper industries, food industries, chemical industries contain residual dyes which are not readily bio degradable is challenging for researchers. Advance oxidation process (AOP) is recently more developed technique in photochemistry and this technique is best for treatment of textile industries effluent (wastewater). We use Methylene Blue Immobilized Resin Dowex-11 for oxidative degradation of dye contaminants. Methylene Blue Immobilized resin is a recently developed photo catalyst and has vast potential of degradation of azo dyes. Methylene Blue can be act as sensitizer for light induced process due to sensitization of catalyst by photo radiation, electron migrates from valance band (VB) to conduction band (CB) and holes are formed in valance band; these holes can generate hydroxyl radicals which are highly oxidizing in nature. Probably hole can react with dye molecule and abstract electron from dye molecule and process of degradation start. Aim of the present work is to gain attention of researchers toward utilization of solar energy for Degradation of dye pollutants by Photo catalyst and find out new photo catalyst for different applications. A new developed cheap and better photo catalyst and its applicability. The potential of degradation of (Sudan IV, is an Azo dye used as model compound for experiment) dyes by (Methylene immobilized Resin Dowex-11) catalyst is better and this catalyst work well in dim light also. We apply it in different condition and find out the effect of different parameters on rate of degradation. These parameters are (1) variation in catalyst loading, (2) variation in dye concentration, (3) variation in pH, (4) variation in light intensity, (5) effect of dissolve oxygen. All the sets are observed for more then 3 hour.

Materials and Methods

Model Dye Information: Sudan-Iv = Loba (Loba Chemicals India)Molecular Formula= $C_{24}H_{20}N_4O$ MolecularWeight:=380.449 λ_{Max}=520 nm. Class=Azo

Photo catalyst: Used Chemicals: We prepare Photo catalyst by following materials Dowex-11 Resin

M.M. Srivastava, L.D. Khemani, S. Srivastava, *Chemistry of Phytopotentials: Health, Energy and Environmental Perspectives*, DOI:10.1007/978-3-642-23394-4_44,

20–50 mesh (Sisco Chemicals India Mumbai), Methylene Blue Hydrate For Microscopy, (C. I . No 52015) (Loba Chemicals India).

Preparation of Photo Catalyst: We prepare approximately M/1000 concentration solution of Methylene Blue in Double distilled water and add Dowex-11 resin in this solution and shake well. Put this mixture for 3 days for complete immobilization of Methylene Blue In side the pores of resin. All the process is carried out in dark place. After three days we can filter Methylene blue immobilized resin from solution, wash this resin by double distilled water twice and used it as photo catalyst.

Role of Methylene blue: Methylene blue is photosensitized dye. When molecule of Methylene Blue immobilized in pores of resin (fill in pores of resin). Methylene blue is photosensitized dye and becomes exited by absorbing photons of light radiations. In first electronic excitation, electron transfers into singlet state and through inter system crossing (ISC) electron can transfer to triplet state of Methylene Blue. Further inter molecule electronic interaction occurs between resin, Methylene blue and solution mixture and resultant is formation of holes, hydroxyl radicals and Supra oxide ions (o^{-}), these are highly oxidative in nature.

Analytical methods: The change in dye concentration is observed simply by Shimadzu-160 UV/Visible spectrophotometer. We suck out 10 ml of solution by

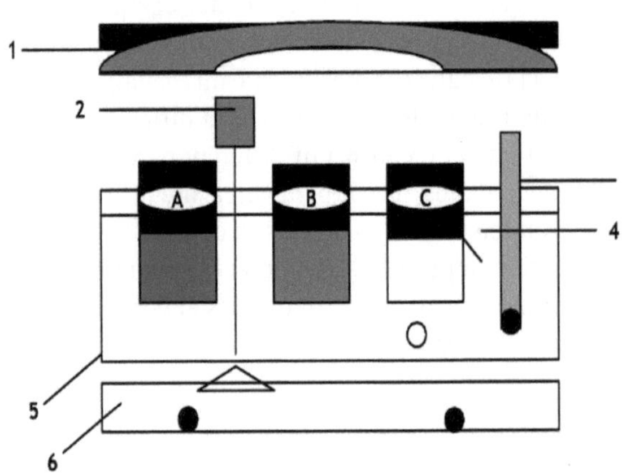

Fig. 1: Schematic diagram of the experimental setup: (1) UV lamp; (2) stirrer; (3) Thermometer; (4) Glass reactor (beaker); (5) water bath; (6) magnetic stirrer. Beaker (a) Dye solution without catalyst shows no change in color (b) shows (degradation of dye molecules) change in color by the action of photo catalyst (c), solution become colorless this shows the complete degradation occurs by the action of Methylene Blue immobilized Resin Dowex-11 (20–50 mesh) in three hour irradiation time duration

pipette at the time interval of 15 minutes and observe changes in percentage transparency of dye solution.

Experimental set up and Experimental procedure: The photo reaction is carried out in glass reactor which contains mixture of Dyes (Sudan IV) and photo catalyst. Solution of reactor is continuously stirred by magnetic stirrer during the experiment. The solution is illuminated by 500 W halogen lamp above the reactor.

At fixed time interval, 10 ml solution is taken out from reaction mixture. Filters catalyst particles and observe variation in transparency of colored water by the help of spectrophotometer (160UV/Visible spectrophotometer).

Following schedule of observation is followed in all experimental process.

(1) In first experiment we observe bio degradability of dyes (without catalyst). We put dye solution in solar/lamp light for 3 days and after 3 days we observe changes in dye concentration. If no changes are found in dye concentration then we carry out next experimental step.
(2) In second experiment which is carried out in dark for test action of catalyst in dark we put reaction mixture (dye solution and catalyst) in dark chamber there is no change observed in dye concentration.
(3) In third experiment we add pure resin without immobilization of Methylene Blue. This experiment is carried out for test action of resin (without immobilized). We observe that absorption process occurs. Absorption efficiency is reduced after first absorption and stop after 2–3 time of use due to fill the pores of resin.
(4) In fourth experiment we use Methylene blue immobilized resin and we observe that polluted and darkly colored mixture transform into transparent water like mineral water. This catalyst can be used many times as there is no effect on efficiency of degradation of dye molecules.

Results and Discussion

Probable Chemical Reaction of This Degradation: Methylene Blue Immobilize Resin Dowex-11. This is newly developed photo catalyst. The dye immobilize in porosity of resin is Methylene Blue. Methylene Blue is photo sensitive in nature, when light radiation is irradiated on it electronic transition occurs from valance band (VB) to conduction band (CB) and through intersystem crossing (ISC) electron reach in to triplet state of Methylene Blue. After it intermolecular electronic transition start between resin, Methylene blue dye molecules, water molecules, dye molecules and dissolved oxygen, resultant through chain process, holes, hydroxyl radicals and Supra oxide ions (o^-) are produced and these are highly oxidizing in nature, by the action of holes, hydroxyl radicals and Supra oxide ions (o^-) on Azo dyes, are transformed in simple organic compounds.

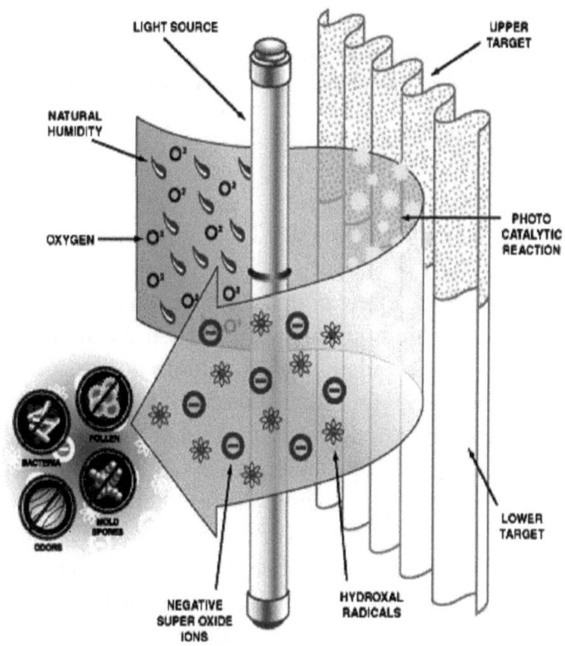

Fig. 2: Shows the Process of generation of holes, hydroxyl radicals and Supra oxide ions (o^-). These possess highly oxidative characteristic

Effect of Amount of Catalyst

The amount of the photo catalyst affects the rate of photo catalytic degradation. We observe effect of variation in amount of photo catalyst on the rate of degradation at constant pH 8. We find out that as concentration of catalyst increases rate of degradation also' increases. Increase in the rate of degradation with increase in amount of catalyst is due to availability of more catalyst surface area for absorption of quanta and interaction of molecules of reaction mixture with catalyst, result is that number of holes, hydroxyl radicals and supra oxide ions (o^-) are increased. These are principle oxidizing intermediate in advance oxidation process and increases the rate of degradation.

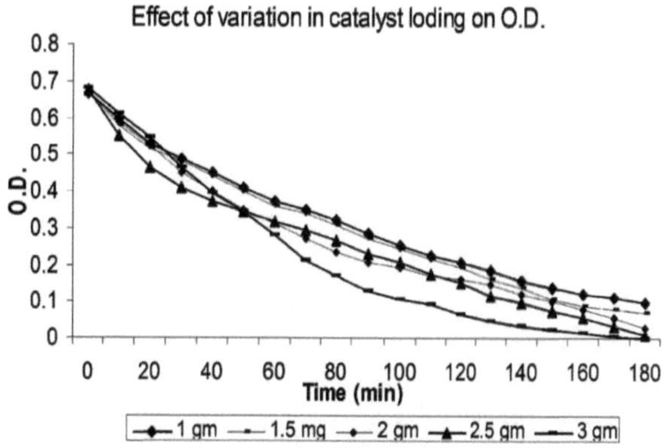

Fig3: Effect of catalyst loading on photo catalytic degradation (Temperature: 303 K, solution volume: 400 ml, initial dye concentration: 40 mg/l, pH 7.5, UV /visible lamp: 10.4 $mWcm^{-2}$.)

Effect of Initial Dye Concentration

We observe the effect of change in dye concentration on photo catalytic degradation and find out that as concentration of dye increases the rate of degradation decreases. This effect may be caused due to following reason

1. This is the most important reason :- As the concentration of dye increases number of photons reaching to catalyst surface decreases result is that less number of catalyst molecules undergoes excitation and due to this effect rate of formation of holes, hydroxyl radicals and supra oxide ions (o^-) is decreased so rate of degradation also decreased.
2. Catalyst surface area is fixed so as the concentration of dye increases rate of degradation decreases because limited number of dye molecules attach at the active site of the catalyst and remaining dye molecules persist in solution until earlier attached molecules are degraded and number of active site of catalyst also decreases due less availability of photons for excitation of catalyst molecules. Competitions between dye molecules to attach to the active site also effect rate of degradation. At higher concentration number of dye molecules are also high so more will be the competition for attachment to active site of catalyst between the dye molecules and result is reduction in the rate of degradation.

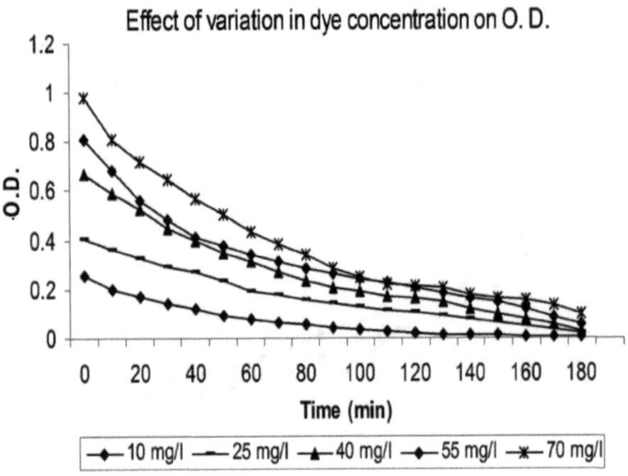

Fig. 4: Effect of initial dye concentration on degradation (Temperature: 303K, solution volume: 400 ml, pH 7.5, UV/visible lamp: 10.4 mW cm^{-2}.)

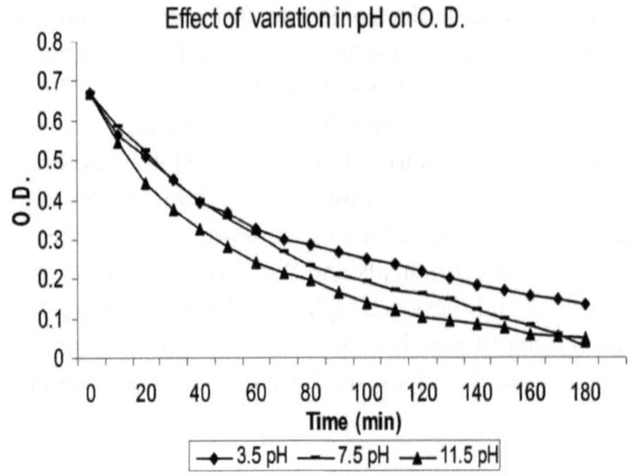

Fig. 5: Effect of pH on degradation (Temperature: 303 K, solution vol.: 400 ml, initial dye concentration: 40 mg/l, UV lamp: 10.4 mW cm^{-2})

Effect of pH

We observe that effect of pH on rate of degradation of dye molecules is very interesting. The results shows that rate of degradation is very low in high acidic pH range, if pH is lower than 3.5 rate of degradation is very less, as pH increases rate of degradation also increases when pH reaches to basic range the rate of degradation increases fast, in pH range 7.5 to 9 rate of degradation is very good. On further increasing pH the rate of degradation also start to decrease after pH range 10 or above rate of degradation is less and decreases as pH increases. So we conclude that rate of degradation in basic medium is higher than acidic medium. The increase in rate of photo catalytic degradation may be due to the more availability of ^{-}OH ions in pH range 7.5 to 9 which will generate more OH radicals by combining with the holes which are formed due to electronic excitation in catalyst. Formation of hydroxyl radicals are responsible more for the photo catalytic degradation than supra oxide ($^{\cdot}O$). At higher pH the rate of degradation decreases. This effect is due to competition between ^{-}OH groups to attach to the active site of catalyst, so rate of attachment of ^{-}OH group decreases. Result is that formation of hydroxyl radicals ($^{\cdot}OH$) decreases and due to this reason rate of degradation also decreases.

Graphical representation of pH effect is shown below

Effect of Light Intensity

We observe the effect of light intensity on rate of degradation. We find out that as light intensity increases the rate of degradation of dye molecules also increases up to certain extent and after it no changes are observed in rate of degradation. These changes in rate of degradation of dye molecules by variation in light intensity is due to the reason that as light intensity increases number of photons reaching the catalyst surface also increases so number of exited catalyst molecules increases and result is increase in the number of holes, hydroxyl radicals and Supra oxide ions (o^{-}) and rate of degradation of dye molecules also increases.

Graphical representation of light intensity variation is shown in Figure 6.

Effect of dissolved oxygen on rate of degradation: We observed the effect of Dissolved oxygen on rate of degradation, as dissolved oxygen increase in dye solution rate of degradation also increase. We observed that when oxygen gas is passed through reaction mixture the rate of degradation increases but when Nitrogen or any other non-reacting gas is passed through this solution no effect is observed on rate of degradation. This effect is may be due to more availability of oxygen for formation of supra oxide ($^{\cdot}O$) and hydroxyl radical. These are highly oxidative in nature and increase the rate of degradation of dye molecules.

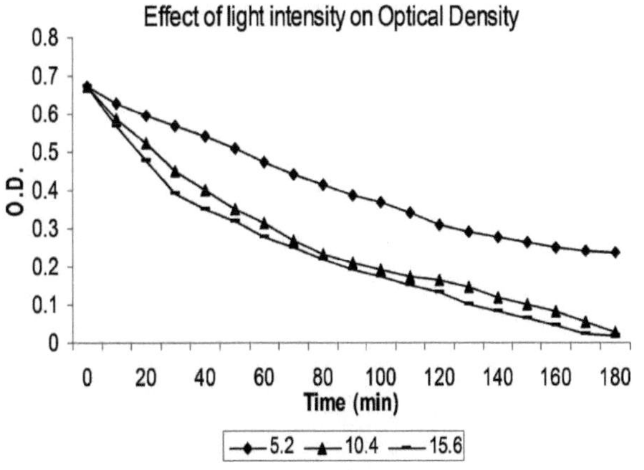

Fig. 6: Effect of variation of light intensity on degradation (Temperature: 303 K, solution volume: 400 ml, initial dye concentration: 40 mg/l, pH 7.5.)

1. Effect of variation in dye concentration: – As concentration of dye increases the rate of degradation of dye deceases.
2. Variation in amount of catalyst: – As concentration of catalyst increases the rate of degradation of dye molecules also increases.
3. Variation in pH: – In acidic range of pH the rate of degradation is very less as pH increases rate of degradation also increases and between pH ranges 7.5 to 9 rate of degradation is faster on further increase in pH the rate of degradation decreases.
4. Variation in light intensity: – On increasing light intensity the rate of degradation of dye molecules increases up to certain limit after it there is no further changes in the rate of degradation.
5. Effect of dissolved oxygen: – Rate of degradation increases up to some extent on increasing dissolved oxygen in dye solution.

Conclusion

After long observation we conclude that this photo catalyst (Methylene Blue immobilized Resin Dowex-11) has good potential of degradation of Azo dyes/ dyes into simple molecules and purify textile effluent (wastewater) which contains large amount of non-fixed dyes mostly Azo dyes. These Azo dyes are non-bio degradable.

We observe the effect of different parameters given in order

References

1. A. Akyol, H. C. Yatmaz and M. Bayramoglu; Appl. Catal. B Environ. 54 (2004) 19
2. Sakthivel, B. Neppolian, M. V. Shankar, B. Arabindoo, M. Palanichamy and V. Murugesan; Sol. C 77(2003) 65
3. C. Hachem, F. Bocquillon, O. Zahraa and M. Bouchy; Dye Pigments 49 (2001)117
4. K. Tanaka, K. Padermpole and T. Hisanaga; Water Res. 34(2000)327
5. C. Hu, J. C. Yu and Z. Hao, Catal.B:Environ. 42(2003) 47
6. C. G. Silva, J. L. Faria and P. K. Wong; Appl. J. Photochem. Photobiol. A: Chem. 155(2003)133
7. S. N. Frank and A. J. Bard; J. Phys.Chem. 81(1977) 1484

45

Biohydrogen Production with Different Ratios of Kitchen Waste and Inoculum in Lab Scale Batch Reactor at Moderate Temperatures

S. K. Bansal, Y. Singhal and R. Singh
Department of Chemistry, Faculty of Science, Dayalbagh Educational Institute,
Dayalbagh, Agra 282110,
E -mail id: bioenergy.dei2011@gmail.com

Abstract

In this study, the effect of ratio of kitchen waste and inoculum like 80:20, 50:50, and 60:40 on biohydrogen production was studied. The volume of each batch reactor was 400 ml and all batch reactors were kept at 37°C in incubator and the incubation time was 10 days. The maximum biohydrogen production was found in the 80:20 (kitchen waste: inoculum) batch reactor i. e. 13.3 % and this was on the 8th day of the start of batch reactor. The reduction in the various physical parameters in the 80:20 batch reactor was studied shows the progress in the reactors and the degradation in this reactor is high as compared to the other reactors i. e. TS-49.25 %; TDS- 68.75 %; VS-55.83 %; TSS-55.83 % and COD was 35.75 %. This study summarizes the comparative study of the biohydrogen production in the different ratios of inoculum and kitchen waste.

Introduction

Dependence on fossil fuels as the main energy sources has led to serious energy crisis and environmental problems, i.e. fossil fuel depletion and pollutant emission. Fossil fuel being one of the major contributors as energy source has been rapidly consumed over the years. Being an unsustainable fuel, it became tedious to meet the global demand. Not only that, it is also one of the significant causes of global warming and air pollution. Based on these reasons the need of finding alternative sources of fuels arose. Hydrogen, being a highly priced commodity to the industry has become a favorite energy carrier and as an alternative fuel. Renewable and clean forms of energy are one of society's greatest need, so we need alternative energy resources which are cheap, renewable and do not cause pollution. Focus is being given to alternate and renewable sources such as solar, wind, thermal, hydroelectric, biomass, etc [1, 2]. In recent years, various advanced strategies for the production of hydrogen from renewable have been explored and are still under investigation. Hydrogen has been considered as an ideal and clean energy carrier for the future since it produces water only upon combustion [3]. Unlike coal and petroleum, hydrogen does not contribute to greenhouse effect, depletion of ozone and acid rain [4]. It is a clean fuel with no CO_2 emission, environmentally benign, safe and also has a high energy content i.e. 120 MJ /kg as compared to gasoline which has only 44 MJ/kg of energy content. Hydrogen can be divided into thermal and non-thermal processes. Biomass gasification is thermal processes, photosynthetic or fermentative processes produce hydrogen in a non-thermal manner. Dark fermentation has a maximum hydrogen yield of 24.3 mol H_2/mol glucose. Dark fermentation is a best way to produce biohydrogen due to its low cost and its high production efficiency [5,6]. Starch containing biomass such as waste agricultural products offer special advantages for biohydrogen production since those raw materials are readily available, cheap and starch can easily be hydrolyzed to carbohydrates [7]. Kitchen waste contains mainly organic material which can easily undergo fermentation. It is also referred to as the organic fraction of municipal solid waste (OFMSW) if combined with irrecoverable paper residues [8]. The different wastes and wastewater are utilized for the production of biohydrogen [9, 16].

M.M. Srivastava, L.D. Khemani, S. Srivastava, *Chemistry of Phytopotentials: Health, Energy and Environmental Perspectives*, DOI:10.1007/978–3-642–23394-4_45, © Springer-Verlag Berlin Heidelberg 2012

Materials and Methods

Kitchen Waste

Kitchen waste obtained from canteen of Dayalbagh Educational Institute, Agra was used for the investigation in the current study. The characteristics of the kitchen waste are given in the Table 1. This waste comprises of various types of food waste materials which are rich in carbohydrate content. The non biodegradable fraction was separated out and the remaining biodegradable portion was mixed enough to make slurry by adding a little amount of water.

Mixed Culture

Bacteria are responsible for the biohydrogen production and these are mainly found in the natural environment such as cow dung, soil, sewage treatment sludge [10]. In the present study, cow dung is used as the mixed culture source. Mixed culture source has various benefits over the pure culture as the former one is more practical in use and are also simple to operate and also easy to control [11]. Mixed culture source i. e. cow dung was pretreated to harness the hydrogen producing bacteria and suppress the hydrogen consuming bacteria. Mixed culture source contains a wide variety of bacteria i. e. Clostridium, Rhodeobacter, Enterobacter, Psuedomonas etc [12]. To enhance the production, nutrient media was added containing 200 mg of KH_2PO_4, 14 mg of $MgCl_2 \cdot 4H_2O$, 2 mg of $Na_2MoO_4 \cdot 4H_2O$, 2 mg of $CaCl_2 \cdot 2H_2O$, 2.5 mg of $MnCl_2 \cdot 6H_2O$, and 10 mg of $FeCl_2 \cdot 4H_2O$ [13].

Experimental Procedure

Hydrogen production experiments in the present study were carried out in the lab scale batch reactors of volume 500 ml. The kitchen waste and inoculum were taken in the different ratio i. e. 80:20, 50:50 and 60:40 and the working volume was adjusted to 400 ml. The batch reactors were purged with the N_2 gas for 10 minutes to create an anaerobic environment and then these were kept in the incubator at 37^{0}C.

Table 1: Various parameters of kitchen waste

Parameter	Unit	Kitchen waste
TS	mg/l	27.06
MOISTURE	%	75–80
ASH	Gm	1.509
C.O.D	mg/l	14,731
pH (slurry)		6

Monitoring

Kitchen waste parameters and whole batch reactors at different ratio of kitchen waste and inoculum parameters like TS, TDS, TSS, COD, ASH were analyzed according to APHA-AWWA methods [14]. The biogas was also monitored after every 24 hours for progressing of the batch reactor.

Analyses

The biogas produced were measured using a gas chromatograph (GC, Rastech) using a thermal conductivity detector having 1:8 m × 3:2 mm stainless-steel column packed with molecular sieve 5A with N_2 as a carrier gas at the flow rate of 30 ml/min. The temperatures of injector, column and detector were kept at 50°C, 80°C, and 90°C, respectively in Gas Chromatography [15]. The sample was taken in the 1 ml Hamilton glass syringe and .5 ml of sample introduced in gas chromatograph.

Results and Discussion

Physical Parameters

Various parameters were studied during the progress of the reactor both initially and finally like TS, VS, TDS, TSS, Fixed Solids and COD. The reduction in various parameters can be seen from the figures. Figures 1,2,3 shows the reduction in the various physical parameters which shows the degradation in carbohydrate content as the reaction proceed and indicates the progress of the batch reactor.

COD values also get reduced compared to 80:20 and 50:50 in which reduction is 35.75 % and 41.96 % respectively [Figure 3, 4]. The COD degradation was seen in 60:40 i. e. 49.18 % and this shows that the carbohydrate is getting degraded and results in the formation of biogas. pH was also measured and found to lowered as compared to initial. This may be due to the formation acids in the reactor due to which pH decreases and all these parameters shows reduction during the study which indicate the progress of the reactors.

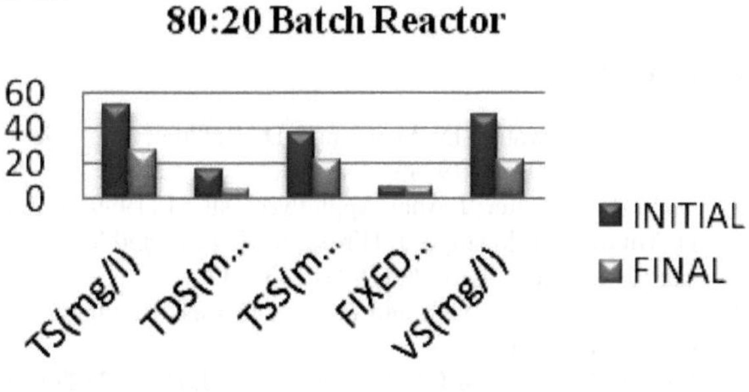

Fig. 1: 80:20 Batch Reactor.

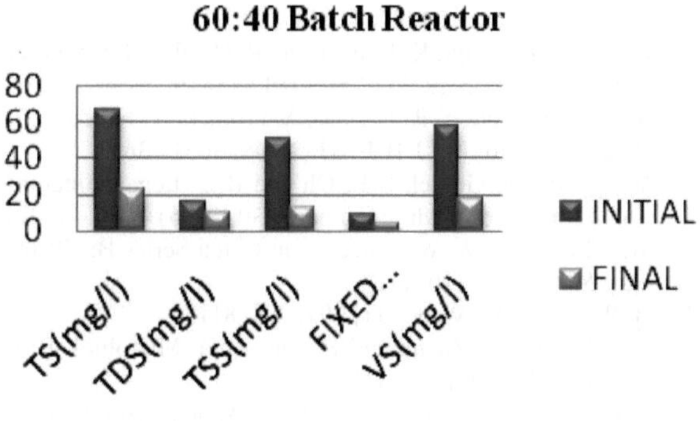

Fig. 2: 60:40 Batch Reactor

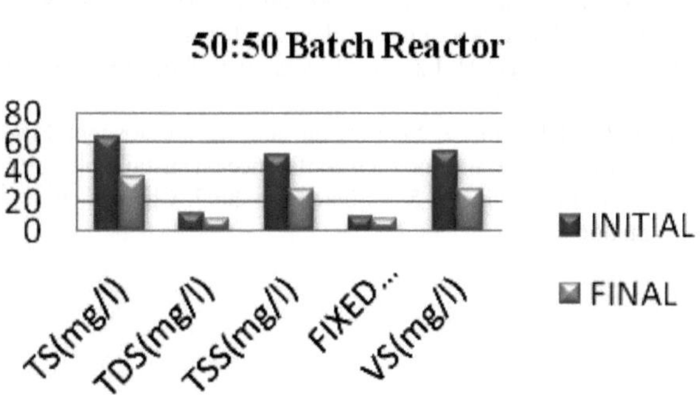

Fig. 3: 50:50 Batch Reactor

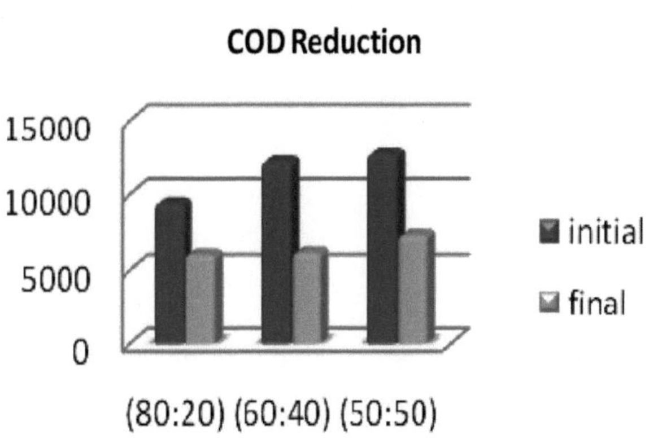

Fig. 4: COD reduction in different ratio batch reactor

Table 2: % Degradation in the various parameters

Ratio	TS (mg/l)	TDS (mg/l)	TSS (mg/l)	COD (mg/l)	VS (mg/l)	Fixed Solid (mg/l)
80:20	49.25	68.75	40.95	35.75	55.38	7.14
50:50	43.02	33.33	45.31	41.96	47.76	16.66
60:40	66.0	37.5	75	49.18	68.27	52.17

Biogas Production

At moderate temperatures biogas produced in different ratio batch reactors. Main constituent of the biogas were CH_4, CO_2 and small amounts of H_2. In 80:20 ratio, the biohydrogen significantly increased with time

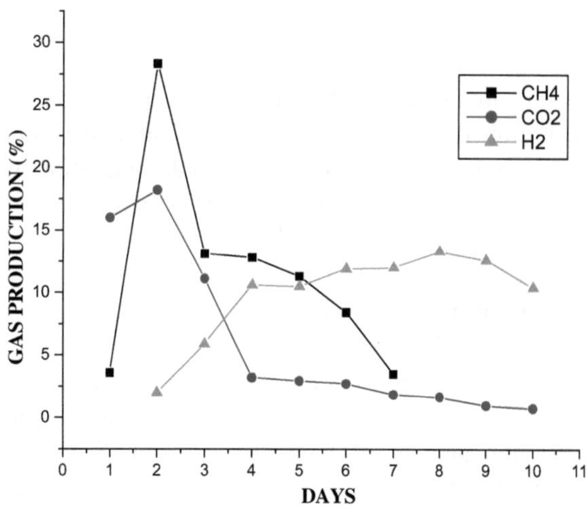

Graph 1: 80:20 ratio biogas production

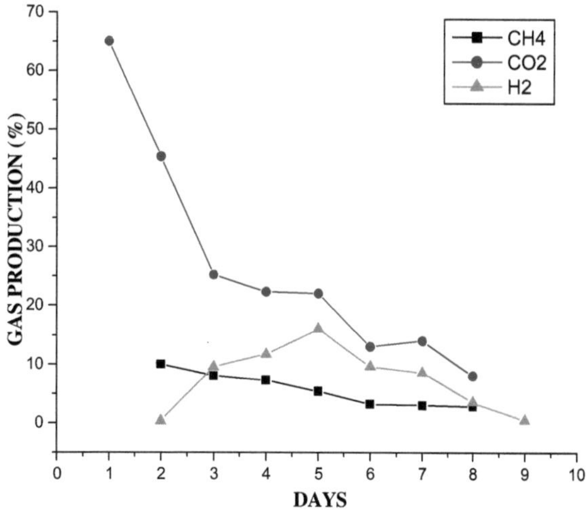

Graph 2: 50:50 ratio biogas production

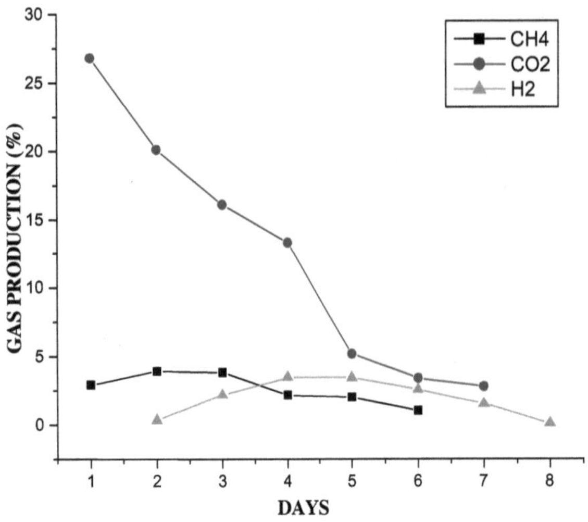

Graph 3: 60:40 ratio biogas production

as compared to the 60:40 and 50:50 ratio. The maximum biohydrogen production in 80:20 at 8th day was 13.3 % and as the reaction proceeded the formation of CH_4 was reduced and of hydrogen was increased.

Conclusion

In the present study, the ratio of kitchen waste to inoculum was carried as 80:20, 50:50 and 60:40. The maximum amount of biohydrogen production was observed in 80:20 at moderate temperatures. The reduction rates in other parameters were: TS-49.25 %, TDS-68.75 %, TSS-55.83 % and COD-35.75 %.

References

1. A. V. Bridgewater, D. Meier and D. Radlein; Org. Geochem. 30 (1999) 1479–93.
2. A. V. Bridgewater; J. Anal. Appl. Pyrolysis 51 (1999) 3–22.
3. H. Yokoi, T. Tokushige, J. Hirose, S. Hayashi and Y. Takasaki; Biotechnol. Lett. 20 (1998) 143–147.
4. G. Chittibabu, K. Nath and D. Das; Process Biochem. 41 (2006) 682–688.
5. P. C. Hallenback and J. R. Benemann; Int. J. H. E 27 (2002) 1185–93.
6. D. B. Levin, L. Pitt and M. Love; International Journal of Hydrogen Energy 29 (2004) 173–189.
7. H. Argun, F. K. Ilgi, K. Kapdan and R. Oztekin; Int. Journal of Hydrogen Energy. 33 (2008) 1813– 1819.
8. H. Zhu, W. Parker, R. Basnar , A. Proracki, P. Falletta, M. Beland, P. Seto; Int. J. H. E. 33 (2008) 3651– 3659.
9. W. Steven, V. Ginkel, S. E. Oh and B. E. Logan; International Journal of Hydrogen Energy 30 (2005) 1535 – 1542.
10. W. J. Long and W. Wei; Science in China Series B: Chemistry 51 (2008) 1110–1117.
11. J. Wang and W. Wan. 34 (2009) 799–811.
12. H. H.P Fang, T. Zhang and H. Liu; Appl. Microbiol. Biotechnol. 58 (2002) 112–118.
13. C. C. Chen, C. Y. Lin and J. S. Chang; Appl. Microbiol. Biotechnol. 57 (2001) 56–64.
14. American Public Health Association (APHA- AWWA), Standard methods of physical parameters.
15. C. Y. Lin and C. H. Lay; Int. Journal of Hydrogen Energy 30 (2005) 285– 292.
16. S. Jayalakshmi, V. Sukumaran and Kurian Joseph; Proceeding of the international conference on sustainable solid waste management. 5–7 September 2007, Chennai, India, pp 356–362.

46

Synthesis and Characterization of Some Schiff Bases and Their Cobalt (II), Nickel (II) and Copper (II) Complexes via Environmentally Benign and Energy-Efficient Greener Methodology

K. Rathore and H. B. Singh
School of Chemical Sciences, Dept. Of Chemistry, St. John's College, Agra
Email: khushboo.kiran@gmail.com,hbs.chemistry@gmail.com

Abstract

Bifunctional tetradentate Schiff bases of o-vanillin and 2-hydroxy-1-naphthaldehyde with ethylenediamine and their complexes of cobalt (II), nickel (II) as well as copper (II) have been synthesized. Schiff base ligands and their metal complexes have been characterized by elemental analysis, molecular weight determination, molar conductance measurements, and UV-Vis., IR, NMR (^1H and ^{13}C) and mass spectral studies. Compounds have been synthesized in an open vessel under microwave irradiation using a domestic microwave oven. The free ligands and their complexes have been tested in vitro against a number of microorganisms in order to assess their anti-microbial properties.

Introduction

The world "Green Chemistry" has emerged as a bright new approach for the traditional field of synthetic chemistry. The concept of green chemistry is enshrined the prevention of waste, the use of safe, environmentally being solvent where possible and use on renewablefeedstocks[1].Many applications of microwaves, as an efficient heating source for organic reactions, have been reported in the literature[2].Microwave heating has proved to be very useful tool to carry out certain organic transformations which not only excludes the use of hazardous non-eco friendly solvents but also enhances the reaction rates greatly. The main advantage of microwave assisted organic synthesis is the shorter reaction time using only small amount of energy. Schiff base and the relevant transition metal complexes are still found to be of great interest in inorganic chemistry although this subject has been studied extensively[3,5].The chelating abilities and analytical and biological applications of these compounds have attracted remarkable attention[6].Tetra dentate Schiff bases are well known for their coordination with various metal ions, forming stable compounds [7,8]. Because of their potential chelating nature symmetrical bis tetradentate Schiff bases of 1, 2-diamines with *o*-hydroxy aldehydes/ketones have been prepared and studied intensively.

In the present work, complexes of Co (II), Ni (II), and Cu (II) with symmetrical bifunctional tetradentate Schiff bases derived from o-vanillin and 2-hydroxy-1-naphthaldehyde with ethylenediamine, have been synthesized, characterized the chemical structure by IR and NMR spectral analysis and to study the anti-microbial activity.

Materials and Methods

O-vanillin and 2-hydroxy-1-naphthaldehyde were purchased from Hi-media. Cupric acetate, cobalt acetate, nickel chloride and ethylenediamine used were of E. Merck grade. Dimethylformamide was obtained from Sisco Chem, Bombay, India. Ethanol and other solvents of A. R. grade were used as received.

Synthesis of Schiff Bases

Unimolar ratios of respective aldehydes (*o*-vanillin or 2-hydroxy-1-naphthaldehyde) with amine (ethyl-

M.M. Srivastava, L. D. Khemani, S. Srivastava, *Chemistry of Phytopotentials: Health, Energy and Environmental Perspectives*, DOI:10.1007/978-3-642-23394-4_46, © Springer-Verlag Berlin Heidelberg 2012

enediamine) were carried out in minimum quantity of ethanol. The reaction mixture was taken in an open beaker and then irradiated inside a microwave oven. The reaction was completed in a short period of 2 minutes. A solid mass separated out on cooling. These are washed with ethanol, diethyl ether and subsequently dried over anhydrous calcium chloride in desiccator.

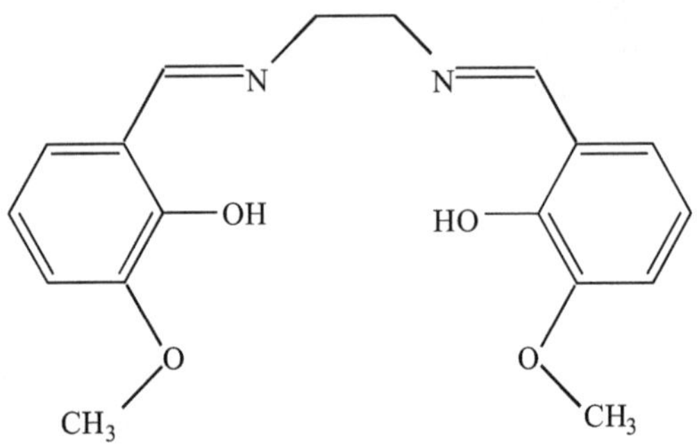

Bis(*o*-vanillinidene)ethylenediamine

L^1H_2

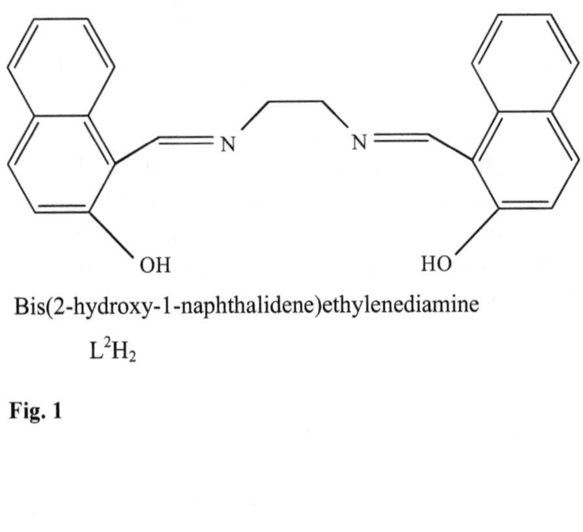

Bis(2-hydroxy-1-naphthalidene)ethylenediamine

L^2H_2

Fig. 1

Synthesis of Metal Complexes

In microwave assisted synthesis, the complexes were synthesized by mixing a hot DMF-ethanolic solution of Schiff bases, with an ethanolic solution (30 ml) of the respectively metal salts. The reaction mixture was taken in an open beaker and then irradiated inside a microwave oven until completion of the reaction. The resulting products were dried over anhydrous calcium chloride in desiccator. The purity of the compounds was checked by TLC.

Physical Measurements and Analytical Data

Microanalyses of carbon, hydrogen and nitrogen of the complexes were carried on a Heraeus Carlo Erba 1108 elemental analyzer. Metal contents were analyzed gravimetrically. IR spectra were recorded on a Perkin-Elmer infrared spectrophotometer in the range 4000–400 cm^{-1}. Electronic spectra of the complexes were recorded on a Helios alpha. ^1H NMR spectra of the ligands and their metal (nickel) complexes and ^{13}C NMR spectrum of the ligand (L^1H_2) were recorded on a Bruker Avance 400 MHz. The ESI mass spectra were recorded on a JEOL-Accu TOF JMS-100LC Mass spectrometer. Molar conductance was measured at room temperature in DMSO using a dip type cell electrode. The molecular weights were determined by the Rast Camphor method.

Antibacterial Activity

Antibacterial activity was evaluated by the paper disc method. The nutrient agar medium (peptone, beef extract, NaCl and agar-agar) and 5 mm diameter paper discs of Whatman No. 1 were used. The compound was dissolved in DMSO in 500 and 1000 ppm concentrations. The filter paper discs were soaked in different solutions of the compounds, dried and then placed in the petriplates previously seeded with the test organisms E. coli and S. aureus. The plates were incubated for 24–30 hours at 28±2°C and the inhibition zone around each disc was measured[9].

Antifungal Screening

The antifungal activity of the ligands and their corresponding complexes was evaluated against Aspergillus niger and Fusarium oxysporum by the agar plate technique. The compounds were dissolved in 50 and 100 ppm concentrations in DMSO and then were mixed with in the medium. These petriplates were wrapped in the polythene bags containing a few drops of alcohol and were placed in an incubator at 25±2°C. The activity was determined after 96 hours of incubation at room temperature (25°C)[10].

Results and Discussion

The observed molar conductance (7.9 to 10.7 ohm^{-1} cm^2 mol^{-1}) of all the complexes in 10^{-3} M dimethyl sulfoxide solutions suggest the non- electrolytic nature[11] of these complexes. The resulting complexes are non-hygroscopic, air stable, colored solids. The physical characteristics of the ligand and its derivatives are enlisted in Table 1.

Infrared Spectra

The IR spectra of the metal complexes show significant changes compared to the ligands. In the IR spectra of the Schiff base ligands a sharp band observed at 1631 and 1638 cm^{-1} is assigned to the √(C=N) mode of the azomethine group. This shifts to lower wave numbers in all the complexes, suggesting the coordination of the azomethine nitrogen to the metal centers. This is further substantiated by the presence of a new band at 490–507 cm^{-1}, assignable to √ (M-N)[12]. The characteristic phenolic √(O-H) mode presences in the ligands were observed at 3444 and 3443 cm^{-1}. The disappearance of phenolic √ (O-H) band in all the complexes under study suggests the coordination by the phenolic oxygen after deprotonation to the metal ion[13]. The bands at 1251and 1257 cm^{-1} due √ (C-O) phenolic were also observed in the ligands. The coordination of phenolic oxygen to the metal atom is further supported by the shifting of √ (C-O) phenolic lower wave number in all the metal complexes[14].

The appearance of a new band at 551–569 cm^{-1} in all the complexes due to √ (M-O)[14]further substantiates it. The overall IR data suggestthe bifunctional tetradentate nature of the ligands.

Electronic Spectra

The absorption bands observed at 17,300 cm^{-1} and 22,000 cm^{-1} in the case of Co(L^1) and at 16,300 and 21,700 cm^{-1} in that of Co(L^2) complexes may be assigned to the transition $^2A_{1g} \rightarrow {}^2B_{1g}$ and $^1A_{1g} \rightarrow {}^2E_g$ respectively. These bands are in accordance with the square planner geometry of the complexes[15]. There are three bands showed at 16,000, 18,500 cm^{-1} in Ni(L^1) and 16, 500, 19,147 cm^{-1} in Ni(L^2) complexes. These bands may be attributed to $^1A_{1g} \rightarrow {}^{1s}A_{2g}$ and $^1A_{2g} \rightarrow {}^1B_{1g}$ transitions[16]respectively, exhibiting to square planner geometry. Further, a strong band observed at 25,000–27,127 cm^{-1} is attributed to the charge-transfer transition[17]. The Cu(L^1) and Cu (L^2) complexes displayed absorption bands in the region 18,880 and 17,739 cm^{-1} in which is assigned to the $^2E_g \rightarrow {}^3T_{2g}$ respectively, corresponding to square planner geometry[17].

Table 1: Physical& analytical data of the ligands and their complexes

Compounds	M.P. (°C)	Yield (%)	Mol. wt. Found/ (Calcd.)	Color	Found/(Calcd.)%			
					C	H	N	M
L^1H_2	142	85	319 (328)	Yellow	65.00 (65.85)	5.83 (6.09)	8.31 (8.54)	–
L^2H_2	275	78	357 (368)	Yellow	78.02 (78.26)	5.33 (5.43)	7.59 (7.61)	–
[Co (L^1)]	260	79	378 (385)	Brown	55.89 (56.10)	4.50 (4.67)	7.11 (7.27)	15.02 (15.32)
[Co (L^2)]	>360	75	419 (425)	Dark purple	67.41 (67.76)	4.18 (4.23)	6.37 (6.59)	13.54 (13.88)
[Ni (L^1)]	340d	82	371 (384.69)	Brown	56.10 (56.18)	4.33 (4.68)	7.27 (7.28)	15.12 (15.21)
[Ni (L^2)]	>360	75	420 (424.69)	Dark red	67.13 (67.81)	4.19 (4.24)	6.53 (6.59)	13.59 (13.82)
[Cu (L^1)]	320	80	383 (389.5)	Black	55.18 (55.24)	4.47 (4.60)	7.02 (7.16)	16.23 (16.62)
[Cu (L^2)]	>360	72	421 (429.5)	Dark green	66.71 (67.03)	4.38 (4.65)	6.19 (6.51)	14.09 (14.66)

^{1}H NMR Spectra

The ^{1}HNMR spectra of the ligands and their [Ni (L)] complexes were recorded in DMSO-d_6. The ligands exhibit a singlet at δ 13.496 and 13.09 ppm respectively due to the phenolic (-OH) proton. The disappearance of phenolic proton singlets in the complexes understudy suggests[18] the coordination of phenolic oxygen to the metal ion after deprotonation of the phenolic group. The ligands show multiplets in the region δ 6.985–8.39 ppm attributable to the aromatic protons, which appear almost at the same position in the metal complexes. The signals at δ 3.719, 3.879–3.67, and 8.523–10.83 ppm are due to OCH_3 group, -CH_2 and azomethine (-CH = N-) protons respectively. These proton signals are observed almost at the same positions in the complexes also.

^{13}C NMR Spectrum

^{13}C NMR spectrum of the Schiff base (L^1H_2) recorded in d_6-DMSO, exhibited signals at δ 39.38, 56.14, 58.82, 115.20–148.47, 152.95 and 67.55 ppm are due to solvent, -OCH_3, CH_2, aromatic carbons, phenolic and azomethine carbon[19]respectively.

Mass Spectra

The molecular ion peak for the nickel complexes of L^1H_2 and L^2H_2 ligands, observed at m/z = 385 and 425, confirm the stoichiometry of metal chelates as ML type.

Antimicrobial Activity

The free ligands and their respective metal chelates were tested against selected fungi and bacteria to assess their potential as antimicrobial agents. The results are quite promising. The antimicrobial activity data reveal that the complexes are more active than the free ligands. Nickel complexes exhibited the highest activity. The antimicrobial activity increased with increase in concentration of the tested compounds.

Thus, on the basis of the above studies the expected structures of the complexes may be represented as shown in Fig. 2.

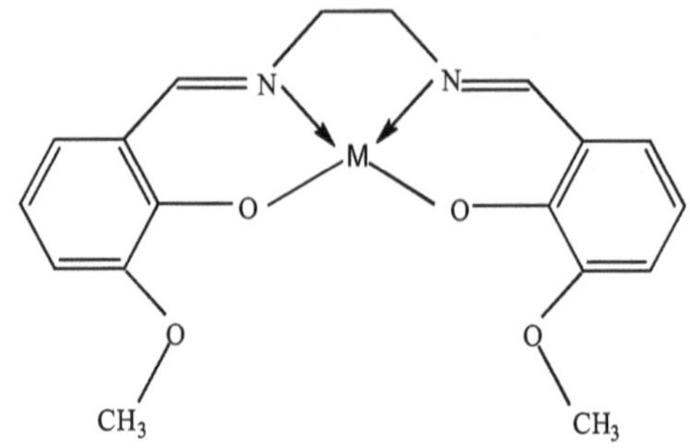

$M(L^1)$

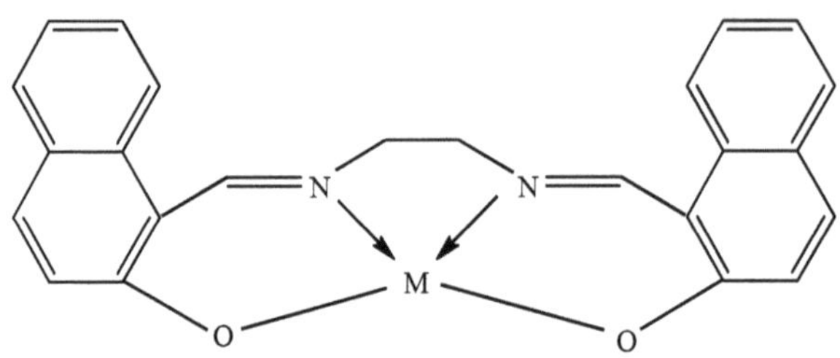

$M(L^2)$

Fig. 2: Where, M = Co (II), Ni (II) & Cu (II)

Conclusion

Microwave irradiation is an efficient and environmentally benign method to accomplish various inorganic synthesizes to afford products in higher yields in shorter reaction periods. The metal complexes synthesized in 1:1 molar ratios with respective ligands were found to be square planar. The ligands behave as a bifunctional tetradentate with the metal atom. Antimicrobial activity showed that the complexes are more active than the parent ligand.

Acknowledgements

The authors are grateful to Principle, St. John's College, for providing laboratory facilities. Analytical and spectral facilities provided by the CDRI Lucknow and BIMR, Gwalior for help in testing antimicrobial activity.

References

1. P. Anastas, and J. C. Warner; Green Chemistry: Theory and practice, Oxford UniversityPress, New York (1998) 30
2. B. Kahveci, M. Ozil, and M. Serdar; Heteroatom Chem.19(1) (2008)38–42

3. N. H. Patel, H. M. Parekh, and M. N. Patel;Transition Metal Chem.30 (2005) 13–17
4. M. Sekerci;XIVth National chemistry congress, Diyarbakir, Turkey (2000) 414
5. M. Sekerci; C. Alkan, A. Cukurovali, and Sayadam;XIIIth National chemistry congress,Samsun, Turkey(1999) 182
6. A. A. Jarrahpour, M. Motamedifar, K. Pakshir, N. Hadi, and M. Zarei;Journal of Molecules9 (2004)815–824
7. S. Achut Munde, N. Amarnath Jagdale, M. Sarika Jadhav, and K. Trimbak Chondhekar; J. Serb. Chem. Soc. 75(3) (2010) 349–359
8. S.A. Sadeek, M. S. Refat; J. Korean Chem. Soc. 2006, 50, 107
9. C. Saxena, D. K. Sharma, and R. V. Singh; Phosphorus, Sulfur and Silicon 85 (1993) 9
10. M. Jain, S. Nehra, P. C. Trivedi, and R. V. Singh; Heterocyclic Communications 9 (2003) 1
11. W. J. Geary; COORD Chem. Rev. 7 (1971) 81
12. R.C. Maurya and P. Patel; Synth React Inorg Met-org Chem. 33 (2003) 801
13. K. Singh, P. Patel, and B.V. Agarwala; Spect. Lett. 28 (1995) 751
14. A. Syamal, and M. R. Maurya; Indian J. Chem. 24A (1985) 836
15. V. Atre, G.V. Reddy, L.N. Sharada, and M.C. Ganorkar; Indian J. Chem. 21A (1982) 79
16. A.B.P. Lever; Inorganic Electronics Spectroscopy, Elsevier Amsterdam (1968) 395.
17. A.K. Tahir, H.S. Shivains, J. Nafees, and K. Shoukat; Indian J. Chem. Sect. 39 (2000) 450
18. R.M. Silverstein, G.C. Bassler, and C.T. Morril; Wiley John and sons, 4th Edn (1981) 241
19. F.W. Wehrli, A.P. Marchand, and S. Wehrli; Interpretation of Carbon-13 NMR Spectra; Wiley:New York, USA (1988)

47
One Pot Preparation of Greener Nanohybrid from Plant Oil

E. Sharmin[1*], D. Akram[1,2], A. Vashist[1], M. Y. Wani[3], A. Ahmad[4], F. Zafar[1] and S. Ahmad[1*]
[1] Materials Research Laboratory, Dept. of Chemistry, Jamia Millia Islamia, New Delhi-110025, India
[2]Dept. of Chemistry, Faculty of Science, Jazan University, PO Box 114, Jazan, Kingdom of Saudi Arabia
[3]Center of Interdisciplinary Research in Basic Sciences, Jamia Millia Islamia, New Delhi-110025, India
[4]Dept. of Biosciences, Jamia Millia Islamia, New Delhi-110025, India
Email: eramsharmin@gmail.com

Abstract

Plants have been the prime resource of food, clothing, medicine and other basic needs of humans since primeval times. Today, much beyond this, the advancements in the knowledge of chemistry, development of sophisticated analytical instruments and techniques as well as the advent of nanotechnology have augmented their use manifolds as biofactories of nature. Plant oils serve as raw materials for polymers, nanohybrids and composites therefrom, for versatile advanced applications. On similar lines, the study presents preliminary account of "one-pot-preparation", characterization and antibacterial behavior of microwave processed plant oil based nanohybrid.

Introduction

Oils derived from any part of plant such as roots, stems, leaves, flowers, seeds or fruits are termed as plant oils. Majority of "fatty" plant oils are derived from seeds such as groundnut, sunflower, soybean, linseed, castor, rapeseed, cottonseed, palm, copra, sesame, maize and others. Plants produce oils from sugars through biosynthetic metabolic pathways involving a series of complex chemical reactions. Oils are accumulated in seeds as triglycerides in their embryo (in cotyledons in sunflower, linseed or rapeseed), endosperm (castor, coriander or carrot), or both (tobacco) [1]. Oils from plant seeds are composed of long aliphatic triglyceride chains that contain carboxyls, double bonds, methylene groups; often, they also bear oxirane rings (e. g., vernolic acid in *Vernonia sp*), hydroxyls (e. g., ricinoleic acid in *Ricinus communis* or Castor, *Lesquerella sp*), furanoid fatty acids (*Exocarpus cupressiformis, Hevea brasiliensis),* both hydroxyls and oxiranes (e. g., ricinoleic and vernolic acids in *Ochrocarpus africanus*), both hydroxyls and cyclopropenoid fatty acids (e. g., in *Scindapsus officinalis*), which may undergo host of derivatization reactions, i. e., epoxidation, hydroxylation, acrylation, vinylation, amidation, trans/interesterification and others to yield value-added polymers through environmentally benign route, with applications as lubricants, adhesives, cosmetics, inks, fuels, antibacterial agents, paints and protective coatings [2, 3].

Plant oil polymers generally exhibit poor mechanical strength, toughness, chemical resistance, thermal stability and low Tg that restricts their high performance applications although they show good flexibility retention characteristic, gloss, moisture resistance (due to hydrophobic oil chains), cost effectiveness, biodegradation, eco- and economy-friendliness. Thus, their modifications as bioblends, bio/nano hybrids and bio/nanocomposites were carried out with plant oil derivatives (epoxies, polyols, polyesters, alkyds, polyurethanes) and some commercial polymers (polyvinylchloride, polyvinyl alcohol), conducting polymers (polyaniline), metal alkoxides, metal nanoparticles, nanoclays, resulting in improved performance characteristics for versatile applications [2–12]. However, their preparation and curing processes are generally complex, multi-step, time consuming, requiring hazardous organic solvents, often also compromising on yield and purity of the final product. Microwave [MW] heating is considered as an energy efficient process for chemical reactions relative to conventional methods. MW chemistry is based on the

M.M. Srivastava, L.D. Khemani, S. Srivastava, *Chemistry of Phytopotentials: Health, Energy and Environmental Perspectives*, DOI:10.1007/978–3-642–23394-4_47, © Springer-Verlag Berlin Heidelberg 2012

efficiency of the reaction mixture to absorb MW radiations owing to MW dielectric heating phenomenon, e.g., dipole interaction or ionic conduction mechanism. By the direct interaction of electromagnetic radiations with the polar molecules of the reaction mixture (reactants, solvents, catalysts), rapid internal (in-core volumetric) heating is produced that results in fast reactions [13–16]. We believe that through MW assisted preparation method, a whole clan of oil polymers may be produced, bearing all the advantages over the aforementioned drawbacks, coupled with cleaner processes, high yield and purity of products, however, the approach is still in cradle stage and scant literature is available on the topic.

Oil from *Ricinus communis* [RCO] seeds or Castor is a non-drying oil (Iodine value= 85g I_2/100 g, Hydroxyl value=160–168 mg KOH/g, Acid value=2.45 mg KOH/g). India is one of the leading producers of RCO. RCO contains 90% ricinoleic acid –a hydroxyl containing fatty acid. Thus, it is a natural polyol, used extensively in polyurethane foams, elastomers, adhesives, coatings and paints [17]. Some researchers have reported the development of organic-inorganic hybrids with RCO derivatives and inorganic reinforcements [18–24]. The present study reports some preliminary findings on structure, morphology and antibacterial behavior of microwave processed RCO biohybrids [RCOH] through "one-pot-preparation", for the first time relative to the several cumbersome steps involved in conventional RCO hybrids. The results have also been compared with their conventionally prepared counterparts [C-RCOH].

Experimental

Materials

RCO [Pioneer, New Delhi] and TEOS (Merck, Germany) were used as received.

Instrumentation and Methods

The synthesis was carried out in a domestic microwave oven model LG MS 1927C operating at 230 V—50 Hz frequency. FT IR spectra were taken on IRPrestige-21, RAffinity-1, FTIR-8400S (Shimadzu corporation analytical and measuring instrument division, F.R. Germany) using a NaCl cell. Morphological studies were carried out by transmission electron microscopy (Morgagni 268-D TEM FEI instrument, Netherlands) on a carbon coated copper grid.

Antibacterial behavior of RCOH (in vitro) was investigated against *E. coli* MTCC 443 and *S. aureus* MTCC 902 by disc diffusion assays in dimethyl sulfoxide [DMSO] [25–27].

Preparation of RCOH and C-RCOH

To prepare RCOH, CO (10.0 g) and TEOS (10% by weight of CO, prehydrolyzed in water/HCl mixture, with pH of TEOS-water-acid mixture ~1.35, at room temperature) were taken in an Erlenmeyer flask. The flask was placed in MW oven for 2 min maintaining the temperature at 60±5°C. Similar reaction flask, equipped with an air condenser, was placed on a preheated magnetic stirrer at 60±5°C for 2 hrs under continuous agitation, to obtain C-RCOH [9]. RCOH and C-RCOH were obtained as free flowing transparent liquids bearing the color of parent RCO.
Aliquots of reaction mixtures were taken at 30s and 30 min, respectively for TEM analyses.

Results and Discussion

One pot preparation of nanohybrids was accomplished by condensation reaction between prehydrolyzed TEOS and secondary –OH groups of RCO, in a MW oven and over magnetic stirrer, respectively (Fig 1) [9,28]. RCOH and C-RCOH were obtained in 2 min and 2 hrs, respectively. The solventless synthesis strategy, as described here, results in enhanced reac-

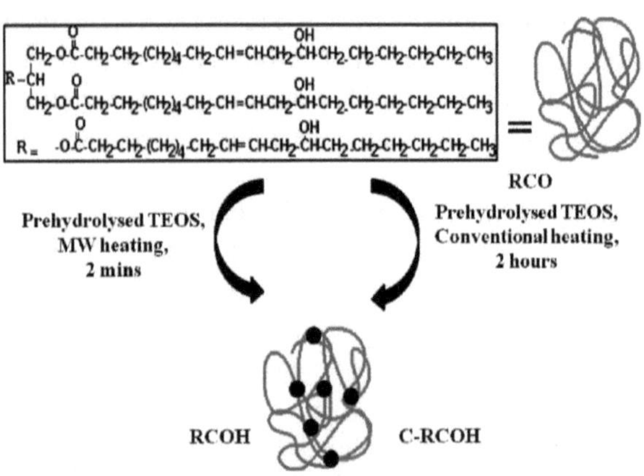

Fig 1. Preparation of nanohybrid

tion rates with different selectivity relative to conventional methods [14]. The salient features of overall approach, such as (i) the use of a domestically abundant, cost effective bioresource, (ii) non-hazardous chemical synthesis, (iii) at lower temperatures, (iv) with no complex derivatization steps using simple basic chemistry, (v) one-pot preparation, (vi) no side reactions or side/waste products, (vii) dry reaction media, and, (viii) MW assisted energy efficient preparation method offer a green protocol to obtain the said biohybrids en route Green Chemistry, hence, RCOH, in particular, may be categorized as greener biohybrid.

FTIR

The spectrum of RCO (Fig 2 a) shows the presence of absorption bands for –OH (3448–3435 cm^{-1}), -CH_2 –, -CH_3 (sym asym str) (2926, 2856 cm^{-1}), -C=C-H (str) (3007 cm^{-1}), -C=O ester (1743 cm^{-1}), -C-C(=O)-O- (str) (1234, 1166 cm^{-1}), -O-C-C- (str) (ester) (1093 cm^{-1}) and -C=C- (1650.0 cm^{-1}). In FTIR spectrum of RCOH (Fig 2b), additional absorption bands appear at 1103 cm^{-1}, 966 cm^{-1}, 914 cm^{-1} and 790 cm^{-1} and, in FTIR spectrum of C-RCOH (not shown here), at 1087 cm^{-1}, 966 cm^{-1} and 801 cm^{-1}, typical for –O-Si-O-. The presence of aforementioned absorption bands confirms the formation of RCO biohybrids.

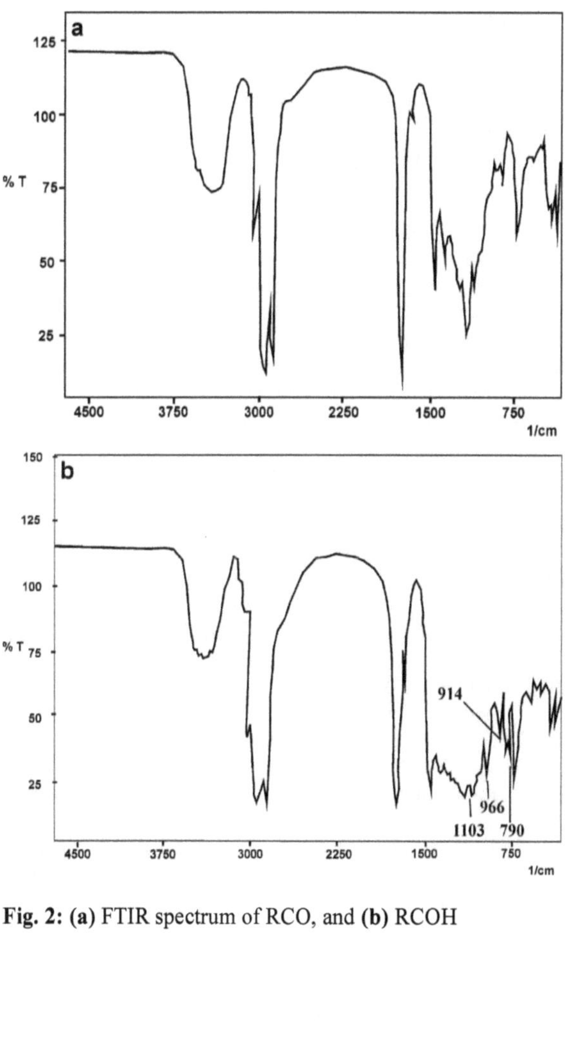

Fig. 2: **(a)** FTIR spectrum of RCO, and **(b)** RCOH

Morphology

TEM micrographs of RCOH and C-RCOH are provided in Fig 3 (a–f). Growth pattern of RCOH (after 30s) reveals the presence of distinct smaller and larger globules of TEOS (Fig 3a) which seemingly merge forming cloudy assemblage of 20–30 nm sized nanosilica particles (Fig b and c), with well defined contours. In C-RCOH (after 30mins), "trampled grass" like morphology can be seen as growth pattern (Fig 3d), finally producing nanosilica aggregates (Fig 3 e and f).

The difference in morphology of RCO nanohybrids may be attributed to the heating methods, i.e., MW assisted and conventional heating. In MW assisted (dielectric) heating, all the molecules of the reaction mixture absorb MW radiations simultaneously via in-core volumetric or bulk and selective heating of polar

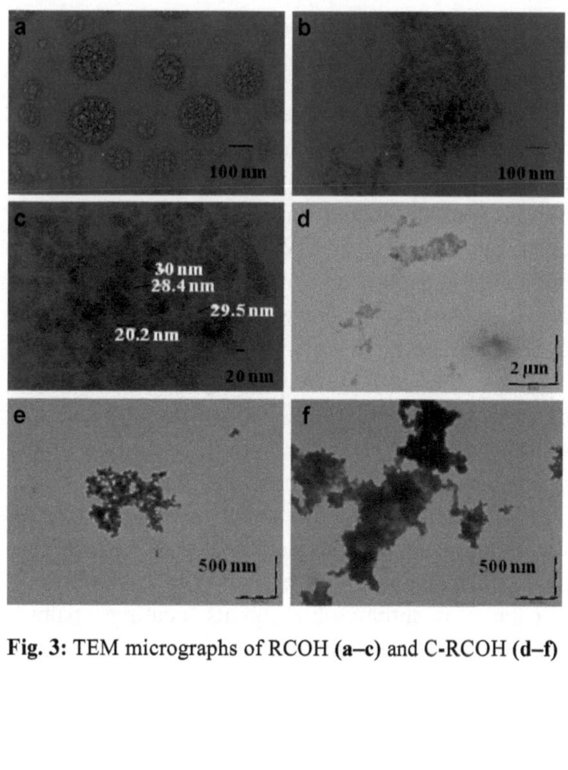

Fig. 3: TEM micrographs of RCOH **(a–c)** and C-RCOH **(d–f)**

entities; they are under the influence of same energy due to equal agitation, thus, they assemble themselves uniformly, producing well-defined morphology of RCOH [29, 30]; in conventional methods, contact-heating occurs through the walls of the reaction vessel, causing inhomogeneous heating from vessel walls to the core, resulting in temperature differences. We believe that this non-uniform heating produces loosely defined nanosilica aggregates in C-RCOH [29, 30]. TEM analysis thus confirms the formation of RCOH nanohybrids.

Antibacterial Behavior

RCOH showed good and moderate antibacterial activity against *E. coli* and *S. aureus* (Zone of inhibition=20 mm and 9 mm, respectively; Minimum inhibitory concentration = 4 mg/mL) while DMSO had no effect on the growth of micro-organisms. The remarkable variations in these results can be attributed to the structural differences in bacterial cells [31]. The flexible nanohybrid chains presumably cut off the required nutrients required by bacterial cell by adhering to its surface via electrostatic interactions, the nano-sized morphology provides large surface area per unit volume facilitating the antibacterial action of RCOH [32, 33].

Conclusion

The synthesis strategy reveals an excellent example of the preparation of greener nanohybrid via simple chemistry. We observed that the morphology of nanohybrids was dependent on mode of heating during chemical reaction, i.e., MW or conventional heating method. The method cuts off the time of reaction and provides a simple and single-step preparation of nanohybrids. The research work further holds scope for extended studies such as the investigation of the effect of varying the content of TEOS on the structure and morphology of the said nanohybrid, assessment of antibacterial and antifungal behavior, thermal stability, Tg, and others. The said nanohybrid may be used as raw material for the production of polyurethane hybrids and composites there from with enhanced applications as antimicrobial agents, coatings, paints, packaging materials and others.

Acknowledgements

Dr Eram Sharmin and Dr Fahmina Zafar acknowledge CSIR, New Delhi, India for Senior Research Associate ships against Grant Nos. 13(8464-A)/2011-Pool and 13(8385-A)/Pool/2010, respectively. Dr Deewan Akram is thankful to CSIR (New Delhi, India) for Senior Research Fellowship against Grant No. 9/466(0122) 2K10-EMR-I. They also thank the Head, Dept of Chemistry, Jamia Millia Islamia, for providing facilities to carry out the research work.

References

1. S. Baud, and L. Lepiniec; Prog. Lipid Sci. 49 (2010) 235
2. F. S. Guner, Y. Yag and A. T. Erciyes; Prog. Polym. Sci. 31 (2006) 633
3. G. Lligadas, J. C. Ronda, M. Galia‘, and Virginia Ca’diz; Biomacromol. 11 (2010) 2825
4. P. Bordes, E. Pollet and L. Avérous; Prog. Polym. Sci. 34 (2009) 125
5. U. Riaz, S. M. Ashraf, and H. O. Sharma; Polym. Degrad. Stab. 96 (2011) 33
6. U. Riaz, A. Vashist, S.A. Ahmad, S. Ahmad and S.M. Ashraf; Biomass and Bioenergy 34 (2010) 396
7. S. Ahmad, S. M. Ashraf, and U. Riaz; Polym. Adv. Technol. 16 (2005) 541
8. H. Deka, N. Karak, R. D. Kalita, and A. K. Buragohain; Polym. Degrad. Stab. 95 (2010) 1509
9. D. Akram, S. Ahmad, E. Sharmin, and S. Ahmad; Macromol Chem. Phys. 211 (2010) 412
10. D. Deffar, and M.D. Soucek; J. Coat. Technol. DOI: 10.1007/BF02698390
11. A. Kumar, P.K. Vemula, P.M.Ajayan, and G.John; Nat. Mater. 7 (2008) 236
12. M. Haq, R. Burgueño, A. K. Mohanty, and M. Misra, Composites Part A: Appl. Sci. Manufacturing 42 (2011) 41
13. J. D.Moselay, C. O.Kappe; Green Chem. 13 (2011) 794
14. R.Sarma, D.Prajapati; Green Chem. 13 (2011) 7183
15. F. Wiesbrock, R.Hoogenboom, and U. S.Schubert; Macromol Rapid Commun. 25 (2004) 1739
16. M. Nuchter, B. Ondruschka, W. Bonrath, and A. Gum; Green Chem., 6 (2004) 128
17. M. Kashif, E. Sharmin, F. Zafar, and S. Ahmad; J. Am. Oil Chem. Soc. (2011) accepted
18. D. M. Becchi, M.A. De Luca, M. Martinelli, S. Mitidieri; J. Am. Oil Chem. Soc. 88 (2011) 101
19. M. Martinelli, M.A. De Luca, D.M. Becchi, S. Mitidieri; J. Sol-Gel Sci. Technol. 52(2009) 202
20. J. Alam, U. Riaz and S. Ahmad; Curr. Appl. Phys. 9 (2009) 80
21. F. Zafar, M. H. Mir, M. Kashif, E. Sharmin and S. Ahmad; J. Inorganic Organomet. Polym. Mater. 21 (2010) 61
22. R. S. Jadhav, D. G. Hundiwale, and P. P. Mahulikar; J. Coat. Technol. 7 (2011) 449
23. N. Jiratumnukul, R. Intarat; J. Appl. Polym. Sci. 110 (2008) 2164
24. H. Uyama, M. Kuwabara, T. Tsujimoto, M. Nakano, A. Usuki, and S. Kobayashi; Chem. Mater. 15 (2003) 2492
25. A. Ahmad, A. Khan, L.A. Khan, N. Manzoor; Microb. Pathog. 48 (2010) 35

26. CLSI/NCCLS. Methods for antimicrobial susceptibility testing of anaerobic bacteria; Approved Standard. 6th ed. M11-A6 (2004)
27. Performance Standards for Antimicrobial Susceptibility Testing; Fifteenth Informational Supplement, Clinical and Laboratory Standards Institute, M100-S15, Vol. 25, No1, 2005
28. M. Fujiwara, K. Kojima, Y. Tanaka, R. Nomura; J. Mater. Chem. 14 (2004) 1195
29. A.I. Gopalan, K.P. Lee, M.H. Hong, P. Santhosh, K.M. Manesh, and S-Ho Kim; J. Nano. Nanotech. 6 (2006) 1594
30. U. Riaz, S.M. Ashraf; Appl. Clay Sci. 52 (2011) 179
31. W.R. Li, X.B. Xie, Q.S. Shi, S.S. Duan , Y.S. Ouyang, and Y.B. Chen; Biomet. 24 (2011) 135
32. S. Pal, Y.K. Tak, J.M. Song; Appl. Environ. Microbiol. 73 (2007) 1712
33. M. Martini, M. Bondi, R. Iseppi, M. Toselli, and F. Pilati; Eur Polym J 43 (2007) 3621

48
Synthesis and Characterization of Fe_2O_3-ZnO Nanocomposites for Efficient Photoelectrochemical Splitting of Water

N. Singh[1], P. Kumar[1], S. Upadhyay[1], S. Choudhary[1],
V. R. Satsangi[2], S. Dass[1] and R. Shrivastav*[1]
[1]Department of Chemistry, Faculty of Science, Dayalbagh Educational Institute, Dayalbagh, Agra, 282 110
[2]Department of Physics & Computer Science, Dayalbagh Educational Institute, Dayalbagh Agra, 282 110
rohitshrivastav_dei@yahoo.co.in

Abstract

It is indispensable to construct clean energy systems in order to solve the current energy crisis issues. In the present work nanocomposite Fe_2O_3-ZnO powders have been prepared. XRD analysis revealed dominant evolution of wurtzite ZnO and α-Fe_2O_3 at Zn/Fe molar ratios being, respectively, ≥ 19.0 and ≤ 0.05. However, at intermediate concentrations, simultaneous growth of two oxides along with cubic $ZnFe_2O_4$ has been observed. Optical characterization led to prominent absorption threshold, which is shifted towards higher wavelength with increasing Fe amount. Results indicate that Fe_2O_3-ZnO nanocomposite at optimized Zn/Fe molar ratio yields significantly increased photoresponse compared to pure ZnO or Fe_2O_3.

Introduction

Fossil fuel reserves are getting depleted continuously with increase in their demand for different life supporting activities. Hence, the need for alternate sources of energy has heightened, leading to the search for an ideal, cost-effective, eco-friendly, sustainable and clean energy source [1]. In the ongoing quest, hydrogen has emerged as an alternative to traditional energy sources such as oil and natural gas [2, 3].

Semiconductor based photoelectrochemical splitting of water employing solar radiations seems to be a good option to produce hydrogen without generating any harmful byproducts [4, 5]. In PEC cell water molecules are reduced by photogenerated electrons to form H_2 providing overall splitting of water utilizing solar energy [6]. PEC cells combine the essential functions of naturally occurring, molecular based energy conversion systems used in plants with those of the more technologically advanced, semiconductor based photovoltaic cells that are used to supply power to spacecraft [7]. Hence water splitting seems to be an attractive reaction which can contribute to an ultimate green sustainable future and help to solve twin issues of energy and environmental pollution [8]. In the present work nanocomposite powders of iron and zinc oxides have been prepared and characterized for morphological and structural features and possible use in photoelectrochemical splitting of water.

Experimental

Iron- zinc oxide nanocomposites were prepared by co-precipitation using, as precursor, iron (III) nitrate nonahydrate [$Fe(NO_3)_3.9H_2O$] and zinc acetate dihydrate [$(CH_3.COO)_2Zn._2H_2O$]. Ammonium hydroxide [NH_4OH] was used for precipitation and ethyl alcohol was employed to wash the precipitate. A series of samples were generated, at varying Zn/Fe molar ratios, starting with the pure zinc oxide at one end and terminating at pure iron oxide at the other end [9]. Aqueous solutions of precursor salts were mixed and stirred for 30 min. along with the slow addition of NH_4OH solution till pH was raised to 7 at which precipitation occurred. Stirring was continued for another 30 min., after which the precipitate was aged for 24 h. After drying for 1 h at 80°C, the content was vacuum filtered and the precipitate was sintered in three stages: Stage I – 30 min at 250°C, Stage II – 30

M.M. Srivastava, L.D. Khemani, S. Srivastava, *Chemistry of Phytopotentials: Health, Energy and Environmental Perspectives*, DOI:10.1007/978–3-642–23394-4_48, © Springer-Verlag Berlin Heidelberg 2012

min at 600°C, and Stage III – 1 h at 800°C. Step sintering was aimed to avoid heat shock to the samples. The sintered powders, thus obtained, were characterized. The crystallographic information, such as, the structure and composition of the evolved phases was gathered through powder XRD analysis using CuK_α radiation in Bruker X-ray diffractometer (Model; AXS D-8 Advanced, Germany). The analysis was made at 2θ varying from 20 to 70° with a step size of 0.02°. UV-Vis absorption spectra were obtained in the region of 200–850 nm on a scanning diffused reflectance UV-Vis spectrophotometer (Perkin Elmer, Lambda 650S). A sample of pure $BaSO_4$ was used as a reflectance standard. Powdered samples prepared were pelletized by pressing in a mechanical press at 7 tones cm^{-2}. Pellets were then sintered at 600 °C for 1h in air. PEC measurements were made by using pellets as working electrode in conjunction with Pt counter electrode and saturated calomel reference electrode and employing a potentiostat (Model ECDA-001, Conserv Enterprises, Mumbai, India) and Xenon arc lamp (150 W, Oriel, USA). PEC cell was filled with 0.1 M NaOH (pH 13) solution, used as electrolyte. The PEC measurements involved determination of current-voltage (I–V) characteristics, both under darkness as well as illumination.

Results and Discussion

The X-ray diffraction pattern (Fig. 1) presents the polycrystalline structure. The high intensity peaks observed at 2θ angles, 31.7, 34.4, and 36.2°, correspond to hexagonal wurtzite ZnO in samples having Zn/Fe molar ratio ≥ 19.0. In samples with Zn/Fe molar ratio ≤ 0.05 peaks at 2θ angles 35.2, 56.6, and 62.2° have been observed indicating the dominant evolution of $\alpha - Fe_2O_3$ (hematite) phase of iron oxide. Despite clear indications on dominant growth of single metal oxides under above conditions, it can be conjectured that such oxides are probably also getting doped by the other metal species present. At intermediate values of Zn/Fe molar ratios (say 0.82), the preferred growth of a mixed oxide phase ($ZnFe_2O_4$) is indicated as evidenced from the peaks observed at 2θ angles 29.9, 35.3 and 56.6°. Employing the diffraction data and the equation 1, Scherrer's calculations were attempted to estimate average crystallite/grain size [10]

$$p = k\lambda/B\cos\theta, \qquad (1)$$

where, p is crystallite/grain size, B the full width at half maxima and λ the wavelength of X-ray used. Average crystallite size in samples varied from 41–68 nm.

Band gap, band edge positions as well as the overall band structure of semiconductors are of crucial importance in photoelectrochemical and photocatalytic applications. The energy position of the band edge level can be controlled by the electronegativity of the dopants, the pH of the solution, as well as by quantum confinement effects [11]. The observed UV-Vis absorption spectra of films (Fig 2) indicated a significant blue shift in absorption edge to higher wavelengths on decreasing the Zn/Fe molar ratios. In samples exhibiting dominant growth of ZnO, sharp absorption edge is seen in the UV region (at around 400 nm), which is a characteristic of hexagonal wurtzite phase of ZnO and may be due to the onset of fundamental absorption corresponding to O:2p → Zn:4s charge-transfer band. On decreasing Zn/Fe molar ratio the absorption shifts to lower energy thresholds and at other extreme where dominant growth of $\alpha - Fe_2O_3$ is seen, it falls in the visible region (around 660 nm). The absorption data was utilized to evaluate optical band gap energy of the samples, based on equation 2,

$$\alpha h\upsilon = C\,(h\upsilon - E_g)^n, \qquad (2)$$

where, hυ is the photon energy, C and n constants, and α the optical absorption coefficient [12]. The observed

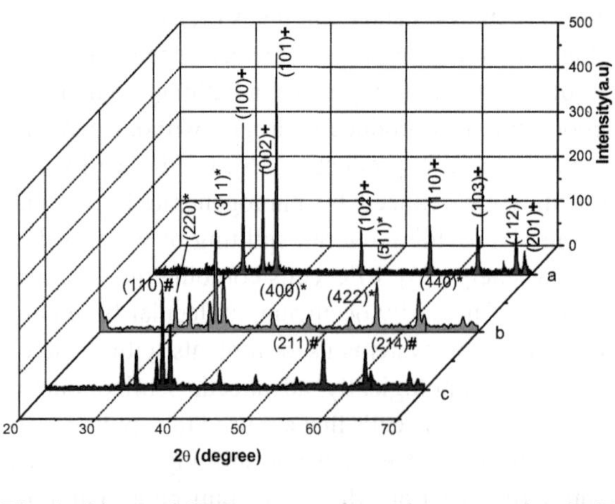

Fig. 1: Observed XRD pattern of samples prepared at Zn/Fe molar ratio: 19.0 **(a)**, 0.82 **(b)** and 0.05 **(c)**. Peaks corresponding to: + ZnO, $*ZnFe_2O_4$, and # α-Fe_2O_3

changes in E_g with variation in Zn/Fe molar ratio in samples is presented in Fig. 3. It is clear that a sharp increase in band gap occurs at Zn/Fe molar ratio 30.49.

The observed I – V curves (Fig. 4) of PEC cell demonstrate photocurrent under anodic bias which is in tune to the expected n-type nature of samples prepared. The variations observed in photocurrent density (i. e. $I_{illumination} - I_{darkness}$) with applied voltage suggest that the samples with Zn/Fe molar ratio 0.42 yield highest photoresponse. Further, the samples with Zn/Fe molar ratio 0.82 and 0.05 also resulted in significant photocurrent generation.

The PEC (I-V) data of few samples showing significant photocurrent was further analyzed to compute "applied bias photon-to-current efficiency" (ABPE) in such cases using equation 3 [13].

$$\%ABPE = \left[\frac{\left|J_{Ph}\left(mAcm^{-2}\right)\right| \times \left(1.23 - |V_b|\right)(V)}{P_{total}\left(mWcm^{-2}\right)}\right] X100, \quad (3)$$

where J_{ph} is the photocurrent density obtained under an applied bias V_b and P_{total} is input power impinged on working electrode from the light source (90 mW/cm^2). Table 1 presents % ABPE values obtained at 700 mV bias against SCE. Besides, the Table also presents short circuit current (J_{sc}) and open circuit voltage (V_{oc}) values under illumination for selected samples.

Since, there is no additional redox couple in the electrolyte, the significant value of photocurrent density can be attributed to electrochemical splitting of water, which was also experimentally indicated by evolution of gases in the form of bubbles on the electrode surface. The gain in photocurrent density in samples, having above stated optimum value of Zn/Fe molar ratio, may be attributed to efficient absorption of light by such samples along with seemingly well aligned band edges with redox levels for hydrogen and oxygen evolution, which is indicated by significantly high short circuit current in such cases. Further with all the samples significant short circuit current is evident even under darkness suggesting a porous/permeable depletion layer.

Table 1: Observed J_{sc}, V_{oc} and % ABPE in the samples having Zn/Fe molar ratios 0.82, 0.42 and 0.05

Sample with Zn/Fe molar ratio	**J_{sc} (mA)**	**V_{oc} (V)**	**ABPE (%)**
0.82	0.91	–0.79	0.8
0.42	0.80	–0.74	2.1
0.05	1.80	–0.78	1.1

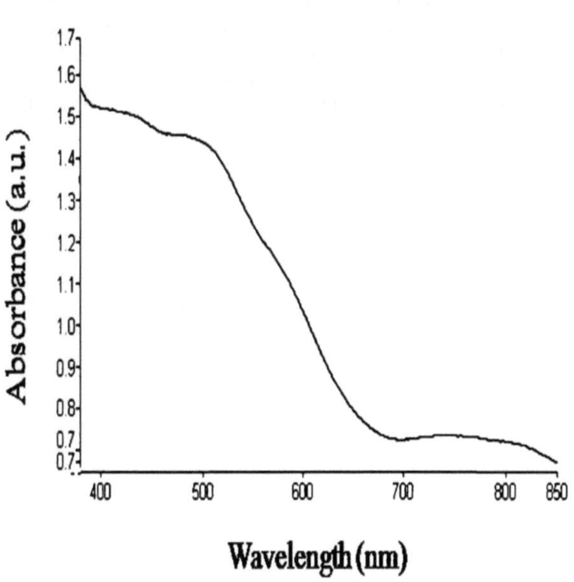

Fig. 2: Observed absorption curve for the sample having Zn/Fe molar ratio 0.82

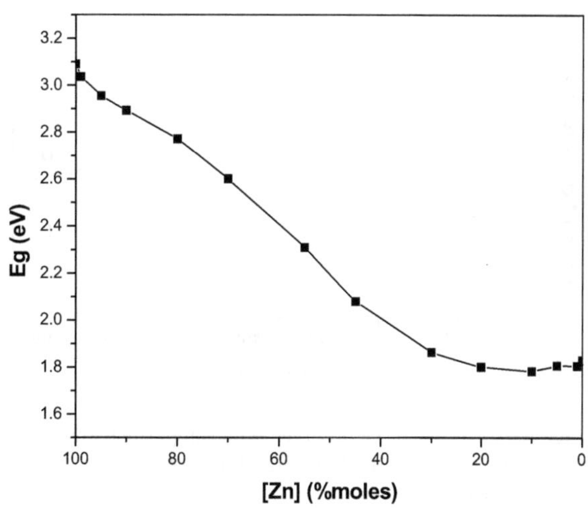

Fig. 3: Observed change in E_g with variation in Zn/Fe molar ratio

Conclusion

The Zn/Fe nanocomposites prepared show dominant evolution of hexagonal wurtzite ZnO, $\alpha - Fe_2O_3$ and $ZnFe_2O_4$ at different Zn/Fe molar ratios. The absorption spectra and the optical band gap of samples suggest that these are efficient – moderate absorber of UV – Visible light, with absorption threshold shifting towards visible light on decreasing Zn/Fe molar ratio. At Zn/Fe molar ratio 0.42 saturation photocurrent has been recorded to be $\approx$ -3.57 mA cm^{-2} and the material can be used as an efficient photoelectrode for photo-splitting of water. The effect can be attributed to better optical absorption and probably well matched band edges.

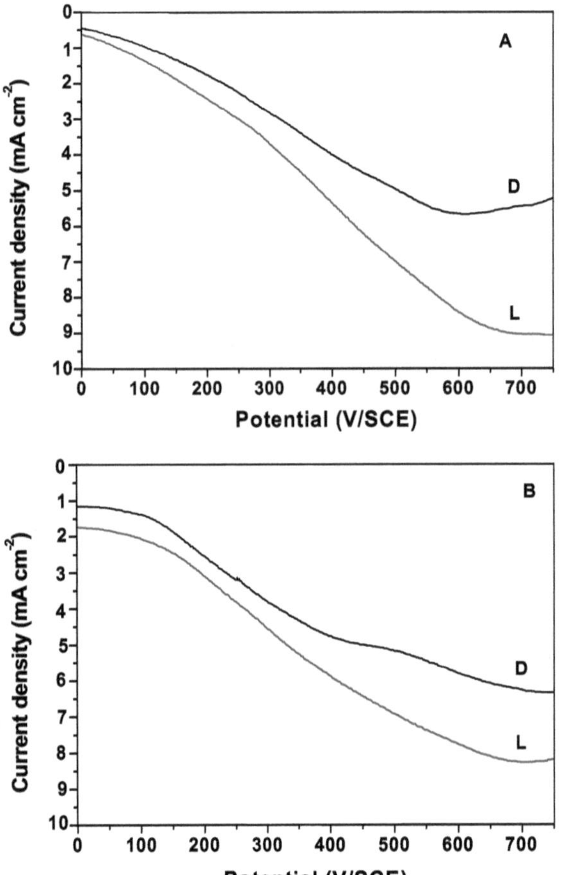

Fig. 4: PEC (I-V) curve observed with samples having Zn/Fe molar ratio: 0.42 **(A)** and 0.05 **(B)**. D: Under darkness, L: Under illumination

Acknowledgements

DAE-BRNS, Govt. of India and University Grants Commission, Govt. of India, is gratefully acknowledged for financial assistance in the form of R&D project to RS. We are grateful to Dr. Sadhana Railu, EMD, National Environmental Engineering Research Institute, Nagpur, India, for DR-UV analysis of samples.

Reference

1. I.E. Paulauskas, G.E. Jellison Jr., L.A. Boatner, and G.M. Brown; Int. J. of Electrochem. (2011) 1–10
2. M. Gupta, V. Sharma, J. Shrivastav, A. Solanki, A.P. Singh, V.R Satsangi, S. Dass and R. Shrivastav; Bull. Mater. Sci. 32 (2009) 23–30.
3. V. Sharma, P. Kumar, J. Shrivastava, A. Solanki, V.R. Satsangi, S. Dass and R. Shrivastav; J. Mat. Sci. 46 (2011) 3792–3801.
4. V. Sharma, P. Kumar, J. Shrivastava, A. Solanki, V.R. Satsangi, S. Dass and R. Shrivastav; Int. J. Hydrogen Ener., 36 (2011) 5236–525
5. M. Guptal, J. Shrivastava, V. Sharma, A. Solanki, A.P. Singh, V.R. Satsangi, S. Dass and R. Shrivastav; Adv. Mat. Res. 67 (2009) 95–102.
6. Y. Fan, D. Li, M. Deng, Y. Luo and Q Meng, Eront; Chem. Cina, 4 (2009) 343–351.
7. M. Gratzel; Nature 414 (2001).
8. A.J. Bard, M.A. Fox; Acc. Chem. Res. 28 (1995) 141–145.
9. A. Hernandez, L. Maya, E. Sanchezmora and E.M. Sanchez; J. Sol-gel Sci. Tech 12 (2007) 71–78.
10. P. Scherrer, Nachr. Ges. Wiss. Gottingn, 2 (1918) 89.
11. A.B. Murphya, P.R.F. Barnesa, L.K. Randeniyaa, I.C. Plumba, I.E. Greyb, M.D. Horneb, J.A. Glasscocka, Int. Jour. of Hyd. Ener. 31 (2006) 1999 – 201.
12. S.C. Ray, Sol Energy Mater Sol Cells, 12 (2001) 68–307.
13. Z. Chen, T.F. Jaramillo, T.G. Deutsch, A.K. Shwarsctein, A.J. Forman, N. Gaillard, R. Garland, K. Takanabe, C. Heske, M. Sunkara, E.W. Mcferland, K. Domen, Eric.L. Miller J.A. Turner and H.N. Dinh; J. Mater. Res., 25 (2010) 3–16.

Section C Environment Perspectives

49
Evaluation of Fluoride Reduction at Different Stages of Sewage Treatment Plant Bhopal, (MP), India

R. K. Kushwah[1], S. Malik[1], A. Bajpai[2], R. Kumar[3]
[1]Department of Chemistry, Sadhu Vaswani College, Bhopal (India)
[2]Makhanlal Chaturvedi University, Bhopal (India)
[3]Sagar Institute of science and Technology, Bhopal (India)
Email: ramkumarkushwah@yahoo.com

Abstract

Fluoride is a normal constituent of natural water samples. Its concentration varies significantly depending on the water sources. The increase in water demand in addition to water shortage has led to growing interest in wastewater reuse. In the present study wastewater samples from different stages of STP, Bhopal was analyzed for Fluoride by using standard methods. During present study the sewage treatment system using different materials showed excellent potential in Fluoride removal from the wastewater. The analysis of treated water for Fluoride indicates that the treated water can be used for industrial cooling and agricultural uses.

Introduction

Water is distributed to a variety of residential, industrial, commercial and municipal clients. Waste water coming from all these sources is collected by sewer system and thereafter transported to the sewage treatment plant. Industrial and human wastes increase fluoride levels in sewage. During treatment of sewage, Primary Treatment Unit removes suspended solid by screening, filtration, flocculation, coagulation, and sedimentation. Several common process using iron salt coagulation, lime or activated alumina may remove fluoride at high concentrations; however, these effects are less effective at the low concentrations of fluoride usually found in municipal sewage. [1, 2] The microorganisms present in most secondary treatment system absorb large quantities of fluoride and may reduce fluoride concentrations in effluent water by up to 50 percent. [3, 4] As water pollutant, elevated concentrations of fluoride may affect a number of organisms, including fish, amphibians, insects, snails, shellfish, protozoa and some aquatic plants. [5] Industrial waste and municipal sewage water may add to the fluoride which is naturally present in all surface water. [6]

Material and Methods

The present Sewage Treatment Plant (Kotra Sewage Treatment Plant) is situated in Bhopal, the capital of central Indian State, Madhya Pradesh, within the geographical coordinates of 23° 15’ 44” N, 77° 28’ 23” E. It receives the wastewater generated in Nehru Nagar, Kotra Sultanabad and adjoining areas. Kotra sewage treatment plant is designed to treat 10 MLD sewage. The Kotra STP is based on waste stabilization technique using anaerobic and facultative ponds. Under present study wastewater samples were collected from influent and effluent of sewage treatment plant (STP) during the period January to December 2010. Samples were analyzed to determine the efficiency of the treatment plants in reducing the fluoride from the influents and effluent samples. Sewage samples were collected in glass containers, precleaned by washing with non-ionic detergents, rinsed in tap water, in 1:1 hydrochloric acid and finally with deionized water before usage. Before sampling, the bottles were rinsed three times with sample water and then filled and Fluoride was analysis in the analytical laboratory according to the methods prescribed in the APHA [7].

M.M. Srivastava, L. D. Khemani, S. Srivastava, *Chemistry of Phytopotentials: Health, Energy and Environmental Perspectives*, DOI:10.1007/978–3–642–23394–4_49, © Springer-Verlag Berlin Heidelberg 2012

Results and Discussion

Monthly samples were collected from different stages of the Sewage Treatment Plant (STP) Kotra, Bhopal. The results obtained for Fluoride is shown in the Table 1.

Variation of Fluoride at different stages of treatment of sewage is shown in following figures.

Fig-1 January - Fluoride

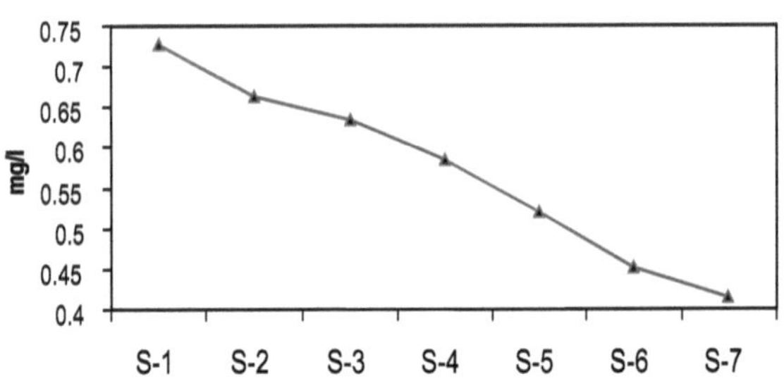

Fig-2 February - Fluoride

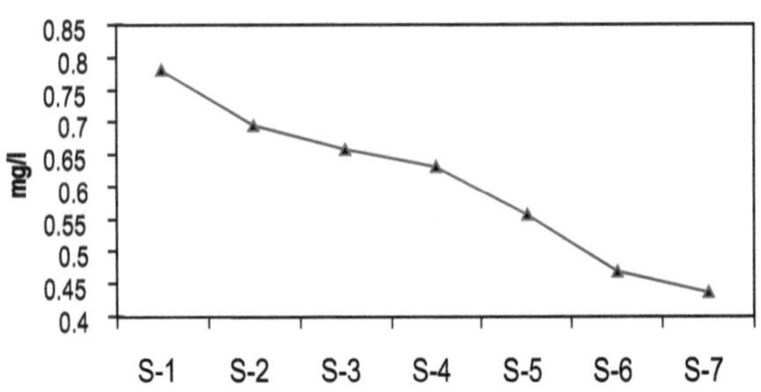

Fig-3 March - Fluoride

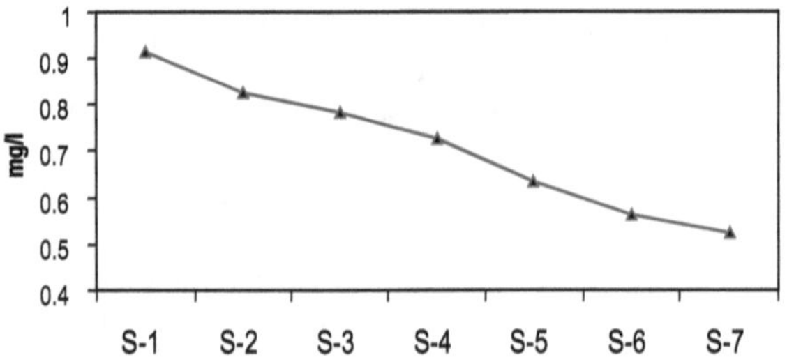

Fig-4 April - Fluoride

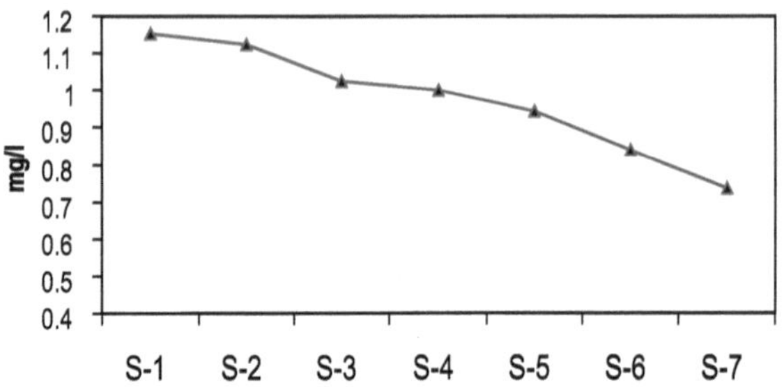

Fig-5 May - Fluoride

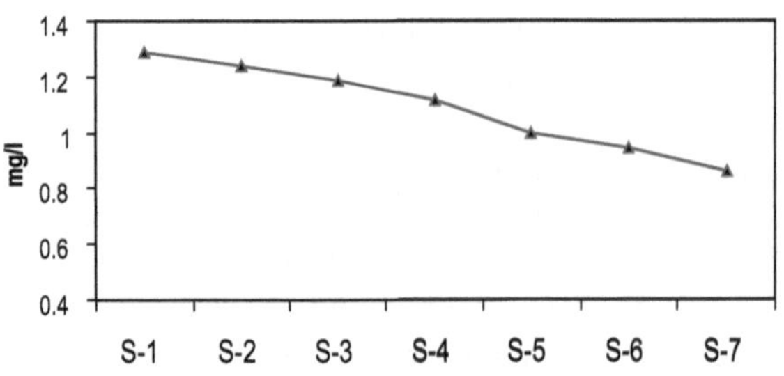

Table 1: Variation of Fluoride in the month of January and February at different stages of STP.

Stations/ Months	Jan	Feb	March	April	May	June	July	Aug	Sept	Oct	Nov	Dec	Average
S -1	0.727	0.782	0.914	1.154	1.287	1.234	0.975	0.886	0.928	0.997	0.844	0.675	0.95
S -2	0.664	0.695	0.827	1.124	1.238	1.185	0.894	0.795	0.865	0.924	0.784	0.615	0.884
S -3	0.635	0.659	0.782	1.025	1.186	1.106	0.854	0.744	0.792	0.865	0.705	0.576	0.827
S -4	0.584	0.631	0.726	1.003	1.114	0.996	0.812	0.653	0.681	0.824	0.657	0.543	0.768
S -5	0.521	0.558	0.634	0.946	0.996	0.954	0.784	0.624	0.662	0.762	0.624	0.512	0.714
S -6	0.452	0.468	0.562	0.842	0.946	0.912	0.744	0.612	0.625	0.678	0.584	0.446	0.655
S -7	0.415	0.436	0.526	0.734	0.864	0.826	0.712	0.584	0.612	0.654	0.562	0.412	0.611

S -1: Inlet of STP **S -2:** Anaerobic Tank-1 **S -3:** Anaerobic Tank-2 **S -4:** Facultative Tank-1
S -5: Facultative Tank-2 **S -6:** Rock Filter **S -7:** Outlet of STP

Fig-6 June - Fluoride

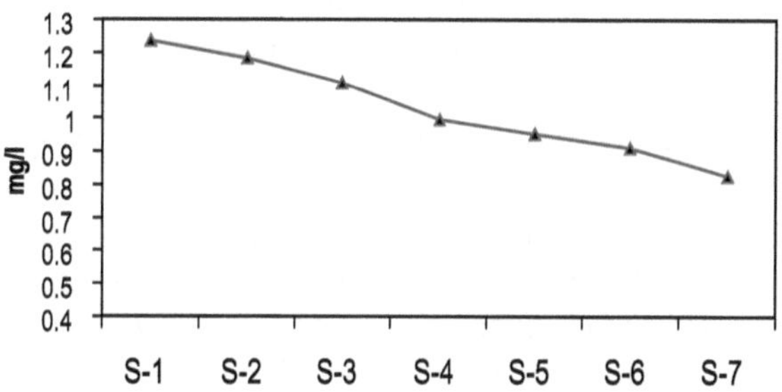

Fig-7 July - Fluoride

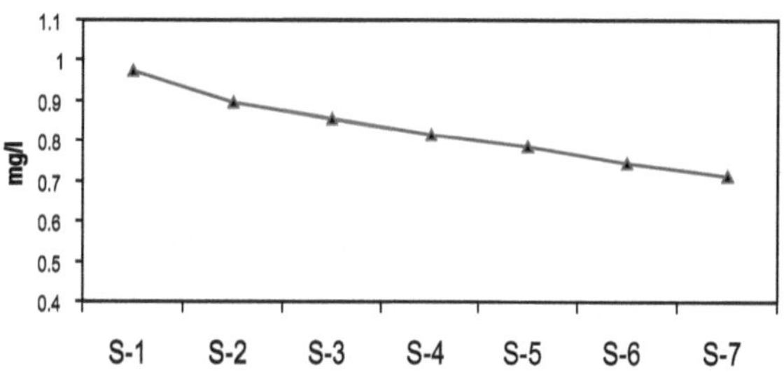

Fig-8 August - Fluoride

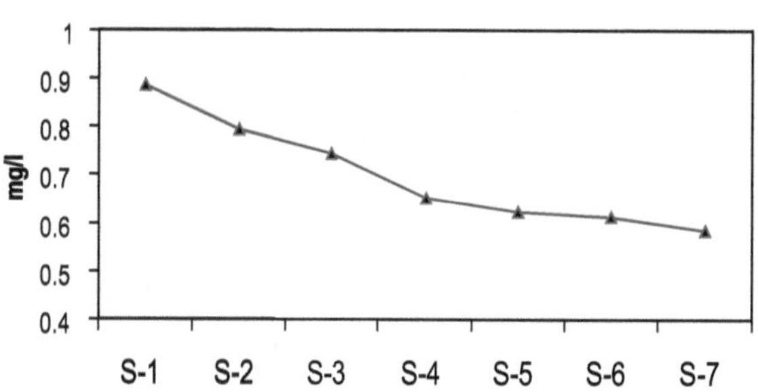

Fig-9 September - Fluoride

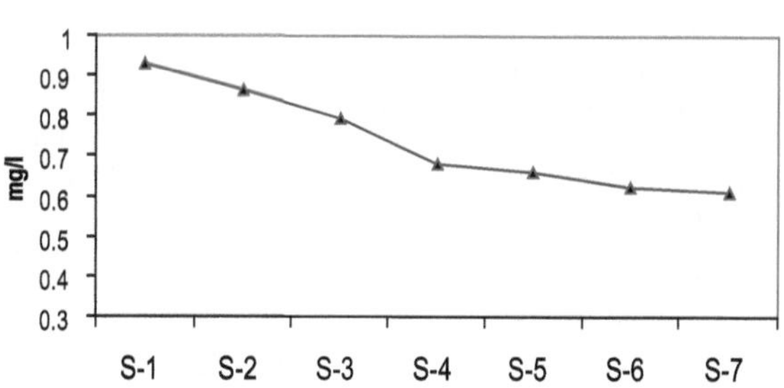

Fig-10 October - Fluoride

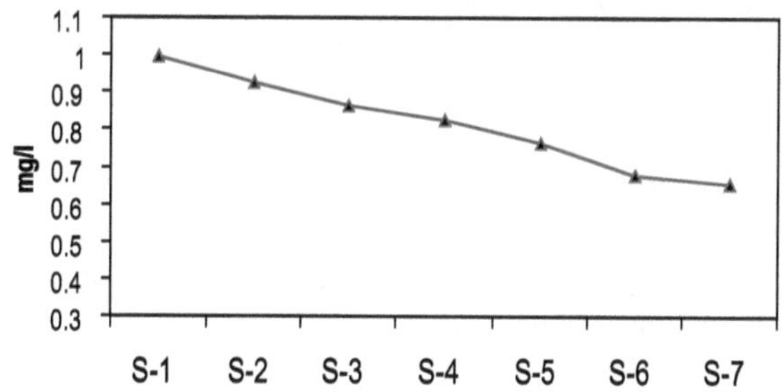

Fig-11 November - Fluoride

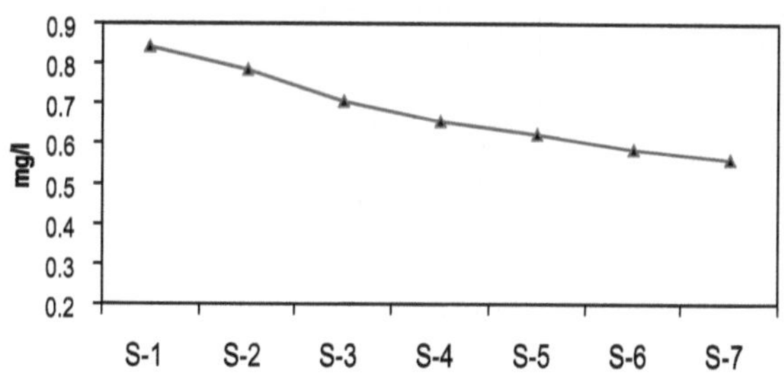

Fig-12 December - Fluoride

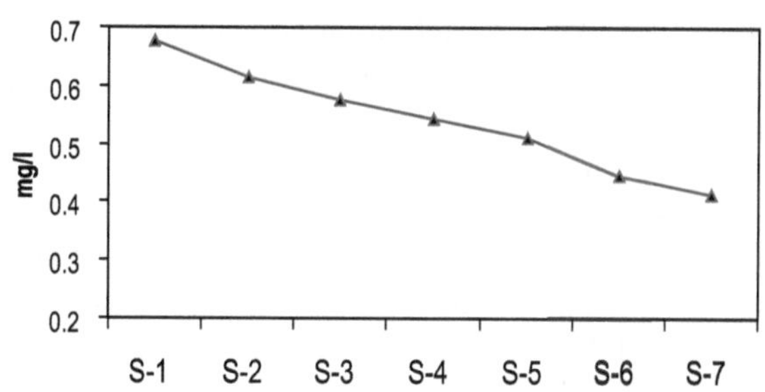

During period investigation in the influent water, Fluoride concentration varied from 0.727 mg/l to 1.287 mg/l and 0.412 mg/l to 0.864 mg/l in the effluent water. The minimum value was observed in the month of January while the maximum value was observed in the month of May in the influent water of sewage treatment plant (STP). The minimum value was observed in month of December while maximum value was observed in the month of May in the final treated effluent water of sewage treatment (Figure 1–12). The present study was conducted to evaluate the effectiveness of the Sewage treatment plant in reducing the fluoride concentration at various stages of treatment. The present Sewage treatment plant is based on oxidation processes. The STP is consists of seven stages having grit removal, two anaerobic ponds, two facultative ponds, a rock filter, outlet of STP. During present investigation maximum reduction in fluoride concentration was observed in facultative tank no. 2. While comparing the percent reduction of fluoride concentration in different months, it was observed that maximum reduction of fluoride was observed in the month of May while minimum reduction was observed in anaerobic tank 1.

Unit	Influent	Effluent	% Reduction
Inlet of STP	0.95	0.884	6.9
Anaerobic Tank-1	0.884	0.827	6.4
Anaerobic Tank-2	0.827	0.768	7.1
Facultative Tank-1	0.768	0.714	7
Facultative Tank-2	0.714	0.655	8.2
Rock Filter	0.655	0.611	6.7

An overview of data reveals that fluoride in drinking water may vary from 0.5 mg/l to 0.5 mg/l [8]. Excessive ingestion of fluoride for a prolonged period (6 month to several years) causes fluoride toxicity in the from of dental, skeletal and gut fluorosis. Fluoride toxicity also affects the soft tissues and enzyme system but its effect on teeth, bones and gut are of practical importance. Out of 6 lakh village in India at least 50% have fluoride content in drinking water exceeding 1.0 mg/l [9, 10].

During the present investigation of Fluoride concentration at effluent water of sewage treatment plant was found within in the prescribed limit of BIS [11]. Burearo of Indian Standard in view of the health problems has laid down the Indian Standard as 1.0 ppm as the maximum permissible limit. This means the body may tolerate fluoride up to a certain limit of 1.0 ppm depending upon the nutritional standard and body physiology. There will be traces of fluoride in any water samples therefore, though BIS has laid down the upper limit as 1.0 ppm it is further specified that lesser the better as the fluoride causes where even 0.4 ppm fluoride in drinking water has caused dental fluorosis [8].

Conclusion

The present study reveals the assessment of Fluoride concentration in waste water, due to various stages of sewage treatment plant (STP) Bhopal. Performance of STP of Kotra was evaluated which has shown its capability to reduce Fluoride from raw sewage. From the above study, it was observed that high concentration of Fluoride was present in the inlet of sewage treatment plant however better water quality was found after treatment in effluent water. Instead of discharging the sewage onto the nearby body of water, it is proposed to let it pass through the sewage treatment plant which would reduce most of the pollutants. So the sewage treatment is essential for maintaining the water quality and the final treated wastewater can be used for secondary purposes like irrigations gardening and industrial cooling.

References

1. L.D. Benefield, J.F. Judkins, B.L. Weand: Process Chemistry for water and waste water treatment. Englewood Cliffs, NJ: Prentice-Hall, (1982) 405–421.
2. W.E. Link, J.G. Rabosky: Fluoride ion removal from waste water employing calcium precipitation and iron salt coagulation. Lafayette, IN (1976).
3. E.D. Atkins, J.R. Hawley: Sources of Metals and Metal Levels in Municipal Waste Waters, Research Report 80. Ottawa (1978).
4. T.T. Masuda; Persistence of fluoride from organic in waste waters. Devel Industr Microbial 5(1964) 53–70.
5. Water Quality Planning Branch, Division of Environmental Management: Naorth Carolina Water Quality Standards Documentation: Toxicity of Fluoride to Freshwater Biota. Report No. 86–01, 1986.
6. E. Groth; Fluoride Pollution along the food chain. Environment 17 (3) (1975) 29–38.
7. American Public Health Association (APHA). Standard methods for the Analysis. 7th Edn, University Press, Washington DC, New York, USA (1989).
8. C.K. Jain, A. Imran and M.K. Sharma; Fluoride contamination in ground water –Indian Scenario. IJEP. 19(4) (1999) 260–266.
9. I. Gupta; Drinking water and fluorosis in Doda. National seminar on water for life, University Jammu (1995).
10. S.K. Gupta, and P. Sharma; An approach to tacking fluoride problem in drinking water. Current Sci., 68 (8) (1995) 706–713.
11. Drinking water specification (1st revision). Amendment no. 1, January, 1993, IS: 10500. Bureau of Indian Standards, New Delhi (1991).
12. S. Pani and S.M. Misra; Assessment of Fluoride Concentration in Some Ground Water Resurgences of Bhopal. IJEP 22(9) (2002) 1003–1006.
13. J.W. Osterman, MD, ScD. Evaluating the Impact of Municipal Water Fluoridation on the Aquatic Environment. AJPH October (1990) 80 (10) (2002) 1230–1235.

50
Adsorption Behavior of *Cedrus Deodara* Leaves for Copper (II) from Synthetically Prepared Waste Water

N. C. Joshi[1], N. S. Bhandari[1] and S. Kumar[2]
[1]Department of Chemistry, Kumaun University, Soban Singh Jeena Campus,Almora, Uttarakhand
[2]Vivekananda Institute of Hill Agriculture, Almora, Uttarakhand
[1]Email: nicks_nicks@rediffmail.com

Abstract

The present work deals with the treatment of copper (II) containing waste water using low cost adsorbent. The readily available waste leaves i. e. Cedrus deodara have been used for removal of copper (II) from synthetic waste water. The adsorption study was carried out by batch operation including contact time, dosage, pH, concentration and temperature. The maximum removal achieved at optimized conditions such as high dose of adsorbent, high pH and low concentration of metal ions. The pH is an extremely important factor than other parameters have been used for removal of metal ion by Deodar leaf powder; 75.42 percent copper (II) has removed after 25 minutes at pH 5. The equilibrium data of adsorption have also been tested with various isotherm models.

Introduction

The presence of heavy metals in industrial waste water is great concern because their toxicity to living organisms. Many of them are soluble in water and become more available for living organism and accumulate. The conventional method for heavy metal removal from waste streams employs various technologies, which are often either expensive or inefficient. Use of inexpensive natural biosorbents such as zeolites, metal oxides, fly ash, clays, coal, peat moss, waste biomass, and chitosan has been considered as a promising alternative for this purpose [1]. Various methods reported for the removal of heavy metals from water and waste water are chemical precipitation, membrane filtration, ion exchange, carbon adsorption and biosorption. Among these biosorption is relatively new and is very promising in the removal of heavy metals in an eco-friendly manner. Copper (II) is one of the toxic heavy metal ion commonly found in an electroplating industrial effluents [2]. Disposal of industrial effluents containing copper (II) ions into natural water beyond limits may harm the living organisms including human [3].The plant C*edrus deodara* is common gymnosperm of Himalayan region. It is evergreen and the tallest plant having needle like leaves. The waste leaves of *Cedrus deodara* were collected from near the SSJ Campus, Almora and Uttarakhand (India).

Experimental

Adsorbent Preparation

The collected waste leaves were washed 2–3 times by distilled water for removing water soluble impurities. Then these leaves were dried for 5–6 days in laboratory and heated at 70°C for next three hours in laboratory oven, for vaporization of water. After grinding and sieved in particle size 63 microns the powder of leaves was preserved in sealed bottles.

Preparation of Synthetic Waste Water

The synthetic waste water containing Cu (II) was prepared from the salt Copper Sulfate ($CuSO4.5H_2O$) in double distilled water. For making 1000 mg/L of Cu (II), 3.921 g of $CuSO_4.5H_2O$ was added in 1000 ml of double distilled water. The pH of this solution was adjusted 3 because most of industrial effluent remains in acid range and acidity favors the solubility of heavy

M.M. Srivastava, L.D. Khemani, S. Srivastava, *Chemistry of Phytopotentials: Health, Energy and Environmental Perspectives*, DOI:10.1007/978-3-642-23394-4_50, © Springer-Verlag Berlin Heidelberg 2012

metals. It is expected that the lower the pH of effluent higher the concentration of heavy metals.

Adsorption Study

The adsorption study was carried out by Batch operation. The batch operation is described as below:-

A 100 ml solution containing desired concentration of metal ions was treated with a certain amount of adsorbent in a 250 ml of conical flask at a constant shake. The solution was then filtered and adsorbent filtered out. The concentration of metal ions before and after adsorption was determined by Atomic Absorption Spectrophotometer (Model; Analytik Jeena, Vario 6). The Removal efficiency of metal ions was calculated by using following equation:-

Removal efficiency = $C_o - C_e / C_o * 100$

Where C_o and C_e are the metal ion concentrations in mg/L initially and at a given time t respectively.

Results and Discussion

Biosorption of copper (II) metal ions onto Deodar leaves has been studied as a function of contact time, dosage, pH, concentration and temperature. The adsorption of metal ions increases with time [6,7,8] for copper (II), the maximum removal efficiency achieved 21.54 after 75 minutes at pH 3 and rpm 170. The adsorbent is able to achieve the percentage removal for copper is18.34after 25 minutes at amount of adsorbent 1 g and pH 3 with rpm 170. The pH is an extremely important factor than other parameters [9,10] have been used for removal of metal ions by Deodar leaf powder; 75.42 percent copper has removed after 25 minutes at pH 5 and rpm 170. The percentage removal of metal ions on Deodar leaves is consistently decreased with concentration but amount of adsorbate per gram of adsorbent (mg/g) increased [11]. The adsorption of metal ions increases [12] on Deodar leaves with temperature but the experimental results indicate it decreases after certain temperature. The removal efficiency of copper achieved 14.03 at 60°C.The equilibrium data of adsorption have been tested with Langmuir [13], Freundlich [14,15] and Temkin isotherm [16] model. The high value of regression (R^2>0.963) for Langmuir isotherm model indicates favorable adsorption on Deodar leaf powder. The value of ,R_L‘ is found less than one in all cases which also confirms Langmuir isotherm model for copper adsorption on Deodar leaves. The high regression value (R^2>0.936) and the value of 1/n is found less than one for metal ions confirms the Freundlich isotherm model. The adsorption capacity (a) and rate of adsorption (b) have also determined by Temkin isotherm model. The high value of regression (R^2>0.914) and other parameters confirms the Temkin isotherm model. The experimental data are found to follow the order Langmuir>Freundlich>Temkin, indicating monolayer adsorption on a homogenous surface.

Conclusion

This work attempts to explore the potential use of Deodar leaves as low cost adsorbent for the effective removal of copper (II). The experimental data indicates that higher the pH of solution, the lower the initial metal ion concentration and higher dose of adsorbent more copper (II) ions are adsorbed on the adsorbent. The equilibrium data of adsorption were evaluated by Langmuir, Freundlich and Temkin isotherm model.

Acknowledgements

The authors are extremely grateful to the Head, Department of chemistry, Kumaun University Nainital, Uttarakhand, India for the support and encouragement provided during the course of this work.

References

1. M. Havelcova, J. Mizera, I. S. korova, M. Pekar; J. of Hazardous Mater. 161(2009) 559–564.
2. S. Babel, T.A. Kurniawan; J. Hazard. Mater. B 97 (2003)219–243.
3. M.D. Mashitah, Z. Zulfadhly, S. Bhatia, J. Artif; Cells, Blood Subst. Immobil. Biotechnol. 27(1999) 429–433.
4. A.Y. Dursun; Biochem. Eng. J. 28(2006) 187–195 .
5. H.L. Liu, B.Y. Chen, Y.W. Lan, Y.C. Cheng; Chem. Eng. J. 97 (2004)195–201.
6. O.I. Oboh and E.O. Aluyor; African J. of Biotech. 7(2008) 4508.
7. M. Ajmal, A.K. Rifaqat, Rao, J. Ahmed, S. Anwar and R. Ahmad; J. Environ. Sci. and Engg. 50(2008)7.
8. M. Alam, M.A. and S. Rais; Rasayan Journal chem. 2(2009) 791–806.

9. A. Sari, D. Mendil, M. Tuzen and M. Soylak, J. Hazard. Mater. 162(2009) 874.
10. J. L. Gardea-Torresday, G dela Rosa and J. R. Peralta Videa, Pure App. Chemistry 76 (2004)801–813.
11. M. Riaz, R. Nadeem, M.A. Hanif,T. M. Ansari, K. U. Rehman;J. of Hazardous Mater.161(2009) 88–94.
12. A. Edwin Vasu; E. Journal of chemistry 5(2008) 1–9.
13. I.Langmuir; J. Am. Chem. Soc. 40(1918) 1361–1403.
14. A. W. Adamson; Physical chemistry of surfaces, 5th edition, John willey and sons Inc., New York (1990).
15. G. Karthikeyan and S. Siva Ilango; E. Journal of chemistry 5 (2008)666–678.
16. A. E. Nemr; J. Hazard. Mater. 161(2009) 132.

51

Zea Mays a Low Cost Eco-friendly Biosorbent: A Green Alternative for Arsenic Removal from Aqueous Solutions

K. R. Raj, A. Kardam and S. Srivastava
Department of Chemistry, Faculty of Science, Dayalbagh Educational Institute, Agra-282110, India
rohitraj.rj@gmail.com, dei.smohanm@gmail.com

Abstract

The development of efficient and ecofriendly biosorbent for the removal of arsenic from water system is a priority in regions where human health is directly affected by elevated arsenic concentrations. Biosorption of arsenic on Zea mays Cob Powder (ZMCP) was investigated for the removal of arsenite and arsenate species from aqueous solutions. Sorption studies, using Batch standard practices result in the standardization of optimum conditions: metal concentration [25 mg/L], contact time [40 min], volume [200 ml] and pH 7.5 for As (III) and 2.5 for As (V) removal. Maximum sorption for As (III) and As (V) species is 70 and 85 %, respectively.

Introduction

The arsenic is one of the most toxic elements that could be found in waters [1]. The main source of arsenic is geological, primarily due to the release of arsenic from arsenic-bearing sediments in groundwater aquifers, but human activities such as agricultural applications (pesticides, fertilizers), mining, smelting and other industrial activities can also cause arsenic pollution [2, 3]. Arsenic contamination in drinking water and groundwater has created serious health problems in countries like India and Bangladesh and in other parts of the world [4]. Long term exposures to arsenic levels can result in permanent and severe damage to human health. Arsenic toxicity causes skin lesions, damage mucous membranes, nervous system, gastrointestinal, cardiovascular, genotoxic, mutagenic and carcinogenic effects [5, 6]. Permissible limit of arsenic in drinking water set by the EPA is 10 mg/L [7].

In general, As (III), which exists predominantly as nonionic H_3AsO_3 in natural waters, is more difficult to remove compared to As (V), which exists as deprotonated oxy anions $H_2AsO^-_4$ or $HAsO^{-2}_4$ [8]. Thus, it requires a pre oxidation step which oxidizes As (III) to As (V). The toxicological, physiological and geochemical behavior depends on its oxidation state. A large number of physico-chemical methods based on coagulation, ion-exchange, reverse osmosis and flocculation with various synthetic coagulants such as aluminum, ferric salts, soda ash, polymers, etc. have been developed for arsenic removal [9, 10]. However, these methods are associated with several disadvantages such as unpredictable metal ion removal, high material costs and the generation of toxic sludge that is often more difficult to manage. As a result, extensive research has been focused on developing eco-friendly method for arsenic removal from water bodies.

The process of adsorption is a good alternative because it take care the disadvantages of the classical chemical destabilization. Numerous inorganic materials have been tested for the removal of toxic metal ion from aqueous solution over the last two decades [11]. However, only a limited number of studies have been carried out on the use of adsorbents derived from plant sources or agricultural wastes. Biosorption has gained important credibility during recent years because of its ecofriendly nature, excellent performance and low cost domestic technique for remediating even heavily metal loaded waste water [12–14].

The present investigation deals with the abatement of arsenite and arsenate ions from aqueous system using *Zea mays* Cob powder (ZMCP). The manuscript also reports the applicability of the biomass for arsenic

M.M. Srivastava, L. D. Khemani, S. Srivastava, *Chemistry of Phytopotentials: Health, Energy and Environmental Perspectives*, DOI:10.1007/978–3–642–23394–4_51, © Springer-Verlag Berlin Heidelberg 2012

recovery as well as to regenerate the exhausted biosorbent thereby making the process more economical, beneficial and cost effective.

Materials and Method

Biosorbent Preparation

Cobs of *Z. mays* were collected from the nearby areas of Dayalbagh Educational Institute, Agra. They were washed repeatedly with water to remove dust and soluble impurities, dried at 65°C for 24 h, crushed and finally sieved through 105, 210, 420 μm mesh of copper sieves.

Biosorption Studies

Sorption studies using standard practices were carried out in batch experiments (triplicates) as functions of biomass dosage (1.0–6.0 g), contact time (20–100 min), metal concentration (0.05–50 mg/L), pH (2–10) and particle size (105–420 μm). The solution of As (III) (sodium arsenite; $NaAsO_2$) was taken in an Erlenmeyer flask. After pH adjustments, a known quantity of biosorbent was added and the metal bearing suspension was kept under magnetic stirring until the equilibrium condition was reached. After shaking, the suspension was allowed to settle down. The residual biomass sorbed with the metal ion was filtered using a Whatman 42 filter paper (Whatman International Ltd., Maidstone, England). The filtrate was collected and subjected for metal ion estimation using Hydride Generation Atomic Absorption Spectroscopy (Perkin Elmer 2380). The metal concentrations before and after adsorption were recorded. The percent metal sorption by the sorbent was computed using the equation:

$$\text{Percent Sorption} = \text{Co—Ce/Co}\cdot 100$$

Where, Co and Ce are the initial and final concentration of metal ions in the solution.

Fourier Transform Infrared Spectroscopy Studies

To access the role of functional groups present on the surface of the biomass that might be involved in metal sorption, FTIR analysis in solid phase was performed using a Shimadzu 8400 Fourier Transform Infrared spectroscopy. Spectra of the sorbent before and after metal sorption were recorded.

Scanning Electron Microscopic Analysis

The morphological characteristics of ZMCP were evaluated using Steroscan 360, Scanning Electron Microscope (Cambridge, UK). The Scanning Electron Micrographs of untreated and As (III) and As (V) treated biomass at bar length equivalent to 10 μm, working voltage 20 KV with 700X magnification were recorded.

BET Surface Area Analysis

The surface area of the biosorbent was measured using a Micromeritics ASAP-2010 BET Surface Area analyzer.

Desorption Studies

Desorption studies (batch process) were conducted to restore the biomass as a function of concentration of different desorption reagents (hydrochloric acid and nitric acid). Metal loaded biosorbent obtained from our sorption experiments, were transferred to Erlenmeyer flasks and shaken with 50 ml of each desorption reagents as a function of time (20, 40, 60, and 80 min) at room temperature. The suspension was filtered using Whatman 42 filter paper and in the filtrate estimation of metal ion concentration was carried out.

Results and Discussions

The effects of various experimental parameters were found to be as follows:

Effect of Metal Concentration on the Sorption Efficiency

Sorption behavior of As (III) and As (V) on plant biomass, carried out in the range of metal concentration (0.05–50 mg/L). Sorption of As (III) and As (V) on the biomass increased with increasing concentration of the metal ion, reaching to an optimal level (25 mg/L) after that percent sorption remained constant.

Effect of Contact Time on the Sorption Efficiency

The effect of contact time on As (III) and As (V) sorption on target biomass was studied for duration of 20–60 min. The percent sorption of metal ion gradually increased with time from 20 to 60 min, finally reaching the optimum value at 40 min.

Effect of Biomass Dosage on the Sorption Efficiency

Biomass dosage used for study varied from 2.0 to 6.0 g. Percent sorption increased with the increase of biomass dosage from 2.0 to 4.0 g. However, no significant increment in the sorption tendency was observed on further increasing the biomass dosage from 4.0 g onwards.

Effect of Initial Volume on the Sorption Efficiency

Percent sorption of As (III) and As (V) on plant biomass was observed under similar experimental conditions in different set of volumes (100–300 ml) of the test solution. Maximum sorption was obtained in the volume (200 ml) of the test solution for As (III) and As (V) species are 70 and 85 % respectively.

Effect of pH on the Sorption Efficiency

The variation of sorption efficiency with pH (2.5–10) in case of As (III) and As (V) has been presented in Fig 1. The biosorption efficiency of As (III) shows a plateau in the pH range of 2.5–6, while maximum at pH 7.5 (70 %). Increase in pH from 7.5 to 10, resulted into decrease in percentage sorption of As (III). The biosorption efficiency of As (V) is maximum (85 %) at pH 2.5, while remain constant in the pH range of 2.5–10.

Based on our experimental findings and pertinent literature available on the topic, we synthesize the possible mechanism to describe arsenic removal by adsorption using plant biomass (*Zea mays* corn cob powder). In the present case, biosorption of arsenic seems to be the resultant of two processes: affinity adsorption and anion exchange process. Affinity adsorption is concerned with the surface behavior of biomass while anion exchange relates with anionic arsenic species and functional groups present on the biomass. The corn cob powder possesses various functional groups predominantly large proportion of polysaccharides (OH groups) on the surface even after the drying process. Further, based on electrochemical theory, carbon (carbohydrates) in contact with water reduces oxygen with the generation of hydroxyl group [15]. Carbon losses electrons and become positively charged, while electrical neutrality is maintained with the hydroxyl ions. When a metal bearing solution containing anions (arsenic species) show relatively higher affinity for carbon than hydroxyl groups and are exchanged.

The biomass–Arsenic interactions can be highlighted on the basis of records of the IR spectrum of native and metal treated ZMCP. In the FTIR spectra of native ZMCP a peak appears at 3466.5 cm^{-1} which corresponds to OH group present in the biomass. Shifting of the OH peak was found in the FTIR spectra of As (III) and As (V) treated ZMCP [3466.5 to 3422.1 cm^{-1} for As (III) & 3466.5 to 3432.6 cm^{-1} for As (V)] as shown in Fig. 2. Shifting of characteristics peak of OH group confirms the biomass-arsenic interaction and their by confirming biosorption of arsenic species.

SEM of the native and metal treated ZMCP was recorded and compared for the difference in morphology and reduction in pore area. SEM of the native biosorbent exhibits large spherical clusters while that of metal treated biosorbent represents dense, agglomerated, irregular type morphology. The change in morphology indicates the accumulation of liquid phase concentration of charge moieties onto ZMCP surface area [Fig. 3]. The fact is further supported

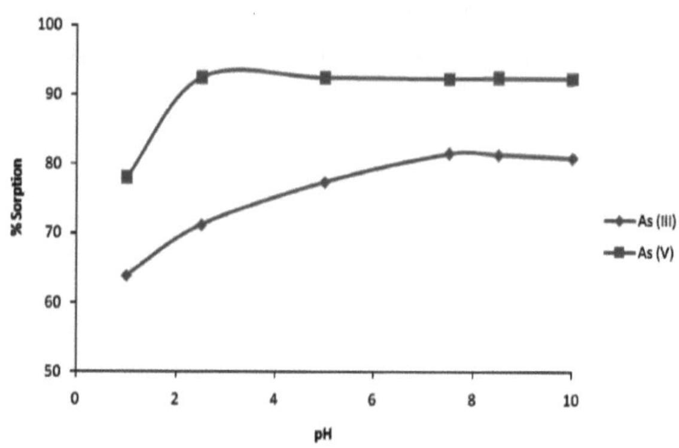

Fig. 1: pH profile and metal ion binding

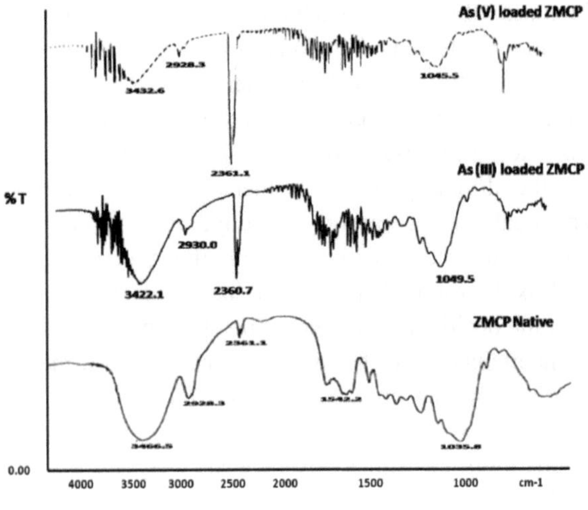

Fig. 2: Comparative FTIR spectra of native, As (III) and As (V) treated ZMCP

by the reduction in pore area from 5.92 μm² [native ZMCP] to 2.53 and 2.10 μm² [As (III) and As (V) respectively] in the BET studies of the biomaterials.

Desorption Studies

Desorption and recovery of spent adsorbent is very important as it will reduce the cost of remediation. The desorption behavior of As (III) & As (V) from metal loaded biomass was observed after desorbing with two mineral acids (hydrochloric acid and nitric acid). The maximum desorption [As (III) 90.55%; As (V) 93.35%] was observed at hydrochloric acid strength (0.05 M) however, better desorption [As (III) 94.40%, As (V) 98.12%] could be achieved with the same strength of nitric acid as desorbing agent. Sorption of As (III) and As (V) on regenerated biomass remained almost constant [As (III) 65.47%; As (V) 81.57%] and then started decreasing in the 3rd cycle.

Conclusion

The laboratory based findings open up new avenues in the abatement of arsenite (70%) and arsenate (85%) by ZMCP which is nontoxic, biodegradable and available at practically little or no cost. Thus, it introduces a less expensive, domestic and environment-friendly green method or at least a pretreatment step before large chemical treatment for metal decontamination and thus presents alternative opportunity for

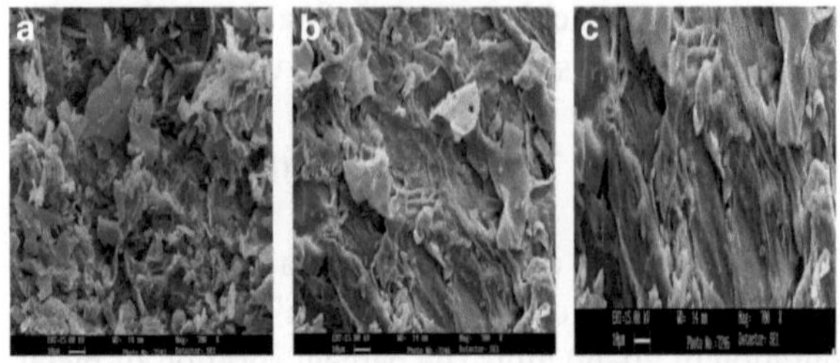

Fig. 3 (a, b, c): SEM of native, As (III) and As (V) treated ZMCP respectively

high-tech methods and synthetic coagulants under the domain of Green Chemistry. The spent biosorbent was regenerated and effectively reused up to three cycles, making the adsorption process further economical.

Acknowledgements

The authors gratefully acknowledge Prof. V.G. Dass, Director and Prof. L.D. Khemani, Head, Department of Chemistry, Dayalbagh Educational Institute, Dayalbagh, Agra. The authors are thankful to UGC SAP program for granting financial assistance.

References

1. A. Katsoyiannis and A.I. Zouboulis; Water Res. 38 (2004) 17.
2. M. Schmöger, M. Oven and E. Grill; Plant Physiology 122 (2000) 793–801.
3. I. Pickering, R. Prince, M. George, R. Smith, G. George and D. Salt; 122 (2000) 1171–1177.
4. J.S. Wang and C.M. Wai; J. Chem. Educ. 81 (2004) 207–213.
5. R.C. Kaltreider, A.M. Davis, J.P. Lariviere and J.W. Hamilton; Environ. Health Persp. 109 (2001) 245–251.
6. B.L. Rivas and M.C. Aguirre; Water Res. 44 (2010) 5730–5739.
7. USEPA; Arsenic Rule Implementation 2003. <http://www.epa.gov/safewater/arsenic.html>.
8. M. Burriel, L. Conde, J. Arribas, M. Hernandez; Quımica Analıtica Cualitativa, Eighteenth ed. Editorial Paraninfo, S.A. (2006).
9. B.L. Rivas and M.C. Aguirre; J. Appl. Polymer Sci. 106 (2007) 1889–1894.
10. D. Ranjan, M. Talat and S.H. Hasan; Ind. & Engin. Chem. Res. 48 (2009)10180–10185.
11. P. Westerhoff, D. Highfield, M. Badruzzaman and Y. Yoon; J. Environ. Eng. ASCE 131(2005) 262–271.
12. A. Kardam, P. Goyal, J.K. Arora, K.R. Raj and S. Srivastava; Natl. Acad. Sci. Lett. 32 (2009) 179–181.
13. M. Malakootian, J. Nouri and H. Hossaini; Int. J. Environ. Sci. Tech. 6 (2009) 183–190.
14. P. Goyal and S. Srivastava. J. Hazard. Mat. 172 (2009) 1206–1211.
15. P. Navarro and F.J. Alguacil; Hydrometallurgy 66 (2002) 101.

52
Removal of Diesel Oil from Water Bodies Using Agricultural Waste *Zea Mays* Cob Powder

M. Sharma, A. Kardam, K. R. Raj and S. Srivastava
Department of Chemistry, Faculty of Science, Dayalbagh Educational Institute, Agra 282110
Email id: dei.smohanm@gmail.com

Abstract

The present studied has been undertaken to study sorption efficiency of diesel oil from water bodies using agricultural waste Zea mays cob powder. The influence of biomass dosage, contact time, particle size and volume of water on the removal process was investigated. Batch studies indicated that maximum sorption efficiency of diesel oil was 72% at contact time (30 min), particle size (105 μm) and volume (100 ml). FTIR & SEM micrographs of native and oil loaded cob powder confirmed diesel oil sorption phenomenon. Petroleum ether was used for regeneration for several cycles with a view to recover the sorbed diesel oil and also to restore the sorbent for its reuse. The findings showed that Zea mays cob powder can easily be envisaged as a new, vibrant, low cost biosorbent for diesel oil clean-up operations.

Introduction

Oil has been used as a source of heat and light but due to the advent and innovations in automobile technology, it has become a source of power for transport. The quantity of oil transported over the sea has enormously increased in volume, encompassing tankers from capacity of 100,000 to 500,000 tones. Accident may occur during each of these transportation and storage steps [1]. The spilled oil contributes an undesirable taste and odor to drinking water and causes severe environmental damage [2].

Currently used water treatment technologies for diesel oil removal, involves gravity separator, centrifugal separator, coalescing separator and vacuum distillation [3]. These available treatment technologies are not effective enough, expensive and require high maintenance. The search for new technologies for the removal of diesel oil has increased attention to biosorption phenomenon. Biosorption has gained important credibility during recent years because of its ecofriendly nature, excellent performance and low cost domestic technique for remediation. Natural materials that are available in large quantities have potential to be used as low cost sorbents [4, 5], as they represent unused resources, various plant materials viz. Salvinia sp. [6], walnut shell [7], groundnut shell husk [8] and rice husk [9] have been studied for oil removal from water bodies.

Corn is the one of the most popular crop of India, which is largely available at low cost. In the present study *Zea mays* cob powder (ZMCP) is used for diesel oil removal from aqueous system.

Experimental

Biosorbent Preparation

Plant *Zea mays* have been identified in the nearby area of the Institute. Their cobs were collected, washed with water to remove the adhering dirt, dried for 24 hours, crushed and finally sieved (105 μm) through copper sieves.

Sorption Studies

Batch experiments (three replicates) were performed in laboratory using standard practices as a function of biomass dosage, contact time, particle size and initial

M.M. Srivastava, L.D. Khemani, S. Srivastava, *Chemistry of Phytopotentials: Health, Energy and Environmental Perspectives*, DOI:10.1007/978–3-642–23394-4_52, © Springer-Verlag Berlin Heidelberg 2012

volume of diesel oil. 10 ml of the diesel oil sample was added to a 500 ml beaker containing 100 ml of water. Water is added to determine the affinity of the sorbent for oil and water. The mixture was stirred for 15 min and 5 g of biomass was added to this mixture. The beaker was placed on magnetic stirrer for 30 min stirring. The oil sorbed biomass was filtered. Petroleum ether was used for regeneration of biomass.

% Oil sorption (diesel) efficiency was calculated by the following formula:

$$\textbf{\% Oil sorption efficiency} = W_{sorb}/W_0 \times 100$$

Where, W_{sorb} = mass of oil sorbed (g)
W_0 = initial mass of oil (diesel, g)

Biomass Characterization

Fourier Transform Infrared Spectroscopy (FTIR)
In order to gain better insight into the shifting of peak that might be involved in oil sorption, FTIR analysis in solid phase was performed using a Shimadzu 8400 Fourier Transform Infrared spectroscopy. Spectra of the sorbent before and after oil sorption were recorded.

Scanning Electron Microscopic Analysis (SEM)
The morphological characteristics of ZMCP were evaluated using Steroscan 360, Scanning Electron Microscopes. The Scanning Electron Micrographs of untreated and oil treated biomass were recorded.

Desorption Studies
Desorption experiments was carried out in order to explore the feasibility of recovering the oil and reuse the exhausted biosorbent for further cycle of sorption. Desorption studies (batch process) was conducted to desorb the target oil from loaded cob biomass as a function of volume of desorption reagent (petroleum ether). Oil loaded biosorbent obtained from our sorption experiments, was transferred to Erlenmeyer flasks and shaken with desorption reagent. The suspension was filtered and in the filtrate estimation of oil volume was carried out.

Results and Discussions

The main parameters influencing diesel oil sorption were investigated: biomass dose, contact time, particle size and volume of the water. Results indicate that ZMCP has considerable potential to be used as biosorbent for diesel oil removal from the water bodies. The effects of various experimental parameters were found to be as follows.

Effect of Biomass Dose (ZMCP)

Batch experiments (three replicates) were performed in laboratory at different biomass dosage 1, 3, 5, and 7 g. Percent sorption of oil increased (45–72 %) with the increased of biomass dose (1–5 g). However, no significant increase in % of oil sorption efficiency was found after increasing the dose from 5 to 7 g. Figure 1 indicate that sorption efficiency is directly related to the number of available sorption sites. Once equilibrium is attained there is no effect on sorption efficiency.

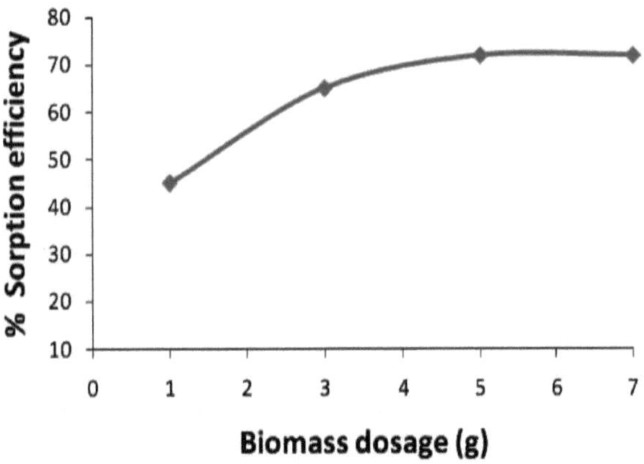

Fig. 1: % Sorption of diesel oil as a function of biomass dosage, fixed contact time (30 min), fixed particle size (105 μm) and fixed volume (100 ml).

Effect of Contact Time

It has been observed that at a fixed biomass dosage (5 g), particle size (105 μm) and fixed volume (100 ml) the percentage of oil sorption efficiency increased as the contact time increased up to certain level. Figure 2 shows that the rate of oil sorption first increased rapidly but after reaching to optimal time value, the sorption efficiencies slightly decreased with increased in the contact time. The effect is due to the saturation of sorption sites.

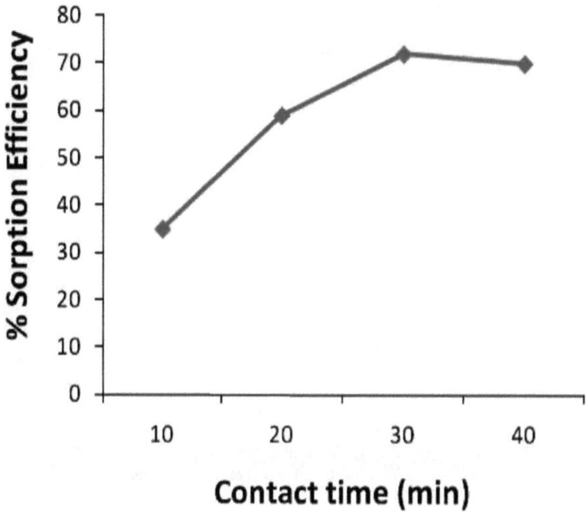

Fig. 2: % Sorption of diesel oil as a function of contact time, fixed biomass dosage (5 g), fixed particle size (105 μm) and fixed volume (100 ml).

Effect of Particle Size

It has been observed that at a fixed biomass dosage (5 g), fixed contact time (30 min) and fixed volume (100 ml) the percentage of oil sorption efficiency is increases as the particle size decreases. Figure 3 shows that smaller particles have relatively higher oil sorption efficiency due to increases in surface area of the biomass.

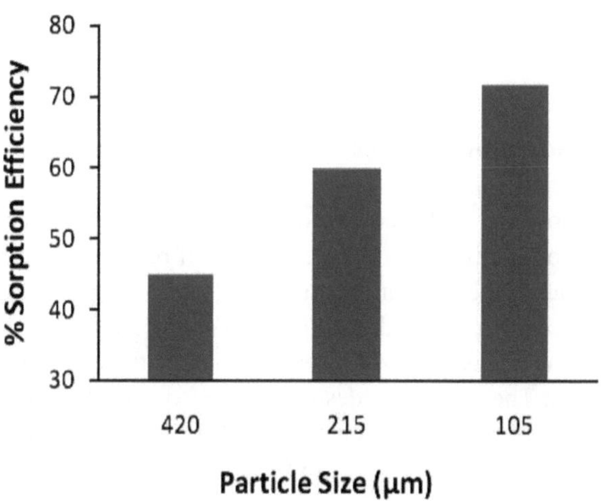

Fig. 3: % Sorption of diesel oil as a function of particle size, fixed biomass dosage (5 g), fixed contact time (30 minutes) and fixed volume (100 ml).

Effect of Test Volume

Batch experiments (three replicates) were performed in laboratory at different volume 50, 100, 150 and 200 ml. Maximum oil sorption efficiency was observed at 100 ml volume of water. Figure 4 shows that the ratio of sorption surface of the cob powder to diesel oil is optimum, exhibiting maximum percentage removal.

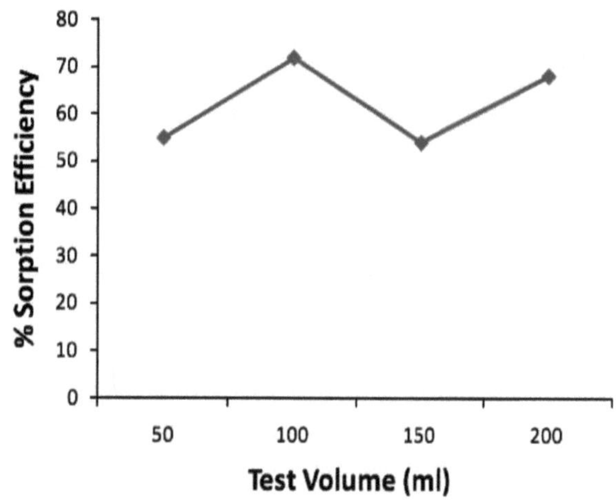

Fig. 4: % Sorption of diesel oil as a function of volume, fixed biomass dosage (5 g), fixed contact time (30min) and particle size (105 μm)

Mechanism

Sorption capacity in a vast range of solids depends on the surface area and pores. Lignocellulosic (ZMCP) materials have more surface area than nonporous materials; therefore, they are good candidates for sorption material. At the initial stage, diesel oil is sorbed by some interaction and van der Waals force between oil in the bath and wax in the ZMCP on the fiber surface. This sorption is due to the fact that both oil and wax are hydrocarbons and there is physical trapping of oil on the fiber surface through its irregular surface morphology. Sorption of oil within the fiber occurs by diffusion through internal capillary movement into sorbent lumens. Cellulose molecules of lignocellulosic (ZMCP) fiber have a hydrophilic nature due to hydroxyl groups. Oil sorption into the pores in the secondary wall maybe negligible.

Biomass Characterization

Fourier Transform Infrared Spectroscopy (FTIR)

The spectra of Fourier transform infrared spectroscopy of native and oil loaded ZMCP are presented in Figure 5 and 6 respectively. This shows shifting of peaks at 3466 cm^{-1}, 1038 cm^{-1} and 2925 cm^{-1} which confined diesel oil sorption phenomenon.

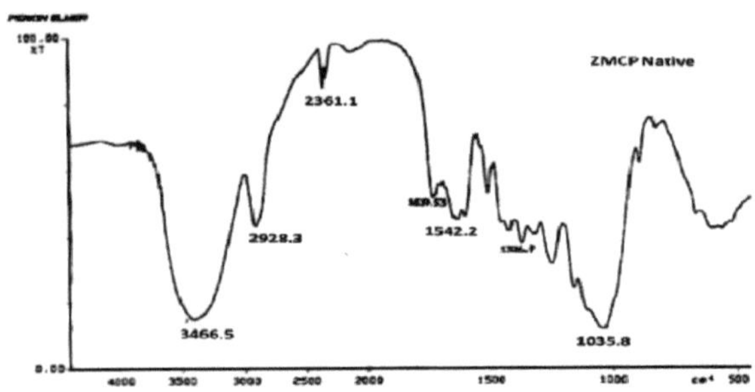

Fig. 5: FTIR image of native ZMCP

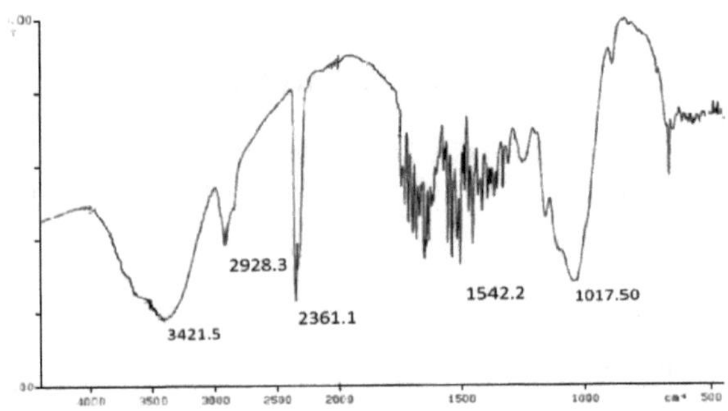

Fig. 6: FTIR spectra of sorbed ZMCP

Scanning Electron Microscopic Analysis (SEM)

The records of scanning electron microscopy of native and oil loaded ZMCP are presented in Figure 7 and 8 respectively. The comparison of scanning electron micrograph of untreated ZMCP represents large spherical clusters as compare to oil treated. Observed reduction in pore area of oil treated ZMCP confirm the biosorption phenomenon.

Fig. 7: SEM image of native ZMCP

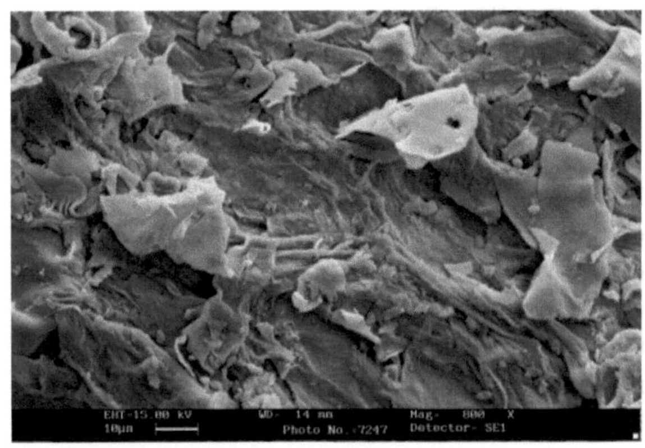

Fig. 8: SEM image of oil loaded ZMCP

Desorption Studies

Desorption and recovery of spent biosorbent is very important as it will reduce the cost of remediation of oil and other contaminants in the environment. With this aim in view, desorption behavior of oil loaded biomass was observed after desorbing with petroleum ether. The maximum desorption [68 %] was observed in the 1st cycle. Sorption of oil on regenerated biomass started decreasing [68 % to 52 %] in the 2nd cycle.

Conclusion

The findings showed that Zea mays cob powder can easily be envisaged as a new, vibrant, low cost biosorbent for diesel oil clean-up operations. Due to its high adsorption capacity, the adsorbent may be structured into continuous belts, sheets or pads so that it could be easily deployed, recovered, and disposed after usage. Thus, the potential of upgrading this method to large scale would be feasible.

Acknowledgements

The authors gratefully acknowledge Professor V.G Das and Professor L. D. Khemani, Director and Head of the Department of Chemistry, Faculty of Science, Dayalbagh Educational Institute, Dayalbagh, Agra for providing the necessary research facilities.

References

1. M.F. Fingas, W.S. Duvall, and G.B. Stevenson, Environ. Protection Serv. (1979).
2. M. Blumer. Oil on the sea .D. P. Hoult, Plenum Press, New York. (1969) 6.
3. N. Biswas. Arch. Environ. Health, 41 (2005) 49–55.
4. A. Kardam, P. Goyal, J.K. Arora, K.R. Raj and S. Srivastava. Nat Acad Sci Lett. 32 (5–6), (2009) 179–181.
5. A. Kardam, K.R. Raj, J.K. Arora and S. Srivastava. Journal of Water Resource and Protection. 2 (2010) 339–344.
6. T.H. Ribeiro, R.W. Smith and J. Rubio, Environ.Sci. Technol.34 (2000) 5201–5205.
7. A. Srinivasan and T. Viraraghavan, J. Environ. Prot. Ecol.43(2010) 78–81.
8. D.B. Nwokoma and U. Anene, Arch. Appl. Sci. Res. 5(2010) 1–19.
9. J. Drelich, J. Hupka, and B. Gutkowski, Environ.Sci.Technol. 457(2004)89–793.

53

Simulation and Optimization of Biosorption Studies for Prediction of Sorption Efficiency of *Leucaena Leucocephala* Seeds for the Removal of Ni (II) From Waste Water

J.K. Arora and S. Srivastava*
Department of Mathematics, Technical College
Department of Chemistry, Faculty of Science,
Dayalbagh Educational Institute, Dayalbagh Agra 282110
Email id: dei.smohanm@gmail.com

Abstract

Simulation and optimization of biosorption studies were carried out using Artificial Neural Network (ANN) modeling. A single layer ANN model was developed to simulate the process and to predict the removal efficiency of Ni (II) ions from aqueous solution using Leucaena Leucocephala seed powder (LLSP). Different NN architecture was tested by varying network topology. The findings indicated that the ANN provided reasonable predictive performance. The influence of each parameter on the variable studied was assessed, and metal concentration, contact time, biomass dosage and initial volume were found to be the most significant factors. Simulations based on the developed ANN model can estimate the behavior of the biosorption phenomenon process under different conditions.

Introduction

The contamination of water by heavy metals is a worldwide environmental problem. Heavy metals are of special concern because they are non-degradable and therefore persistent [1]. Nickel (II) containing wastewaters are common as it is used in a number of industries including electroplating, batteries manufacturing, mining, metal finishing and forging. Removal and recovery of heavy metals are very important with respect to environmental and economical considerations [2]. Traditional treatment techniques include chemical precipitation, membrane filtration, electrodialysis and ion exchange. Nevertheless, the application of such processes is sometimes restricted because of technical or economic constraints [3]. With increasing environmental awareness and legal constraints being imposed on discharge of effluents, a need for cost-effective alternative technology is essential. In this endeavor, biosorption seems to be a promising alternative for treating metal contaminated water [4].

To achieve an optimum management for any control measure, the concept of modeling for an efficient operation and design should be developed. A high quality representative model can provide a favorable solution to the process control. Adsorption processes are usually modeled using the mechanistic or empirical based kinetic expressions. Nonetheless, in many cases, these empirical models fall-short to represent the biosorption phenomena and its in-behind physical meaning [5]. In addition, predictive conclusions are hardly drawn from systems operating at different conditions. Because of reliable, robust and salient characteristics in capturing the non-linear relationships of variables in complex systems, application of Artificial Neural Network (ANN) has been successfully employed in environmental engineering [6–8] and bioremediation [9–12].

In continuation of our work on development of green processes for decontamination of toxic metals from waste water [13–17], the present paper proposes the sorption of Ni (II) ions from aqueous system using *Leucaena Leucocephala* Seed Powder (LLSP). It also reports the applicability of a single layer ANN model using a back propagation (BP) algorithm to predict the removal efficiency of LLSP for Ni (II) ions. Pursuing benchmark comparisons of BP algorithms, a study was conducted to optimize network structure.

M.M. Srivastava, L.D. Khemani, S. Srivastava, *Chemistry of Phytopotentials: Health, Energy and Environmental Perspectives*, DOI:10.1007/978-3-642-23394-4_53,

Experimental

Biosorbent Preparation

The *Leucaena leucocephala* tree has been identified in the nearby area of Dayalbagh Educational Institute and seeds were collected during April 2009. The seeds were washed repeatedly with water to remove dust and soluble impurities, dried in sun light for 24 hours, crushed and finally sieved through (420, 210, 105μm) mesh copper sieves. No other chemical or physical treatments were used prior to adsorption experiments.

Biosorption Studies

Sorption studies using standard practices were carried out in batch experiments (triplicate) as a function of biomass dosage (1.0–6.0 g), contact time (10–60 min), volume of the test solution (100–300 ml), metal concentration (1–100 mg/l), particle size (105μm) and pH (2–8.5). A required amount of Ni (II) (Nickel Sulfate, AR grade) was taken in an Erlenmeyer flask and after pH adjustments, a known quantity of dried biosorbent was added and metal bearing suspensions were kept under magnetic stirring until equilibrium conditions were reached. After shaking, the suspension was allowed to settle. The residual biomass sorbed with metal ion was filtered using Whatman 42 filter paper (Whatman International Ltd., Maid stone, England). Filtrate was collected and subjected for metal ion estimation using Flame atomic absorption spectrometer. Percent metal uptake by the sorbent has been computed using the equation: % Sorption = Co – Ce / Co · 100, where Co and Ce were the initial and final concentration of metal ions in the solution.

ANN Structure

Network structure has significant effects on the predictive results. As per the network topology the neural network employed has four input nodes corresponding to the process variables namely metal concentration, biomass dosage, contact time, initial volume and for output nodes corresponding to the sorption efficiency of Ni (II) ions. However, the optimal number of hidden layers and the optimal number of nodes in each layer are case dependent and there is no straight forward method for the determination of them. Neural Network Toolbox Neuro Solution 6.0 ® mathematical software was used to predict the sorption efficiency. One hundred twenty experimental sets were used to develop the ANN model. A Single layer ANN with sigmoid axon transfer function was used for input and output layers. The data gathered from batch experiments were divided into input matrix and desired matrix. The single layer sigmoid network represents functional relationship between inputs and output, provided sigmoid layer has enough neurons. Levenberg-Marquardt algorithm is fastest training algorithm for network of moderate size, therefore, used in the present study.

Backpropagation Neural Network

The back propagation network is a multilayer feed forward network with a different transfer function in the artificial neuron and a more powerful learning rule. The learning rule is known as back propagation, which is a kind of gradient descent technique with backward error (gradient) propagation (Fig. 1). The training instances have been set in order to interconnect weights between the neurons to settle into a state of correct classification of input patterns. Once the network is trained with different architectures, it has the ability to generalize over similar features found in different patterns. The four steps in the training processes viz. assemble the training data, create the network object, train the network and simulate the network were used to respond to new inputs.

Results and Discussions

Sorption Studies

Sorption studies lead to the standardization of the optimum conditions as: metal concentration (25 mg/l), contact time (40 min), particle size (105 μM) and volume (200 ml) at pH 6.5 for maximum Ni (II) removal (76.23%).

Optimization of the ANN Architecture

The removal of Ni (II) ion using LLSP was calculated in the laboratory batch experiments as a function of biomass dosage, contact time, volume of the solution and metal ion concentration in terms of percentage sorption. As biosorption influenced by number of process variables which holds complex non-linear relationship among them so as to affect the sorption process, any simple feed forward network may not be sufficient to handle the prediction efficiency [18]. ANN model based on single layer recurrent back propagation algorithm for the experimental data, generated from the above batch experiments was applied to train the Neural Network. During training, the output vector is computed by a forward pass in which the input is propagated forward through the network to compute the output value of each unit. The output vector is then compared with the desired vector which resulted into error signal for each output unit. In order to minimize the error, appropriate adjustments were made for each of the weights of the network. After several such iterations, the network was trained to give the desired output for a given input vector. The single layer network structure included fifteen hidden neurons, describing the dynamics of Ni (II) in effluent (Fig. 1).

The sigmoid axon was considered transfer function with 0.7 momentums. Series of experiment resulted into the evaluation of performance based on 60 % data for training, 20 % data for testing and 20 % data for cross validation at 3000 Epoch with 0.70000 momentums. The performance of network simulation was evaluated in terms of mean square error (MSE) criterion. The minimum MSE in the group of four variables was determined for training and cross validation are 2.11455790712831E-08 and 0.00327509470454907 respectively. Fig. 2 (a,b,c) shows the result obtained by the Neural Network simulation for both the training and cross validation data sets. The Ni (II) ion concentrations were precisely predicted for the training data sets.

Testing and Sensitivity Analysis

The developed network model was examined for its ability to predict the response of experimental data not forming the part of the training program. The network was finally tested for training, cross validation and testing data sets. The comparison results of desired outputs and network outputs are shown in fig.4. During testing a linear correlation coefficient of (R^2 = 0.99, 0.95 and 0.91) were obtained for the training, cross validation and testing data sets. A sensitivity analysis was conducted to determine the degree of effectiveness of variables. Performance of the group of input vectors included biomass dosage, Ni (II) ion concentration, contact time and volume. The degree of effectiveness of variables was found in the order of biomass dosage > metal concentration > contact time > initial volume.

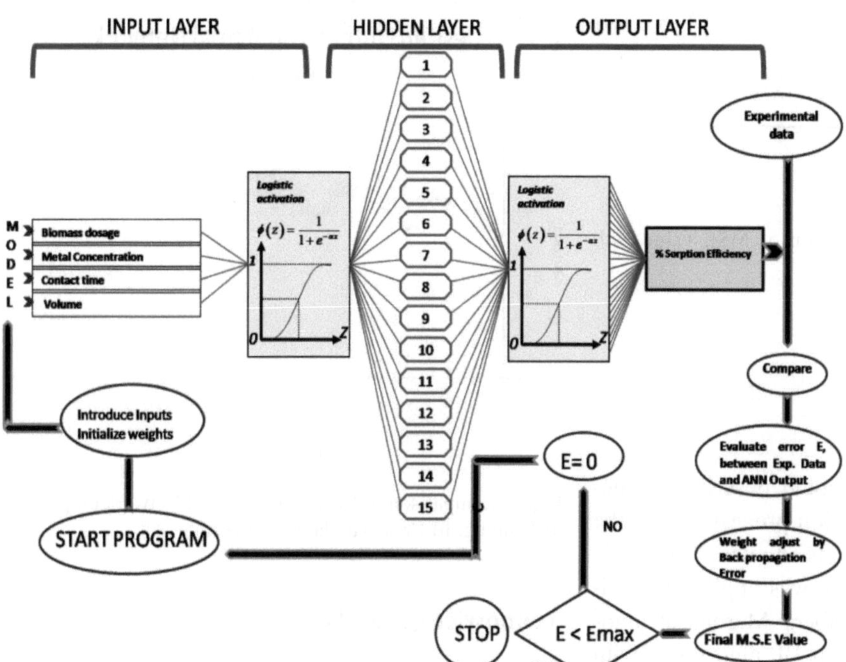

Fig. 1: Single layer Neural Network structure for the simulation and prediction of the biosorption efficiency.

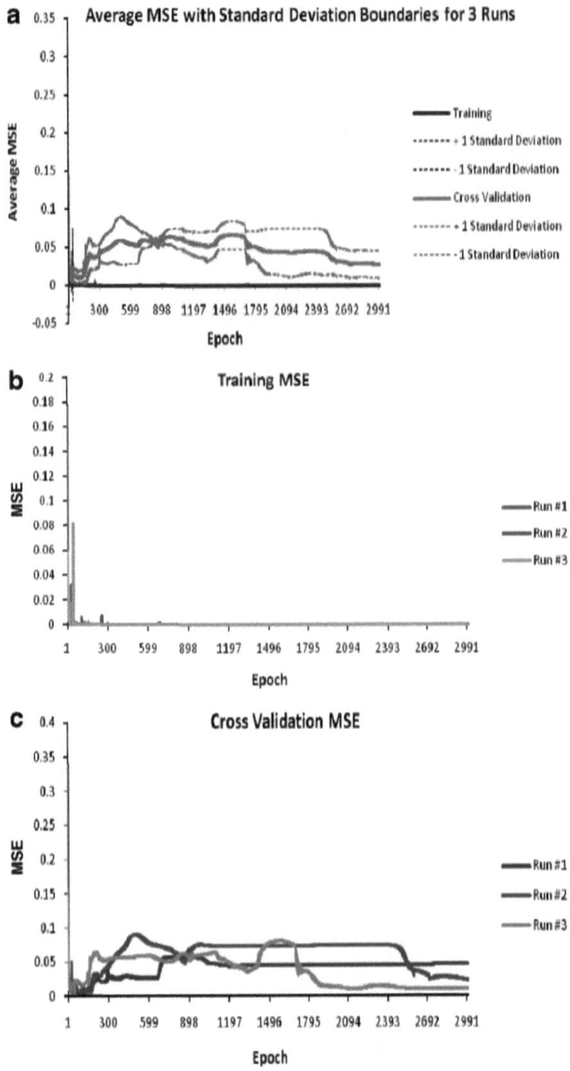

Fig. 2 (a–c): Graphical representation of MSE value with 3000 Epoch

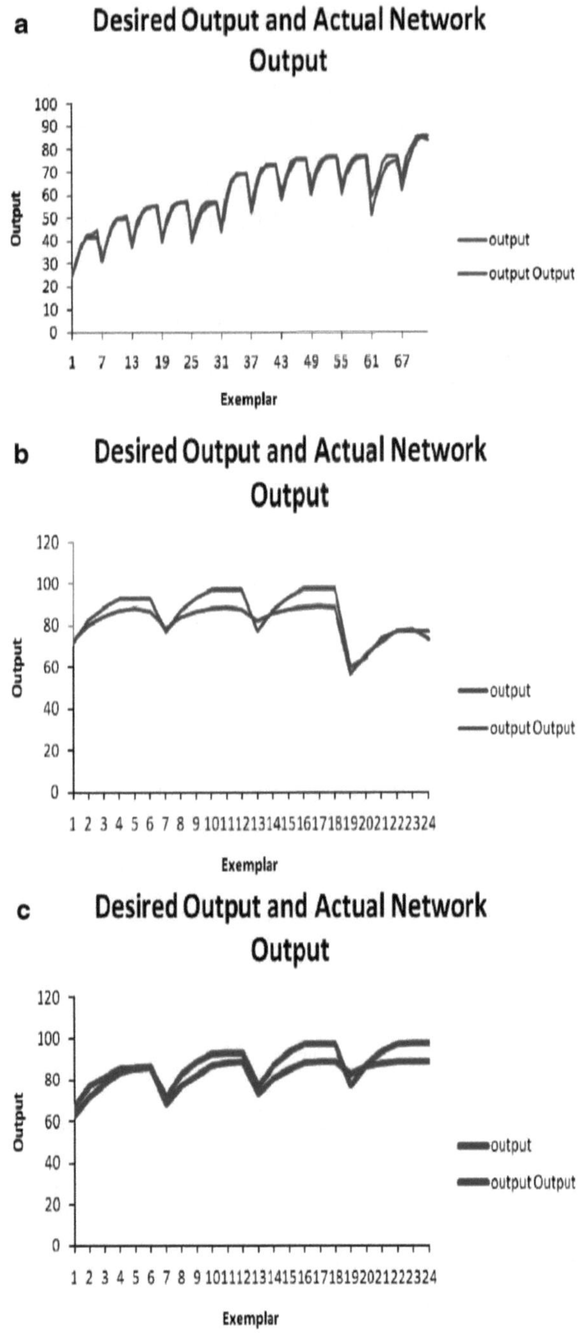

Fig. 3: Comparison of Desired output and Network output for: **a)** Training; **b)** Cross validation; **c)** Testing

Conclusion

The developed ANN model could describe the behavior of the complex interaction process within the range of experimental conditions adopted. The single layer ANN modeling technique was applied to optimize this process. The Levenberg–Marquardt algorithm (LMA) was found best of BP algorithms with a minimum mean squared error (MSE) for training and cross validation as 2.11455790712831E-08 and 0.00327509470454907 respectively.

Acknowledgements

The authors gratefully acknowledge Professor V.G Das and Professor L.D. Khemani, Director and Head of the Department of Chemistry, Faculty of Science, Dayalbagh Educational Institute, Dayalbagh, Agra for providing the necessary research facilities.

References

1. R. A. K. Rao, M.A. Khan, Colloids and Surfaces A: Physicochem. Eng. Aspects 332 (2009) 121–128.
2. M.N. Nourbakhsh, S. Kilicarshan, S. Ilhan, H. Ozdag, Chem. Eng. J. 85 (2002) 351–355.
3. T.A. Kurniawan, G.Y.S. Chan, W.H. Lo, S. Babel, Chem. Eng. J. 118 (2006) 83–98.
4. H.K. Alluri, S.R. Ronda, V.S. Settalluri, J.S. Bondili, V. Suryanarayana, P. Venkateshwar, African J. Biotech. 6 (2007) 2924–2931.
5. J. Febrianto, A.N Kosasih, J. Sunarso, Y.H Jua, N. Indraswati, S. Ismadji, J. Haz. Mat. 162 (2009) 616–645.
6. Y.S. Park, T.S. Chon, I.S. Kwak, S. Lek, Science of the Total Environ, 327, (2004) 105–122.
7. L. Belanche, J.J. Valdes, J. Comas, I.R. Roda, M. Poch, Artif. Intell. Eng, 14, (2000) 307–317.
8. G.R. Shetty, S. Chellam, J. Membrane Sci, 217 (2003) 69–86.
9. A. Kardam, K.R. Raj, J.K. Arora, S. Srivastava, Natl. Acad. Sci. Lett. 33 (3&4) (2010) 83–87.
10. K.R. Raj, A. Kardam, J.K. Arora, S. Srivastava, J. Radioanal. Nucl. Chem. 283 (2010) 797–801.
11. A. Kardam, K.R. Raj, J.K. Arora, S. Srivastava, J. Water Resource and Protection 2 (2010) 339–344.
12. K.R. Raj, A. Kardam, J.K. Arora, S. Srivastava, J. Water Resource and Protection 2 (2010) 331–338.
13. A. Kardam, P. Goyal, J.K. Arora, K.R. Raj, S. Srivastava, Natl. Acad. Sci. Lett. 32 (2009) 179–181.
14. P. Goyal, P. Sharma, S. Srivastava, M.M. Srivastava, Int. J. Environ. Sci. Technol. 5 (2008) 27–34.
15. P. Goyal, S. Srivastava, Arch. Environ. Prot. 34 (2008) 35–45.
16. P. Goyal, S. Srivastava, Natl. Acad. Sci. Lett. 31 (2008) 347–351.
17. P. Goyal, S. Srivastava, Journal of Hazardous Materials, 172 (2009) 1206–1211.
18. D.R. Baughman, Y.A. Lieu. Neural network in bioprocessing and chemical Engineering, Academic Press. San Diego, 1995.

54
Treatment of Saline Soil by Application of Cyanobacteria for Green Farming of Rice in Dayalbagh

S. Yadav and G. P. Satsangi
Department of Botany, Dayalbagh Educational Institute,
Agra- 282 110, U. P., India.
Email: sanjaydei84@gmail.com

Abstract

In the present investigation attempts were made to understand the stages of succession of cyanobacteria in paddy fields and treatment of saline soil by the application cyanobacterial bio-fertilizer for green farming of rice. A chronological sequence of appearance of valuable cyanobacterial genera in nearby area was recorded through out the paddy cropping session. On the addition of cyanobacterial bio-fertilizer pH and E. C. of soil were decreased 12.16% and 78.94% respectively. Interestingly at the end of 4-weeks soil properties were profoundly changed. There was a considerable improvement in the total N_2 content and organic carbon of the soil therefore C/N ratio increased from 9.3 to 10.6 (13.98%) as a result of cyanobacterial treatment.

Introduction

The increasing cost of synthetic nitrogenous fertilizers has made it imperative to find alternate sources of N_2. On the other hand heavy use of chemical fertilizers, herbicides and pesticides may cause environmental problems in near future[1]. In this respect, biological nitrogen fixation attracts the focus of workers as a nitrogen supplement i.e., *Azolla* and Nitrogen fixing cyanobacteria, especially for sustainable rice crop management. The significant role of BGA in maintaining soil fertility and their distribution in Indian rice fields have been studied by various workers during past half century[2–4]. The most common genera found *Anabaena, Aulosira, Calothrix, Cylindrospermum, Gloeocapsa, Nostock, Rivularia, Sytonema and Tolypothrix*[5]. The cyanobacterial succession in saline soil was reported by Pandey *et al.*[6].

The alkaline soils are characterized by impermeability, extreme hardness and occasional presence of undesirable salts on the surface all of which affect adversely the rice plant growth. Therefore it is important to reclamation of such soil so these soils make fertile[7]. Soil pH is an important factor, which determines the distribution and dominance of cyanobacteria. Indeed cyanobacteria can decrease soil pH from 9.2–7.5[8–9]. In soil the oxidizable matter sulfide, phosphorus and iron significantly reduced by inoculation of cyanobacteria. Under such condition cyanobacterial bio-fertilizers could provide essential nutrition and organic matter for better soil health as observed earlier by Singh and Bisoy [10].

Material and Methods

The study was carried out from month of July 2010 to March 2011 to develop a chronological of succession of cyanobacterial genera in the agricultural fields of Dayalbagh, Agra (U. P.). Identification of cyanobacterial strains were carried out using the Taxonomic publication of Desikacharya[11]. To observe the effect of cyanobacteria on the soil health, the experiment was performed under natural conditions/*In-vivo* inoculated with and with- out cyanobacterial bio-fertilizer. To carry out this experiment soil based cyanobacterial bio-fertilizer from both N_2 fixing and nondiazotrophic was prepared in open air shallow cultures by following the method of Venktaraman[11]. Soil without inoculation of cyanobacterial fertilizer was taken as control. All soil samples were analyzed for some physical and chemical properties viz., pH, electrical

M.M. Srivastava, L. D. Khemani, S. Srivastava, *Chemistry of Phytopotentials: Health, Energy and Environmental Perspectives*, DOI:10.1007/978–3–642–23394–4_54, © Springer-Verlag Berlin Heidelberg 2012

conductivity (EC), total N_2, organic C and C/N ratio by following the method of Jackson [13] as pre and post cyanobacterial treatment at an interval of 2–4 weeks.

Results and Discussion

During the summer (April–June) fields become dry. At the end of month June and beginning of July small patches *Anabeana* sps along with *Nostoc* sps, *Cylindrospermum* sps, *Aphanothece* sps, *Scytonema* sps, *Croococus* sps and *Oscillatoria* sps appeared on the surface due to first shower of rain (Table 1). In the month of July-August most of fields were logged due to heavy rains. At the end of month of August sps of *Tolypothrix* was frequently found from the study area and the paddy fields were covered with a thin greenish layer due to cyanobacterial colonization.

After 2 weeks of cyanobacterial treatment of saline soil, pH and EC were started to decrease, although there is no change was recorded in total N_2 and organic C of the soil (Table 2). Interestingly at the end of 4 weeks of treatment soil properties were profoundly changed. It was observed that soil pH and EC decreased from 9.45–8.3 (12.16%) and 1.33–0.28 (78.94%) respectively. There was a considerable improvement in the total N2 content (39.28%) and organic C (38.46%) of the soil (Table 2). Therefore C/N ratio increased from 9.3 to 10.6 (13.98%) as a result of cyanobacterial treatment. More time and field studies are required to understand the cyanobacterial effect on the complete characteristics of soil under local environmental conditions.

Table 1: Identified Cyanobacterial genera and their succession period (July 2010-March 2011)

Cyanobacterial genera	Succession Period
Anabeana sps	July – February
Nostoc sps	July – February
Cylindrospermum sps	July – October
Aphanothece sps	July – December
Scytonema sps	July – March
Croococus sps	July – Feb
Oscillatoria sps	August – January
Tolypothrix sps	September – March

References

1. I.M. Chung, K.H. Kim, J.K. Ahn and H.J. Ju; Korean Weed Sci. 17 (1997) 52–58.
2. A.C. Shukla; Rev. Algal. 10 (1971) 257–270.
3. R.S. Aiyer, S. Shlahuddin and G.S. Venkataraman; J. Agric. Sci. 42 (1972) 382.
4. G.L. Tiwari, R.S. Pandey; Nova Hedwigia. 27 (1976) 701–730.
5. G.S. Venkataraman; The role of blue–green algae in tropical rice cultivation. In Nitrogen Fixation by Free Living Micro-organisms (ed. Stewart, W.D.P.), Cambridge Univ. Press. Cambridge, (1975) 207–218.
6. K.D. Pandey, P.N. Shukla, D.D. Giri and A.K. Kashyap; Biol. Fert. Soils, 41(2005) 231–235.
7. R.N. Singh; Pub. ICAR, New Delhi (1961).
8. G.E. Fogg; Annu. Rev. Plant Physiol. 7 (1956) 51–70.
9. B.A. Whitton, M. Potts; Kluwer, Derdacht, (2000) 233–255.
10. P.K. Singh, R.N. Bisoy; Bio-fertilizers for restoration of soil fertility. In: Singh, J.S. (ed). Restoration of degraded land; concept and strategies. Rastogi Pub. Meerut, (1993) pp 25–47.
11. T.V. Desikachary; Cyanophyta. Indian Council of Agricultural Research. New Delhi, India (1959).
12. G.S., Venkataraman. Algal bio-fertilize for rice, IARI, New Delhi (1981) 1–10.
13 M.L. Jackson; Soil Chemical analysis. Prentice Hall of India Pvt. Ltd., New Delhi, India (1973).

Table 2: Effect of cyanobacterial inoculation on the characteristics of saline soil under natural Conditions (Mean±SD)

Soil Parameters	Pre-cyanobacterial inoculation	Post-cyanobacterial inoculation		Increase/decrease After 4 weeks (%)
		After 2 weeks	After 4 weeks	
pH	9.45±0.072	9.12±0.059	8.3±0.016	12.16[d]
EC	1.33±0.370	1.1±0.082	0.28±0.043	78.94[d]
Total N_2	0.028±0.002	#	0.034±0.003	39.28
Organic C	0.26±0.016	#	0.36±0.036	38.46
C/N Ratio	9.3	#	10.6	13.98

 # – No change, d – Decrease

55
Effect of Anionic and Non-ionic Surfactants in Soil-Plant System Under Pot Culture

A. Mohammad* and A. Moheman
Department of Applied Chemistry, Faculty of Engineering and Technology, Aligarh Muslim University, Aligarh 202 002 (India)
*Email: alimohammad08@gmail.com

Abstract

A study was carried out to investigate the effects of anionic (sodium dodecyl sulfate, SDS), and nonionic (TritonX-100) surfactants on dry biomass, and nutrients uptake in tissues of wheat (Triticum aestivum) under pot. Concentrations of nutrients (i. e. K, Ca, Zn, and Fe) in shoots and roots were determined respectively. The results showed that nutrient concentrations in shoots were less affected with increase in concentration of surfactant (0.1 % aqueous SDS or TritonX-100) in soil as compared to roots. Similarly, shoot dry biomass seemed less affected than root dry biomass while anionic surfactant (SDS) was found to be more toxic to plant growth and nutrient uptake.

Introduction

Surfactants are used on large scale in detergents for domestic cleaning and various industrial processes due to their efficiency to clean dirt and greasy materials. The first study on the use of surfactant for clean-up of contaminated soil was reported from the American Petroleum Institute in 1979. Surfactants are known to possess powerful bactericidal properties and are capable to modify the physical and chemical properties of the soils [1]. Long-term treatment with surfactants causes pollution [2] of soil and water, which affects growth, development, and metabolism of soil microorganisms [3] and plants [4–7].They solubilize nonpolar plant substances such as waxy critical or the lipoidal part of the cell wall that facilitate the rapid absorption of the toxic as well as beneficial chemicals. They enter into the intercellular spaces and affect plant growth systems [8].

Literature survey reveals that the available information about the use of surfactants in relation to plant growth and nutrient uptake are few. Hence, during present study an experiment was designed to investigate the effect of an anionic (SDS) and nonionic (TritonX-100) surfactants on growth and nutrients uptake by wheat plants.

Experimental

The soil (0–30 cm depth) used in the experiment was collected from the field of Aligarh Muslim University, Aligarh (India). The soil was air dried, passed through a 5-mm nylon mesh sieve. The physico-chemical properties of the soil were as follows: pH (H_2O) 8.56; organic matter 0.49 %; cation exchange capacity 16.2 meq/100 g soil and the exchangeable cations (K^+, 0.5 and Ca^{2+}, 3.7 meq/100 g soil) as measured by the standard method. The available Zn, and Fe contents of the soil were 0.37, and 10.5 ppm (DTPA method) [9].

Before potting, soils were treated with 0.1 % aqueous solutions of SDS, and TritonX-100 at different concentrations levels (0 (control), 5, 10, 20, 40 and 80 ml kg^{-1} soil). The basal fertilizers applied (0.45 g kg^{-1} soil) were N (as urea), P_2O_5 (as $CaHPO_4$) and K_2O (as K_2SO_4) in 1:1:1 ratio by weight. These were mixed homogeneously with the soil. The resulting soil was then adjusted with distilled water to 60 % of water holding capacity (WHC) and maintained at this moisture content for 2 days. The subsamples were mixed homogeneously again and aliquots (1.6 kg oven dry weight) were transferred to glazed earthenware pots and sown with seeds of winter wheat (10 seeds pot^{-1}). After seedlings emerged, the pots were thinned

M.M. Srivastava, L. D. Khemani, S. Srivastava, *Chemistry of Phytopotentials: Health, Energy and Environmental Perspectives*, DOI:10.1007/978-3-642-23394-4_55,

to five seedlings/pot. There were three replicates of each treatment. The pots were randomly arranged in a glass house and rearranged several times during the growth period. The growth temperature was 22 ± 3°C. During the growth period the plants were watered with distilled water every three days and treated with above surfactants doses every 15 days till 75 days. The plants were harvested after 90 days.

At harvest, shoots were cut at the shoot-root junction. The roots were carefully washed from the pots with tap water and a 5-mm sieve to collect root samples as much as possible. Both shoots and roots were washed with distilled water and dried in an oven at 70 ± 1°C for 48 hour. The dry biomass of shoots and roots were weighed and ashes at 180°C for 4 h and finely powdered in a stainless steel grinder. About 1 g of powdered plant material (shoots or roots) was digested in 10 mL of acid mixture containing conc. HNO_3 and $HClO_4$ (4:1, v/v). The digested mixture was treated on a hot plate till brown fumes ceased and converted to a syrupy liquid along with some white fumes. The samples of the syrupy liquid were dissolved in 5 mL of conc. HCl and diluted with distilled water. To obtain the clear solutions, they were filtered and then adjusted to a volume of 25 mL in each case. The resulting solutions were analyzed for K, Ca, Zn, and Fe. The K, and Ca were estimated by ‚Systronic' flame photometer. The concentration of Zn was estimated with the aid of double beam atomic absorption spectrophotometer (GBC-902) and Fe by spectrophotometry.

Results and Discussion

Effects of anionic (SDS) and non-ionic (TritonX-100) surfactants are evident in this experiment in terms of both dry biomass and nutrients uptake in plant tissues. In the present pot experiment, shoots and roots biomass were increased with the increase of TritonX-100 from 0 to 10 mL/kg, followed by decrease on further increase in concentration level of TritonX-100 from 10 to 80 mL/kg soil (Fig. 1). However, in case of SDS shoots and roots biomass seemed less affected by SDS treatments from 0 to 10 mL/kg, then decreased as SDS increased from 10 to 80 mL/kg (Fig. 1). The beneficial effect of TritonX-100 at lower doses (0 to 10 mL/kg) may be attributed to the increase in nutrient contents in plant tissues (Figs. 2 and 3). Calcium level in shoots does not change much by surfactants treatments (Fig. 2). However, calcium concentration in roots was found to decrease significantly with the increase in the concentration of surfactants (Fig. 3). It is obvious that Ca uptake is affected by the surfactants, but it is not the translocation of Ca within the plants. A decrease of Ca concentration in roots under SDS or TritonX-100 treatments may be a symptom of a damaged intercellular defense system. The optimum concentration of the TritonX-100 brought the interfacial tension to the most suitable degree resulting in maximum absorption of nutrients by plants leading to the beneficial effect. The negative effects of SDS (0 to 80 mL/kg soil) and TritonX-100 (20 to 80 mL/kg soil) doses may be attributed to the reduction in microbial activity of *Bacillus siliceus* and *Aspergillus niger*, which ultimately reduce the solublizing effects. And also may be due to the fixation of nutrients into non-exchangeable form. This result is consistent with some previous studies [5–7, 10] where application of surfactant to soil has shown

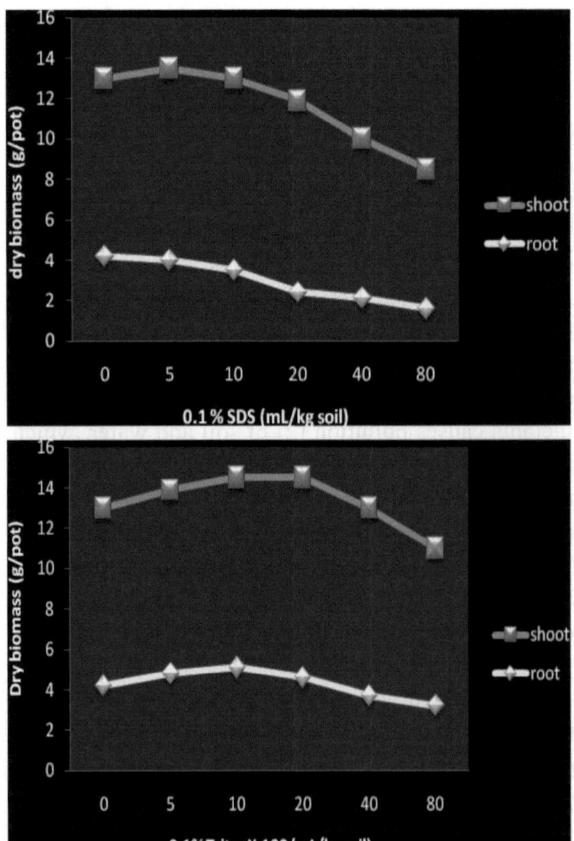

Fig. 1: Effects of SDS and TritonX-100 on Dry shoot and root biomass of wheat (*Triticum aestivum*)

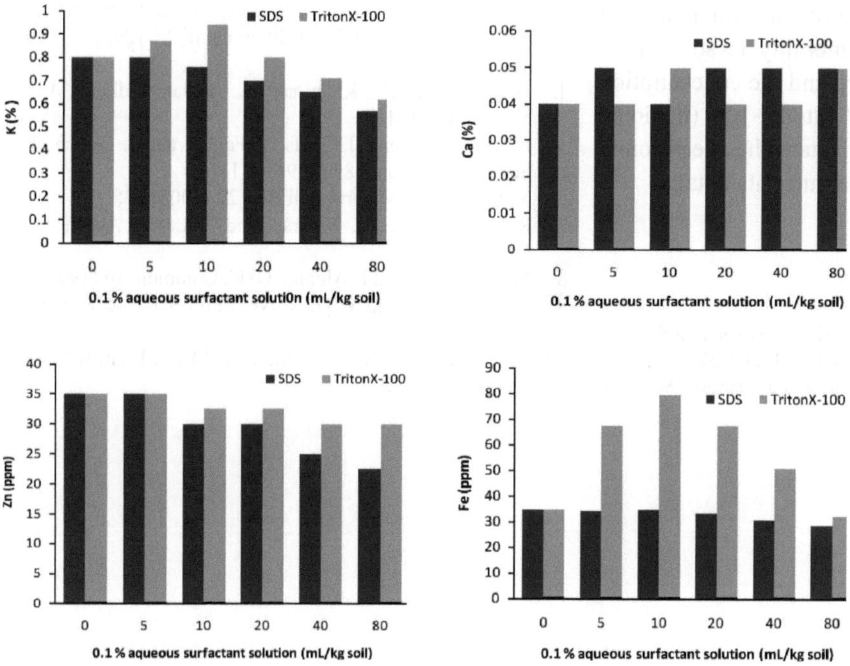

Fig. 2: Effects of SDS or TritonX-100 on K, Ca, Zn and Fe uptake in the shoots of wheat (*Triticum aestivum*)

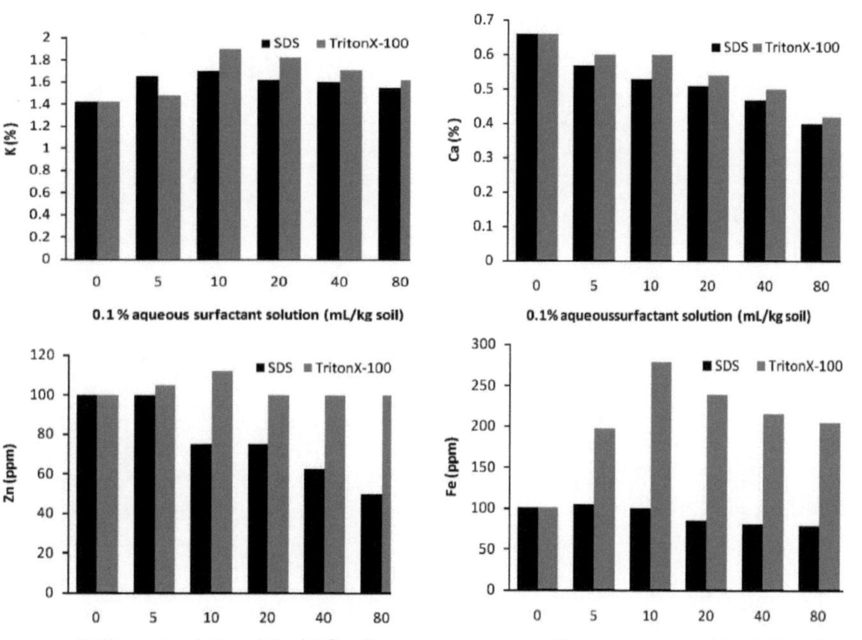

Fig. 3: Effects of SDS or TritonX-100 on K, Ca, Zn and Fe uptake in the roots of wheat (*Triticum aestivum*)

detrimental effects on plant growth, soil microbial activity as well as nutrients mobility in soil. These effects depend upon the nature and the concentration of surfactant. Compared to TritonX-100 (nonionic surfactant), SDS (anionic surfactant) has been found more toxic to plant growth and nutrients uptake.

References

1. G.C. Klingman, *A Text Book of Weed Control as a Science*, Wiley Eastern Pvt. Ltd., New Delhi, 81, (1973)
2. N. Valoras, J. Letey, J.P. Martin, and J. Osborn; Soil Sci. Soc. Am. J. 40 (1976) 60.
3. J.H. Kowalezyk, *Effect of Synthetic Surfactants on Microbial Activity in Soil,* Umweltkommam. 2nd 1292, 2, (1993) 365.
4. J.P. Singhal, S.U. Khan, and D. Nandan; Indian J. Agric. Sci. 49 (1979) 423.
5. P. Kloepper-Sams, F. Torfs, T. Feijtel, and J. Gooch; The Sci. total environ. 185 (1996) 171.
6. N. Lal, and R. Mishra; Poll Res. 22 (2003) 335.
7. R.J. Luxmoore, N. Valoras, and J. Letey; Agron J. 66 (1974) 673.
8. M.Y. Husain, and F. Ahmad; Oxid. Commun. 16 (1993) 62.
9. W.L. Lindsay and W.H. Norvell; Soil Sci. Soc. Am. J. 42 (1978) 421.
10. S.U. Khan, J.A. Khan, S. Jabin, and R.K. Bhardwaj; Clay Res. 19 (2000) 27.

56
Studies on Efficacy of Eco-Friendly Insecticide Obtained from Plant Products Against Aphids Found on Tomato Plant

S. Dubey, S. Verghese P., D. Jain and Nisha
Department of Chemistry, St. John's College Agra
Email: sngt.dubey@gmail.com

Abstract

A number of synthetic pesticides have been used for the control of aphids in agriculture, but increasing public concerns over their adverse effects on the environment have required more environmentally friendly methods for pest management. Tomato is one of the most consumed and widely grown vegetable crops in the world, but their yield is suppressed by the diseases like tomato yellow leaf (TYL) caused by aphids. Aphids are the most common insects found on tomato plants. Aphids suck sap and juice from the new growth on tomato plants, causes wilting, loss of vigor and distorts the shape of the plant, and leave a characteristic sticky excrement called honeydew the leaves will have sticky shiny spots on them when honeydew is present. Aphids may even transmit viral diseases from one plant to another like tomato yellow leaf curl virus (TYLCV) and tomato spotted wilt virus (TSWV) and tomato yellow top. Aphids can be significantly reduced by the application of soap based insecticide containing Pyrethroid which is eco-friendly in nature as pyrethroid is obtained from the extract of Chrysanthemum plant. This insecticidal spray of potassium laurate and pyrethroid has reduced immensely the disease on tomato plant named tomato yellow leaf (TYL) spread by aphids .Hence an attempt has been made to develop an eco friendly economical product for integrated pest management strategy.

Introduction

Aphids are one of the major pests in the agriculture crops. It is a popular vegetable in India and Bangladesh and also a popular source of vitamins and minerals in human diet[1]. Its production is seriously hampered by many diseases like tomato yellow leaf (TYL). This disease is transmitted by aphids[2]. In extreme conditions they even cause stunting of the overall growth of plant.

The yield loss is reported to be 63–95 %[3]. These aphids (insects) suck sap and juice from new growth on tomato plants, and also act as vectors of many diseases like leaf curl and yellow spot on plant leaf, due to which the crop suffer heavy losses[4] like wilting and the loss of vigor and also distorts the shape of the plant. Pyrethroid is eco-friendly and safe to use against aphids. Pyrethroid has shown very good efficacy against aphid diseases on tomato plants, and thus highly desirable for controlling these diseases and improving grain yields[5].

Tomato aphids are important insects and period of April to September is quite favorable for their infestation, insecticides are important in controlling tomato plant diseases and farmer should scout their field for aphids regularly during the flowering period and use insecticides for controlling aphids[6].

Generally, farmers use insecticides to control the aphids. But mere dependence on insecticides causes environmental pollution and pesticides resistance.

Study has been done[7] for the management of tomato plant diseases by pyrethroid in past. Therefore, the present study was carried out to evaluate the efficiency of soap based insecticide containing potassium laurate and pyrethroid on aphids found on tomato plants as there is an urgency in the search for eco-friendly alternative approaches for the plant diseases.

Material and Method

Pyrethroid is extracted from Chrysanthemum. Chrysanthemum flowers were harvested, dried, crushed and conserved in a freezer[8]. Pyrethrins are extracted with Kerosene method[9]. Potassium laurate was pre-

M.M. Srivastava, L.D. Khemani, S. Srivastava, *Chemistry of Phytopotentials: Health, Energy and Environmental Perspectives*, DOI:10.1007/978–3-642–23394-4_56,

pared by refluxing equivalent amount of corresponding fatty acids and aqueous solution of potassium hydroxide for 6–8 hours on a water bath. Potassium laurate was purified by recrystallization with benzene-methanol mixture and dried under reduced pressure. The purity of potassium laurate checked by the determination of their melting point (120°C).

The conductivity measurements of the solution of potassium laurate in distilled water was made with "BIOCRAFT" direct reading conductometer and a dipping type glass conductivity cell with platinized electrodes at a room temperature and the CMC (criticle micelle concentration) of potassium laurate is 0.0045 dm^3/l on plotting specific conductance Vs concentration at various concentrations.

Many samples of various concentration having different P^H have been prepared by mixing potassium laurate of concentration (LCMC) and pyrethroid of different dilution (%), and then sprayed on tomato plant to check the efficacy of this insecticidal spray on weekly and bi-weekly interval.

Results and Discussion

A field experiment was laid out in Agra during Jan-May and a mixture of potassium laurate and Pyrethroid of different concentration having different P^H value being sprayed weekly and biweekly on tomato plants grow in various rows (1–13). All the agronomic practices were followed as and when required. Considering the pest population in experimental area. There were thirteen rows having three plants of tomato each. The first row having three tomato plants found healthy but the remaining rows were afflicted with tomato yellow leaf diseases either on leaf or on fruit or on stem. These were the initial signs of disease of yellow leaf of tomato plant caused by aphids. Each plant in each row subjected to insecticidal spray contains potassium laurate (LCMC) and pyrethroids (%) of various P^H (9.10–7.52) on bi-weekly in the beginning when the symptoms started appearing and converted into weekly application when the plants (leaf, fruit and stem) started growing healthy. The mixture of the potassium laurate and pyrethroid of P^H 9.10 has been sprayed on second row and similarly the liquid spray of decreasing P^H have been sprayed on subsequent rows. The impact and efficacy of the mixture of potassium laurate (LCMC) and pyrethroid (%) on tomato plant for tomato yellow leaf disease having given in Table 1.

It was noted that the plants of second, third, fourth, fifth and sixth row need frequent application as the symptoms were recurring soon. The plants of seventh, eight, ninth, tenth and twelfth row have been sprayed with P^H 8.40–7.52 found in better condition, than the plants present in previous rows. The plants of eleventh row, which was sprayed with solution of P^H 7.78 was found healthy even after weekly application.

Conclusion

In our experiment we conclude that, the liquid spray insecticide contains potassium laurate and pyrethroid having P^H 7.78 is found to be more effective than other solutions. Other solutions were found to be less effective in curing tomato yellow leaf (TYL) transmitted/spread by aphids on tomato plants.

Table 1: Impact of mixture of potassium laurate and pyrethroid on tomato plant

S. No.	Number Of Rows	Concentration Of Pyrethroid (%)	P^H Value Of Spray	Effect On Plants
1.	First row (healthy)	——	——	——
2.	Second row	1	9.00	Needs frequent application
3.	Third row	2	9.04	Needs frequent application
4.	Forth row	3	9.10	Needs frequent application
5.	Fifth row	4	8.96	Needs frequent application
6.	Sixth row	5	8.90	Needs frequent application
7.	Seventh row	6	8.40	Effective
8.	Eight row	7	8.00	Effective
9.	Ninth row	8	7.95	Effective
10	Tenth row	9	7.88	Effective
11	Eleventh row	10	7.70	More effective
12	Twelfth row	11	7.61	Effective
13.	Thirteen row	12	7.52	Effective

References

1. B.A. Nault, J (III). Speese, D. Jolly and R.L. Groves; J. Crop Prot. 22 (2003) 505- 512.
2. J.C. Desai; Thesis, Gujarat Agric. Univ., Gujarat (India). 2000.
3. N.D. Gupta; Thesis Department of Plant Pathology, BSMRAU, Bangladesh. 2000. 77p.
4. U.C. Singh, Singh Reeti and K.N. Nagaich; Entomology, 6(2) (1998) 181–183.
5. E.K. Chatzivassiliou, D. Peters and N.I. Katis; Phytopathology 92(2002) 603–609.
6. E.K. Chatzivassiliou, D. Peters and N.I. Katis; J. Phytopathol. 11(2007) 699–705.
7. J.K. Mac Intyre Allen, C.D. Scott-Dupree, J.H. Tolman and C. Ron Harris; J. Pest Manage. Sci. 61 (2005) 809–815.
8. C.B. Gnadinger, I.E. Evans and C.S. Corl; Bulletin 1933. 401. 19 pp.
9. M. Keen; Du solala Table, 1988 (October). No. 166:21.

References

57
Studies on Cr (III) and Cr (VI) Speciation in the Xylem Sap of Maize Plants

S. J. Verma[1] and S. Prakash[2]
[1]Dept. of Chemistry, T.S Chankaya (Institute of Nautical Sciences), Karave, Navi Mumbai 400706
[2]Emeritus Professor, Dept. of Chemistry, Faculty of Science, Dayalbagh Educational Institute, Dayalbagh Agra 282001.
Email: shikhaj@rediffmail.com

Abstract

The toxicity, mobility and bioavailability of chromium, a persistent pollutant depends on its chemical form. Cr (III) is innocuous whereas Cr(VI) is toxic. In vitro studies were carried out using radiotagged Cr (III) and Cr(VI) in the xylem sap of mature maize plants using anion and cation exchange elution chromatography. Xylem sap transports ions to the aerial parts of the plant after absorption by roots. Both Cr (III) and Cr (VI) were present as mobile and soluble anionic organic complexes, probably Cr (III)-citrate in the xylem sap. This indicates a mechanism used by plants for detoxification of toxic Cr (VI) reducing the phytotoxicity via the food chain and providing an opportunity for Phytoremediation.

Introduction

Chromium Contamination

The extensive use of chromium, an indispensable metal but a persistent contaminant has resulted in its accumulation in various environmental compartments. The major contributor of Cr pollution to the biosphere is the leather industry accounting for 40% of the total industrial use [1]. In India, about 2000–32,000 tons of elemental Cr annually escapes into the environment from tanning industries. In recent times, there is thrust to explore the mechanism of absorption, uptake, translocation, tolerance, bonding and especially the chemical form of chromium in plants to understand its phytoavailability from contaminated soils and waters, their further transfer through agricultural food chains and to evaluate the potential of plants to exploit them for phytoremediation to treat Cr contaminated soil [3,4].

Chemical Speciation of Chromium

Chemical speciation is significantly applicable in the case of chromium as its two important and stable states viz. hexavalent and trivalent species show contrasting behavior with respect to reactivity, solubility, mobility and toxicity. Cr (III) is considered an essential micronutrient needed by the body for the maintenance of normal glucose tolerance levels [5] whereas anionic Cr (VI) is toxic as an oxidizing agent and even carcinogenic [6].

Chromium Uptake and Speciation in Plants

The uptake of Cr by plants depends on Chemical Speciation changes in soil and water, in the plant root environment (rhizosphere), in the root, in the xylem sap and in the leaves and fruits [7]. There are several studies on the uptake, accumulation and phytotoxicity of the two Cr species in different plant parts in various cereal crops and vegetables [8–11]. The few studies on the chemical form of Cr in different plant components of different plant species indicate the conversion of toxic Cr (VI) to the harmless Cr (III) species in root and the presence of Cr as Cr (III)-organic acid complex in the roots and leaves [4, 12–15].

Chromium Speciation in Xylem Sap

There is a dearth of studies on the translocation and speciation of Cr in the xylem sap [16–18]. The xylem

M.M. Srivastava, L. D. Khemani, S. Srivastava, *Chemistry of Phytopotentials: Health, Energy and Environmental Perspectives*, DOI:10.1007/978-3-642-23394-4_57,

sap translocates water, nutrient and non-nutrient ions after absorption by roots, to the aerial parts i. e. shoot and fruits. This component of the plant can provide a unique insight into the basic mechanisms employed to transport nutrient as well as pollutant metals in plants. Besides inorganic ions, xylem sap has various organic ligands of which the carboxylic acids are the major complexants [19, 20]. Heavy metals that exist in many oxidation states and/or are prone to hydrolysis are translocated after interaction with the organic ligands of the xylem sap [7, 16–18, 21–22].

In a series of studies aimed on chemical speciation of Cr in plants, in the present study, the chemical form in which Cr(III) and Cr(VI) is translocated by the xylem sap in 90 days old maize plants (*Zea mays. L;* Ganga 5) representing the most important stage of the crop, the cob stage was determined. Simultaneous cation exchange (Dowex-50 X8) and anion exchange (Dowex-1 X8) column elution chromatography was used for Speciation analysis using radiotagged Cr. The latter was detected and quantified by Gamma Ray Spectrometer.

Experimental

Plant Material

Seeds of maize (*Zea mays L.*; seed type: Ganga 5; sowing season: October–November) procured from National Seed Corporation Ltd., Indian Agricultural Research Institute, New Delhi, India were sterilized and soaked for 24 hours. They were sown in polyethylene pots filled with washed dried quartz sand (pH, conductivity of washing: 6.5 – 7.0 and 1.4 – 1.6 mmho respectively) and the pots placed in a green house. Till the seeds germinated, water was sprinkled twice a day. There after, they were irrigated alternately by water and nutrient solution. The nutrient solution [23] contained the following nutrient millimolar concentrations: K, 5.0; N, 15.0; Ca 5.0; P, 1.33; Mg, 3.0; S, 1.5; Fe, 0.1; Mn, 0.01; Cu, 0.001; Zn, 0.002; B 0.03; Mo, 0.0002; Co, 0.0001; Na, 0.1, Cl, 0.04. Citrate was replaced by EDTA to maintain the solubility of nutrient Fe.

Collection of Xylem Sap Samples

Plant stems were cut perpendicular to the stem axis with sterilized stainless-steel razor blades at approximately 12 cm above the root. The first drop of exudate from the cut surface was discarded and the cut surface washed and wiped gently. The subsequent drops that emerged were sucked by a micropipette and collected as xylem sap samples in ice cooled sterilized borosil glass vials. Plant stem was severed in a basipetal fashion and fresh cuts were made after every 1.5 – 2.0 hours at a distance of 1 – 2 cm from the previous cut till the junction of stem and root. The sampling was done for a period of 24 hours and the xylem sap of various plants was pooled. The sap was filtered with Millipore GV nylon filters (0.22 mm), purged with pure nitrogen (N_2) gas and stored at freezing temperatures.

Separation of Carboxylic Acid Fraction of Xylem Sap

Separation of carboxylic acids (major complexants) from the other constituents of xylem sap mainly, amino acids was carried out using cation exchange resin (Dowex-50 X8). pH of the sap was adjusted to 7.0, and loaded on a 1 X 20 cm column with a bed height of 6 cm. The carboxylic acids were eluted with deionized distilled water (pH 6.95) [24]. This separated fraction was utilized for in vitro studies as well as quantification of the major carboxylic acids by ion suppressed reversed phase HPLC.

Preparation of Reagents

All solutions were prepared in deionized distilled water using ‚Analytical Grade Reagents‘. The radioactive ^{51}Cr isotope was obtained as aqueous CrO_4^{2-} (specific activity 50µci/mg) from Board of Radiation and Isotope Technology (BRIT), Mumbai, India. 200µM radiotagged Cr (III) stock solution was prepared. 1 ml of 1000 µM Cr (VI) stock was mixed with active Cr (VI) (41.7µci) and reduced with Na_2SO_3 such that the pH of the resultant solution was 4.5. The prepared oxidation state of radiotagged Cr (III) was checked by batch cation and anion exchange experiments. 0.0192 gm citric acid was made up to 50 ml at pH 5.0 to prepare 2000µM stock solutions.

0.01M, 0.1M, 1M and 2.0M $NaNO_3$ at pH 4.5 were the eluents used in anion exchange elution chromatography and 0.01M, 0.1M and 1M $HClO_4$ and 0.25M, 0.5M and 1M Ca $(NO_3)_2$ solutions prepared in 0.01M $HClO_4$ solution were the eluents for cation exchange elution chromatography.

Cr (III) and Cr (VI) Complexation and *in Vitro* Studies

Cr (III) and citric acid was mixed in the ratio of 1:10 (100μM: 1000μM) at pH 4.5–5.0 at 25 °C. The complexation analysis was carried out on the 4th, 7th, 10th and 30th day of mixing. Only the results of 10th and 30th day of complexation are shown.

For the *in vitro* studies, 0.5 ml of Cr (III)/Cr (VI) solution containing 100μM of Cr (III)/Cr (VI) was mixed with 0.5 ml of carboxylic acid fraction of sap or 0.125 ml of pure sap corresponding to 1000μM of total carboxylic acid as determined by HPLC. The total volume was made up to 1 ml. The pH of these solutions was in the range of 4.5 and 5.0 and they were kept at 25 °C. For Cr (III), the analysis was carried out just as in synthetic Cr (III) complexation studies whereas for Cr (VI), analysis was carried out after a period of 20 days at 25 °C.

Ion Exchange Elution Chromatography and Preparation of Elution Curves

Dowex-1 and Dowex-50 ion exchange columns with bed heights 3 cm and 2 cm respectively were prepared by pouring the slurry of the respective resin in the first eluent in columns of dimensions 1 X 20 cm. The elution was carried out at a flow rate of 0.5 ml/min. 5 ml fractions of each eluent were collected from the column in scintillation vials. Each fraction was counted using Gamma Ray Spectrometer at 0.320 MeV corresponding to the photopeak of ^{51}Cr. The Gamma ray Spectrometer model used for the present work was a NaI (Tl) Detector coupled to a 4K Multichannel Analyser (Canberra Accuspec Card PC386AT). The elution pattern was obtained by plotting the volume (ml) of eluent against the percentage of total counts obtained in each fraction using MS Office Excel Software.

The elution profiles of pure Cr (III), pure Cr (VI) and Cr (III)-citric acid complexation were compared with elution curves obtained for interaction of pure Cr (III) with the carboxylic acid fraction and the pure sap in the in vitro studies. The Cr (III) complexes which are chiefly anionic, were separated by anion exchange elution chromatography. The extent of reaction or the unreacted Cr (III) and formation of any positive complexes was ascertained by comparison with the elution curve of pure Cr (III) from the cation exchanger.

Results and Discussion

Cr (III) In Vitro Studies Results

Time based interaction of radiotagged Cr (III) with the carboxylic acid fraction of maize sap (WIIIx) indicates gradual formation of anionic complexes (Fig. 1a). The anionic complexes elute as peaks at 30 and 65 ml, with the dominance of the 65 ml peak on 30th day. The elution curve profiles for Cr (III) complexation with pure xylem sap (WIIIy) are comparable to the carboxylic acid fraction with similar dominant anionic complexes being eluted at 30 ml and 65 ml (Fig. 1b).

The above elution curves obtained are similar to Cr (III) complexation studies carried out with citric acid (Fig. 2). Three major peaks elute from the anion exchange column at 30, 65 and 80 ml. The 30 ml peak is considerably reduced whereas the 65 ml peak becomes dominant on the 30th day.

These peaks along with the other minor peaks represent various 1:1 citrate complexes.

The extent of complex formation is substantiated by the bar diagram illustrated for elution from the cation exchanger on the 10th day (Fig 3a,b) for Cr(III)-

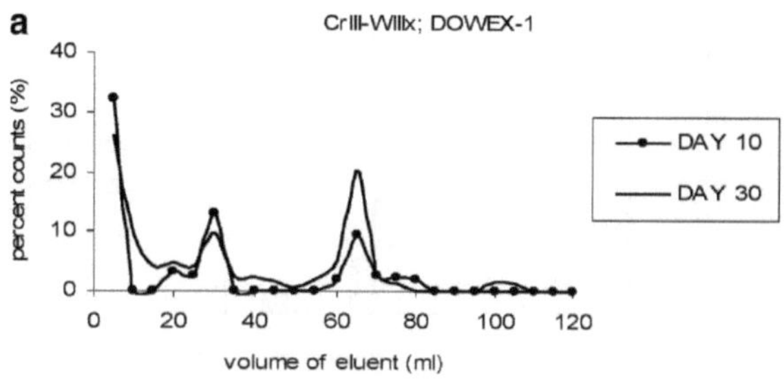

Fig. 1a: Cr (III)-Carboxylic acid fraction (WIIIx) elution curve on Dowex-1

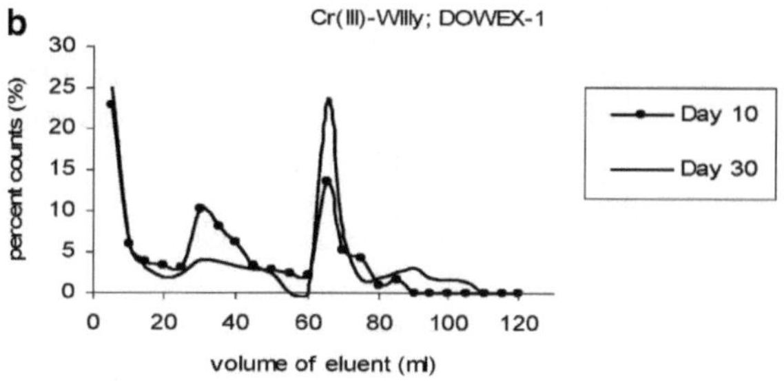

Fig. 1b: Cr(III)-Pure sap (WIIIy) elution curve on anionic Dowex-1

WIIIx and Cr(III)-WIIIy respectively as compared to pure Cr (III).

Cr(III) is present as different hydrolytic species of which $[Cr(H_2O)_6]^{3+}$ is the most dominant and gets eluted at 65 ml in the D fraction. Most species in the xylem sap samples are anionic as they elute in the first fraction from the cation exchanger. The D fraction is also reduced considerably. This signifies that Cr (III) forms anionic complexes or species with the pure sap (WIIIy) as well as the carboxylic acid fraction (WIIIx), though the extent of complexation is more with the former.

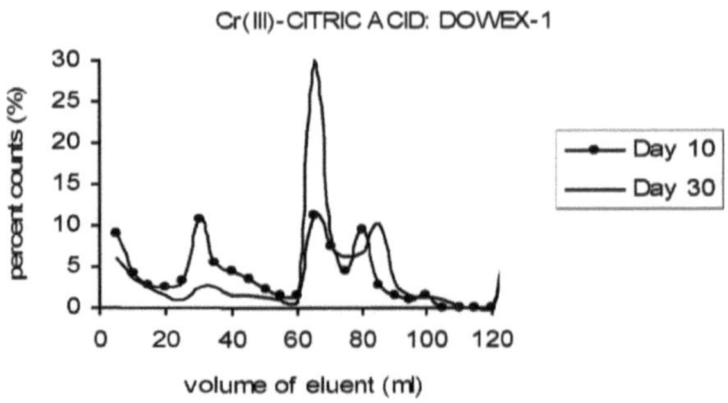

Fig. 2: Cr (III)-citric acid complexation elution profile

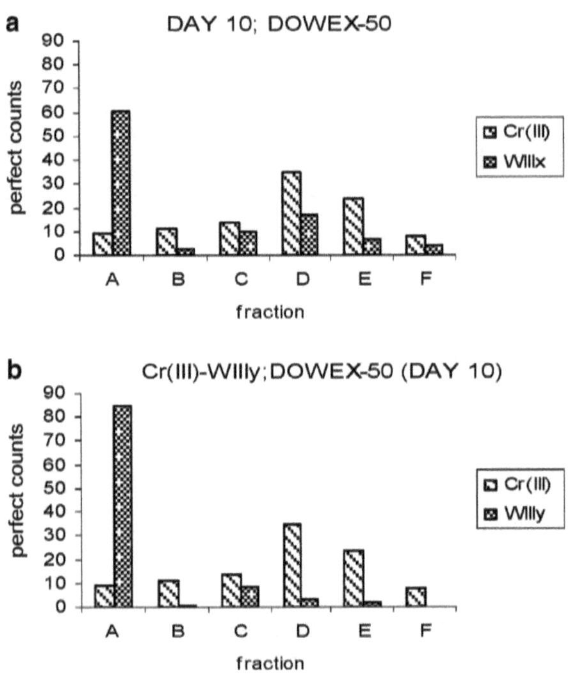

Fig. 3a: Comparison of elution of pure Cr(III) and Cr(III)-WIIIx from cation exchanger Dowex-50; **Fig. 3b** Comparison of elution of pure Cr(III) and Cr(III)-WIIIy from cation exchanger Dowex-50

Cr (VI) In Vitro Studies Results

The elution profile of Cr (VI)-WIIIy (pure sap) interaction from the anion exchanger Dowex-1 shows shallow peaks at 65 ml and 80 ml (Fig. 4a) as compared to a single sharp peak for pure Cr (VI) at 65 ml (Fig. 5a). Thus this profile is similar to Cr(III)-xylem sap curve (Fig. 1b) whereas the elution profile of this sample on cation exchanger Dowex-50 (Fig. 4b) is akin to pure Cr(III) curve (Fig. 5b). Thus there is a definite speciation change of Cr (VI) i. e. reduction as well as complexation by the organic ligands of the sap.

Interestingly, the interaction of Cr (VI) with the carboxylic acid fraction does not reveal a significant reduction of Cr (VI) to form Cr (III) or Cr (III) complexes (Fig. 6a, b). The same was observed for Cr (VI)-citric acid complexation. The reduction of Cr (VI) to Cr (III) is not feasible. Reduction of Cr(VI) by oxygen-containing functional groups at pH values 4–8 is a slow process, though it occurs at significant rates in presence of catalysts specially Fe(II) [23]. Iron is absent in the carboxylic acid fraction, being held up in the cation exchange resin while separating the carboxylic acids in the separation procedure. Fe is

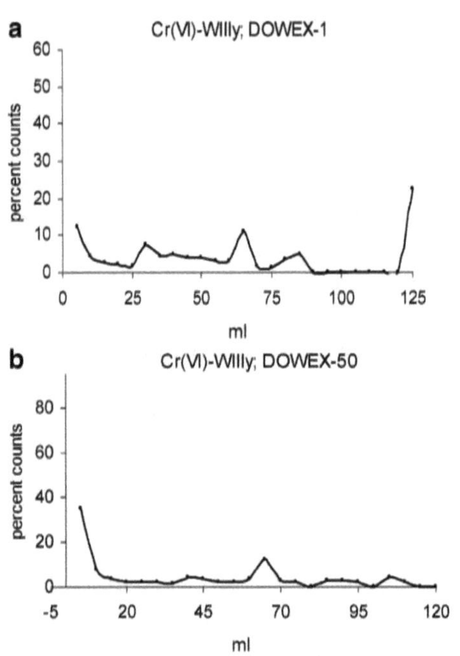

Fig. 4a: Cr (VI) – (WIIIy) pure sap elution curve on anionic Dowex-1. **Fig. 4b:** Cr (VI) – (WIIIy) pure sap elution curve on cationic Dowex-50

a major ion present in the xylem sap and is present primarily as an organic acid complex [7, 21]. Fe (II) and Fe-(II) carboxylates are known to enhance Cr (VI) reduction and lead to the formation of complexed soluble Cr (III) [24]. This aspect needs to be further explored in the plant system. The reduction may also be attributed to other organic constituents (e. g. amino acids, sugars) [7, 13] that are present at a higher concentration in this mature stage [20].

Evaluation of Plant Role in Alleviation of Toxic Cr (VI)

The in vitro results are corroborated by similar studies carried out using electrophoresis technique [18] as well Cr (VI) in vivo analysis. Thus Cr(III) is present as an anionic carboxylate complex probably Cr(III)-citrate and Cr(VI) is also likely to be carried in the same form after reduction to Cr(III) in the xylem sap of mature maize plants.

The above findings correspond to other speciation studies on Cr in plants. Cr (VI) speciation analysis [16] in cabbage xylem sap, indicated that 95 % of Cr (VI) was present in the trivalent form and it has been shown that once in the vascular system, low concentration of Cr (VI) was transported as a Cr (III)-organic complex in subterranean clover [14]. The Cr ion is found in an octahedral conformation bound to six oxygen atoms, with an absence of Cr (VI) in the shoots and leaves due to reduction to Cr (III) in the roots [11–18].

This study also highlights the role of organic acids in the detoxification mechanism inside the plant. The detoxification starts in the roots, where the toxic Cr (VI) gets reduced within 4 h in the fine lateral roots [13] and is chelated and compartmentalized in the vacuole by low-molecular-weight organic acids (LMWOA) [9]. On translocation to the aerial parts, Cr(VI) gets reduced and bound by chelates of xylem sap [16–18] and further gets stored as less toxic Cr(III) complexes in the aerial parts [4,13–15]. The detoxification mechanism thus works at every step. The food chain also seems to be well protected from excess and toxic Cr by the Soil–plant Barrier [25] unless the conditions are drastic. In the in vivo study, no toxic symptoms were visible in the plants at a Cr (VI) concentration of 5.2 ppm in hydroponics. In field conditions, Cr (VI) is readily immobilized in soils by adsorption, reduction, and precipitation processes.

Thus the detoxification of Cr (VI) at each step enables that the phytotoxicity of chromium in ed-

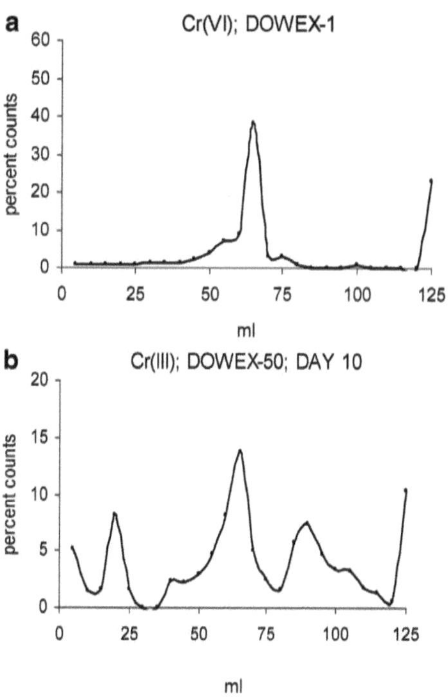

Fig. 5a: Pure Cr (VI) elution profile on Dowex-1. **Fig. 5b:** Pure Cr (III) elution profile on Dowex-50

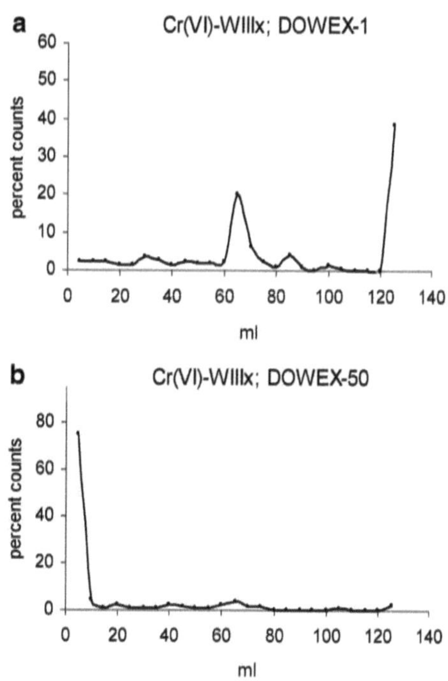

Fig. 6a: Cr (VI)-WIIIx elution curve on anionic Dowex-1. **Fig. 6b:** Cr(VI)- WIIIx elution curve on cationic Dowex-50

ible plant tissues occurs at concentrations below that are injurious to animals or humans. In fact, this biotransformation of Cr(VI) to the less toxic and mobile Cr(III) complexes presents a significant approach for the *Phytodetoxification* or the *in situ* detoxification of Cr(VI) through plant-based reduction and chelation mechanisms for bioremediation of chromium contaminated wastelands.

References

1. J. Barnhart; Regul Toxicol Pham 26 (1997) S3–7.
2. P. Chandra, S. Sinha and U.N. Ray; *ACS Symposium Series #664* Eds. E.L. Kruger, T.A. Anderson and J.R. Coats; Washington DC, (1997) p 274–282.
3. A.K. Shanker, C. Cervantes, H. Loza-Taverac and S. Avudainayagam; Environ. Int. 31 (2005) 739– 753.
4. S. Bluskov, J.M. Arocena, O.O. Omotoso and J.P. Young, Int. J. Phytorem. 7(2) (2005)153–165.
5. W.Mertz; J. Nutrit. 123 (1993) 626–63.
6. S.A. Katz and H. Salem; VCH Publishers New Jersey, USA, (1994).
7. D.A. Cataldo, R.E. Wildung, and T.R. Garland; J. Environ. Qual. 16(4) (1987) 289–295.
8. M.M Srivastava, A. Juneja, S. Dass, R. Srivastava, S. Srivastava, S. Mishra, S. Srivastav, V. Singh, and S. Prakash; Chem. Spec. and Bioavail. 6(2)(1994) 27–30.
9. L.S. di Toppi, F. Fossati, R. Musetti, I. Mikerezi, and M.A. Favali; J. Plant Nutri. 25(4) (2002) 701–717.
10. F.X. Han, B.B.M. Sridhar, D.L. Monts and Y. Su New Phyto. 162(2)(2004) 489 – 499.
11. J.L. Gardea-Torresdey, J.R. de la Rosa G, M. Peralta-Videa, G. Montes, Cruz-Jimenez and I. Cano-Aguilera. Arch. of Environ. Con. Toxicol. 48(2) (2005) 225 – 232.
12. X.Z. Yu, J.D. Gu and L.Q. Xing; Ecotoxicol. 17(8) (2008) 747–55.
13. C.M. Lytle, F.W. Lytle, N. Yang, J.H. Qian, D. Hansen, A. Zayed, and N. Terry; Environ. Sci. Technol. 32 (1998) 3087–3093.
14. J.A. Howe, R.H. Loeppert, V.J. DeRose, D.B. Hunter, D.B. and P.M. Bertsch, Environ. Sci. Technol. **37** (2003) 4091 – 4097.
15. Y. Zhao, J.G. Parsons, J.R. Peralta-Videa, M.L. Lopez-Morenoa and J.L. Gardea-Torresdey; Metallomics 1 (2009) 330–338.
16. R. Milacic and J. Štupar; Analyst 119 (1994) 627–632.
17. S. Juneja and S. Prakash; Chem. Spec. and Bioavail 17(4) (2005) 161–169.
18. S. Verma nee Juneja, and S. Prakash; Chem Spec and Bioavail, 20(2) (2008) 55–63.
19. M.C. White, A.M. Decker, and R.L. Chaney; Plant Physiol 67 (1981a) 292–300.
20. S. Juneja and S. Prakash; DEIJSER,14(1&2)(2007)91–100.
21. L.O. Tiffin; Plant Physiol 48 (1971) 273–277.
22. V.G. Mihucz, E. Tatar, I. Virag, E. Cseh, F. Fodor, G. Zaray; Anal Bioanal Chem 383 (2005) 461–466.
23. B.I. Deng and A.T. Stone; Environ. Sci. Technol. 30(8) (1996) 2482–2494.
24. S.J Hug, H.U. Laubecher and B.R. James; Environ. Sci Tech. 31(1) (1997)160–170.
25. M.M. Gatti, C. Baffi and S. Silva; Geophy Res Abs. SRef-ID: 9 (2007)1607–7962.

58
Cobalt and Zinc Containing Plant Oil Based Polymer: Synthesis and Physicochemical Studies

T. Singh and A. A. Hashmi
Department of Chemistry, Jamia Millia Islamia (Central University)
New Delhi-110025
Email: dr.aahashmi@yahoo.co.in

Abstract

Plant derived oils bear a large potential for the substitution of currently used petrochemicals. Polymer containing transition metal elements in their backbone continue to be of considerable interest. Cobalt and Zinc containing oil based polyesteramide resins were developed by condensation polymerization reaction. Spectroscopic techniques such as FTIR and ^{1}HNMR have been used to establish the structure of the polymer. Presence of metal has been confirmed through AAS. Standard laboratory methods were used to study the physicochemical characteristics like acid value, hydroxyl value, saponification value, iodine value, specific gravity, and viscosity. The thermal behavior of the polymer was analyzed by TG/DTA.

Introduction

Replacement of petroleum derived raw materials with plant-based products are quite significant from both a social and an environmental viewpoint in the production of natural polymers. Plant oils, polysaccharides (mainly cellulose and starch), sugars and wood are some of the most widely applied renewable raw materials in the chemical industry for non-fuel applications. Biopolymers offer the advantages of low cost, ready availability from renewable natural resources, and potential biodegradability. Development of polymeric materials from renewable resources in academic research and chemical industry open new vista towards sustainable development [1].

Plant oils have attracted attention as alternative raw materials for the preparation of resins and polymers, to replace or augment the traditional petro-chemical based polymers [2,3]. Vegetable oils, possessing a triglyceride structure with saturated and unsaturated fatty acid side chains, which in its pure form are considered as platform chemicals for the production of bio-based polymers [4,5]. Soybean oil contains natural antioxidants which help to prevent the oxidative rancidity. Soybean oil is also very nourishing because it contains vitamins and other nutrients [6]. Transition metals incorporation lead to biological and further modification of polymer [7–11]. The importance of zinc for normal growth and survival of plants and animals was recognized a long time ago. Zinc metal finds application as biomedical, antibacterial, antifungal and antiviral agents [12–16]. Cobalt is also recognize as an essential trace element for all animals as the active center of coenzymes called cobalamins. Among the key support of cobalt involved in preventing and treating pernicious anemia, helps red blood cell production and supports normal nervous system function [17–18].

We intent to report the synthesis, characterization and physicochemical studies of bimetallic oil based polymer. Single metal containing oil based polymers have also been reported [19–24]. In this bimetallic polymer synthesis we are using soybean oil as polymer and Zinc and Cobalt as metals.

Experimental

Materials

Soybean oil was obtained from Jayant Oil mill Vadodara, India. diethanolamine, sodium ethoxide (Rankem chemicals); zinc acetate, sebacic acid, cobalt sulfate, diethyl ether, xylene, sodium chloride

M.M. Srivastava, L. D. Khemani, S. Srivastava, *Chemistry of Phytopotentials: Health, Energy and Environmental Perspectives*, DOI:10.1007/978–3-642–23394-4_58, © Springer-Verlag Berlin Heidelberg 2012

and all the solvents were used as obtained from Merck chemicals, India.

Characterization

Structures of PEA, ZPEA and CZPEA were determined by spectroscopic techniques like FTIR and ^{1}HNMR. FTIR spectra of the polymers were recorded on Perkin–Elmer 1750 FTIR spectrophotometer using NaCl cell. ^{1}HNMR spectra were recorded on JEOL JNMECX 400P FTNMR system using TMS as an internal standard. AAS was recorded on Perkin Elmer 3100 using HCl Lamp. Thermo Gravimetry/Differential Thermal Analyzer (**TG/DTA**) was recorded on EXSTAR 6000 from 50°C to 600°C in nitrogen atmosphere at a healing rate of 10°C per minute with alumina as reference sample. Solubility of PEA, ZPEA and CZPEA in various solvents was checked at room temperature. Inherent viscosity, Iodine value, saponification values, specific gravity and refractive index of PEA, ZPEA and CZPEA were also determined through standard laboratory methods.

Synthesis

Hydroxyethyl Soybean Oil (HESA)

30 g of soybean oil was mixed in 25 g of diethanolamine in a beaker. The solution was yellow in color and viscous in nature. After this 15 ml sodium ethoxide was mixed in the solution. It was noticed that color of the solution remained same. The solution was kept in microwave for 1 min and 30 s. At the end a yellowish brown viscous solution was obtained. The progress of the reaction was monitored by thin layer chromatography (TLC).

Polyesteramide (PEA)

HESA was dissolved in 100 ml diethyl ether. 15 % aqueous sodium chloride was mixed in it. The solution was yellow in color. The solution was kept overnight for separation in separating funnel. Two layers were obtained after separation. Upper yellow layer was retained and lower off white layer was discarded as it contains impurities. Upper layer was taken in a beaker and kept on stirrer for 3 hr and 30 min at 40°C for proper dissolution and evaporation of ether. A viscous gel of yellow color was obtained at room temperature. 17 g of this solution along with .02 M sebacic acid and 3 ml xylene was dissolved in a beaker. The resultant solution was kept in microwave for 4 min. The product obtained was solid and off white in color.

Zinc Polyesteramide (ZPEA)

10 g PEA, 3.38 g zinc acetate and 5 ml xylene was taken in a beaker. The mixture was subjected to microwave radiation at 100 % power for 6min to finally obtain the dark brown and viscous solution of ZPEA. The progress of the reaction was monitored by running TLC.

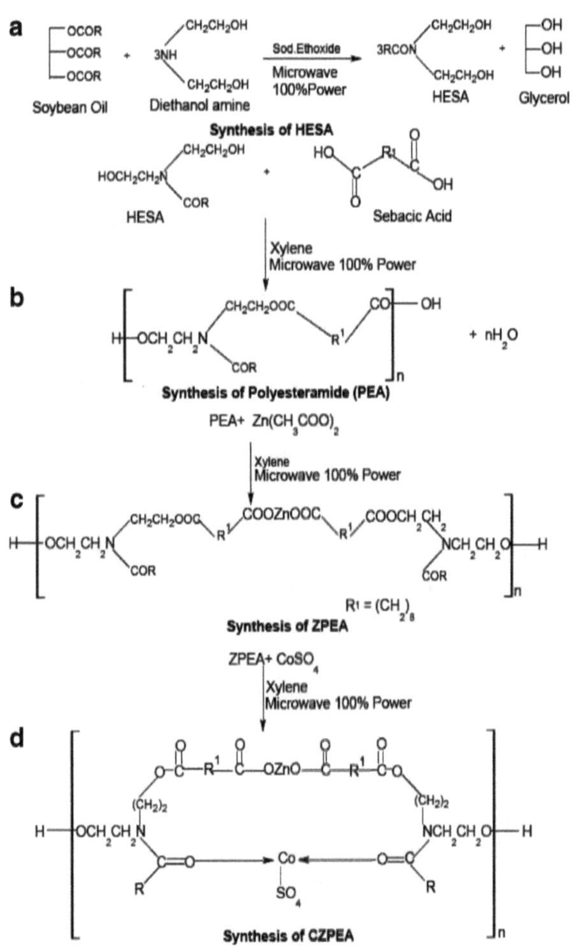

Reaction Scheme: Synthesis of Bimetallic Polymer CZPEA

Cobalt Zinc Polyesteramide (CZPEA)

2 g ZPEA was mixed in 1 g cobalt sulfate and kept in microwave for 7 min. CZPEA was prepared it was brown viscous solid. TLC was conducted to investigate the progress of the reaction.

Results and Discussion

The physical properties, analytical data and spectral data of the PEA, ZPEA and CZPEA support their proposed structure. The physico-chemical characteristics data (Table 1) indicates a decrease in saponification and iodine values from PEA to ZPEA and then to CZPEA. These trends indicate the disappearance of OH group of PEA during the condensation reaction. Hydroxyl value, inherent viscosity and specific gravity shows a gradual increase from PEA to ZPEA and then to CZPEA due to increase in molar mass and crosslinking.

Solubility of CZPEA was checked with various solvents. It was found to be partially soluble in acetone, benzene, butan-2-one, chloroform, ethanol, methanol, tetrahydrofuran, dimethylsulfoxide and insoluble in carbontetrachloride, dichloromethane, diethylether, toluene and xylene. Synthesized polymer was not soluble in water which is due to the hydrophobic nature of triglycerides present in oils.

Spectral Analysis of PEA

FT-IR (cm^{-1}) of PEA shows peaks at 3600(OH stretching), 2937 (>CH_2 symmetrical stretching), 1737 (COOR stretching), 1650 (CO amide grp), 1455 (CH_2 bending), 1171 (RCOR stretching), 950 (CH_2 out of plane bend).

Table 1: Characterization of PEA, ZPEA and CZPEA

Characteristics	PEA	ZPEA	CZPEA
Saponification Value	140	134	132
Hydroxyl Value	179	147	144
Iodine Value	37	22	21
Specific Gravity	.921	.923	.927
Viscosity	–	.587	.592

The ^1H-NMR (ppm) spectrum of PEA shows proton of terminal CH_3 at 0.8316, chain CH_2 at 1.24, =CH-CH_2 adjacent to terminal olefinic group at 2.1063. The sharp peak of methylene adjacent to ester group is observed at 2.70, CH_2 attached to amide nitrogen at 3.54, and alcoholic OH at 4.13.

Spectral Analysis of ZPEA

FT-IR (cm^{-1}) of ZPEA shows peaks at 3485 (OH stretching), 2925 (>CH_2 symmetrical stretching), 1731 (COOR stretching), 1632 ($CONR_2$ grp stretching), 1469 (CH_2 bending), 1119 (RCOR bending), 618 (CH_2 out of plane bend).

The ^1H-NMR (ppm) spectrum of ZPEA shows proton of terminal CH_3 at 0.8190, chain CH_2 at 1.223, =CH-CH_2 adjacent to terminal olefinic group at 2.1031. The sharp peak of methylene adjacent to ester group is observed at 2.682 and alcoholic OH at 5.301.

Spectral Analysis of CZPEA

FT-IR (cm^{-1}) of CZPEA show peaks at 3459 (OH stretching), 2923 (>CH_2 symmetrical stretching), 1714 (COOR stretching), 1633($CONR_2$ grp stretching), 1455 (CH_2 bending), 1071 (COO stretching).

The ^1H-NMR(ppm) spectrum of CZPEA shows proton of terminal CH_2 at 0.88, chain CH_2 at 1.25, =CH-CH_2 adjacent to terminal olefinic group at 2.00, CH_2 attached to amide nitrogen at 2.30 and 3.51, CH_2CH_2OH at 3.60 and alcoholic OH at 4.20.

The IR spectrum of ZPEA & CZPEA show no separate peaks for Zn-O but shifting of peaks due to C=O ester, along with pronouncement of these peaks, is observed in the case of ZPEA. This shows that Zn has interacted with PEA and withdraws electrons toward it, causing decrease in the bond strength of C=O and R-C-O, leading to the aforementioned shifts in their peaks. Co interaction with oxygen was observed by shifting of CH_2 bond.

^1H-NMR spectrum of ZPEA & CZPEA shows the reduction in the intensity of the peak and shifting to downfield is due to the effect of metal on the methylene group adjacent to ester linkage.

Atomic Absorption Spectroscopy and Thermal Analysis

Heavy metal analysis was carried out using atomic absorption spectroscopy. The detection limits for Zinc and Cobalt in CZPEA were 0.599 and 0.244 mgL^{-1}. Thermogravimetry (TG) curve of bimetallic polymer CZPEA indicates first degradation at 102.5°C, second degradation at 343.8°C, third degradation at 434.8°C and fourth degradation at 543.1°C. Differential thermal analysis (DTA) data revealed the thermal degradation steps of thermogram and shows melting temperature at 116°C, 346.8°C, 455°C and 527.6°C.

Conclusion

Plant oil becomes potential starting material for the preparation of polymers. Polymers prepared from soybean oil are ideal replacement for petroleum based polymers. Bimetallic plant oil based polymer have been prepared by condensation polymerization using soybean oil based fatty amide polyole and sebacic acid in a process initiated by sodium ethoxide. The copolymers were prepared by microwave irradiation technique with minimal usage of organic solvents. The improved method can save reaction time and gave high percentage yield of the products. CZPEA was found to be thermally stable.

References

1. M. Flieger, K. M, A. Prell, T. Rezanka and J. Votruba; Folia Microb. 48 (2003) 27–44.
2. H. Uyama, M. Kuwabara, T. Tsujimoto, M. Nakano, A. Usuki and S. Kobayashi; Macrom. Bios. 4 (2004) 354–360.
3. A. Guo, I. Javni and Z. Petrovic; J. A. Polym. Sci. 77 (2000) 467–473.
4. C. Geoffry and M. A. Hillmyer; Macromolecules 42 (2009) 7987–7989.
5. J. C. Ronda, G. Lligadas, M. Galia and V. Cadiz; Europ. J. Lipid Sci. and Tech. 113 (2011) 46–58.
6. K. M. Doll and S.Z Erhan; Green Chemistry 7 (2005) 849–854.
7. E. Sharmin, S. M. Asharaf and S. Ahmad; Europ. J. Lipid Sci. & Tech. 109 (2007) 134–146.
8. F. Zafar, S. M. Asharaf and Sharif Ahmad; Prog. in Organ. Coat. 50 (2004) 250–256.
9. H. Matsuda; J. App. Polym. Sci. 22 (1978) 2093–2108.
10. F. Zafar, S. M. Asharaf and S. Ahmad; React. Funct. Polym. 67 (2007) 928–935.
11. H. Matsuda and S. Takechi; J. Polym. Sci. Part A Polym. Chem. 28 (1990) 1895–1908.
12. R. Jaykumar and S. Nanjundan; J. Polym. Sci. Part A Polym. Chem. 42 (2004) 1809–1819.
13. R. Jayakumar, and S. Nanjundan; Europ. Polym. J. 41 (2005) 1623–1629.
14. K. H. Brown, Janet M, Peerson, J. Rivera and L.H Allen; Amer. J.Clin. Nut.75 (2002) 1062–1071.
15. Z. H. Chohan, H.Pervez, A.Rauf, K. M. Khan and C. T. Supuran; J. Enzyme Inhib. Med.Chem. 21 (2006) 193–201.
16. Mokhles, A. Elzaher; Appl. Organom. Chem. 18 (2004) 149–155.
17. Enghag and Per; "Cobalt" Encyclopedia of the elements: technical data, history, processing, applications. (2004) 667.
18. Kobayashi, Michihiko; Shimizu and Sakayu Europ. J. Biochem. 261 (1999) 1–9.
19. M. Alam, A. R. Ray, S. M. Ashraf and S. Ahmad; J. Americ. Oil Chem. Soc. 86 (2009) 573–580.
20. F. Zafar, S. M. Asharaf and S. Ahmad; Prog. in Organ. Coat. 59 (2007) 68–75.
21. N. P. Bharathi, M. Alam, S. Shreaz and A. A. Hashmi; J. Inorg. and Organom. Polym. and Mat.19 (2009) 459–465.
22. N. P. Bharathi, M. Alam, S. Shreaz, A.A. Hashmi; J.Inorg. and Organom.Polym. and Mat. 20 (2010) 839–846.
23. N. P. Bharathi, M. Alam, S. Shreaz and A.A. Hashmi; J.Inorg. and Organom.Polym.and Mat. 20 (2010) 833–838.
24. N. P. Bharathi, N.Umar, S. Shreaz, A.A. Hashmi J. Inorg. and Organom. Polym. and Mat. 19 (2009) 558–565.

59
Cation Exchange Resin (Amberlyst® 15 DRY): An Efficient, Environment Friendly and Recyclable Heterogeneous Catalyst for the Biginelli Reaction

S. Jain, S. R. Jetti, N. Babu G, T. Kadre and A. Jaiswal
School of Studies in Chemistry & Biochemistry, Vikram University,
Ujjain, Madhya Pradesh-456010
Email: srinujetti479@gmail.com

Abstract

Ion exchange resin catalyzed multicomponent Biginelli reaction has been studied for the synthesis of 3, 4-dihydropyrimidin-2-ones. Among the various solid acid catalysts Amberlyst® 15 DRY was found to be the most efficient, recyclable and environmentally benign heterogeneous catalyst regarding reaction time, yield and ease of work up procedure.

Introduction

Replacement of conventional, toxic and polluting Bronsted and Lewis acid catalysts with eco-friendly reusable solid acid heterogeneous catalysts like acidic zeolites, clays, sulfated zirconia and ion exchange resins is an area of current interest [1, 2]. The use of solid acid catalyst instead of liquids includes many advantages, such as reduced equipment corrosion, ease of product separation, recycling of the catalyst and environmental acceptability. In the recent past ion exchange resins in general and styrene-DVB matrixed resin sulfonic acid (Amberlyst® 15 DRY) in particular, which are strongly acidic and chemically as well as thermally stable have been found to be excellent catalysts for a variety of the major organic reactions like esterification, alkylation, acetalization, acylation and condensation [3–6].

The Biginelli reaction is one of the most important multi-component reactions for the synthesis of dihydropyrimidinones. Dihydropyrimidinone are known to exhibit a wide range of biological activities such as antiviral, antitumor, antibacterial, and anti-inflammatory properties [7]. In addition, these compounds have emerged [8] as potential calcium channel blockers, antihypertensive, α1a–adrenergic antagonists and neuropeptide antagonists. Furthermore the 2-oxodihydropyrimidine-5-carboxylate core unit is also found in many marine natural products [9], including the batzelladine alkaloids, which have been found to be potent HIV gp-120-CD4 inhibitors.

The clinically important antiretroviral like agents like AZT, DDC, DDI possess the pyrimidine scaffold. Another related framework of very easily accessible via Multi Component Reaction involving Urea, active methylene compounds and aldehydes in the presence of a catalyst as originally reported [10] by Biginelli. In recent years type 1b pyrimidine scaffold has been under intensive investigation [11] as it has a very high pharmalogical profile.

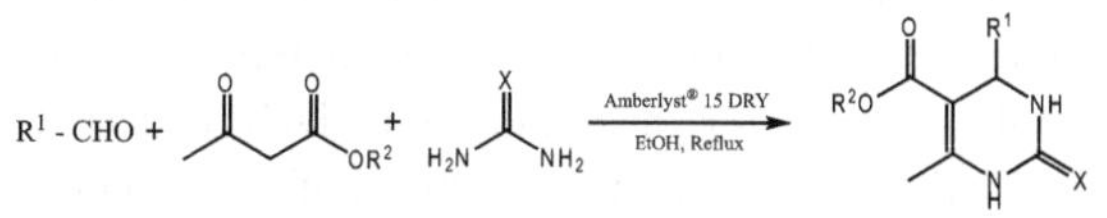

Scheme-1

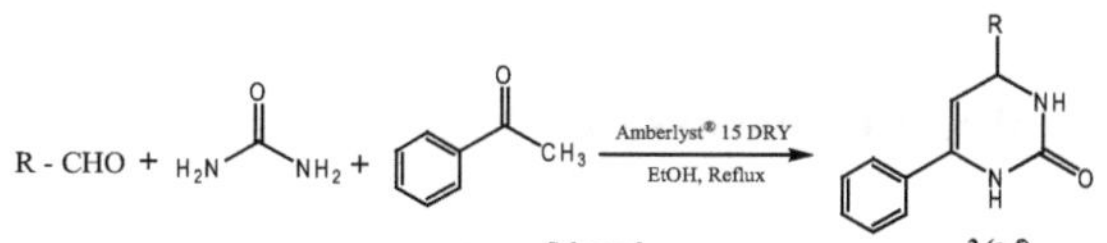

Scheme-3

The chemistry of C_5-C_6 double bond has been extensively explored in type 1a skeleton and careful manipulation of this bond have led to interesting chemistry [12] and many useful new structures. In contrast

M.M. Srivastava, L. D. Khemani, S. Srivastava, *Chemistry of Phytopotentials: Health, Energy and Environmental Perspectives*, DOI:10.1007/978–3-642–23394-4_59, © Springer-Verlag Berlin Heidelberg 2012

C_5-C_6 double bond in Biginelli scaffold is relatively less explored and only a few useful transformations are attempted [13] involving very careful manipulations. This less developed chemistry of C_5-C_6 bond in Biginelli compounds appears to be due to difficulties in manipulating the methyl group at C_6 and ester group at C_5 which are traditionally placed in these positions.

A plethora of reagents/methods have been reported for the synthesis of 3, 4-dihydropyrimidin-2(1H)-ones such as ceric ammonium nitrate under ultrasonication [14], Lewis acids (such as BF3. Et2O) in combination with transition metal and suitable proton source [15, 16], lanthanide triflates [17], lanthanide chloride [18], indium chloride [19], antimony chloride [20], bismuth nitrate under microwave irradiation [21], copper iodide [22], molecular iodine [23], heteropolyacids [24], ionic liquids [25]. However, in spite of their potential utility many of the existing methods suffer from the drawbacks, such as the use of strong acidic conditions, longer reaction times, tedious workup, environmental disposal problems and lower yields of the products, leaving scope for further development of an efficient and versatile method for Biginelli reaction.

Growing concern about environmental damage leads to an urgent requirement for the development of eco-friendly technology and economic processes. It is of great practical importance to synthesize DHPM derivatives by the Biginelli reaction by using a solid acid catalyst, because of the ability to modify the acid strength, ease of handling, recycling of the catalyst and environmental compatibility. In view of the above requirement, and as a part of our program towards green synthesis, we herein report a single-step and eco-friendly protocol for the synthesis of DHPM derivatives by the multicomponent reactions of β-dicarbonyl compound, aldehydes and urea (Scheme 1) over Amberlyst® 15 DRY with good yields and selectivity.

Results and Discussion

To evaluate the catalytic effect of various ion exchange resins we started with the model reaction of ethylacetoacetate (1.0 mmol) with benzaldehyde (1.0 mmol) and urea (1.2 mmol) in refluxing ethanol without and with use of various acidic ion exchange resins as catalysts to afford dihydropyrimidine 1a in various yields (Table 1). It can be seen from Table 1 that Amberlyst® 15 DRY was the most efficient (Table 1, entries 3) among the five solid acidic ion exchange resins studied. It was found that 50 mg of Amberlyst® 15 DRY is sufficient to carry out the Biginelli reaction successfully. An increase in the amount of Amberlyst® 15 DRY to more than 50 mg showed no substantial improvement in the yield, whereas the yield is reduced by decreasing the amount of Amberlyst® 15 DRY.

The effect of solvent on the reaction was studied (Table 2, entries 1–6) and ethanol was found to be the best solvent when considering the reaction yields and environmental damage.

All aforementioned reactions proceeded expeditiously and delivered good yields with range of aromatic aldehydes containing electron donating and electron withdrawing groups. This three-component reaction also proceeds efficiently with a broad range of structurally diverse 1, 3-dicarbonyl compounds, aldehydes and urea under this protocol to produce the corresponding DHPMs (Table 3). A variety of dicarbonyl compounds could be used successfully. Thiourea was also used with similar success. The results are presented in Table 3. All the substrates were smoothly converted to their corresponding DHPMs in excellent yields.

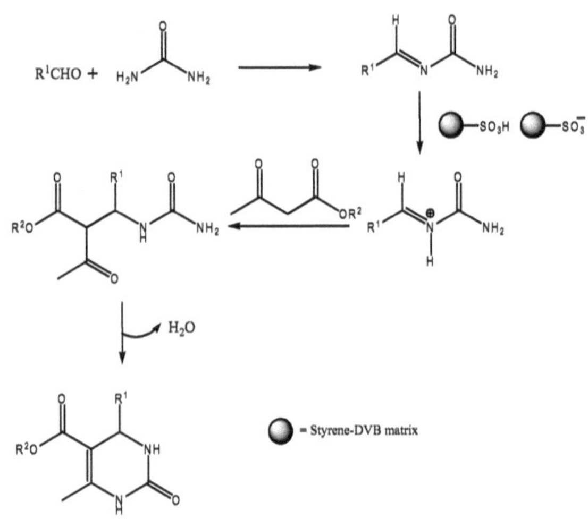

Scheme-2: The proposed mechanism for the formation of pyrimidine derivative

The probable mechanism (Scheme-3) of the reaction appears to involve the activation of the carbonyl function by Amberlyst® 15 DRY, thereby making the methyl group readily enolizable, which in turn reacts with aldehyde and urea derived imine in a Michael type step to produce product 2 as represented in (Table 4).

Table 1: Catalytic Activity of Different Ion Exchange Resins in Biginelli Condensation

S. No	Ion Exchange Resin	Reaction Time (h)	Yield [b] (%)
1	–	10	Trace
2	Amberlyst-70	3	81
3	Amberlyst® 15 DRY[a]	5.5	94
4	Indion-130	3	92.5
5	Indion-190	3.5	92
6	Nafion-H	4.5	85

[a] Reaction conditions: ethyl acetoacetate (1.0 mmol), Benzaldehyde (1.0mmol) and Urea (1.2mmol) in dry ethanol (10 ml), Ion exchange resin (50 mg) at refluxing temperature

Table 2: Optimization of the Reaction Conditions for the Synthesis of 1a[a]

Entry	Catalyst	Solvent	Time (h)	Yield (%)
1	Amberlyst ® 15 DRY	Water	4	90
2	Amberlyst ® 15 DRY	EtOH	5.5	94
3	Amberlyst ® 15 DRY	CH_3CN	6.5	85
4	Amberlyst ® 15 DRY	THF	6	87
5	Amberlyst ® 15 DRY	Benzene	10	Trace
6	Amberlyst ® 15 DRY	Toluene	10	Trace

[a] All Reactions were conducted at reflux temperature of the solvent used.

Table 3: Amberlyst ® 15 DRY catalyzed Synthesis of Dihydropyrimidine-2-(1H)-ones/Thiones

S. No	R[1]	R[2]	X	Products[a]	Yield [b] (%)	M.P (°C)
1	C_6H_5	Et	O	1a	88	206–208
2	4-(CH_3O)-C_6H_4	Et	O	1b	90	201–202
3	4-(NMe_2)-C_6H_4	Et	O	1c	80	255–257
4	4-NO_2-C_6H_4	Et	O	1d	94	211–213
5	4-(Cl)-C_6H_4	Et	O	1e	90	215–216
6	3-(Cl)-C_6H_4	Et	O	1f	88	192–193
7	3-(Br)-C_6H_4	Et	O	1g	81	185–186
8	2,4-(Cl)-C_6H_3	Et	O	1h	92	249–250
9	4-Cl-C_6H_4	Me	O	1i	89	204–205
10	4-(NO_2)C_6H_4	Me	O	1j	95	236–238
11	4-(CH_3O)-C_6H_4	Me	O	1k	85	192–194
12	C_6H_5	Me	O	1l	86	209–211
13	4-F-C_6H_4	Et	O	1m	92	182–184
14	3-NO_2-C_6H_4	Et	O	1n	91	227–229
15	2-NO_2-C_6H_4	Et	O	1o	96	208–210
16	C_6H_5-CH=CH	Et	O	1p	90	230–232
17	2–4-$(Cl)_2$-C_6H_5	Me	O	1q	93	252–253
18	C_6H_5	Et	S	1r	89	208–210
19	2-NO_2-C_6H_4	Et	S	1s	92	205–207
20	4-(CH_3O)-C_6H_4	Et	S	1t	89	153–155

Experimental Section

All solvents and reagents were purchased from Aldrich and Merck with high-grade quality, and used without any purification. The Indion-130 and Indion-190 were purchased from Ion Exchange India Ltd. Nafion-H, Amberlyst-70 and Amberlyst® 15 DRY were purchased from Aldrich. Melting points were determined on electro thermal apparatus by using open capillaries and are uncorrected. Thin-layer chromatography was accomplished on 0.2-mm precoated plates of silica gel G60 F254 (Merck). Visualization was made with UV light (254 and 365 nm) or with an iodine vapor. IR spectra were recorded on a FTIR-8400 spectrophotometer using DRS prob. ^{1}H-NMR and ^{13}C-NMR spectra were recorded in DMSO-d6 solutions on a Bruker AVANCE 400NMR spectrometer operating at 400 (^{1}H) and 100 (^{13}C) MHz LCMS analysis (EI, 70V) were performed on a Hewlett-Packard HP 5971 instrument.

Physical Properties of Amberlyst ® 15 DRY

Physical form ________________ Opaque beads
Ionic form as shipped ________________ Hydrogen
Concentration of acid dites ________________ 4.7eq/Kg
Water content ________________ ≤1.5 % (H^+ form)

Table 4: Amberlyst® 15 DRY catalyzed Synthesis of 5-Unsubstituted 3.4-dihydropyrimidin-2(1H)-ones

S. No	R	Products	Yield [b] (%)	M.P (°C)
1	C_6H_5	2a	90	233–36
2	4-(Cl)-C_6H_4	2b	92	267–69
3	4-(CH_3)-C_6H_4	2c	86	248–50
4	4-(CH_3O)-C_6H_4	2d	84	259–61
5	2-(Cl)-C_6H_4	2e	91	260–63
6	3-(CH_3O)-C_6H_4	2f	88	256–58

Shipping weight ______________ 610 g/L (38lbs/ft)
Fines content______________ < 0.300 mm: 1.0 %max
Surface area__________________________ 45 ^{m2}/g
Average pore diameter__________________ 250A^0
Swelling ___60 to 70 % (dry to water)
10 to 15 % (dry to hexane)
10 to 15 % (dry to toluene)
15 to 20 % (dry to ethylene dichloride)
30 to 40 % (dry to ethyl acetate)
60 to 70 % (dry to ethyl alcohol, 95 %)
15 to 20 % (dry to benzene)

General Procedure for the Synthesis 4-aryl Substituted 3, 4 Dihydropyrimidinones

A mixture of β-diketone (1.0 mmol), aldehyde (1.0 mmol), urea/thiourea (1.2 mmol), and Amberlyst ® 15 DRY (50 mg) in anhydrous ethanol (10 mL) were refluxed for an appropriate time as indicated by TLC. The catalyst was filtered and washed with ethyl acetate until free from organic material. The solvent was evaporated at reduced pressure and obtained solid was crystallized from ethanol to afford pure 3, 4-dihydropyrimidin-2-one/thione **1** in excellent yields.

General Procedure for the Synthesis 3, 4-dihydro-4, 6-diphenylpyrimidin-2(1H)-ones

To a solution of acetophenone (1.0mmol) in ethanol (10mL) was added benzaldehyde (1.0mmol), urea (1.5mmol) and Amberlyst ® 15 DRY (50mg) and was refluxed for an appropriate time as indicated by TLC. The catalyst was filtered and washed with ethyl acetate until free from organic material. The solvent was evaporated at reduced pressure and solid obtained was recrystallized from ethanol to afford pure 3, 4-dihydro-4, 6-diphenylpyrimidine-2(1H)-one **2** in excellent yields.

Conclusions

In conclusion, we have developed a simple, efficient, environmentally benign and improved protocol for the synthesis of 3, 4-dihydropyrimidin-2-ones/thiones over Amberlyst ® 15 DRY as the catalyst with excellent yields. The simplicity of the system, ease of separation/reuse of the catalyst due to its heterogeneous nature, excellent yields of the products and ease of work-up fulfill the triple bottom line philosophy of green chemistry and make the present methodology environmentally benign.

Acknowledgements

One of the authors (Srinivasa Rao Jetti) is grateful to Madhya Pradesh Council of Science and Technology (MPCST), Bhopal for award of fellowship. Authors are also thankful to Deputy Director and Head, SAIF, Central Drug Research Institute (CDRI), Lucknow for providing elemental analysis and spectral data and the Department of Chemistry, Vikram University, Ujjain for extending laboratory facilities and IR data.

References

1. J.H. Clark (Ed.); *Catalysis of organic reactions by supported reagents*, VCH Publishers, New York, USA,(1994), pp. 35.
2. R.A. Sheldon and H. Van Bekkam; Catalysis through heterogeneous catalysis, Wiley-VCH Publishers, Weinheim, Germany, (2002).
3. G.D. Yadav and M.S. Krishnan; *Org.Process Res. Develop.2* (1998) 86–95.
4. G.D. Yadav and P.K. Goel; *Green Chem.* 2 (2000) 71–78.
5. S.M. Mahajani and M.M. Sharma; 1 97–105.
6. S.B. Patil, R.P. Bhat and S.D. Samant; *Synthetic Commun.* 36 (2006) 2163–2168.
7. C.O. Kappe; *Eur.J.Med.Chem.* 35 (2000) 1043–1052.
8. K.S. Atwal, B.N. Swanson, S.E. Unger, D.M. Floyed, S. Moreland, A. Hedberg and B.C. O'Reilly; *J. Med.Chem.34* (1991) 806.
9. B.B. Snider, J. Chen, A.D. Patil and A. Freyer; *Tetrahedon Lett.* 37 (1996) 6977–6980.
10. P. Biginelli; *Chem Ber.* 24 (1891) 1317.
11. C.O. Kappe; *Tetrahedron.* 56 (1993) 6937- 6963.
12. P.J. Bhuyan, R.C. Borah and J.S. Sandhu; *J.Org.Chem.* 55 (1990) 568–571.
13. T. George, R. Tahilramani and D.V. Mehta; *Synthesis. 6* (1975) 405–407.

14. J. S. Yadav, B. V. S. Reddy, K. B.Reddy, K. S. Raj and A. R. Prasad; *J Chem Soc Perkin Trans:1* (2001) 1939–1941.
15. E. H. Hu, D. R. Sidler and U. H. Dolling; J.Org.Chem. 63 (1998) 3454–3457.
16. C. Liu, J. Wang and Y. Li; J.Mol.Catal.A. 258 (2006) 367–370.
17. Y. Ma, C. Quan, L. Wang and M. Yang; J.Org.Chem. 65 (2000) 3846- 3849.
18. J. Lu, Y. Bai, Z. Wang, B. Yang and H. Ma; Tetrahedron Lett. 41 (2000) 9075–9078.
19. B. C. Rannu, A. Hajra and U. Janu; J.Org.Chem. 65 (2000) 6270–6272.
20. I. Capanec, M. Litvic and I. Grungold; Tetrahedron. 41 (2007) 11822.
21. B. K. Banik, A. T. Reddy and A. Datta; Tetrahedron Lett. 48 (2007) 7392–7394.
22. H. Kalita and P. Phukan; Cul. Cat. Commun.8 (2007)179–182.
23. R. S. Bhosale, S. V. Bhosale, T. Wang and P. K. Zubaidha; Tetrahedron Lett. 45 (2004) 9111–9113.
24. S. P. Maradur and G. S. Gokavi; *Cat. Commun.* 8 (2007) 279–284.
25. J. Peng and Y. Den; Tetradedron Lett. 42 (2001) 5917- 5919.

60

An Efficient Method for the Extraction of Polyphenolics from Some Traditional Varieties of Rice of North-East India

A. Begum[1], A. Goswami[2], P. K. Goswami[2] and P. Chowdhury[1]

[1]Natural Products Chemistry Division

[2]Chemical Engineering Division

CSIR North-East Institute of Science and Technology, Jorhat-785006

Email: pritishchowdhury@yahoo.com Fax: +91-376-2370011

Abstract

Polyphenolics are secondary metabolites having numerous health benefits. One of the potential sources of phenolics is rice. Despite being the staple food of India, phenolic extraction from rice has not gained much significance due to its concentration in the aleurone layer which is lost during processing. Two procedures were studied for the effective extraction of total phenolics from rice: ultrasound-assisted aqueous methanol extraction and aqueous methanol extraction. The results indicated that ultrasound aided extraction improved the levels of phenolic content by 1.3–2.8 times.

Introduction

Polyphenolics are widely distributed in plants as aromatic secondary metabolites. They contribute to the sensory quality (appearance, taste, and odor) as well as the antioxidant activity of fruits, vegetables, beverages and grains. Polyphenolic substances have been reported to induce health-promoting effects such as antimicrobial and anti-inflammatory action, reduce risk of cardiovascular disease, and inhibit the oxidation of human low-density lipoproteins and anticarcinogenic effects. The beneficial effects of phenols have been extensively attributed to their antioxidant properties based on the scavenging of active oxygen species and free radicals [3].Phenolic acids are an important group of polyphenolics and include hydroxybenzoic (C_6-C_1), hydroxyphenylacetic (C_6-C_2) and hydroxycinnamic (C_6-C_3) acids. Hydroxycinnamic acid is the most widely distributed in plants. The important ones include p-coumaric, caffeic, ferulic and sinapic acids.

Cereal consumption is an excellent way to increase phenolic compound intake. However, consumption of these phenolic compounds from cereals is currently neglected. The major reason for this is that these compounds are concentrated in the aleurone layer and are lost with the separation of seed coat during processing. By the same token, most phenolic compounds in rice, which is a major staple cereal all over the world, particularly in Asia, are also lost with the aleurone layer of rice [5]. This layer is very rich in bioactive molecules, mostly polyphenols.

Traditional rice varieties exhibit tremendous nutritional characteristics than the high-yielding varieties (HYV) and hence can be used as potent nutraceuticals.

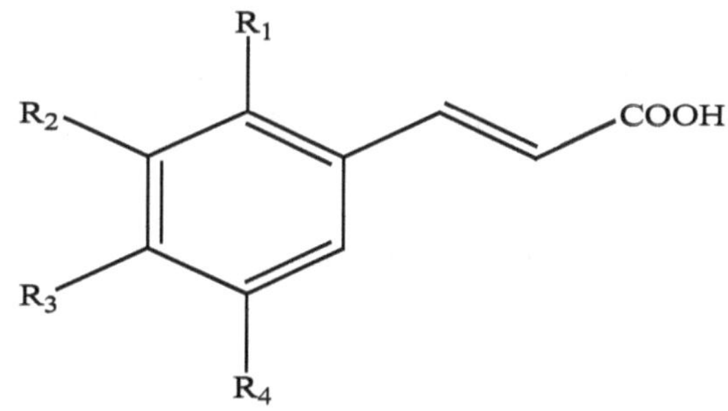

R_1=R_2=R_4=H, R_3=OH	p-Coumaric acid
R_1=R_4=H, R_2=R_3=OH	Caffeic acid
R_1=R_4=H, R_2=OCH_3, R_3=OH	Ferulic acid
R_1=H, R_2=R_4=OCH_3, R_3=OH	Sinapic acid

Fig. 1.: Chemical structure of hydroxycinnamic acid

M.M. Srivastava, L.D. Khemani, S. Srivastava, *Chemistry of Phytopotentials: Health, Energy and Environmental Perspectives*, DOI:10.1007/978–3-642–23394-4_60, © Springer-Verlag Berlin Heidelberg 2012

The main step in the recovery and isolation of polyphenolics and evaluation of individual and total polyphenolics from various plant-based sources is their extraction. Solvent extraction has been established as an effective method [7]. Polyphenolics are often most soluble in lesser polar solvents than water. Effective extraction generally involves a proper solvent selection, elevated temperature and mechanical agitation to maximize polyphenol recovery.

The objective of the present study was to develop a proper extraction method for polyphenolics from some traditional rice varieties of North-East India for their maximum recovery.

Materials and Methods

Chemicals

Gallic acid was purchased from Fluka. Folin-Ciocalteau reagent was obtained from Nice Chemicals.

Rice Samples

Four traditional rice cultivars (Jengoni, Sokua, Jahingia and Betguti) of *Oryza sativa* L. were collected from the North-Eastern region of India during the period from November to February of 2010–2011.

Preparation of Rice Samples

Rough rice was dehusked using a Rice Sheller (Indosaw, Haryana, India) and milled in a Rice Polisher (Indosaw, Haryana, India) to obtain the aleurone layer of rice. Moisture was determined by drying at 103°C to constant mass according to AACC Methods of Analysis. Analysis was performed in triplicate and results were expressed on a dry basis.

Extraction of Rice Total Phenolics

Two methods were examined for the extraction of total phenolics from rice samples:

(a) Aqueous Methanol Extraction
Rice sample (10 g) was extracted with 100mL of 80 % aqueous MeOH for 2h in an orbital shaker (200rpm) at room temperature. The extracts were then filtered through a Whatman filter paper. The residue was re-extracted twice. The filtrates were then collected, evaporated to dryness under reduced pressure at 40°C by a rotary-evaporator (Buchi, Switzerland), dried overnight in a desiccator and weighed.

(b) Ultrasound-assisted Aqueous Methanol Extraction
Rice sample (10 g) was mixed with 100mL of 80 % aqueous MeOH in a 250mL Erlenmeyer flask. The flask was immersed into an ultrasonic bath (Emcolite Ultrasonics) and sonicated for 15min at room temperature with occasional shaking. The mixture was filtered through a Whatman filter paper and the residue re-extracted twice. The filtrates were then evaporated to dryness in a rotary-evaporator (Buchi, Switzerland) at 40°C, dried overnight in a desiccator and weighed.

The extraction procedure was done in triplicate for each of the extracts. The extraction yield was expressed as weight percentage on a dry basis. All extracts were stored in amber-colored bottles at –4°C until further analysis.

Determination of Total Phenolic Content (TPC)

The total phenolic content (TPC) of the rice extracts was determined by Folin-Ciocalteau method [6]. The rice extracts were redissolved in 10mL of methanol and then 0.5mL of appropriate diluted extract was mixed with 2.5mL of 5 % Folin-Ciocalteau reagent. The reaction was carried out for 5min in dark. Then 2mL of sodium carbonate solution (7.5 %) was added. The mixture was kept in dark at ambient temperature for 60min. The absorbance was then measured at 765 nm using a UV-Vis spectrophotometer (Perkin-Elmer). All measurements were conducted in triplicate. Gallic acid was used as the standard and the total phenolic content was expressed as mg gallic acid equivalents (GAE)/100 g of rice.

Results and Discussions

Ultrasonic extraction affected the total phenolic content of rice as indicated in Tables 1 and 2.

The total phenolic extract ranged from 12.1–19 % in ultrasound-assisted aqueous methanol extraction to 7.4–12.5 % in aqueous methanol extraction. Similarly, the total phenolic content was shown to be in the range 93.5–327.9mgGAE/100 g rice for ultrasound-assisted aqueous methanol extracts compared to 42.3–235.3mgGAE/100 g rice for aqueous methanol extracts. This may be ascribed to the cavitational phe-

Table 1: Total phenolic extract of rice (% wt, dry basis)

Samples	Ultrasound-assisted methanol extraction	Methanol extraction
Jengoni	15.36±1.02	8.47±0.45
Sokua	19.05±0.73	12.55±0.51
Jahingia	12.33±0.15	8.37±0.37
Betguti	12.16±0.15	7.48±0.50

Table 2: Total phenolic content of rice (mg GAE/100 g rice)

Samples	Ultrasound-assisted methanol extraction	Methanol extraction
Jengoni	327.93±2.1	235.33±2.51
Sokua	155.29±2.56	55.33±2.51
Jahingia	97.46±2.15	67.33±2.49
Betguti	93.51±1.5	42.33±2.4

nomenon, which is the rapid formation and collapse of countless microscopic bubbles in a solvent produced by the alternating low- and high-pressure waves generated by ultrasonic sound. This enhances the mass transfer rate and solvent penetration into cellular materials. In addition, the disruption or damage of biological cell walls by ultrasound results in facilitated release of the intracellular contents. Hence phenolic extraction is enhanced by ultrasound-aided solvent extraction.

Conclusion

Methanolic extraction with ultrasound gave a significant increase in total phenolic yield of traditional rice varieties compared to conventional reflux extraction. Thus these traditional varieties may be used as potent sources of polyphenols for the development of nutraceuticals for health benefit.

References

1. J. Sungsopha, A. Moongngarm and R. Kansakoo; Aus. J. Basic & Applied Sci. 3 (2009) 3653–3662.
2. M. W. Zhang, R. F. Zhang, F. X. Zhang and R. H. Liu; J. Agric. Food Chem. 58 (2010) 7580–7587.
3. P. G. Kapasakalidis, R.A. Rastall and M.H. Gordon; J. Agric. Food Chem. 54 (2006) 4016–4021.
4. R. Yawadio, S. Tanimori and N. Morita; Food Chem. 101 (2007) 1616–1625.
5. S. Tian, K. Nakamura and H. Kayahara; J. Agric. Food Chem. 52 (2004) 4808–4813.
6. S. Jang and Z. Xu; J. Agric. Food Chem. 57 (2009) 858–862.
7. T. Régnier and J.J. Macheix; J. Agric. Food Chem. 44 (1996) 1727–1730.
8. AACC Approved Methods of Analysis, 11th Ed.

61
Determination of Heavy Metal Ions in Selected Medicinal Plants of Agra

A. Khanam and B. S. Singh
Department of Botany, St. John's College, Agra
Email – arishkhanam@yahoo.in , aashi.sultana1@gmail.com

Abstract

Therapeutic plants have always been valued as a mode of treatment of variety of ailments in folk cultures and have played a very important role in discovering the modern day medicines with novel chemical constituents. Now, is it has been established fact that over dose or prolonged ingestion of medicinal plants leads to the chronic accumulation of different elements which causes various health problems. In this context, elemental contents of the medicinal plants are very important and need to be screened for their quality control. Essential metals can also produce toxic effects when the metal intake is high concentrations, whereas nonessential metals are toxic even in very low concentration for human health. During the present study the levels of Copper, Iron and Zinc in five most important therapeutic plants i. e. Achyranthus aspera, Azadirachta indica, Adhatoda vasica, Ricinus communis and Withania somnifera, of Agra city were determined through AAS. The results were compared with the permissible limits (PL) set by WHO. Accordingly, all plants showed increased levels of Copper and Iron which is alarming. Maximum 35.8 ppm Copper in Azadirachta indica and 674.2ppm Iron in Achyranthus aspera were being observed. Thus, the tested plants showed high content of toxic pollutant due to industrial pollution. Therefore, special care must be taken during the administration of routinely used control practices for herbal medicines screening in order to protect consumers from toxicity.

Introduction

Among ancient civilizations, India has been known to rich repository of medicinal plant. India has a long history and tradition as well as rich heritage of using medicinal plants and aromatic plant for health care and beauty in improving the quality of life. Medicinal plants have been age of long remedies for human diseases because they contain components of therapeutic value (Nostro et al., 2000). In recent years, several authors all across the world, reported many studies on the important of element constituents of the herbal drug plants which enhanced the awareness about trace elements in these plants (Sharma et al., 2009). Rapid Industrialization and urbanization, in the area, are threat to the local medicinal flora in context of heavy metals pollution (Salman and Fida, 2009). Thus, from ongoing studies it can be inferred that heavy metal accumulation even utilized in micro quantity may cause serious hazards to human life and it should be mandatory for the pharmaceutical industries to detect the heavy metal concentration in raw drug before processing (Rai et al., 2001).Therefore, it is important to have a look on good quality control of medicinal herbs in order to protect consumers from contamination. In the present study, attempts have been made to study the presence of copper, iron and zinc in selected medicinal plants.

Materials and Methods

Sample Preparation

Plant samples were washed with deionized water and oven dried at 60 °C for 24 hrs. They were pounded in mortar to obtain powder form.

Digestion

1gm powder of each plant sample was dissolved in 5 ml of HNO3 for12h and then heated until the red-

M.M. Srivastava, L.D. Khemani, S. Srivastava, *Chemistry of Phytopotentials: Health, Energy and Environmental Perspectives*, DOI:10.1007/978–3-642–23394-4_61, © Springer-Verlag Berlin Heidelberg 2012

dish brown fumes disappear. Before drying, then cool it for 30 min and 5 ml perchloric acid was added to the above solution and heated for 5 min then 5 ml of aquaregia was add and heated to small volume and up to marked 250 ml by adding deionized water. Filtered samples were stored and analyzed for heavy metals by Atomic absorption spectrophotometer (Perkin Elmer ANALYST 100).

Results and Discussion

Copper

The maximum concentration was estimated 35.8 ppm in *A. indica* and minimum concentration was estimated in *R. communis* (Table 2 and Figure 1). The permissible limit set by WHO (2005) in medicinal plants was 20 ppm. After comparison, metal limit in the medicinal plants with those proposed by WHO (2005) it is found that all plants accumulate Cu above this limit.

Table 1: Pharmacognostic Features of the Herbs

S. no	Plant species	Local name and family	Part use	Disease cure
1	*Achyranthes aspera*	Puthkanda (Amaranthaceae)	Whole plant	Cough, asthma, kidney stone, anti inflammatory, diuretic,
2	*Azadirchta. indica*	Neem (Meliaceae)	Whole plant	Bitter; tonic, antimicrobial, antifungal, astringent, antiperiodic; useful in wounds and skin infections.
3	*Adhatoda vasica*	Vasaka (Acanthaceae)	Leaves	The drug vasika is used in bronchial troubles and consumption; also used in diarrhea, dysentery, glandular, tumors and skin affections.
4	*Ricinus communis*	Arand (Euphorbiaceae)	Whole plant	Constipation, stomach disorder swelling, Chambal, against scorpion sting.
5	*Withania somnifera*	Asghand (Solanaceae)	Whole plant	Aphrodisiac, diuretic, bronchitis, Ulcer.

Iron

The range of Fe in the studied plants was high with a maximum of 674.2 ppm in *A. aspera* and minimum of 215.2 ppm in *A. vasica*. The permissible limit set by FAO/WHO (1984) in edible plants was 20 ppm. After comparison, metal limit in the medicinal plants with those proposed by FAO/ WHO(1984) it is found that all plants accumulate Fe above this limit. However, for medicinal plants the WHO (2005) limits not yet been established for Fe. Fe is necessary for the formation of hemoglobin and also plays an important role in oxygen and electron transfer in human body (Kaya and Incekara, 2000).

Zinc

In studied plants Zn concentration ranged between 20.60ppm in *W. somnifera* and 63.90 ppm in *A. aspera*. The permissible limit set by FAO/WHO

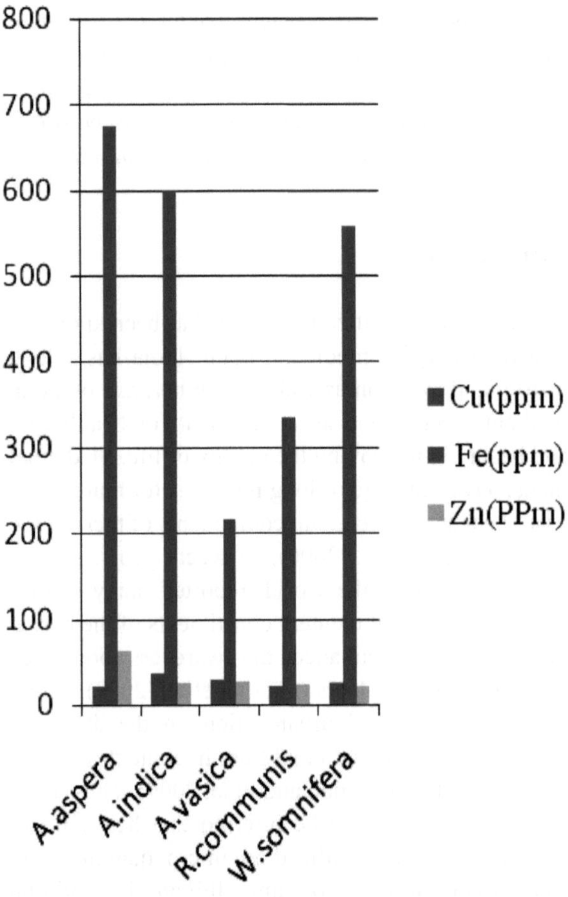

Fig. 1: Concentrations of Cu, Fe, and Zn in selected medicinal plants

Table 2: Concentrations (ppm) of heavy metal ions in selected medicinal plants

S. no	Name of plant species	Cu (ppm)	Fe (ppm)	Zn (ppm)
1	*A. aspera*	20.9	674.2	63.90
2	*A. indica*	35.8	600.2	23.93
3	*A. vasica*	28.5	215.2	26.68
4	*R. communis*	20.3	333.5	23.30
5	*W. somnifera*	24.2	556.8	20.60

(1984) in edible plants was 27.4 ppm. After comparison, metal limit in the studied medicinal plants with those proposed by FAO/WHO (1984) it is found that only *A. aspera* accumulate Zn above this limit while all others plants are within limits. However, for medicinal plants the WHO (2005) limits not yet been established for Zn.

Conclusion

The selected medicinal plants have been recommended as remedies for myriad of conditions in the traditional system of medicine. In the field of phytotherapy, tremendous progress has been documented regarding scientific evaluation of medicinal plants across the globe. The practical repercussion of the changing situation may be witnessed in the WHO monographs, National Pharmacopoeias and herbs processing industries. The concentration of Zn determined in selected medicinal plants were well below the critical limit except *A. aspera*. Zn is an essential trace nutrient to all plants it is required in large number of enzymes and plays an essential role in DNA transcription (Mengal and Kirkby, 1982). The results suggest that medicinal plants used for human consumption or for preparation of herbal products and standardized extracts should be collected from an unpolluted natural habitat.

References

1. FAO/WHO Contaminants. In Codex Alimentarius, vol. XVII, Edition 1. FAO/WHO, Codex Alimentarius Commision, Rome (1984).
2. I. Kaya, N. Incekara; J. Turkish Weed Sci. 3 (2000) 56–64.
3. K. Mengel, E.A. Kirkby; Principles of plant nutrition. International Potash Institute, Bern, Switzerland (1982).
4. A. Nostro, M.P. Germano, V. D‘ Angelo, A. Marino, M.A. Cannatelli; Plant Medica 25 (2000) 24.
5. V. Rai, M. Agarwal, S. Khatoon, A.K.S. Rawat, S. Mehrotra; Estimation of Co and Mn in some medicinal plants. Bull. Environ. Contam. Toxicol.66 (2001) 427–432.
6. S.M. Salman, M. Fida;Taibah Int. Chem. March 23 –25. Al-Madinah Al-Munawwarah, Saudi Arabia. (2009).
7. K.R. Sharma, M. Agrawal, M.F. Marshal; Food Chem. Toxicol. 47 (2009) 583–591.
8. R.A. Sial, M.F. Chaudhary, S.T. Abbas, M.I. Latif, A.G. Khan; J. Zhejiang Uni. Sci. B. 7(12) (2006) 974–980.
9. WHO Quality Control Methods for Medicinal Plant Materials Revised, Geneva, Switzerland (1998, 2005).

62
Electro Chemical Determination of Pb (II) Ions by Carbon Paste Electrode Modified with Coconut Powder

D S Rajawat, S Srivastava and S P Satsangee
USIC, Remote Instrumentation Lab, Dayalbagh Educational Institute, Agra – 282110, India
Department Of Chemistry, Dayalbagh Educational Institute, Agra – 282110, India
Email : deiusic@gmail.com

Abstract

Coconut powder was used as modifier for carbon paste electrode for voltammetric determination of lead ions in aqueous samples. Different parameters like electrode composition, different electrolytes, electrolyte concentration, different pH values, deposition time and deposition potential were optimized for application of Coconut powder modified electrodes. The use of 0.2M HCl as supporting electrolyte and 20% (w/w) modifier, deposition time 390 sec and deposition potential −1800mV gave an optimum current response for lead (II) solution. Results showed that the modified carbon paste electrode can determine Lead with better sensitivity in aqueous solutions.

Introduction

Heavy metals are highly toxic and dangerous pollutants [1, 2]. Lead is one such deleterious metal, detected in waste streams from mining operations, tanneries, electronics, electroplating, petrochemical industries [3]. Now days, the importance of controlling the level of environmental pollutants in natural waterways and potable water has generated increasing interest in the development of novel sensors for the detection of heavy metals [4]. Frequently used methods for lead determination are atomic absorption spectrometry, neutron activation and inductively coupled plasma atomic emission spectroscopy. However these are very expensive and often do not offer sensitivity usually required for environmental samples. Electrochemical determination of metal ions is relatively inexpensive and is one of the most sensitive and selective techniques in the determination of trace amounts of metals at natural levels [5]. Stripping voltammetry at the mercury electrode is a well established technique for the determination at the low levels of different metal ions. However, mercury electrodes have the disadvantage of being mechanically unstable during various steps of the assay procedure, thus they are less desirable than solid state sensors in routine field applications [6]. In addition, mercury based electrodes have issues related to the use and disposal of mercury.

Hence electrochemical determination of heavy metals at low concentration with the use of modified electrodes has been in great demand. A common approach to increase selectivity of a electrode is to attach host molecules which selectively interact with specific guest molecules. Such type of electrodes utilizes chemical and biological modifying moieties such as tissues, extracts, ligands, redox mediators, algae and enzymes that can accumulate or complex metal ions. These sensors offer several advantages i. e. fast analytical response, ease of fabrication, low cost, suitability for miniaturization and Ecofriendly nature [7].

Materials and Methods

Preparation of Modified Electrodes

Green coconut was obtained from the local market of Agra. Mesocarp was used for the analysis of metals. First the mesocrap tissues were washed with acetonitrile and then the tissues were dried. Powder was obtained by grinding of dried coconut tissues. Then the powder was sieved to get particles of <150 μm. Modified electrodes were prepared by thorough mixing of graphite powder and mineral oil in a ratio of 3:1 w/w with different proportion of coconut powder

M.M. Srivastava, L.D. Khemani, S. Srivastava, *Chemistry of Phytopotentials: Health, Energy and Environmental Perspectives*, DOI:10.1007/978–3-642–23394-4_62,

varying from 5 to 30% w/w using mortar and pestle. Then the homogenized paste was packed in a glass tube (id 3 mm) where a copper wire was inserted for electrical contact. The electrode surface was renewed by squeezing out a small amount of paste, scrapping off the excess amount of paste and the surface was smoothed by polishing the tip on a photopaper.

Procedure

Voltammetric measurements were carried out in a three electrode cell connected to a Metrohm 663 VA stand, module IME 663 and PGSTAT302N (Metrohm, Switzerland). The experiment was controlled by a PC connected to potentiostat using an Interface (Metrohm, Switzerland). The three electrode cell is consisted of Ag/AgCl as reference electrode, platinum as auxiliary electrode and the modified carbon paste electrode described below is used as the working electrode. All potentials were given Vs Ag/AgCl reference electrode. Different parameters like supporting electrolytes, electrolyte concentration, amount of modifier, pH of electrolyte solution, deposition time and deposition potential were optimized for metal determination using Differential Pulse Anodic Stripping Voltammetric (DPASV) analysis.

Results and Discussion

Anodic Stripping Voltammetry in differential pulse mode was performed for the Voltammetric analysis of lead ions. The effect of different amounts of modifier, different electrolytes (HCl, NaOH, NaCl), electrolyte concentration, deposition time and deposition potential and pH was investigated.

The first step for characterization of a prepared carbon paste electrode is Cyclic-voltammetry. Cyclic Voltammetry is used to determine the potential window of the modified CPE. A modified CPE can only be utilized for analysis in a potential region that exhibits a constant or minimal current response. It is clear from Fig. 1 that plant modified CPE showed inertness in the potential region –1 to +1 making it useful in evaluating electro active species within this range.

The fresh electrode surface was first electrically activated by 10 replicated direct current sweeping from –1000 to +1000 mV with a scan rate of 100 mV/s, immersed in 0.2M HCl solution. The solution is then exchanged by a sample solution containing 1 ppm of Lead(ll) in HCl electrolyte then purged by pure nitrogen gas for 5 min. In order to evaluate the performance of the new modifier, the working electrodes containing 0, 5, 10, 15, 20, 22.5, 25, 30% Coconut powder are prepared and examined under identical conditions.

Using DPASV a narrow peak was observed around –0.5V for Lead using coconut modified Carbon Paste Electrodes.

Differential pulse Anodic stripping voltammetry (DPASV) confirmed that lead can be detected using the Coconut powder modified CPE and was used for the optimization of different parameters.

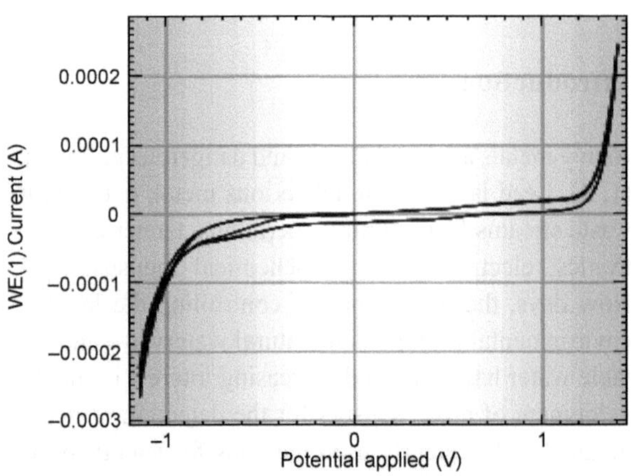

Fig. 1: Cyclic Voltammogram of plant modified CPE in HCl electrolyte with in potential range +1.2 to –1.1

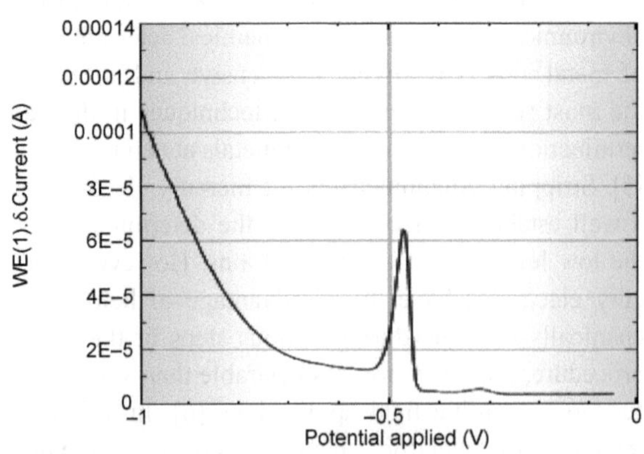

Fig. 2: Differential Pulse Anodic Stripping Voltammogram of 1 ppm Lead in 0.2 M HCl

Optimization of Different Parameters

Electrode Composition

Effect of different composition of coconut powder on the electrode response of coconut powder modified CPE was studied using different ratios of Carbon paste to coconut powder. It is obvious from Fig. 2 that the modified electrode accumulate lead ions from the solution onto the electrode surface and give stripping peak at a potential of –500 mV. Fig. 3 shows that peak height for metal determination increases with the increase in the amount of Coconut powder and then decreases.

The maximum peak height is observed when the plant modifier is 20%. It is mainly due to the fact that when we increase the plant modifier percentage a sharp increase in the peak height is observed due to a great number of sites available for cation exchange with Pb (ll) ions, but further increase in the modifier concentration results in decrease of the peak height, which is mainly due to the less conductivity at the surface of the CPE since there is a larger amount of the plant material and it is non conducting. Results show that the current value for the 20% modifier is greater than the bare electrode shown in Fig. 3. Thus the modifier enhances the sensitivity of the electrode compared to the bare CPE.

Supporting Electrolyte

The role of supporting electrolyte in electrochemical analysis is to facilitate electrical conduction in solution. They are used to decrease the resistance of the solution, to eliminate electro migration effects and to maintain a constant ionic strength in controlled potential experiments. It must be carefully selected in order to avoid unwanted interferences during voltammetric runs. Different Supporting electrolytes (Acid, Base and salt solutions) have been used in determination of appropriate supporting electrolytes. Results are shown in Fig. 4. The highest peak is observed in HCl followed by NaOH and then by NaCl. HCl gave sharp, well defined and reproducible analytical signals for anodic stripping voltammetry compared to others. Similar results were also found in Voltammetric determination of Pb(ll) using CPE modified with pineapple peelings.[8]

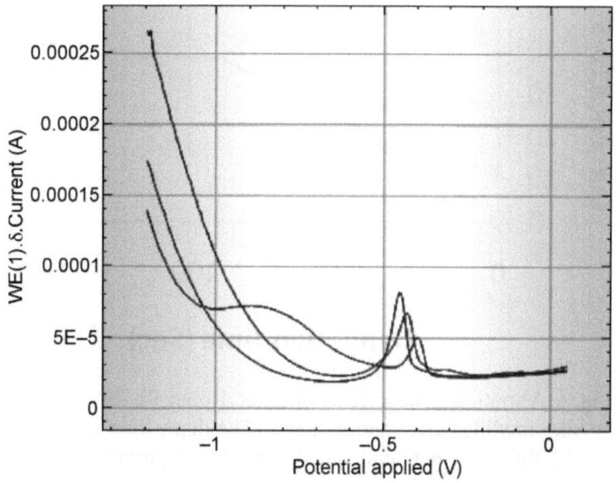

Fig. 4: Differential pulse voltammograms of 1 ppm Pb using different electrolytes for coconut powder modified electrodes. (The deposition potential E: –1200mV, step potential: 5 mV and pulse amplitude: 30 mV)

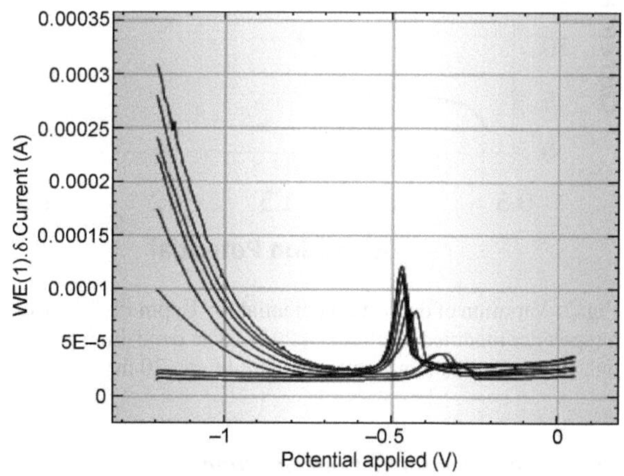

Fig. 5: Differential pulse voltammograms of 1 ppm Pb for different electrolyte concentration of HCl for coconut powder modified electrodes of 20%. (The deposition potential E: –1200mV, step potential: 5 mV and pulse amplitude: 30 mV)

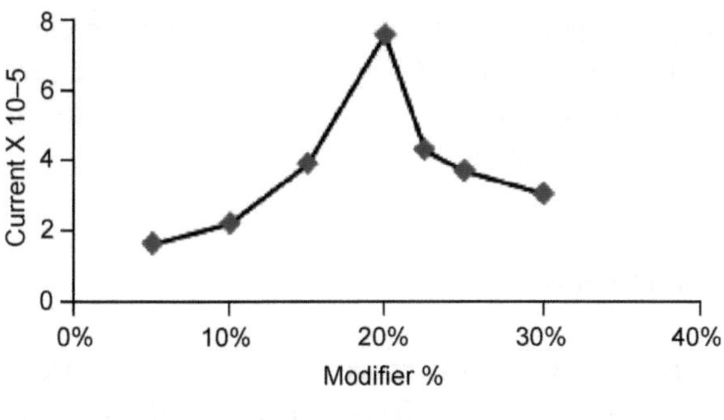

Fig. 3: Variation of modifier % for 1 ppm Pb for Carbon Paste Electrodes of Different coconut powder modified electrodes. (The deposition potential E: –1200mV, step potential: 5 mV and pulse amplitude: 30 mV)

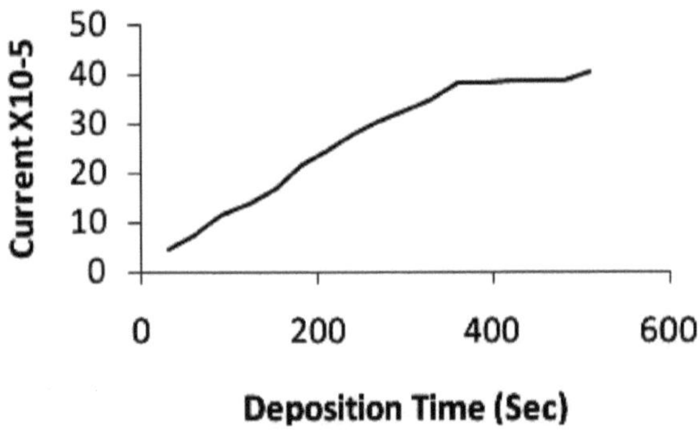

Fig. 6: Variation of deposition time for 1 ppm Pb for coconut powder modified electrode of 20%. (The deposition potential E: –1200mV, step potential: 5 mV and pulse amplitude: 30 mV)

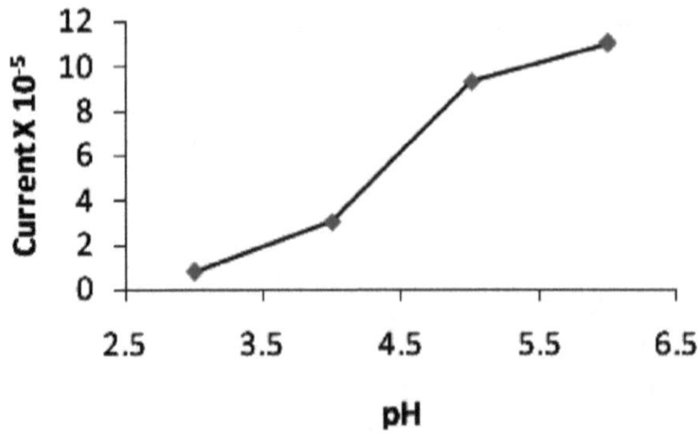

Fig. 8: Variation of pH of Acetate buffer for 1 ppm Pb for coconut powder modified electrode of 20%. (The deposition potential E: –1200mV, step potential: 5 mV and pulse amplitude: 30 mV)

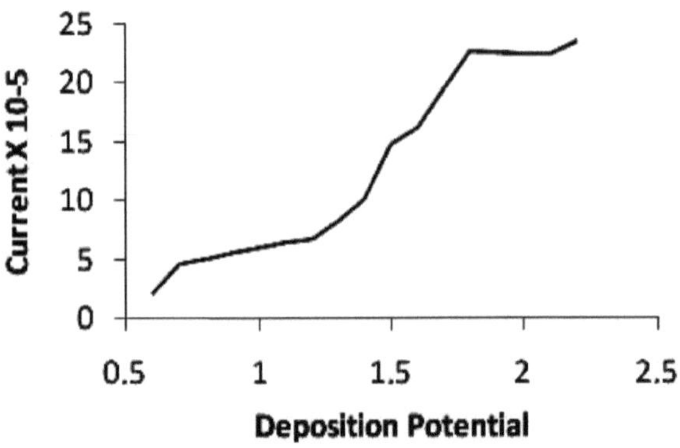

Fig. 7: Variation of deposition potential for 1 ppm Pb for coconut powder modified electrode of 20%. (The deposition potential, step potential: 5 mV and pulse amplitude: 30 mV)

Supporting Electrolyte Concentration

The effect of supporting electrolyte concentration on the activity of the modified CPE is shown in Fig. 5. HCl was used as a supporting electrolyte in a range varying from 0.01M to 0.5M. When the concentration of the supporting electrolyte increased from 0.01M to 0.5 M, the current value increase up to 0.25M concentration of supporting electrolyte and decreases on further increase of the electrolyte concentration.

Deposition Time

Deposition time is the time required to reduce Pb^{2+} to Pb^0. The effect of deposition time on peak current is shown in the Fig. 6. Peak current increases with the increase in the deposition time. This is because when deposition time is long, maximum amount of the accumulated analyte are reduced to its elemental form, consequently giving higher peak current response.

Deposition Potential

Deposition potential is the applied potential at which the accumulated metal on the electrode surface is deposited or reduced. The change in the peak height with change of deposition potential is shown in the Fig.7. It was found that activity of modified CPE was maximum when the deposition potential was 1800 mV. There is continuous increase in the peak height with the increase of the peak potential.

pH Variation

The effect of pH on the electrode response was determined using Acetate Buffer. It can be observed from Fig. 8 that the peak current increases from pH 3 to 6. Thus optimum peak height was achieved at pH = 6.

Regeneration of the Electrode Surface

The success of quantitative voltammetric analysis with CPEs depends upon the quality of the carbon paste electrode surface. Modified CPE surface was renewed by simple removal of the used carbon paste layer and polishing it on photo paper. Modified electrodes were stored in refrigerator. The electrode surface was activated by doing cyclic voltammetry in electrolyte solution.

Conclusion

Coconut powder was successfully used as environment friendly, cost effective, non-hazardous modifier for voltammetric determination of Lead using modified CPE. Coconut powder modified CPE was able to detect lead (II) ions in aqueous samples using DPASV. Optimization of parameters like amount of modifier, supporting electrolyte, electrolyte concentration, deposition time, deposition potential and pH was performed. The peak height of bare carbon paste electrode and modified carbon paste electrode shows that modified carbon paste electrode shows two times increase in the peak height as compared to bare carbon paste electrode.

References

1. A. Kardam, K. R. Raj, J. K. Arora and S. Srivastava. Journal of instrumentation society of India, 40 (3) (2010) 175–176.
2. A. Kardam, P. Goyal, J. K. Arora, K. R. Raj and S. Srivastava. Nat Acad Sci Lett. 32 (5–6) (2009) 179–181.
3. P. Vasuderan, V. Padmavathy and S. C. Dhingra. Bioresour. Technol., 89 (3) (2003) 281–287.
4. M. R. Ganjalia, N. Motakef-Kazamia, F. Faridboda, S. Khoeec and P. Norouzia; Journal of Hazardous Materials 173 (2010) 415–419.
5. B. Jannat, N. Sadeghi, M. R. Oveisi, A. Behfar, H. Komeilizadeh and A. Shafaati;, Iranian Journal of Pharmaceutical Research, 8(3) (2009) 159–162.
6. E. R. E. Mojica, J. M. Vidal, A. B. Pelegrina and J. R. Micor. J. Applied Sci. 7(9) (2007) 1286–1292.
7. E. E. Mojica, J. R. L. Micor and C. C. Deocaris. J. Appl. Sci. Res. 1(1) (2005) 99–102.
8. E. R. E. Mojica, S. P. Gomez and J. R. L. Micor, Philipp. Agric. Sci., 89 (2006) 134–140.

63
Assessment of Surface Ozone levels at Agra and its impact on Wheat Crop

V. Singla[1], T. Pachauri[2], A. Satsangi[3], K. Maharaj Kumari[4] and A. Lakhani[5]
Department of Chemistry, Faculty of Science, Dayalbagh Educational Institute,
Dayalbagh, Agra 282 110.
Email: anitasaran2003@yahoo.co.in

Abstract

Ozone is currently the most important air pollutant that negatively affects growth and yield of agricultural crops in most parts of the world, and wheat is arguably the most important food crop in the Northern India. The higher ozone concentration in different regions is posing threat to food production. Measurement of surface ozone was made during the growing season (January-March 2011) of wheat crop at Agra. The daytime maximum was found to vary from 62–72 ppb and minima varied from 20–23 ppb. The effect of ozone on crop yield has been examined using exposure indices AOT40 and SUM06. The calculated AOT40 (2562 ppbh) and SUM06 (7470 ppbh) were found below the critical levels thereby indicating that wheat crop grown in Agra is safe from the threats of existing surface ozone levels.

Introduction

Surface ozone (O_3) is a secondary air pollutant produced from photochemical reactions involving the oxidation of volatile organic compounds (VOC) and carbon monoxide (CO) in the presence of nitrogen oxides (NOx) [1]. Enhanced concentrations of ozone are ubiquitous to most of the populated regions of the world, where the burning of biofuels and/or fossil fuels used for cooking, heating, transportation and the generation of energy cause the emission of VOC, CO and NOx and thus the photochemical generation of ozone. Rise in concentration of surface ozone is most likely to threaten food production across the globe due to its phytotoxicity and prevalence over important agricultural regions of North America, Europe and Asia [2–4].

Ozone affects plant production by diffusing into the leaf via the stomata and then in intercellular air spaces where it dissolves in water contained in mesophyll cell walls and leads to production of reactive oxygen species. Early symptoms of chronic ambient air exposure are decreased rate of photosynthesis, accelerated senescence and decreased leaf area. The ability of elevated surface ozone to damage agricultural crops has been well documented by research projects conducted in the United States (US) and Europe in the 1980s and 1990s, i.e. the US National Crop Loss Assessment Network (NCLAN) [5] and European Crop Loss Assessment Network (EUCLAN) programs using Open Top Chamber (OTC). NCLAN results indicate a reduced annual soybean yield of 10% and a reduced cotton yield of 12% for seasonal mean O_3 mixing ratios greater than 50 ppbv [6]. Fuhrer et al., [7] modeled a reduction of yield with increasing O_3 over a 40 ppbv threshold, resulting in a 10% reduction in spring wheat yield for O_3 mixing ratios in southern Europe. AOT40 and SUM06 are the different indices that account for threshold effects of O_3. Pleijel et al., [8] compared a number of these indices for wheat and potatoes and concluded that the threshold-based flux best captured O_3 damage to crop yield. Nevertheless, Ozone-crop-effects studies have shown that a reasonably robust estimate of the impact of ozone exposure on net crop yields can be obtained through the use of indices that track exposure over an entire season [9,10]. Such indices are SUM06; AOT40; W126 and N100 [11].

In India, studies conducted on test crop plants wheat (*Triticum aestivum*), mustard (*Brassica campestris*), mung bean (*Vigna radiata*) and spinach showed significant decrease in yield at a rural site experiencing

M.M. Srivastava, L.D. Khemani, S. Srivastava, *Chemistry of Phytopotentials: Health, Energy and Environmental Perspectives*, DOI:10.1007/978-3-642-23394-4_63, © Springer-Verlag Berlin Heidelberg 2012

high ozone concentration as compared to low O_3 site [12]. At Dayalbagh, wheat crop is harvested in the month of April. Therefore in this study, we present the diurnal variation of O_3 measured at Dayalbagh, Agra during the growing season of wheat crop i. e. January-March. Preliminary assessment of the effects of O_3 pollution on wheat crop is discussed using exposure indices like AOT40 and SUM06.

Experimental

Sampling Site

Agra is situated at latitude of 27°10 ON and longitude of 78°05 OE with an altitude of 169 m above the sea level in the semi-arid zone of India. It is positioned with the Thar Desert of Rajasthan to the West, central hot plains to the South, Gangetic plains to the East and cooler hilly regions to the North. Study was carried out at Dayalbagh Educational Institute Campus. It is a small suburban site with no industrial activity around. Dayalbagh is a small residential community lying immediately outside the city where agricultural activities predominate. The sampling site lies by the side of a road that carries mixed vehicular traffic, moderate (of the order of 1000 vehicles) during the day and minimal (of the order of 100 vehicles) at night. The campus lies about 2 km north of the National Highway-2 (NH-2) which has dense vehicular traffic (10^6 vehicles) throughout the day and night. Agra has a continental type of climate characterized by extreme dryness in summer and cold winters with calm periods. The summer in Agra is hot with intense solar radiation and is dominated by strong southeasterly winds (wind speed ranges from 10–16 kmh^{-1}). Intense solar radiation varies from 19–23 Wm^{-2}. The winters are associated with calm periods of about 40% (wind speed ranges from 0.5–5 kmh^{-1}).

Instrumentation

Surface O_3 concentrations were recorded using continuously operating O_3 analyzer (Thermo Fischer Model 49i) from January-April 2011. The ozone concentration measurement is based on ultraviolet absorption photometry, resting upon absorption of radiation of wavelength 254 nm by ozone in the analyzed sample. The radiation source is a UV-lamp and clean air (zero) and the sample itself are alternately measured in cells. The minimum detection limit and precision of the analyzer is about 0.5 ppb and 1 ppb respectively.

Results and Discussion

Diurnal Variation of O_3

Fig. 1 gives the representative diurnal variation of O_3 observed during the study period. The variation in ozone concentration exhibits marked diurnal variability, with high concentrations during the day and low concentrations during night. With the onset of sunshine, ozone concentration starts increasing gradually, becomes maximum (62–72 ppb) in noon when the solar radiation and temperature reaches at their maximum. The concentration further starts decreasing with the diminishing of sunshine and becomes minimum (20–23 ppb) at night and early morning hours. The maximum concentrations during noontime are attributed to photochemical production of O_3 in presence of precursor gases such as CO, CH_4 and NMHCs in presence of sufficient amount of NOx and proceed via following set of reactions:

$$\mathbf{NO_2 + \mathit{h}\nu\ (\lambda < 420\,nm) \rightarrow NO + O}$$
$$\mathbf{O + O_2 + M \rightarrow O_3 + M}$$

Boundary layer height also rises gradually after sunrise and reaches maximum during local noontime due to convective heating and starts descending as the temperature starts decreasing in the evening after sunset. The decrease in night-time O_3 concentrations is mainly due to titration of O_3 by surface emission of NO and ground level destruction of O_3 in a shallow boundary layer involving the loss of O_3 by NO ($O_3 + NO \rightarrow O_2 + NO_2$). Similar patterns of seasonal variation have also been observed at other sites like Gadanki and Anantapur in India during the past decade [14, 15].

During the study period, hourly O_3 concentrations were often in excess of 80–90 ppb and reached as much as 100 ppb in the noontime. High temperature favors O_3 formation. The mean temperature remains low (23–27 °C) during January and February but rises rapidly in March and April (35–38 °C) and so the ozone concentrations. Fig. 2 shows the variation of O_3 as a function of time as a contour during the study period.

Exposure Indices

Apart from its important climatic implications, O_3 is also known to have phytotoxic effects. Ozone is regarded as the most important component associated with agricultural crop damage. Ozone has strong oxidizing properties and leads to inhibited photosynthesis, respiration, nutrient uptake which leads to reduced yields of agricultural crops. Several approaches have been tried to define robust exposure indices. Two such exposure indices, AOT40 and SUM06 had been used in the present study to discuss the ozone exposure to wheat crop. AOT40 is used as the main exposure index for ozone effects on plants, including crops [15] and SUM06 is used in applied studies and for setting air-quality standards [16].

AOT40 is calculated as the sum of differences between the hourly mean concentrations (O_3) and 40 ppb for hours when $O_3 > 40$ ppb, for each daylight hour with global radiation ≥ 50 Wm^{-2} over a 3 month period:

$$\text{AOT 40} = \sum_{n=1}^{n}([O_3] - 40)_i \text{ for } [O_3] > 40 \text{ ppb},$$

where n is the number of hours (i) that the threshold of 40 ppb is exceeded. For studies of wheat in Europe it has been shown that AOT40 gives better performance as an exposure index [17].

An AOT 40 value of 10,000 ppb h for daylight hours (radiation > 50 W m^{-2}) over a 6 month period has been established as a critical level for the protection of forests. While, for the protection of agricultural crops of 5 % loss in yield, an AOT 40 value of 3000 ppb h for daylight hours over 3 months growing season and 5300 ppb h specially for cereal crops has been established as the critical level [18,19]. The calculated AOT40 during the growing season of wheat crop (January-March) was observed to be 2562 ppbh which is below the critical levels (3000 ppb h) of O_3 (Fig. 3). This indicates that the crop growing in the vicinity of sampling site is not affected by enhanced levels of O_3 and therefore the yield loss of crop might not take place.

SUM06 is defined as the sum, over a 3-month period, of the hourly ozone concentrations for daylight hours (0700–0900 h) when the concentration (O_3) is at or above 60 ppbh (0.06 ppmh):

$$\text{SUM06 (ppmv-hour, ppmh)} = \sum_{n=1}^{n}(CO_3)_i \quad \text{for } CO_3 \geq 60 \text{ ppb in months},$$

where n is the number of hours (i) that the threshold is exceeded. Table 1 gives the critical levels for different species of wheat.

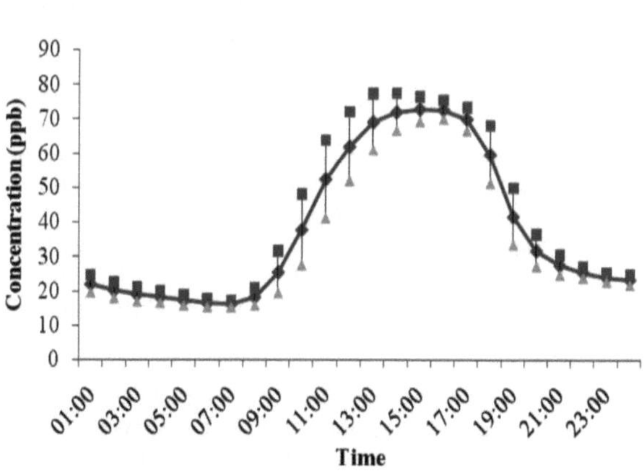

Fig. 1: Diurnal variation of Ozone

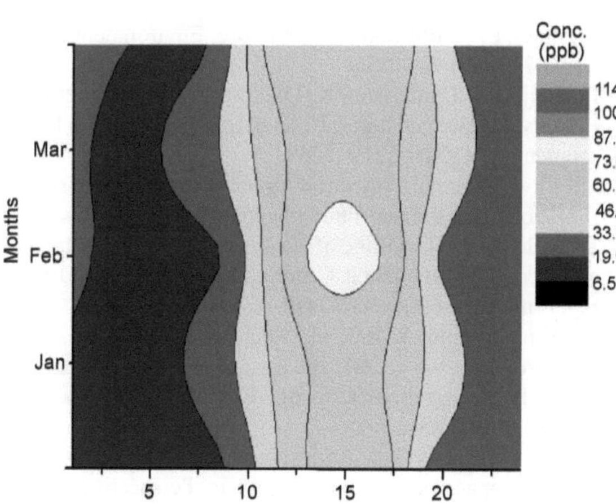

Fig. 2: Contour plot of Ozone

Table 1: Critical Levels for Different Species of Wheat

Crop	Species	12 hour SUM06
Wheat	ABE	25,100
	ARTHUR	21,300
	ROLAND	7,400
	ABE	34,800
	ARTHUR	27,700
	VONA	2,900
	VONA	7,700

Source: NAAQO, 1999

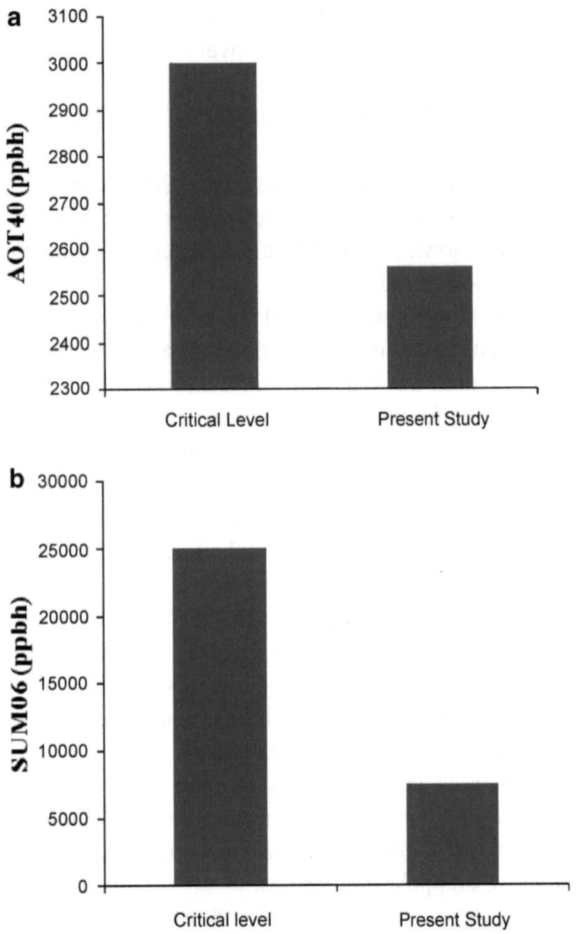

Fig. 3: Bar diagram showing critical levels and calculated (a) AOT40 and (b) SUM06

The observed 3-month SUM06 for wheat at the present site was found to be 7470 ppbh which is much lower than the critical levels (21,300–34,800 ppb h) of O_3 reported for wheat [20].

This lower 3-month SUM06 value further supports the study citing that wheat crop grown in the surrounding area of this site does not experience any harm.

Conclusion

Measurement of O_3 during the growing season of wheat crop (January-March 2011) showed maximum concentration during daytime (62–72 ppb at 1200–1500 h) and minima at night (20–23 ppb at 2100–2300 h) due to maximum solar radiation and high temperature during noon hours. The 3-month AOT40 value (2562 ppbh) was observed to be less than equal to critical level (3000 ppbh) and therefore the wheat crop yield is not likely to be affected by the current O_3 concentration. The 3-month SUM06 value (7470 ppbh) was found to be lower than the critical levels of SUM06 set for wheat crop.

Acknowledgements

We thank Prof. V. G. Das, Director and Prof. L. D. Khemani, Head, Department of Chemistry, Faculty of Science, Dayalbagh Educational Institute, for the facilities provided. The authors also gratefully acknowledge ISRO-GBP Project for providing funds to carry out the experimental work.

References

1. W. Huixiang, C. S. Kiang, T. Xiaoyan, Z. Xiuji and W. L. Chameides; Atmospheric Environment, 39 (2005) 3843–3850.
2. J. Fuhrer and F. Booker; Environment International. 29 (2003) 141–154.
3. Royal Society. Ground-level Ozone in the 21st Century: Future Trends, Impacts and Policy Implications. The Royal Society, London. Science Policy report 15/08 (2008).
4. L. D. Emberson, P. Buker, M. R. Ashmore, G. Mills, L. S. Jackson, M. Agrawal, M. D. Atikuzzaman, S. Cinderby, M. Engardt, C. Jamir, K. Kobayashi, N. T. K. Oanh and Q. F. Quadir, A. Wahid. Atmospheric Environment. 43 (2009) 1945–1953.
5. W. W. Heck, W. W. Cure, J. O. Rawlings, L. J. Zaragoza, A. S. Heagle, H. E. Heggestad, R. J. Kohut, L. W. Kress and P. J. Temple; Journal of Air Pollution Control Association. 34 (1984) 729–735.
6. A. S. Heagle; Annual Review of Phytopathology. 27 (1989) 397–423.
7. J. Fuhrer, L. Skarby and M. R. Asmore; Environment Pollution. 97 (1997) 91–106.
8. H. Pleijel, H. Danielsson, K. Ojanpera, L. D. Temmerman, P. Hogy, M. Badiani and P. E. Karlsson. Atmospheric Environment. 38 (2004) 2259–2269.
9. US Environmental Protection Agency. Air Quality Criteria for Ozone and Related Photochemical Oxidants. Research Triangle Park, NC: Office of Health and Environmental Assessment, Environmental Criteria and Assessment Office; report no. EPA-600/p-93/004aF-cF (1996).
10. C. Wang, J. Guo, Y. Bai, M. Wen, J. Liu and L. Li. Acta Meteorologica Sinica. 60 (2002a) 238–241 (in Chinese).
11. W. W. Heck and E. B. Cowing; Environmental Management. 3 (1997) 23–33.
12. M. Agrawal; Trend in tropospheric ozone concentration and its impact on agriculture: Indian Perspective. EnviroNews Archives. (April 2007) 13(2).

13. M. Naja and S. Lal; Surface ozone and precursor gases at Gadanki (13.58°N, 79.28°E), a tropical rural site in India. Journal of Geophysical Research. 107 (2002) (D14).
14. Y. N. Ahammed, R. R. Reddy, K. R. Gopal, K. Narasimhulu, D. B. Basha, L. S. S. Reddy and T. V. R. Rao; Atmospheric Research 80 (2006) 151–164.
15. L. Karenlampi and L. Skarby; Critical levels for Ozone in Europe: Testing and Finalizing the concepts. UN-ECE Workshop Report. University of Kuopio, Department of Ecology and Environmental Science, Finland (1996) 363.
16. US Environmental Protection Agency. Air Quality Criteria for Ozone and Related Photochemical Oxidants. Research Triangle Park, NC: Office of Health and Environmental Assessment, Environmental Criteria and Assessment Office; report no. EPA-600/p-93/004aF-cF (1996).
17. K. Aunan, T. Berntsen and H. Seip; Ambio. 29 (2000) 294–301.
18. WHO. Update and revision of the WHO air quality guidelines for Europe, Ecotoxic Effects, Ozone Effects on Vegetation. European Center for Environment and Health, Bilthoven, the Netherlands (1996).
19. G. S. Satsangi, A. Lakhani, P. R. Kulshrestha and A. Taneja; Journal of Atmospheric Chemistry. 47 (2004) 271–286.
20. National Ambient Air Quality Objectives for Ground-Level Ozone-Summary Science Assessment Document. Cat. No. En42–17/7–1–1999E (1999).

64
Synthesis and Characterization of an Eco-Friendly Herbicides Against Weeds

N. Sidhardhan, S. Verghese.P, S. Dubey and D. Jain
Department of Chemistry, St. John's College, Agra
Email: sidharthnisha@gmail.com

Abstract

Herbicides that kill plants by inhibiting specific vital functions do not distinguish between crop plants and weeds. Such non-selective herbicides are generally applied before sowing/emergence of crop plants and their residual effects may affect crop performance. There is limited flexibility in the schedule of their application and their use requires caution. However, some crop plants enjoy naturally endowed resistance to specific herbicides. It is important to recall that although a large number of chemicals have been approved for weed control, their widespread and continuous use is not desirable owing to their toxicity and long-term effects on the environment. Hence an environmental friendly degradable, non-persistent, non-accumulative herbicides which can kill weeds, found in the roof top footpath and agricultural field have been synthesized by metathesis of ammonium salt and fatty acid characterized by IR, pH and Conductometric studies, as ammonium myristate is inherently a non-toxic substances and disappear after the desired period of herbicides activity and it is very safe to use as a weedicide. It is also found out the effective concentration to remove the weeds are at above CMC on daily application for fifteen days before the crop sowing.

Introduction

"Weeds are problem in Indian agriculture, as elsewhere in the world. Weeds compete with crops for moisture and nutrients. Loss of yield due to weed infestation is variable and is more pronounced in crops grown under rainfed conditions. Some parasitic weeds drawn water and nutrients from crop plants and can inflict severe damage. Further, weeds serve as alternate hosts to pathogens and also harbor pests. Control of weeds during early stages of crop growth, when the young seedlings of crop plants are unable to compete with hardy weeds, is crucial for capturing yield potential. For this reason, labor demand for weeding operation is high during early phase of crop cycle and manual weed control over large areas is not feasible from the point of labor supply and monitory costs. Some weeds that are wild relatives of crop plants are difficult to distinguish from crop plants at early stages and pose challenge for manual weeding. Under these situations, chemical weed control is relevant for realizing higher productivity and production". In India, about 6000 tons of herbicides are currently used for weed control, mainly in irrigated crops (about 77% on wheat and rice) and on plantations (about 10%) [3]. However, herbicides form only 12% of the pesticides used on crops in India. A wide variety of weeds (perennial and annual are generally encountered in crop fields. However, specific weeds predominate different cropping systems and zones. Both broad spectrum/non-selective and selective herbicides are in use. Continuous use of some herbicides has led to development of resistant weeds and has exacerbated weed problems. For example, in rice-wheat cropping system of Punjab and Haryana, Phalaris minor has developed resistance against isoproturon [(3)]

Methodology

$$\underset{\text{Ammonium Cholride}}{NH_4Cl} + \underset{\text{Myristic acid}}{C_{14}H_{28}O_2H} \longrightarrow \underset{\text{Ammonium myristate (SOAP)}}{C_{13}H_{27}COONH_4} + H_2O$$

Ammonium myristate was prepared by refluxing equivalent amounts of corresponding fatty acids and

M.M. Srivastava, L. D. Khemani, S. Srivastava, *Chemistry of Phytopotentials: Health, Energy and Environmental Perspectives*, DOI:10.1007/978-3-642-23394-4_64,

aqueous solution of NH_4CL for 6–8 hours on a water bath. The ammonium myristate was purified by recrystallization with benzene-methanol mixture and dried under reduced pressure. The purity of the ammonium laurate was checked by the determination of their melting point. The melting point of purified ammonium myristate was 144.8ºC. The conductivity measurements of the solution of ammonium myristate in distilled water was made with "BIOCRAFT" direct reading Conductometer and a dipping type glass conductivity cell with platanized electrodes at a room temperature.

Results and Discussion

Infra-red Absorption Spectra

The infrared absorption spectra of stearic acid and of corresponding Ammonium myristate was obtained with a Perkin-Elmer grating spectrophotometer in the region 4000–400 Cm^{-1}, using potassium bromide technique. In the IR spectrum of ammonium myristate the absorption bands of C-H stretching vibrations Viz. the symmetrical vibration of CH_2 at 2860–2850 cm^{-1}, the asymmetrical stretching vibration of CH_2 at 2920–2910 cm^{-1}, the asymmetrical stretching vibration of CH_3 at 2960–2940 cm^{-1} and the deformation of CH_2 at 1498–1320 cm^{-1} are observed in the spectra of ammonium myristate as well as in myristic acid .The evenly spaced progressive bands near 1350–1188 cm^{-1} which are characteristic of the hydrocarbon chain of acid remain unchanged on preparing the carboxylate from the corresponding fatty acid. The absorption bands observed near 2660–2640, 1700, 930–900, 575–530 cm^{-1}, in the spectra of fatty acid have indicated the presence of localized –COOH group in the form of dimeric structure and the existence of intermolecular hydrogen bonding between two molecules of the acid. The absorption bands observed near 2660–2640, 1700 and 950–930 cm^{-1} corresponding to the –OH group in the spectra of fatty acids have completely disappeared in the spectra of corresponding ammonium myristate. The appearance of two absorption band of carbonyl group corresponding to the symmetric and asymmetric vibrations of carboxylate ion near 1470–1410 and 1560–1540 cm^{-1} respectively in the spectra of potassium stearate indicate that there is a complete resonance in the C-O bonds of carbonyl group of the carboxylates molecules and the two bonds become identical with the force constant assuming the value intermediate between those of normal double and single bonds. It is therefore concluded that the resonance character of the ionized carboxyl group is retained in these metal carboxylates and the fatty acids exist with dimeric structure through hydrogen bonding whereas the metal-to-oxygen bonds in these metal carboxylates are ionic in character. It is also confirmed that the molecules of carboxylates retain the resonance character of the carboxyl group.[(6)]

Specific Conductance

The specific conductance, k of the solution of ammonium myristate soap in a mixture of NH_4CL+Myristic acid increases with the increase in the soap concentration, C (Table: 1). The increase in the specific conductance with the increase in the soap concentration may be due to the ionization of ammonium myristate soap into a simple metal cations and fatty acids anions in solutions and also due to the formation of micelles at higher soap concentration. The plots of specific conductance k, against the soap concentration, C are characterized by an intersection of single straight lines at a definite soap concentration, which corresponds to the CMC of the soap. The result shows that the solute-solvent interaction is larger than the solute-solute interaction of dilute soap solutions. It is therefore concluded that the soap molecules do not show appreciable aggregation below the CMC and there is a marked increase in the aggregation of the soap molecules at this definite soap concentration.[(1)] The CMC (critical micelle concentration) of ammonium myristate is

Table 1

S.NO	Specific conductance x10^2	pH	Impact on weeds
1	9.0x10^2	8.93	No Impact
2	7.8x10^2	7.10	N0 Impact
3	4.3x10^2	6.14	N0 Impact
4	4.3x10^2	5.73	N0 Impact
5	1.7x10^2	5.78	Less Impact
6	8.7x10^2	5.75	Effective
7	3.0x10^2	5.30	Effective
8	2.0x10^2	5.14	Effective
9	1.0x10^2	5.00	Effective
10	5.0x10^2	4.33	Most Effective

$2.5X10^{-3}dm^3/l$ on plotting specific conductance Vs concentration at various concentrations. [2] Many samples of various concentration having different P^H have been prepared by mixing ammonium myristate of concentration of different dilution (%), and then sprayed on weed plant to check the efficacy of this insecticidal spray on daily and bi-weekly interval. (Fig. 1).

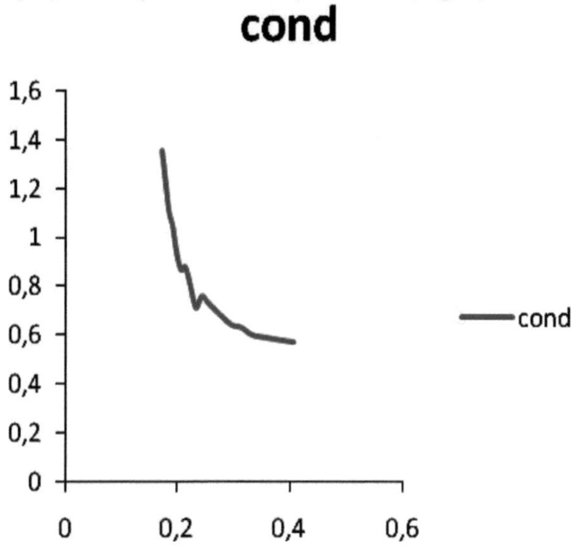

Fig. 1: Conductivity graph

Conclusion

Ammonium myristate have been found to be an effective herbicide to kill Weeds (Parthenium hysterophorus) found on agriculture field, roof top on daily application for its continue days.

References

1. H. G. Hetzal; "Safe Use of Pesticides in Agriculture", Vergenatech.com (1996).
2. K. Amman; Trends biotechnology. 23 (2005) 388–394.
3. K. P.Gupta; "Pesticide in the Indian Environment-Environmental Science Series. Interpret New Delhi" (1986) 15–16.
4. R. Wiilliam, Buck and B. Goff "Morphology and Classification of mosses," in A.Jonathan Shaw and Bernard Goffnet (edn), Bryophyte Biology Cambridge University (2003) 71–123.
5. S.Saksena, "Managing weeds: accent on chemical control". Pestic. Inf., XXVIII, 6–11 (2003).
6. Steve, Master Gardener "Moss control in lawns" (Web). Gardening in western Washington. Washington State University (1996).
7. Weed act 1959" Department for Environmental, Food and Rural affairs (Defra), UK., 2009.

65
Role of Phenolics in Plant Defense Against Insect Herbivory

F. Rehman[1], F.A. Khan[1] and S.M.A. Badruddin[2]
[1]Department of Botany, Aligarh Muslim University, Aligarh
[2]Department of Zoology, Aligarh Muslim University, Aligarh
E. mail: farharehman2@gmail.com, fareedkhan.amu2007@rediffmail.com, alibadruddin7@gmail.com

Abstract

Several secondary metabolites synthesized in plants have significant defensive role against herbivores, pests and pathogens. The defensive role played by such secondary metabolites include deterrence/antifeedant activity, toxicity or acting as precursors to physical defence systems. Many specialist herbivores and pathogens being one step more evolved circumvent the deterrent effects of secondary metabolites but actually utilize these compounds as either host recognition cues or nutrients. Phenylpropanoids are a group of phenolics, it is a chemically diverse family of compounds ranging from simple phenolic acid to large and complex polymers such as tannins, lignin and flavonoid. Phenolics derived from amino acids and their precursors and some compounds which derived from shikimic acid pathway. This group includes metabolites derived from the condensation of acetate units (e. g., terpenoids) and produced by the modification of aromatic amino acids (e. g., phenylpropanoids; cinnamic acids, lignin, precursors, hydroxybenzoic acids, catechols and coumarins), flavonoids, isoflavonoids and tannins like dihydroxyphenols and flavonols polymerized by the action of peroxidases and polyphenoloxidases. This review present an overview of biosynthesis and role of phenolics in plants by which they protect themselves against herbivory.

Introduction

An enormous variety of secondary metabolites are derived from shikimic acid or aromatic amino acids, many of which have important roles in defense mechanism against herbivory and wounding [1]. The term, phenolics, has been used to describe a group of structurally diverse plant secondary metabolites [2]. Plants produce a high diversity of secondary metabolites having a prominent protective function against predators and microbial pathogens for having a toxic nature and repellence to herbivore and microbes. Some of these metabolites are also known for defense against abiotic stress (e. g., UV-B exposure) as well as means for the communication of the plants with other organisms [3, 4], insignificant for growth and developmental processes [5, 4]. Secondary metabolites are therefore, most important part of the plants defense system against pests and diseases including root parasitic nematodes [6]. Plants produce a large variety of secondary products that contain a phenol group, a hydroxyl functional group on an aromatic ring called phenol, a chemically heterogenous group also.

Phenolic compounds are aromatic compounds bearing one or more hydroxyl group on an aromatic ring. Phenylpropanoids are the phenolic compounds derived from phenylalanine. There are >8000 known phenolic compounds, roughly categorized into 14 compound classes based on the number of carbons and their arrangement [7]. Only some of them have been implicated in plant herbivore interactions. Amongst these are the benzoic acids, hydroxycinnamic acids (and their conjugates), furanocoumarins, coumarins, stilbenes, flavonoids (especially flavonols), hydrolysable tannins, condensed tannins and lignin [8]. The role if phenolic compounds in plant herbivore interaction has been described by different worker such as hydroxycinnamic acids may act as cell wall crosslinks that fortify and protect plant cell walls against chewing damage [9,10]. Tannins accumulate in many plant species specially in trees, in response to herbivory [11,12]. In insects, tannins produced lesions in the midgut of the animals feeding on plants. Insects that normally feed on tannins have been found to have a relatively thick protective peritrophic membrane lining the midgut epithelium. It is presumed to be an

M.M. Srivastava, L.D. Khemani, S. Srivastava, *Chemistry of Phytopotentials: Health, Energy and Environmental Perspectives*, DOI:10.1007/978-3-642-23394-4_65, © Springer-Verlag Berlin Heidelberg 2012

evolutionary evidence that tannins play a role in plant insect interactions [11,12,13].

Biosynthesis of Phenolic Compounds

Phenolics are widely distributed in plants, and accumulate during normal growth and development. Thus, phenolics are constitutive products in plants and are present prior to insect or mammalian herbivory-induced damage (i. e. wounding). Indeed, roles for phenolics are pre-formed (constitutive) defense against herbivory [14, 9, 10]. The phenolics act as physical barriers present as cell wall bound phenolics, lignin suberin, and cuticle-associated phenolics as well as stored compounds that have deterring (antifeedent) or directly toxic (insecticidal) effect on herbivores [15].

Phenylpropanoid metabolism has been divided into two main stages, 'general phenylpropanoid metabolism' (i. e. biosynthesis of hydroxycinnamic acids) and 'pathway specific metabolism' (i. e. hydroxycinnamic acids as precursors to specific classes of phenylpropanoids such as flavonoids or coumarins [7]. But later, it has been found that the interconversion between hydroxycinnamic acid does not necessarily lie within a defined general pathway [16] and in fact differentiate into compound class specific pathways as early as cinnamate (or cinnamoyl-CoA), after its hydroxylation to form p-coumarate. Consequently 'general phenylpropanoid metabolism' consists of two common steps: the deamination of phenylalanine by phenylalanine ammonia lyase (PAL) and the 4-hydroxylation of cinnamate by cinnamate-4-hydroxylase (C4H) [13]. Shikmic acid pathway is a key intermediate in the synthesis of phenylpropanoids and amino acids [13] [Fig. 1].

Shikimate Pathway

The biosynthesis of most phenylpropanoid begins with amino acid phenylalanine tyrosine, and tryptophan. The first enzyme in the shikimate pathway is 3-deoxy-D-arabino-heplulosonate-7-phosphate (DAHP) synthase, which catalyses the condensation of one molecule each of phosphoenolpyruvate and erythrose-4-phosphate [13]. Both of these starting material are derived from primary metabolism. DAHP synthase has been shown to be induced, at the transcript protein, and enzymes activity levels as noted, in wounded potato tubers and tomato fruits [17].

The main classes of soluble phenolic compound involved in the response of plants to wounding and herbivory are hydrocinnamic acids (and their conjugates and derivatives) and tannins(via flavonoids)[Fig.1]. However, phenylpropanoids are derived from phenylalanine, which also has to be synthesized *de novo* as part of the overall wound/herbivory response, and it is important to consider this as an early stage of induced phenylpropanoids metabolism. Phenolic compounds generally divided into different subcompound such coumarinfurano-coumarins, lignin, isoflavonoid tannins, flavonoids, there roles in plants are described below [Fig. 1].

A. Flavonoids

Flavonoids are C_{15}compounds and form the building block of condensed tannins [13]. It is one of the largest classes of plant phenolic, perform different functions in plant system including pigmentation and defense [18]. The first key enzyme of flavonoid pathway is chalcon synthase (CHS) catalyses the first step in this pathway [19] [Fig. 1].

Isoflavonoids are derived from a flavonone intermediate, naringenin ubiquitously present in plants and play a critical role in plant developmental and defense response. The metabolic fate of naringenin is an important branch point in flavonoid biosynthesis giving rise to the flavonols (e. g. kaempferol, quercitin), dihydroflavonols, isoflavonoids (e. g. genistein) and flavones (e. g. apigenin), these all are collectively called as flavonoids [13][Fig.1].

B. Coumarins

Coumarins are simple phenolic compounds, widely distributed in vascular plants and appear to function in different capacities in various plant defense mechanism against insect herbivores and fungi. It is product of shikimic acid pathway [4] [Fig. 1]. Halogenated coumarin derivatives work very effectively *in vitro* to inhibit fungal growth. For example, 7-hydroxylated simple may play a defensive role against parasitism of *Orobanche cernua* by preventing successful germination, penetration and connection to host vascular system [20].

C. Lignin

Lignin is highly branched polymer of three simple phenolic alcohol known as monolignols, its physical toughness deter feeding by herbivorous animals and its chemical durability makes it relatively indigestible to herbivores and insects pathogens[21]. Lignifications block the growth of pathogens and are a frequent response to infection or wounding [24, 4]. But the principle function of lignin is structural, it has also been implicated as a defensive chemical. Lignin itself is not readily digested by herbivores and, because it is covalently linked to cellulose and cell wall xyloglucans, its presence decreases the digestibility of these polymers as well [23].

D. Tannins

The biological role of tannins is not clear, but tannin do appear to deter feeding by many animals when tannins free alternative are available [23]. Tannins are general toxins that significantly reduce the growth and survivorship of many herbivores and also act as feeding repellents to a great diversity of animals. They cause a sharp, astringent sensation in mouth as a result of their binding of salivary proteins. Due to this, mammalian herbivores, mammals such as cattle, dear and apes characteristically avoid plant with high tannin contents [24][Fig.1].

There are several examples of constitutive phenolics acting as feeding deterrents for herbivores and inhibitors of enzymes [25]. The evidence for the role in resistance against fungi, bacteria and nematodes is more circumstantial [26]. In this relations it appears to be the speed and duration of *de novo* biosynthesis of phenolics that is more important for resistance than the constitutive concentration.

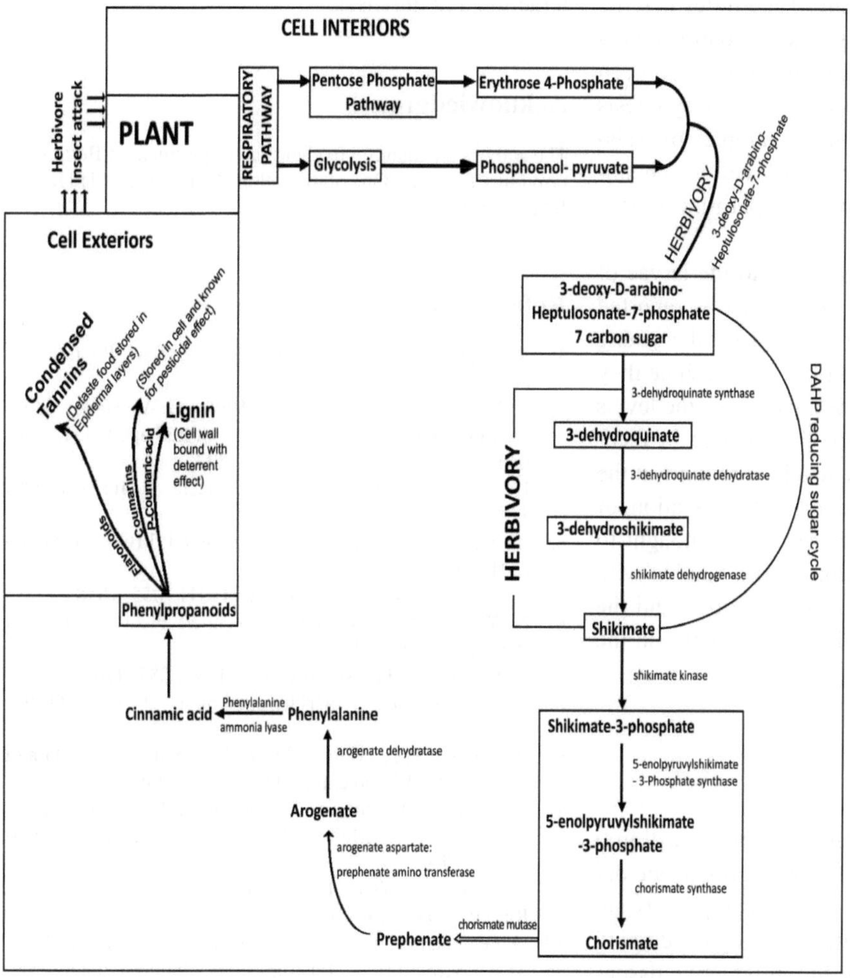

Fig. 1.: Biosynthesis, release and storage of defensive phenolics in plants attacked by herbivores. (adapted and modified in parts from Razal et al. 1996; Strack 1997, Croteau et al. 2000)

Role of Plant Phenolics and Resistance to Insects

The more complex interaction between insect pests, insect predators and phenolics in the host plant has been reported in two species of Mimosoideae [27]. Strong correlation between the constitutive concentrations of catechol based phenolics in strawberry leaves and resistance to the two spotted spider mites namely, (*Tetrancychus urticae*) [28]. A cotton phenolic pigment (gossypol) also had a deterrent effect against numerous insect pests [29]. This is strong evidence for the role of gossypol in resistance against nonspecialist insect pests, specialist feeders have developed strategies to overcome the gossypol toxicity [1]. The development of the mites on cultivars containing high concentration of phenolics was clearly suppressed, especially in cultivars with high catechol concentrations. The delayed development of the mites may be due to the phenolics covalently binding to mite digestive enzymes and inactivating them. It has also been shown that mites damage induce the *de novo* synthesis of phenolics in plants [30]. Other examples of resistance to the two spotted spider mites involving phenolics include their interaction with peppermint (monoterpenes and phenolics) [31].

The role of tannins and resistance to the larvae of the oak moth (*Opheropthera brumata*) was reported by Feeny [32]. The moth larvae feed on the young leaves of the trees in the spring but by mid-June they suddenly stop. No environmental factors or the levels of predators could account for this change in feeding. Feeny [32] found that the levels of tannins in the leaves significantly increased prior to cessation of feeding by the larvae The deterrence was thought to be because of the complexation of the tannins with the host proteins (making them indigestible) and the reaction of the tannins with digestive enzymes in the gut of the larvae.

Conclusion

The distribution of a secondary metabolite within a plant, both between tissues and during growth and development, is rarely uniform. Many compounds are synthesized by, and accumulate in, young developing tissues, particularly leaves, or in reproductive tissues such as flowers and seeds. Some secondary metabolite systems are dynamic, responding to attack, infection or stress, and that enhanced synthesis/ accumulation of secondary metabolites is part of an integrated defence mechanisms.

In this article we described the metabolism of phenylpropanoids or phenol (including coumarins, isoflavonoids, lignin and tannins in plant against herbivory. This review of induced phenylropanoids metabolism, at the transcript and/or enzyme level, is by no means exhaustive. The Figure 1 describing hydroxyl synthesis of flavonoids, coumarins and *p*-coumaric acid and condensed tannins vis-à-vis their role in plant insect interactions may serve as model.

From all of these we conclude that the study of wound and/or herbivory induced phenylpropanoid metabolism also provides the opportunity to discover novel biochemistry involved in the plant herbivore interaction and defenses.

Acknowledgment

The authors are thankful to chairman Department of Botany for providing facilities. One of the authors (F. R.) acknowledges the fellowship from UGC.

References

1. R.N. Bennett and R.M. Wallsgrove; New Phytol. 127 (1994) 617–633.
2. E. Wong. Plant phenolics. In: Butler G.W., Bailey R.W. (eds.) Chemsitry and biochemistry of herbage vol. 1, London: Academic Press (1973) 265–322.
3. H. Schafer and M. Wink; Biotech. J 4(12) (2009) 1684–1703.
4. M. Mazid, T.A. Khan and F. Mohammad; J Biol and Med 3 (2011) 232–249.
5. G.A. Rosenthal; Phytochem. 30 (1991) 1055–1058.
6. N. Wuyts, D. Waela and R. Swenner; Plant Physiol and Phytochem 44 (2006) 308–314.
7. D. Strack; Acad press New York (1997) 387–416.
8. C.P. Constable; AmerPhytopathSoct St Paul (1999) 137–166.
9. R. Santigo, R.A. Malvar, M.D. Baamonde, P. Revilla and X.C. Souto; J Econ entomol 98 (2005) 1349–1356.
10. R. Santigo, A. Butron, J.T. Arnason, L.M. Reid, X.C. Souto and R.A. Malvar; J Agric Food Chem 54 (2006) 2274–2279.
11. E.A. Bernays, G.C. Driver and M. Bilgener; Adv. Ecol. Res. 19 (1989) 263–302.
12. D. Peters and C.P. Constabel; Plant J 32 (2002) 701–712.
13. M.A. Bernards and Bastrup-Spohr; Induced plant resistance to Herbivory. 9thEds Schaller A Springer Stuttgart Germany (2008) 189–211.

14. H. Ding, R.L. Lamp and N. Ames; J ChemEcol 26 (2000) 969–984.
15. L.L. Walling; J plant Growth Regul 19 (2000) 195–216.
16. R.A. Dixon, F. Chen, D. Guo and K Parnathi; Phytochem 57 (2001) 1069–1084.
17. W.E. Dyler, J.M. Henstrand, A.K. Handa and K.M. Herrmann; ProcNatlAcadSciUSa 86 (1989) 7370–7373.
18. T. Kondo, K. Yoshida, A. Nukagawa, T. Kawai, H. Tamura and T. Goto; Nature 358 (1992) 515–518.
19. J.A. Lake, K.J. Field, M.P. Davey, D.J. Beerling and B.H. Lomax; Plant Cell and Environment 32 (2009) 1377–1389.
20. K. Serghini, A De Lugue Perez, M.M. Castejon, T.L. Garcia and J.V. Jorrin; J Exp Bot 52 (2001) 227–234.
21. M./Madar and F.V. Amberg; Plant Physiol 70 (1982) 1128–1131.
22. J.M. Gould; Physiol 14 (1983) 25–91.
23. Secondary metabolites in Introduction to Plant physiology, 4thedition., Eds. W.G. Hokins and N.P.A. Huner USA (2009) 459–479.
24. J.F. Oates, P.G. Waterman and G.M. Choo; Oecologia 45 (1980) 45–56.
25. P.R. Cheeks, Vol IV. Phenolics Boca Raton CRC Press 1989.
26. J.B. Harborne; Introduciton to Ecological Chemistry. London: Acad Press (1988).
27. S. Koptur; Ecology 66 (1985) 1639–1650.
28. A. Luczynski, M.B. Isman and D.A. Raworth; J Eco Ento. 83 (1990) 557–563.
29. M.B. Abou-Donia; Vol. IVthPhenolics Boca Raton: CRC Press (1989) 2–22.
30. M. Inoe, S. Sezaki, T. Sarin and T. Soquira; Applied Entol and Zol 20 (1985) 348–349.
31. P.O. Larsen. The biochemistry of plants, vol. 7, Ed.s P.K. Stumpf, E.E. Conn Acad Press New York (1981) 501–525.
32. P.P. Feeny; Ecology 51 (1970) 565–581.
33. R.A. Razal, S. Ellis, S. Singh, N.G. Lewis, G.H.N. Tower; phytochem 41(1996) 31–36.
34. Biochemistry and molecular biology of plants, Eds. R. Croteau, T.M. Kutchan, N.G. Lewis; Amercn soc plant biolgst Rochville;(2000)1250–1318.

66
Water and Wastewater Treatment using Nano-technology

N. A. Khan [1], K. A. Khan[2] and M. Islam[1]
[1]Department of Civil Engineering AMU, Aligarh
[2]Girls polytechnic AMU, Aligarh,
Email: er.nadim@gmail.com

Abstract

The growth history of usage of reversible energy (from 1973) and development of nanotechnology (from 1994) shows that world's society searching for usage of modern technology in practical development of reversible energy in great amount in global village. So the third world countries and development countries according to great profit it of free reversible energy they must use those sources when fossil sources has being finished (for independence in providing of energy. Today nanoparticles, nanomembrane and nanopowder used for detection and removal of chemical and biological substances include metals (e. g. Cadmium, copper, lead, mercury, nickel, zinc), nutrients (e. g. Phosphate, ammonia, nitrate and nitrite), cyanide, organics, algae (e. g. cyanobacterial toxins) viruses, bacteria, parasites and antibiotics. Basically four classes of nanoscale materials that are being evaluated as functional materials for water purification e. g. metal-containing nanoparticles, carbonaceous nanomaterials, zeolites and dendrimers. Carbon nanotubes and nanofibers also show some positive result. Nanomaterials reveal good result than other techniques used in water treatment because of its high surface area (surface/volume ratio)

Introduction

Usage of Nano in Building of Dam with the Purpose of Energy divided in four groups in the world that are countries of nano materials production, they go towards pollutant energy and in reversible, Pollutant energy and the production of practical thing in dam's part. Following reversible, Energy whit out pollutant and in reversible items is related to this: and finally Energy whit out pollutant and unlimited Occurring of three factors in 1995, because a reference C Unlimited age of dam. Point of water reversible energy that is followings:

- Weather changing because of the gathering of green dam). Houses gases in atmosphere.
- Extreme sensitivity with different usage
- Open perspective about reversible energy.

In 2005, world consuming light energy of water reaches to 2627 tetra watt. In this year, 21.8% North America, 23.9% Europe, 21.7% of Asian America country20% southern and Central America, 5.7% previous belongs consuming of light energy of water to themselves. Among the countries of world, the most using, belong to Canada, Brazil, china, America with the order, 12.6%, 10.3%, 9.8%. 8.1%.

Through production and use of nanotechnology-based products, disposal of engineered nanomaterials (eNMs) into sewage is currently occurring [1]. Concentrations of eNMs detected at municipal wastewater treatment plants (WWTPs) will inevitably increase as the nanotechnology industry grows. WWTPs are important sources of contaminant discharge into the environment through treated effluent, biosolids, and plant-generated aerosols. Therefore, quantification and characterization of the release of eNMs from WWTPs will be an important contribution to understanding eNM environmental fate and exposure. Nanoscale titanium dioxide (TiO_2), silver (Ag), and fullerenes (C60) are currently the most-produced eNMs and found in a wide range of commercial products [2]. Recent eNM exposure modeling indicated that of nano-TiO_2, nanosilver, and carbon nanotubes, only predicted concentrations of nano-TiO_2in WWTP effluents (0.7–16 μg/L) were higher than the predicted no-effect concentration level (1 μg/L) [3]. Furthermore, TiO_2has been used in consumer products for decades. Thus, titanium (Ti) may serve as an indicator of the fate of nanomaterials passing through a WWTP.

M.M. Srivastava, L. D. Khemani, S. Srivastava, *Chemistry of Phytopotentials: Health, Energy and Environmental Perspectives*, DOI:10.1007/978-3-642-23394-4_66, © Springer-Verlag Berlin Heidelberg 2012

However, different types of eNMs, such as nanosilver or fullerenes, have different properties, so their release from WWTPs may differ from that of TiO2. The objectives of this study were to:

(1) measure current Ti concentrations in a full-scale municipal WWTP;
(2) quantify Ti concentrations in lab-scale reactors and compare to full-scale concentrations, and
(3) compare degrees of sorption of nanoscale TiO2, Ag, and C60to wastewater biomass.

They have balance in different parts contrast to micro silica is fast and very little amount like nano metal oxides and nano non metal particles of this materials have the same effect of pozzolanic (nano silica, nano zirconium) [1–10].

High Performance Concrete and Multiple Purposes

The new solutions for solving the problem of controlling (leashing) water behind the waterpower house, is using of high performance concrete. This kind the corrosion, keeping among the natural destroying of is a kind of composite material and for solidity factors. Concludes of composite material and it has multi phase and comply. The feature of the concrete depends on nano structure of the concrete that makes a unity, tenacity. Development of high performance concrete for construction of dams with the purpose of waterpower house almost contains of different parameters like mixing usual concrete with concrete of different ravel. This concrete has more features that mechanical feature, like; electro magnetic feature and it can be used in atomic and waiting until it dries with neat. We see that the structure (maintenance the ray) of that for keeping the energy system [8].

Materials and Methods and Discussion

Samples from each process unit and finished biosolids were collected from a conventional activated sludge wastewater reclamation facility in Aligarh civil engineering department brought from kashipur Nainital. Ti concentrations in both filtered (0.7-μm glass-fiber filter) and non-filtered samples were measured. Nano-TiO2in filtered tertiary effluent was imaged by scanning electron microscopy (SEM). Sequencing batch reactors (SBRs) of heterotrophic bacteria were set up to represent WWTP aeration and settling (10-hour hydraulic residence time, 6-day solids retention time).The SBRs were dose with nano-TiO2, and completely mixed and settled supernatant solutions were removed once per day and analyzed for Ti concentrations. Batch sorption experiments were conducted to compare sorption of TiO2, Ag, functionalized Ag (f-Ag), fullerene (aq-nC60), and functionalized C60 (nC60(OH)24) nanoparticles to rinsed activated sludge (wastewater biomass). Metal and carbon concentrations were determined by inductively-coupled plasma optical emission spectroscopy and total organic carbon measurement, respectively. Nano – Silica – Amor of in concrete industry, silica is the (HPC), that product is silicofiom or micro-silica that it has 0.1 – 1 mm diameter and it has about 90 % of oxide silica we can say that micro-silica is a product that is been used for feature of the composite.

Nanotechnology is regarded as one of the key technologies of the future and associated with high expectations by politics, science and economy. On the basis of the definition by the Office of Technology Assessment at the German Parliament (TAB), the term of nanotechnology is referred to by the Federal Environment Agency (UBA) as the manufacturing, analysis and use of structures – for example particles, layers or tubes – of less than 100 nanometers (nm) in at least one dimension. Artificially produced nano-sized particles and nanoscale system components have new properties which are of importance for the development of new products and applications. Such new properties of materials and substances result from the special properties of surfaces and interfaces and in part, from the geometric shape of the material. Based on the available literature data, it has been assumed by the UBA that in the decades to come, nanotechnology will have a strong influence on essential industries such as the automotive, chemical and pharmaceutical industries as well as mechanical engineering, medicine, biotechnology and environmental engineering, and that it has a potential for fundamentally changing whole fields of technology. In 2006, some 550 companies have been active in the field of nanotechnology in Germany, employing a total of ca. 50 000 persons. Industry is expecting great market potentials reaching up to one trillion US dollars worldwide in 2015.

In the opinion of a great number of experts, nanotechnology has a positive potential not only for economic development. Considerable improvement is also expected with regard to the protection of the environment and human health. Thus, nanotechnology development may increase the efficiency of resources and improve the overall performance of environmental protection.

However, despite the vehement development of nanotechnology in recent years and an increasing number of products made by means of nanotechnology, knowledge about the exposure of humans and the environment to nano-sized particles has been very scarce so far. Questions arising as to the implications of the exposure to nanoparticles for humans environment have not yet been sufficiently elucidated. Due to the novel properties of nanoparticles, the technological development should therefore be accompanied by a corresponding risk assessment in order to identify and subsequently avoid potential damage and cost caused by the new technology, a procedure that has meanwhile become common for any new technology. In particular, the Federal Ministry of Education and Research (BMBF) and the European Commission have responded to this in recent years by supporting a number of research projects. Examples of action taken include the "Innovation and Technology Analysis on Nanotechnology" (2002 – 2004) and the Initiative, "NanoCare" (2006 – 2008), both conducted by the BMBF, and the 6th and 7th Framework Programmes for Research, "NanoSafe 1" (2003 – 2004) and "NanoSafe 2" (since 2005) conducted by the European Union.

The UBA has been involved in this discussion (see Annex: Activities of the Federal Environment Agency). The discussion about the opportunities and risks of nanotechnology should take place in an unemotional and objective atmosphere beyond technoscepticism and technomania.

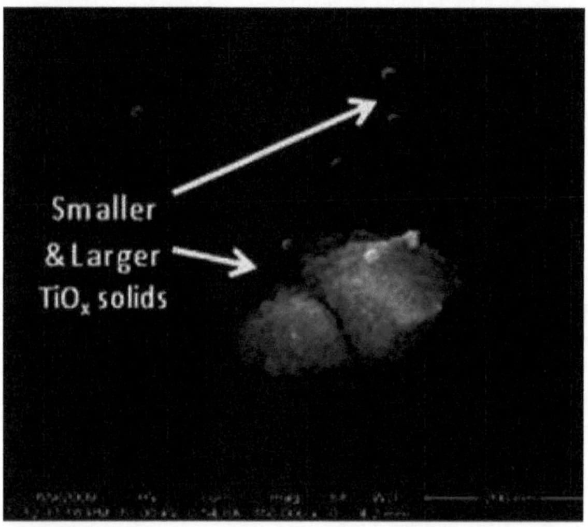

Figure 1: Titanium concentrations of arizona Wwtp nonfiltered and filtered headworks and process effluent samples; (b) SEM analysis of nanoscale TiO_x in Wwtp tertiary effluent; scale bar is 200 nm

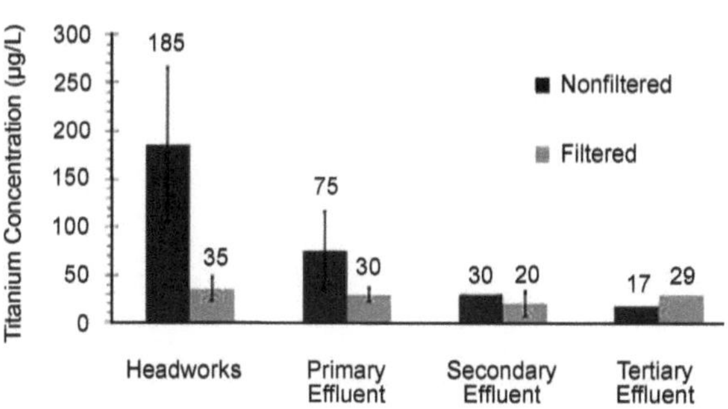

Figure 2: Titanium concentrations vs. treatment process

Conclusion

According to survey and that technology and work are going forward with each other, we understand that developing counties especially third word countries for freeing of problems of providing energy they should use nano products in this part in order to find complete information should these items:

1. Potential of each of restoration energy.
2. Defining and choosing of proper area (finding site)
3. View for the future of nanotechnology in restoration energy.
4. Economical justification use of nanotechnology in part of energy according to various factors. C Planning, stocking volume with choosing of each energy.
5. Ranked plan for developing of modern technology in relation of restoration energy.
6. Volume of replace ability and to answer.

This study defines current environmentally-relevant concentrations for examining nano-TiO2toxicity.

WWTP biosolids represent a potentially important source of TiO2release into the environment.TiO2may serve as an indicator or model of the environmental fate of a widely-used eNM.Lab-scale experimental data were valuable for substantiating Ti removal observed in the full-scale WWTP. Nonfunctionalized eNMs are more effectively removed from wastewater than functionalized eNMs, which may persist to a greater degree in effluent.

References

1. Council, Rio De Janerio, Energy in Brazil. Brazilian National Committee of the World Energy arming State of the world. Wordwatch Institute, CFlavin Slowing Global, York, USA (1990).
2. Global Energy Perspectives, World Energy Conference, Canada (2005) 2010–2030.
3. D. Bowman, and G. Hodge; Columbia Science Technology Law Rev. 8 (2007) 1–32.
4. Attitudes toward nanotechnology and federal regulatory agencies.
5. L. Paz, P. Comba, F. Bianchi, L. Martina, M. Menegozzo, F. Mitis, L. Pizzuti, R. Santoro, M. Trinca and M. Martuzzi, Conservation International nano product, Annals of the New York Academy of Sci., 1076 (2006) 449–461.
6. A. Maynard, and E. Michelson; Nanotech Project, 29 (2006) 5- 27.
7. B. S. Federici, and R. Handy; Aquatic Toxicol. 84 (2007) 415–430.
8. Bi, X., G. Jones, K. Qu, W. Sheng, G. Martin and J. Fu; Sci. Technol. 41 (2007) 5647–5653.
9. W. Bastos, J. Gomes, R. Oliveira, R. Almeida, E. Nascimento, J. Bernardi, D. Lacerda, L. Da Silveira and W. Pfeiffer; Sci. Total Environ. 368 (2006) 344–351.
10. D. Bowman, and G. Hodge; Columbia Sci. Technol. Law Rev. 8 (2007) 1–32.
11. T. M. Benn, P. K. Westerhoff; Environ. Sci. Tech. 42 (2008) 4133–4139.
12. Woodrow Wilson International Center for Scholars. 2009.
13. B. Nowack, T. D. Bucheli. Occurrence, behavior, and effects of nanoparticles in the environment. Environ. Pollution 150 (2007) 5–22.

67
Role of Plants in Removing Indoor Air Pollutants

A. S. Pipal, A. Kumar, R. Jan and A. Taneja*
Department of Chemistry, Dr. B. R. Ambedkar University, Agra, India
*Email: atanejas5@hotmail.com, aspippal@gmail.com

Abstract

The quality of the indoor environment has become a major health consideration, since people spend 80–90% of their time indoors. Research by a number of authors is reviewed here, demonstrating capacities of indoor plants to improve Indoor air quality. The studies show that plants can reduce indoor air pollutants by 75% in different conditions. An evaluation of these studies done throughout the world clearly indicates that potted-plants can provide an efficient, self-regulating, low-cost, sustainable, bioremediation system for indoor air pollution, which can effectively compliment engineering measures to reduce indoor air pollution, and hence improve human wellbeing and productivity.

Introduction

The quality of the indoor environment has become a major health consideration in the developed world, since urban-dwellers generally spend 80–90% of their time indoors [1–4]. Indoor air can often contain 5 to 7 times the contaminant concentrations of outdoor air [5–6]. The harmful effects of these mixtures have been recognized as components of 'sick building syndrome' or 'building-related' [7–8], with symptoms of headache, dizziness, nausea, sore eyes and throat, or loss of concentration. A number of studies have shown that potted-plants have a capacity to contribute to the improvement of indoor air quality, by reducing air-borne contaminants such as VOCs, nitrogen oxides and dust [9–12], as well as by aiding humidity, temperature and noise control [13]. It has also been shown that staff wellbeing (as measured by questionnaire surveys and interviews) and productivity (as reductions in sick-leave rates) is improved where indoor plants have been installed [14–15]. The aim of this paper is to provide a review of research on indoor plants to improve indoor air quality, to outline findings, and to present new data that further demonstrate the ability of potted-plants to remove indoor air pollutants and enhance IAQ.

Potted-plants Improve a Number of Aspects of IAQ

The studies undertaken to demonstrate the role of plants in controlling IAQ will be discussed in detail like Yoneyama et al. 2002 [16] reviewed absorption and metabolism of NO_2 and NH_3 in 220 species (sun- and shade-loving plants, the latter of which can be used indoors). In a UK homes with flueless gas appliances, Coward et al. 1996[12] found that houses with six or more potted-plants showed reductions of over one third in NO_2 levels. In 1999, Lee and Sim [17] in a study in Korea showed that indoor plants absorb and metabolize SO_2. In the USA, Lohr and Pearson-Mims 1996 [11] showed that indoor plants significantly reduce dust (particulate) levels. Costa and James 1999 [13] found that potted-plants also reduce indoor noise levels.

Plants as Decontaminants

Table 1 shows some plants which play an important role in removing indoor air pollutants, literature also reveals that staff wellbeing is improved with sick leave absences reduced over 60% [14–15]. An ability of foliage plants to absorb chemical compounds from the indoor air has been demonstrated in many studies. In the report of the National Aeronautics and Space Administration (NASA) Wolverton et al. 2005 [18]

M.M. Srivastava, L. D. Khemani, S. Srivastava, *Chemistry of Phytopotentials: Health, Energy and Environmental Perspectives*, DOI:10.1007/978–3-642–23394-4_67, © Springer-Verlag Berlin Heidelberg 2012

showed that low-light-requiring houseplants such as *Bamboo palm*, *English ivy* have the potential for improving indoor air quality by removing trace organic pollutants from the air in energy-efficient buildings. The removal of airborne toluene by means of the phyllosphere of *Azalea indica* augmented with a toluene-degrading enrichment culture of *Pseudomonas putida* TVA8 was studied by Kempeneer et al. 2004 [19]. This technique is promising and could be practical implemented in the field of indoor air pollution control.

Table: 1 Air pollutant removal by some house plants

House Plant	Chemical Pollutant	Initial ppm	% Removed
English Ivy	Benzene	0.235	90%
	Trichlorethylene	0.174	11%
Peace Lily	Benzene	0.166	80%
	Formaldehyde	10.0	50%
	Trichlorethylene	20.0	50%
Spider Plant	Formaldehyde	14.0	86%
	Carbon Monoxide	128.0	96%
Chrysanthemum	Benzene	58.0	54%
	Formaldehyde	18.0	61%
	Trichlorethylene	17.0	41%
Mother-in-law tongue	Benzene	0.156	53%
	Trichlorethylene	0.269	13%
Golden Pathos	Benzene	0.156	53%
	Formaldehyde	18.0	67%
	Carbon Monoxide	113.0	75%
Madag Dragon Tree	Benzene	0.176	79%
	Formaldehyde	15.0	60%
	Trichlorethylene	0.136	13%
Waneckii	Benzene	0.182	70%
	Formaldehyde	8.0	50%
	Trichlorethylene	17.0	24%
Heart Leaf	Formaldehyde	27.0	71%
Corn Plant	Formaldehyde	20.0	70%
Chinese Evergreen	Benzene	0.204	48%

Conclusions

The numerous studies reported here, and a number of different sources around the world, show conclusively that the potted-plant microcosm (PPM) can greatly improve IAQ by removing many major pollutants. Thus the PPM represents an adaptive, self-regulating, portable, flexible, low-cost, sustainable and beautiful biofiltration and bioremediation system for IAQ. This innovative technology can complement any engineering measures and can be used in any building.

Indoor potted-plants can remove air-borne contaminants such as volatile organic compounds (VOCs), CO, CO_2, Benzene, formaldehyde, trichloroethylene, NOx over 300 of which have been identified in indoor air. This paper reviews the capacity of the potted-plant microcosm to contribute to cleaner indoor air, and lay the foundation for the development of the plant/substrate system as a complementary biofiltration system. To ensure sustainability of the urban environment, satisfying the ‘triple bottom line’ of environmental, social and economic considerations, it is expected that indoor plants will become standard technology – a vital building installation element, for improving IAQ.

References

1. G. Abbritti, and G. Muzi; *Proceedings of Healthy Buildings, 95 an International Conference on Healthy Buildings in Mild Climate,* University of Milano and International Centre for Pesticide Safety, Milano, Italy, (1995). pp. 185–195
2. M. Krzyanowski; *Proceedings of Indoor Air 99, the 8th International Conference on Indoor Air Quality and Climate*, Edinburgh, Scotland, (1999). pp. 230–232
3. D.O. Carpenter; Environ. Monit. Assess. 53 (1998) 245–258
4. American Lung Association; Air Quality, Australia (2001) pp. 10–13
5. A. Taneja, R. Saini and A. Masih; Ann. N.Y. Acod. Sci. 1140 (2008) 228–245
6. A. Kulshrestha, D. S. Bisth, J. Masih, D. Massey, S. Tiwari and A. Taneja; J. Atmos. Chem 62(2009) 121–138
7. D. Massey, J Masih, a. Kulshrestha, M. Habil, A. Taneja; Building and Environment, 44 (2009) 2037–2045
8. P. Carrer, D. Alcini, D. Cavallo, F. Visigalli, D. Bollini, and M. Maroni; *Proceedings of Indoor Air 99, The 8th International Conference on Indoor Air Quality and Climate*, Edinburgh, Scotland, August, (1999) 129–134

9. B.C. Wolverton, and J.D. Wolverton; *J. Mississippi Acad. Sci.* 38 (2) (1993) 11–15
10. M. Giese, U. Bauer-Doranth, C. Langebartels, and J.H. Sandermann; *Plant Physiol.* 104 (1994) 1301–1309
11. V.I. Lohr, and C.H. Pearson-Mims; *Atmos. Environ.* 30 (14) (1996) 2565–2568
12. M. Coward, D. Ross, S. Coward, S. Cayless, and G. Raw; Building Research Establishment Note N154/96, Watford, UK (1996)
13. P. Costa, R.W. James; In Proceedings of Indoor Air 99, Edinburgh, International Conference on Indoor Air Quality and Climate, Vol. 3, (1999) 234–239
14. J. Bergs; Proceedings of International Plants for People Symposium, Floriade, Amsterdam, NL, June, Flower Council of Holland, NL (2002)
15. T. Fjeld; In *Proceedings of International Plants for People Symposium*, Floriade, Amsterdam, NL, June, Flower Council of Holland, NL (2002)
16. T. Yoneyama; Prospects for Phytomonitoring and Phytoremediation, Springer, Tokyo, Japan (2002) 221–234
17. H. Lee and W.K. Sim; Contributions from International People-Plant Symposium", Sydney (1999) 101–108
18. B.C. Walveston, A.Johnson and K.Bounds, Final Report, NASA Stennis Space Centre, MS.USA (1989)
19. L.D. Kempeneer, B. Sercu, W. Vanbrabant, V. Langenhove; Appl Microbial Biotechnol. 64 (2004) 284–288

68

Decolorization and Mineralization of Commercial Textile Dye Acid Red 18 by Photo-Fenton Reagent and Study of Effect of Homogeneous Catalyst Uranyl Acetate

M. Surana[1] and B. V. Kabra[2]

[1]Department of Chemistry, Faculty of Science, Mewar University, Chittorgarh (Raj.),

[2] Department of Chemistry, Faculty of Science, M. L. V.Govt. College, Bhilwara (Raj.) – 311001

Email: monasurana2711@yahoo.com

Abstract

Degradation of commercial textile dye ACID RED 18 (AR18) by Photo-Fenton reagent has been investigated under irradiation of visible light in aqueous medium. The degradation process was carried out by hydroxyl free radicals which were the most reactive chemical species in the whole process. The influence of different experimental parameters such as concentration of Dye (AR18), photocatalyst (Fenton reagent), H_2O_2 and pH of the experimental solutions on the initial rate and photodegradation of the dye were assessed and optimized. Complete mineralization of dye AR18 (~100%) was achieved by Photo-Fenton reagent. The photobleaching process follows first order kinetics. A simple, efficient and environmental friendly method for the photodegradation of dye is developed.

Introduction

In the last two-three decades, azo dyes are produced and consumed annually in large quantities in textile industries. Most of the azo dyes are difficult to be degraded due to their complex structure. Some of them are toxic and suspected carcinogens. The untreated dyes in effluent from dyeing factories and leather industries are a group of hazardous chemicals as well as major sources of water pollution [1, 2]. The color in the waste water is an obvious indicator of water pollution due to dyes and pigments [3, 4]. Also the dyes in water affect photosynthetic activity in environment due to reduced light penetration and may be toxic to some aquatic lives [5]. The environmental concern of these potentially carcinogenic pollutants in contaminated water has drawn the attention of many research workers.

Currently, the major methods of textile wastewater treatment are advanced oxidation processes which are widely used are UV irradiation, H_2O_2/ UV [6], TiO_2/ UV [7], Ozonation [8], Fenton reagent [9] or UV-Fenton [10] process etc. Fenton reagent exhibits a number of features that make its use advantageous as compared to other methods such as high degree of catalytic efficiency, high degree of specificity, absence of side reactions and cost effectiveness. Experimental observations indicate that it can be used to degrade chemicals and dyes [11, 12].

For the present investigation, water soluble disazo dye Acid Red 18 is selected. AR18 is toxic and carcinogenic in nature. Thus it is considered worthwhile to see how textile dye AR18 could be degraded in aqueous medium by photo-Fenton reagent.

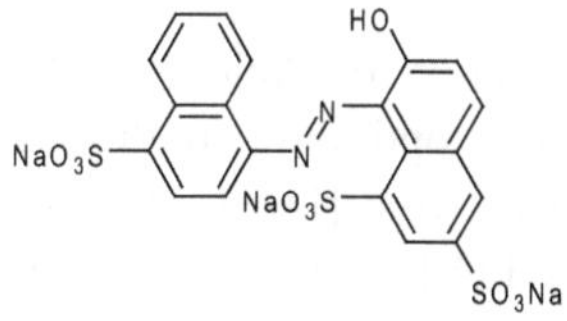

Materials

For the present studies the commercial disazo dye Acid Red 18 (C.I No. 16255, Colortex). Photocatalyst $FeSO_4$ (Merck, 99% purity) and H_2O_2 (Merck, 6% purity) were used for photocatalytic degradation. 1×10^{-3}M (0.604 g/l) stock solution of dye was prepared in double distilled water and diluted as required. The desired pH of the solution was adjusted by the

M.M. Srivastava, L.D. Khemani, S. Srivastava, *Chemistry of Phytopotentials: Health, Energy and Environmental Perspectives*, DOI:10.1007/978–3-642–23394-4_68, © Springer-Verlag Berlin Heidelberg 2012

addition of previously standardized sulfuric acid solutions. All laboratory reagents were of analytical grade.

Procedure and Analysis

The reaction mixture was prepared by taking 3 ml of stock solution of dye (1×10^{-3} M), 3.5 ml of $FeSO_4$ (1×10^{-3} M), 1.6 ml of H_2O_2 (6%) in a round bottom flask. The total volume of the reaction mixture was made 100 ml by adding double distilled water. The concentration of different ingredients in the reaction mixture was [Dye] $=3 \times 10^{-5}$ M, [$FeSO_4$] $=3.5 \times 10^{-5}$ M and [H_2O_2] $=3.1 \times 10^{-2}$ M.

To carry out the photobleaching, the reaction mixture was irradiated under light source (2 × 200 W Tungsten lamps) with stirring. A water filter was used to cut off thermal radiations. The pH of the solution was adjusted by adding acid solution. The progress of the reaction was observed at definite time intervals by measuring absorbance using spectrophotometer (Schimadzu, UV- 1700) at 506 nm. The rate of decrease of absorbance with time was continuously monitored. After complete mineralization, the presence of NO_2^-, NO_3^-, SO_4^{2-} ions and evolution of CO_2 were tested by standard procedure.

Results and Discussion

Control experiments (in absence of photocatalyst, light) confirm the necessity of photocatalyst, light to follow the photocatalytic path for the photobleaching of dye.

The photocatalytic degradation of Acid Red 18 was observed at 506 nm. The optimum conditions for the photobleaching of dye were [Dye] $=3 \times 10^{-5}$ M, [$FeSO_4$] $=3.5 \times 10^{-5}$ M, [H_2O_2] $=3.1 \times 10^{-2}$ M and pH=3.5. The results are presented graphically in Fig-1.

It was observed that absorbance decreases with time, indicating that the dye is degraded on irradiation. A graph between *2+log Abs* and time has been plotted. The linearity of the plot indicates that the photocatalytic bleaching follows a first order kinetics. The rate constant of this photobleaching process was determined using the expression.

Rate (k) = 2.303 x Slope = 10.47×10^{-5} sec^{-1}

The effect of variation in various reaction parameters has been studied, e. g. pH, concentration of the dye, $FeSO_4$ and H_2O_2.

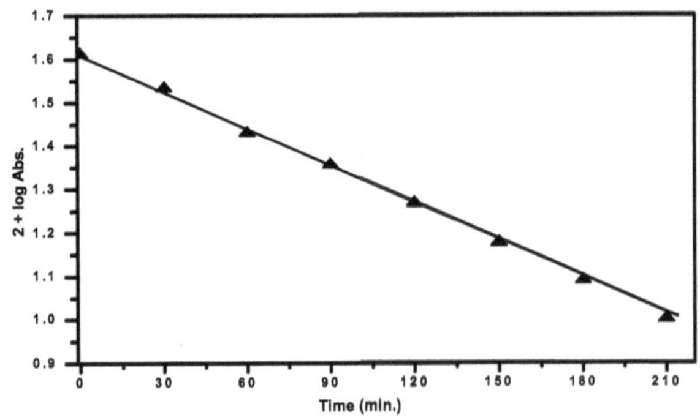

Fig. 1: A plot showing a typical run of photocatalytic degradation of Acid Red 18 observed at 506 nm under optimum conditions

Effect of Variation in pH

The effect of pH on the rate of photocatalytic bleaching of dye was observed. The photodegradation was performed at different pH (1.5–5). The results (Fig-2) reveals that the rates of photobleaching of dye increases with an increase in pH up to 3.5, after which it decreases with increasing pH. At pH >3.5, Fe^{2+} gets converted into Fe^{3+} which decomposes H_2O_2 into water and oxygen, instead of forming ·OH radicals. Thus, all subsequent experiments were carried out at pH 3.5.

Effect of Variation in Dye Concentration

The effect of [Dye] on the degradation of Acid Red 18 was studied at different concentrations varying from 1.0×10^{-5} M to 8.0×10^{-5} M keeping all other factors identical.

The result (Fig-3) reveals that the rate of photobleaching of dye decreases with the increase in the concentration of dye. It can be explained on the basis

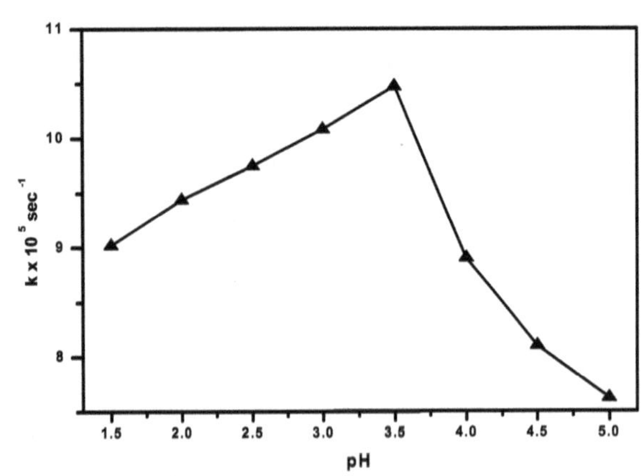

Fig. 2: A plot showing effect of pH on photocatalytic degradation of Acid Red 18 by Fenton reagent

of the observations that as the concentration of the dye is increased, the dye itself may act as a filter for the incident light, preventing sufficient intensity of light from reaching the dye molecule in the solution.

Effect of Variation in Catalyst Concentration

Keeping all other factors identical, the concentration of catalyst was changed and its effect on the rate of photochemical degradation was observed. The result of Fig-4 reveals that the rate of photobleaching of dye increases with the increase in the concentration of catalyst up to 3.5×10^{-5} M. The increase in ferrous ions in the reaction mixture is accompanied by enhanced generation of ·OH radicals, consequently increasing the rate of photodegradation. After the optimal Fe^{+2} addition, the higher dose of Fe^{+2} resulted in a brown turbidity that causes the recombination of ·OH radicals and Fe^{+2} reacts with ·OH as a scavenger. Therefore, on further increase, the rate becomes almost constant.

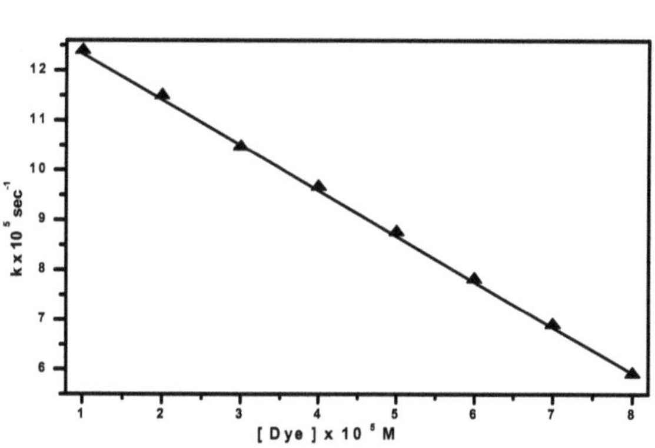

Fig. 3: A plot showing effect of [Dye] on photocatalytic degradation of Acid Red 18 by Fenton reagent

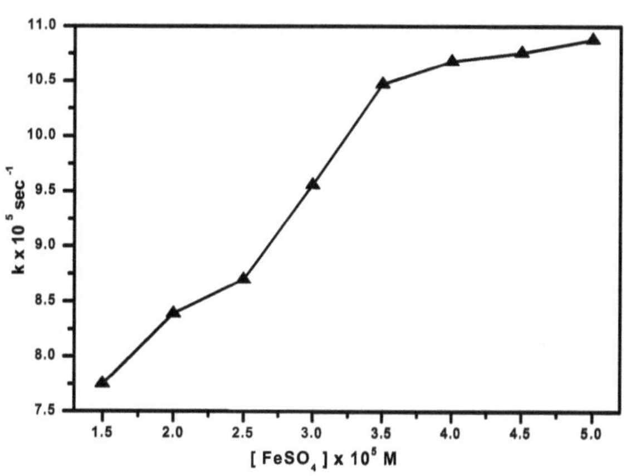

Fig. 4: A plot showing effect of catalyst concentration on photocatalytic degradation of dye by Fenton reagent

Effect of H_2O_2 Concentration

Keeping all other factors constant, the concentration of H_2O_2 was changed and its effect on the rate of photobleaching was studied. The result reported in Fig-5 reveals that the rate of photobleaching of dye increases with the increase in the concentration of H_2O_2 up to 3.10×10^{-2} M. Further increase in concentration has negligible effect as the reagent itself reacts with ·OH radicals to produce $^{\cdot}O_2H$ which has much lower oxidation capacities.

Mechanism

The mechanism of photo-Fenton oxidation is based on the generation of ·OH radicals by the catalytic decomposition of H_2O_2 in acidic media [13]. In presence of Fe^{2+}, the peroxide breaks down to ·OH and OH^-, according to the following reactions (Eqn. 1, 2, 3).

$$Fe^{2+} + H_2O_2 \rightarrow Fe^{3+} + {}^{\cdot}OH + OH^- \quad (1)$$

$$Fe^{3+} + H_2O \xrightarrow{h\nu} Fe^{2+} + {}^{\cdot}OOH + H^+ \quad (2)$$

$$2\,H_2O_2 \rightarrow H_2O + {}^{\cdot}OH + {}^{\cdot}O_2H \quad (3)$$

The incorporation of ·OH radicals with H_2O_2 also produces ·OOH radicals (Eqn. 4).

$${}^{\cdot}OH + H_2O_2 \rightarrow {}^{\cdot}O_2H + H_2O \quad (4)$$

Ferrous ions will undergo oxidation to ferric ions by the addition of ·OH radicals (Eqn. 5), while ferric ions

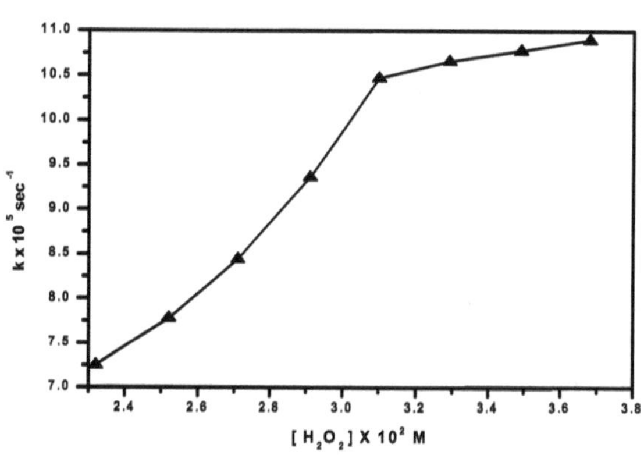

Fig. 5: A plot showing effect of $[H_2O_2]$ on photocatalytic degradation of Acid Red 18 by Fenton reagent

are reduced to ferrous ions by the incorporation of $^{\cdot}OOH$ radicals producing H^+ ions (Eqn. 6).

$$Fe^{2+} + {}^{\cdot}OH \rightarrow Fe^{3+} + OH^- \quad (5)$$

$$Fe^{3+} + {}^{\cdot}O_2H \rightarrow Fe^{2+} + O_2 + H^+ \quad (6)$$

$^{\cdot}OOH$ radicals are highly unstable in water and undergo facile disproportionation rather than reacting slowly with the dye molecules. The participation of the $^{\cdot}OH$ radical as an active oxidizing species was confirmed. The hydroxyl radical scavengers like 2-Propanol or t-Butyl Alcohol were used, which drastically reduced rate of degradation.

The hydroxyl radical (oxidation potential 2.8 eV) attacks dye by either abstracting a hydrogen atom or adding itself to double bonds. Initially the hydroxyl radicals attack on the azo linkage of the dye molecule and break down to produce Sodium 4-nitronaphthalene-1-sulfonate (A) and Disodium 7-hydroxy-8-nitronaphthalene-1, 3-disulfonate (B). These further undergo continuous attack of $^{\cdot}OH$ radicals and oxygen to produce toluene and naphthalene derivatives and finally oxidize to CO_2 After continuous irradiation, the complete mineralization of dye occurred via converting into CO_2, H_2O, NO_2^-, NO_3^-, SO_4^{2-} .

The end products are simple molecules and harmless to the environment (Eqn.7).

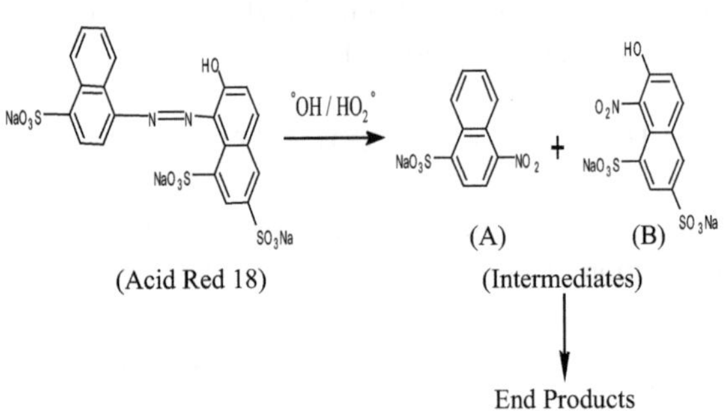

$$(CO_2 + H^+ + H_2O + NO_2^- + NO_3^- + SO_4^{2-}) \quad (7)$$

The end products were detected and their presence in the reaction mixture was ascertained by chemical physical method.

Nitrate ions were detected by nitrate ion selective electrode having a solid-state PVC polymer membrane. Nitrite ions were confirmed by using H_2SO_4 and $FeSO_4$ (a dark brown solution, arising from the iron nitric oxide complex). Sulfate ions were detected by gravimetric analysis (barium chloride solution). CO_2 was confirmed by introducing the gas to freshly prepared limewater.

Effect of Addition of Homogeneous Catalyst Uranyl Acetate

The effect of addition of homogeneous photocatalyst U (VI) on photodegradation has been investigated. A search was made to suggest the combination of ferrous ions by some other active metals, which have special oxygen transfer properties to generate more highly reactive hydroxyl radicals.

So the effect of UO_2^{2+} (3.5×10^{-5} M) with H_2O_2 was studied (Fig.6) which shown appreciable increase in photodegradation of the dye at optimum conditions. The reaction rate was faster as it generates two hydroxyl radicals per uranyl ion (Eqn.8, 9).

$$2UO_2^{2+} + 2H_2O_2 \rightarrow 2UO^{2+} + 4{}^{\cdot}OH + O_2 \quad (8)$$

$$2UO^{2+} + O_2 \xrightarrow{h\nu} 2UO_2^{2+} \quad (9)$$

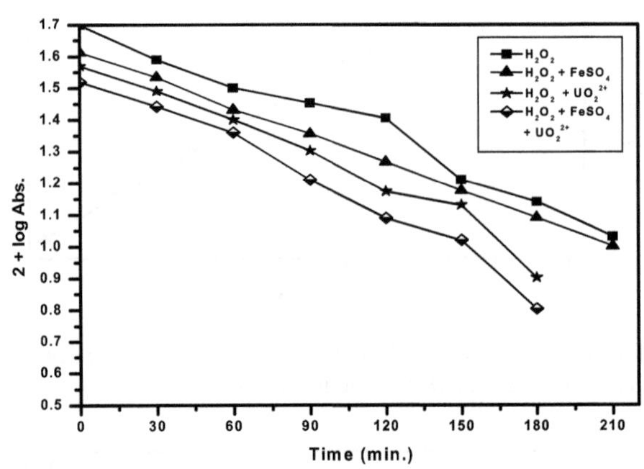

Fig. 6: A plot showing effect of $[UO_2^{2+}]$ on photocatalytic degradation of Acid Red 18 by Fenton reagent

References

1. C. Radulescu, A. M. Hossu, I. Ionita, and E.I Moater; Dyes and Pigments 76(2) (2008) 366
2. A. S. Stasinakis; Global Nest Journal 10(3) (2008) 376
3. P. Mehta, R. Mehta, M. Surana and B.V. Kabra; J. Ind. Council Chem. 26(2) (2009) 158
4. A. Kumar, M. Paliwal, , R. Ameta, and Ameta, S. C; J. Iran. Chem. Soc. 5(2) (2008) 346
5. Li. Huang, D. Gu, L. Yang, L. Xia, R. Zhang, and H. Hou; J. Environ. Sci. 20 (2008) 183

6. M. Ugurlu and I. Kula; Environ. Sci. Poll. Res. 14 (2007) 319
7. D. Chatterjee and S. Dasgupta, J. Photochem. Photobiol. 6 (2005) 186
8. C. J. Haperman, B. G. Anderson, A. Torrents and A. J. Acher Aric; Food Chem. 45 (1997) 1006
9. M. Surana, P. Mehta, K. Pamecha and B. V. Kabra; Der Pharma Chemica. 3 (2) (2011) 39
10. H. Gabriel and J. Hon; Research J. App. Sci. 3 (2008) 216
11. M. M. Rahman, M.A. Hasnat and K. Sawada; J. Sci. Res. 1(1) (2009) 108
12. T. R. Sundararaman, V. Ramamurthi and N. Partha; Modern Applied Science 3(8) (2009) 15
13. R. J. Bigdha; Chem. Eng. Prog. 12 (1995) 62

69

A Green Approach for the Synthesis of Thiazolidine-2,4-dione and its Analogues Using Gold NPs as Catalyst in Water

K. Kumari,[1,3] P. Singh,[2,*] R. C Shrivastava,[4] P. Kumar,[5] G. K. Mehrotra,[1] M. Samim,[3] R. Chandra,[5] Mordhwaj[6]
[1] MNNIT, Allahabad, Uttar Pradesh, India
[2] A. R. S. D. College, University of Delhi, Delhi- 110021, India
[3] Department of Chemistry, Jamia Hamdard, New Delhi-110062, India
[4] Department of Chemsitry, Amity institute of Applied Sceinces, Amity university, Noida, Uttar Pradesh, India
[5] Department of Chemistry,University of Delhi, Delhi-110007,India
[6]ACBR, University of Delhi, Delhi-11007, India
Email: **arsdchemistry@yahoo.in**

Abstract

The condensation reaction of chloroacetic acid and thiourea derivatives were carried out with gold nanoparticles in water to afford thiazolidine-2,4-dione (TZD) and its analogues, in shorter times with higher yields is described. Well characterized gold nanoparticles shows remarkable activity.

Introduction

Thiazolidine-2,4-diones (TZDs) are a new class of antidiabetic agents and include the compounds which are in clinical use – DRF-2189, ciglitazone, rosiglitazone and pioglitazone as well as several others that have been limited to pre-clinical study. All TZDs in clinical use presently act on PPAR and result in significant reduction in plasma glucose, peripheral glucose uptake and insulin levels. They also improve some of the abnormalities of lipid metabolism and decrease triglycerides levels. (Shoda et al, 1984; Garrino et al, 1986; Shoda et al, 1990; Cuzzocreaa et al, 2004) This group of compounds stimulates adipocyte differentiation, generating small adipocytes that are more insulin-sensitive than large adipocytes. (Ajeet et al, 2008)

Some of the major advances in chemistry over the past decades have been found in the area of green solvents, catalysis and catalytic processes, which play a major role in establishing the clean technology revolution in the pharmaceutical industry.

Nanoparticles as the name suggests have a size range of 1–100 nm in diameter. Accordingly they show size dependent properties. (Stange et al, 2004; Bruss 2008; Wang et al, 2008) Small size of nanoparticles makes them attractive in catalysis due to their large surface-to-volume ratio. (Damal et al, 2007; Raffa et al, 2008; Serna et al, 2009) Metal nanoparticles can be recovered from the reaction mixture easily and can be reused for further reactions with almost same efficiency, as the reaction reported till now do not affect the oxidation state and the integrity of metal nanoparticles.

Recently the green Solvents (water and ionic liquids) are being used successfully in organic reactions in place of volatile organic solvents and they do not cause health and environmental problems. In view of this, the search for alternatives to the damaging solvent is of highest priority. This is particularly important as solvents are used in huge amounts in organic synthesis and these are mostly volatile liquids, which are difficult to contain. (Zhang et al, 2005; Weihua et al., 2008; Yawei et al, 2008) The reactions in ionic liquids are easy to perform and need no special apparatus or methodologies. Also, the ionic liquids can be recycled and this leads to reduction of the cost of the processes. With these applications ionic liquids provide more dispersion of metal nanoparticles and more stability in comparison of volatile organic solvents.

In quest for biologically potent antidiabetic compounds, we envisage to synthesize thiazolidine-2,4-dione and its derivatives using different types of

M.M. Srivastava, L. D. Khemani, S. Srivastava, *Chemistry of Phytopotentials: Health, Energy and Environmental Perspectives*, DOI:10.1007/978–3-642–23394-4_69,

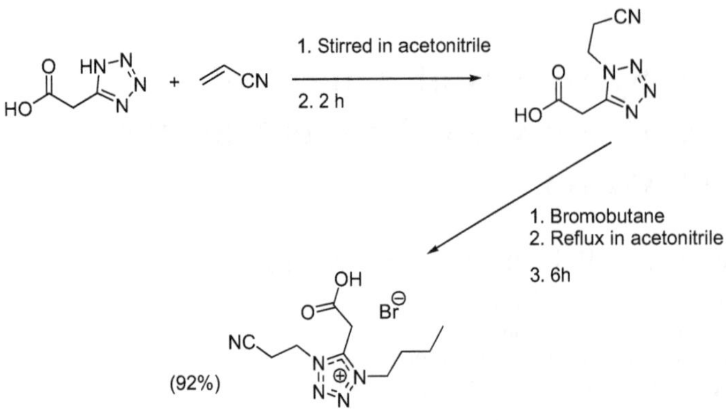

Scheme 1: Synthesis of ionic liquid (1-Butyl-5-carboxymethyl-4-(2-cyano-ethyl)-4H-tetrazolium bromide)

catalysts in green solvents. The synthesis of thiazolidine-2,4-dione derivatives had been tried earlier using organic solvents and catalysts. In the present study, we synthesized ionic liquid, gold nanoparticles and utilize in synthesis of thiazolidine-2,4-dione and its derivatives. Metal nanoparticles are synthesized by many methods reduction in ionic liquid, (Bruss et al, 2006) by reverse micelle method, (Kidwai et al, 2007) simple reduction on alumina (Kantam et al, 2007) reduction method in PVP, (Li et al, 2005) reduction method in polyethylene glycol (PEG), (Yan et al, 2006) etc. but the preparation in liquid is preferred because it provides stability to metal nanoparticles with no oxidized metal particles. Here we used tetrazolium ring based ionic liquid for the preparation of metal nanoparticles, which are highly stable in comparison of other ionic liquids.

Experimental

Synthesis of tetrazolium ring based ionic liquid (1-Butyl-5-carboxymethyl-4-(2-cyano-ethyl)-4*H*-tetrazolium bromide)

In a round bottom flask, 1-*H*-tetrazole-5-acetic acid (10 mmol) and acrylonitrile (10 mmol) in acetonitrile (2 mL) were stirred. Completion of the reaction was checked by thin layer chromatography (TLC). On completion of the reaction, the solvent was removed under reduced pressure to get the product. Then, above compound in acetonitrile was taken and bromobutane was added to the round bottom flask in 1:1.1 ratio respectively. The reaction mixture was refluxed. Then solvent and excess of bromobutane were evaporated under reduced pressure to afford 1-Butyl-5-carboxymethyl-4-(2-cyano-ethyl)-4*H*-tetrazolium bromide as in Scheme 1.

Analytical Data of Ionic Liquid

$C_{10}H_{16}N_5O_2$; Pale green liquid; IR (u=cm^{-1}; KBr) 3338 (O-H), 3146 (N-H), 2962 (C-CH), 2361 (CN), 1749 (C=O); O-H (9.39, 1H); ^{1}H-NMR (d) aliphatic hydrogen (5.426, 4H), (4.71, 2H), (1.431, 9H); ^{13}C-NMR (d) carbonyl carbons (168), nitrile (166), alkenic carbons (146), aliphatic carbon (66, 61, 51, 49, 33, 30).

Synthesis of gold nanoparticles in ionic liquid (1-butyl-5-carboxymethyl-4-(2-cyano-ethyl)-4*H*-tetrazolium bromide).

In a round bottom flask (10 mL), 2 ml of the ionic liquid and 10 mg of salt of gold i. e., $HAuCl_4$ were stirred for 10 minutes (yellow color) and the mixture was treated with excess of methanolic solution of sodium borohydride (20 mg in 10 ml of methanol). A red colored solution was obtained from yellow color solution of $HAuCl_4$ indicating the formation of gold in zero oxidation state. Stirring was continued for another 6 hours for reduction of Au (III) to Au (0). Then the solution obtained was centrifuged for 10 minutes at 10,000 rpm and the supernatant was discarded. Washed the centrifuged pellet with ethanol. The nanoparticles were analyzed for characterization using powder X-ray diffraction (XRD), transmission electron microscopy (TEM), quasi elastic light scattering (QELS) and UV-Visible techniques to determine shape, size and oxidation state of gold nanoparticles as in Scheme 2 (Singh et al 2009).

Characterization of the Gold Nanoparticles

Powder X-ray Diffraction Analysis

The scaling of powder X-ray diffraction plot are 2 theta and intensity counts plotted on x- and y-axes respectively. Peak at 38.1 degree indicates the presence of gold clearly in zero oxidation state as in **Fig. 1**. The intensity peak counts indicate the crystallinity of sample. More the intensity peak counts more will be crystalline. Powder X-ray plot indicates the formation of crystalline gold nanoparticles in zero oxidation state.

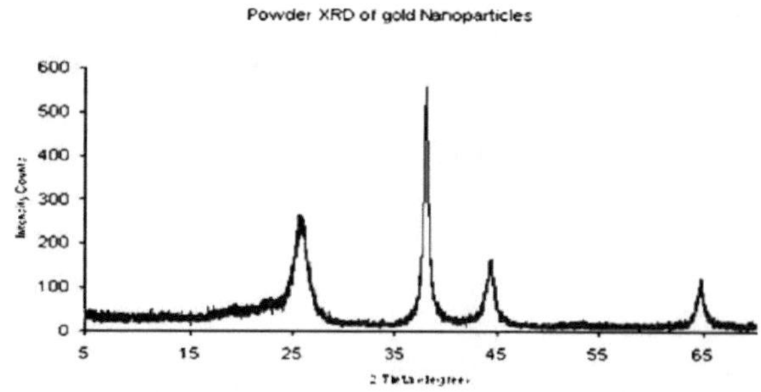

Fig. 1: Powder X-ray diffraction of Gold nanoparticles

Transmission Electron Microscopy (TEM) Analysis

It determines the shape and size of the particles. We found the range of size of gold nanoparticles is 20–30 nm and the particles are spherical in nature as in Fig. 2.

Quasi Elastic Light Scattering (QELS) Analysis and UV-Visible Spectrum of Gold Nanoparticles

It calculates the particles size as well as the dispersivity of the particles present in the sample. From the QELS picture, we observed the sample is monodispersive and the average particles size of the gold nanoparticles is 22.16 nm, which is in correlation with TEM picture as in Fig. 3a and the same is confirm by UV-Visible spectrum. As the particles size decreases the l_{max} also increases or shifts to higher wavelength. The l_{max} at 545 nm indicates the particles size of 20–30 nm as in Fig. 3b.

Synthesis of Thiazolidine-2,4-dione and its Derivatives

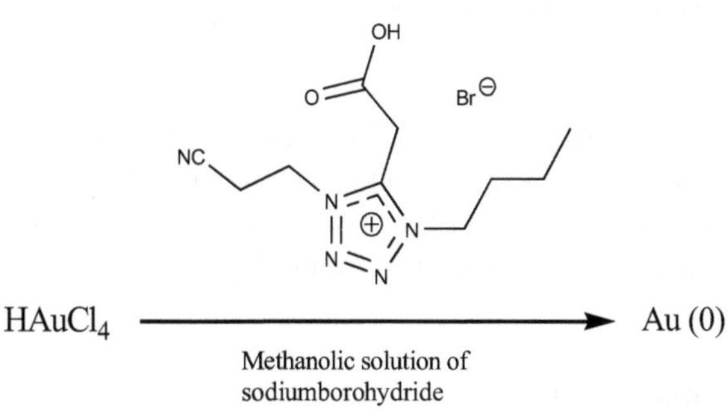

Scheme 2: Synthesis of gold nanoparticles in ionic liquid

Optimization of Solvent for the Synthesis of Thiazolidine-2,4-dione (Compound 3a)

As per Table 1, synthesis of thiazolidine-2,4-dione (Compound 3a) by the reaction between chloroacetic acid (Compound 1a) and thiourea (Compound 1b) was examined in the presence of polar solvents. In this we used polar solvents only like ethanol, methanol, water and acetonitrile because thiourea and chloroacetic acid derivatives are soluble in only polar solvents not in non polar solvents like hexane. We found that, water is the best solvent for the synthesis of thiazolidine-2,4-dione amongst all the solvents present in Table 1, because it gives high yields in shortest duration of time. Along with these qualities, water is cheap and a green solvent. We cannot run this reaction in ionic liquid because of decomposition of ionic liquid start at 90 °C.

Optimization of Catalyst for Synthesis of Thiazolidine-2,4-dione (Compound 3a)

As per Table 2, we compared sulfuric acid, acetic acid, gold nanoparticles and tungstic acid as catalysts in the synthesis of thiazolidine-2,4-dione in the above reactions. Amongst these catalysts sulfuric acid and acetic acid are highly toxic in nature and can't be recovered from the product while gold nanoparticles and tungstic acid can be recovered and can be reused after washing with the appropriate solvents. But gold nanoparticles were found out to be the best catalysts

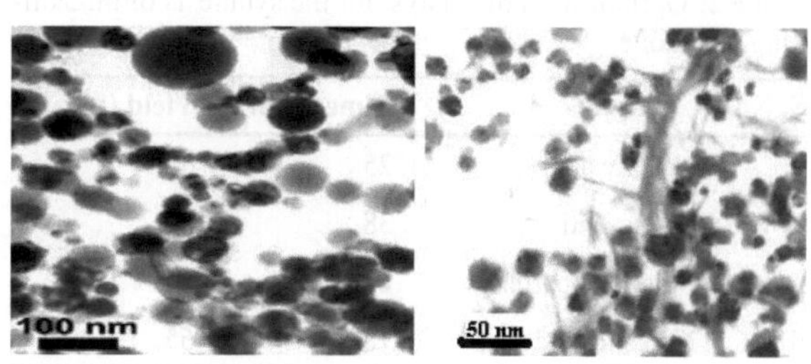

Fig. 2: Representative transmission electron microscopic pictures of gold nanoparticles at 120 kV and 80 kV

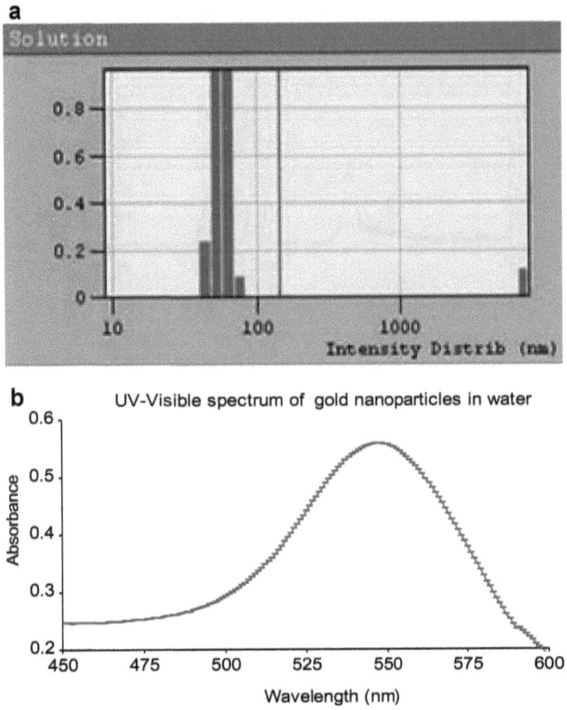

Fig. 3: Pictures of (a) quasi elastic light scattering and (b) UV-Visible spectrum of gold nanoparticles in water

Table 1: Optimization of the solvent for the synthesis of thiazolidine-2,4–dione [a, b]

S. No.	Solvent	Time (hrs)	Yield (%)
1	Ethanol	63	40
2	Water	42	60
3	Methanol	45	45
4	Acetonitrile	60	42

[a] Reaction condition: Chloroacetic acid (10 mmol) and thiourea (10 mmol) were refluxed in above solvents in absence of catalyst for the appropriate time; [b] Characterized by FT-IR, NMR and Mass spectra.

Table 2: Optimization of catalyst for the synthesis of thiazolidine-2,4-dione [a, b]

S. No.	Catalyst	Time (hours)	Yield (%)
1	Sulfuric acid	25	68
2	Acetic acid	38	60
3	Gold nanoparticles	8	80
4	Tungstic acid	13	65

[a] Reaction condition: Chloroacetic acid (10 mmol) and thiourea (10 mmol) were refluxed in water using above catalysts for the appropriate time; [b] Characterized by FT-IR, NMR and Mass spectra.

Table 3: Synthesis of thiazolidine-2,4-dione and its analogs [a,b]

S. No.	Structure of the product	Compound Number	Yield (%)	m.p. (^{0} C)
1	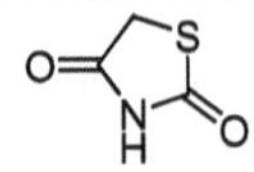	**3a**	80	162–164
2	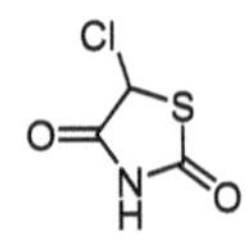	**3b**	72	134–136
3	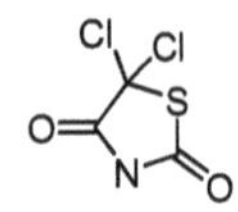	**3c**	60	104–106
4	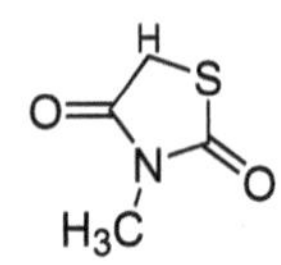	**3d**	85	176–178
5	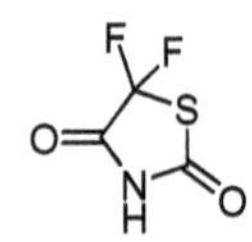	**3e**	50	98–100

[a] Reaction condition: Chloroacetic acid derivatives (10 mmol) and thiourea derivatives (10 mmol) were refluxed in water using gold nanoparticles; [b] Characterized by FT-IR, NMR and Mass spectra

for the synthesis of thiazolidine-2,4-dione because duration of time for the reaction is minimum as well as yield is maximum. Along with these applications, gold nanoparticles do not have toxicity.

Scope of the Reaction

In the scope of the reaction, new compounds were synthesized by the reaction between chloroacetic acid derivatives and thiourea derivatives using gold nanoparticles in water as in Table 3.

Conclusion

In summary, we have shown that the synthesized Au nanoparticles catalyze the condensation between chloroacetic acid and thiourea derivatives to yield thiazolidine-2,4-dione and its analogues in high yields and in short duration of time. Our protocol avoids the use of expensive reagents, high temperatures and the reaction being performed in water serves as an green and efficient method It is remarkable that the reaction in water which makes the procedure quite simple, more convenient and environmentally benign. Au nanoparticles were well characterized by transmission electron microscopy (TEM), powder X-ray diffraction (powder XRD) and quasi elastic light scattering (QELS). Further catalytic applications of Au nanoparticles for addition-elimination reactions with complex structures of biological significance are currently under investigation.

Acknowledgements

This work is dedicated to late Dr. N.N. Ghosh, my mentor (Dr. Prashant Singh) and we are highly thankful to University Grant Commission for the financial assistance.

References

1. A Kumar, P Singh, A Saxena, A De, R Chandra, S Mozumdar; Cat. Comm. 10 (2008) 17–22
2. J.A. Bruss and M.A. Gelesky; J. Mol. Cat. A: Chemical. 252 (2006) 212–218
3. S. Cuzzocreaa, B. Pisanob and L. Dugoa; Eur. J. Pharmaco. 483 (2004) 79–93
4. J. Damel, J. Cejka and P. Stepnicka; J. Mol. Cat. A: Chemical. 274 (2007) 127–132
5. M.G. Garrino, H.P. Meissner and J.C. Henquin; Eur. J. Pharmacol. 124 (1986) 309–316
6. M. Kidwai and N.K. Mishra; Tet. Lett. 48 (2007) 8883–8887
7. M.L.Kantam and V.S. Jaya; Cat. Comm. 8 (2007) 1963–1968
8. P.Li, L. Wang and H. Li; Tetrahedron 61 (2005) 8633–8640
9. P. Raffa and V. Evangeilisti; Tet. Lett. 49 (2008) 3221–3224
10. P. Serna, P. LopezHaro; J. Cat. 263 (2009) 328–334
11. T. Sohda, K. Meguro and Y. Kawamatsu; Chem. Pharm. Bull. (Tokyo). 32 (1984) 2267–78
12. T. Sohda, Y. Momose and K. Meguro; Arzneimittelforschung. 40 (1990) 37–42
13. K.J. Stange and R. Angelici; J.J. Mol. Cat. A: Chemical. 207 (2004) 59–68
14. M. Svensson, J.W. Eriksson and G. Dahlquist; Diabetes Care. 27 (2004) 955–962
15. G.Weihua, W. Congmin, Y. Xiao, H. Xingbang, W. Yong and L. Haoran; Cat. Comm. 9 (2008) 1979–1981
16. L. Wang and C. Cai J. Mol. Cat. A: Chemical. 306 (2009) 97–101
17. W.Yan and R.Wang; J. Mol. Cat. A: Chemical. 255 (2006) 81–85
18. H.Yawei, G. Yingying, Z. Sumei, Y. Fan, D. Hanming, S.J. Yongkui; Mol. Liquids. 143 (2008) 154–187
19. H. Zhang, A. Zhang and D.E. Kohan; Proc. Natl. Acad. Sci. U.S.A. 102 (2005) 9406–11
20. P. Singh, K. Kumari, A. Katyal, R. Kalra, R. Chandra. Spectrochim. Acta A: Mol Biomol. Spect. 73 (2009) 218–220

70

Synthesis of Potential Phytochemicals: Pyrrolylindolinones and Quinoxaline Derivatives using PEG as an Environmentally Benign Solvent

A. V. K. Anand*, K. Dasary and A. Lavania
School of Chemical Sciences, Department of Chemistry, St. John's College, Agra University
Agra-282002, U. P. India
Email: ananita_7@yahoo.co.in

Abstract

3-pyrrolylindolines and pyrrolylindeno[1,2-b]quinoxaline derivative, having pyrrole, indole and quinoxaline moieties known for antibacterial, antifungal and antiprotozoal properties, used in agricultural fields as fungicides, herbicides and insecticides, present in hormones responsible for plant growth have been synthesized in good yields, using Polyethylene glycol (PEG-400) which is an effective, inexpensive, non-toxic and an environmentally benign solvent used in drug designing

Introduction

Five-membered, nitrogen-containing heterocycles such as pyrrolyl-indolinones are important building block in number of biologically active compounds [1]. Such type of heterocyclic moieties is of exceptional interest in pharmaceutical applications as they represent the core part of several drugs [2]. Pyrrole based compounds are very important among the five membered heterocycles as its nucleus occurs in various natural products, the green pigment chlorophyll (coloring matter of leaves) and hemin (coloring matter of blood) are vital components of the living organisms. Some pyrrole derivatives from Lycium Chinese fruits [3], have been used as a tonic in traditional Oriental medicine and were reported to exhibit hypotensive, hypoglycemic and antipyretic activities and to prevent stress-induced ulceration in experimental animals [4, 5]. pyrrole derivatives extracted from datura leaves exhibit antifungal activity [6].

Indole [7] and its derivatives have captured the imagination of organic chemists for more than a century they have important application in the industry of plant growth, indoleacetic acid (auxin) is a plant growth hormone, tryptophan an essential α-amino acid, as such, is the constituent of most proteins. Futhermore, indole alkaloids isolated from medicinal plants have been used as antibiotics, anti-inflammatory, antihypertensive and antitumor agents, the indole nucleus has taken on considerable pharmacological importance, they help in the development of roots in plant and are used as herbicides. Indoline, dihydroindole, is used for making agrochemicals and medicines. Quinoxalines [8] have attracted considerable attention in the past years they exhibit biological activity such as antibacterial and antiprotozoal, the protection of plants from foliar and soil borne phytopathogens is accomplished by the treatment of the plants with the specified class of tetrazolo[1,5-a]quinoxalines [9]. 2-Chloro-3-methylquinoxaline nucleus have been reported to possess optimized antimicrobial activity [10].

The importance of protecting plants from phytopathogens is extremely important as every ornamental crop plant suffers from the ravages of phytopathogens which cause plant diseases. Many economically important crops cannot be successfully raised without a chemical agent for the protection of plants from phytopathogens.[11] The search for new and improved plant protective agents therefore continued to be pursued vigorously.

In context to the above protocol 3-pyrrolylindolines and pyrrolylindeno[1,2-b]quinoxaline derivatives have been prepared, Poly Ethylene Glycol (PEG-400),[12] has been found to be an efficient reaction medium in the preparation of these compounds.

M.M. Srivastava, L.D. Khemani, S. Srivastava, *Chemistry of Phytopotentials: Health, Energy and Environmental Perspectives*, DOI:10.1007/978-3-642-23394-4_70, © Springer-Verlag Berlin Heidelberg 2012

It is a biologically acceptable polymer, inexpensive, thermally stable, non-toxic media and has been extensively used in drug designing and as a tool for diagnostics. It has been efficiently used as a support and phase transfer catalyst in various transformation and synthesis.

Results and Discussion

In the present work Polyethylene Glycol (PEG-400) has been successfully used as a reaction medium for the preparation of 3-pyrrolylindolin-2-ones 2(a-i) and pyrrolylindeno [1,2-*b*]quinoxaline (4a) by the condensation of trans-4-hydroxy-L-proline with isatin derivatives under neutral condition. The reaction proceeded efficiently at 60–70°C without the use of any catalyst. A series of these compounds have been prepared by the condensation of *trans*-4-hydroxy-L-proline with isatin derivatives as shown in Table 1 by the above method. No additional catalyst was required and the condensation was completed within 40–45min. Different types of isatin derivatives underwent the condensation smoothly.

The formation of an inclusion complex, green in color was observed after 25 minutes of heating, this inclusion complex was insoluble in water and isolable in some cases as it was very short lived. This was probably due to the complexation of isatin with PEG which in turn accelerates the condensation reaction with trans-4-hydroxy-L-proline. The catalytic activity of PEG-400 was established by the fact that no reaction was observed in the absence of PEG. The products were detected and analyzed, the structures have been established from their spectral (IR; ^{1}H-NMR and MS) data.

A series of compounds have been prepared by the condensation of *trans*-4-hydroxy-L-proline with isatin derivatives as shown in figure 1 and 2

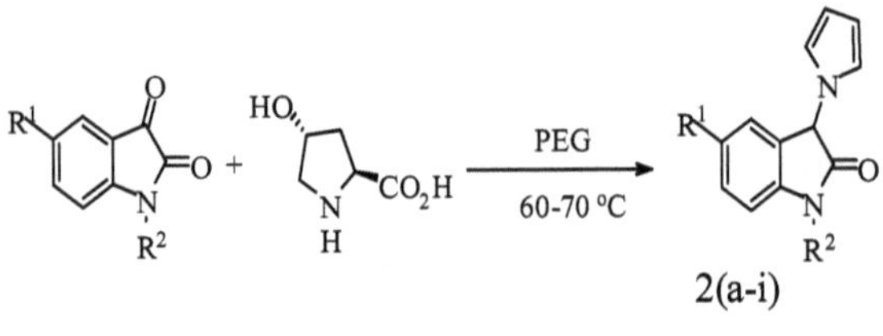

Fig. 1

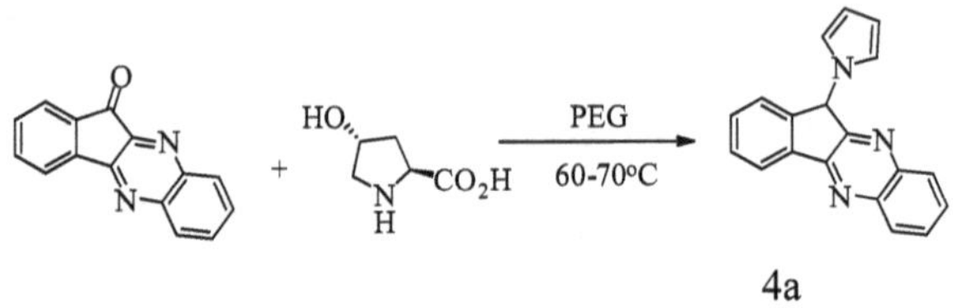

Fig. 2

Reaction Mechanism

The proposed reaction mechanism has been suggested during the synthesis of pyrrole derivatives, the presence of PEG increases the activity of both keto and amino group of isatin and trans-4-hydroxy-L-proline respectively due to the formation of intermolecular hydrogen bonding, hence facilitating the sequence of the reaction.

The formation of an inclusion complex, green in color was observed after 25 minutes of heating, this inclusion complex was insoluble in water and isolable in some cases as it was very short lived. This was probably due to the complexation of isatin with PEG which in turn accelerates the condensation reaction with trans-4-hydroxy-L-proline.

Table 1: Preparation of 3-(1H-pyrrol-1-yl)indolin-2-ones 2(a-i) and 11-(1H-pyrrol-1-yl) 11H-indeno[1,2b] quinoxaline (4a) in the presence of (PEG)

S. No.	R^1	R^2	reaction Time (min)	M.p (°C)	Yield (%)
(2a)	H	H	60	140	91
(2b)	Br	H	60	170	92
(2c)	CH_3	H	60	162	90
(2d)	F	H	60	132	89
(2e)	CN	H	65	124	85
(2f)	NO_2	H	65	180	**85**
(2g)	H	CH_3	60	132	**90**
(2h)	H	CH_2Ph	70	124	91
(2i)	H	Ph	60	160	92
(4a)	–	–	70	142	91

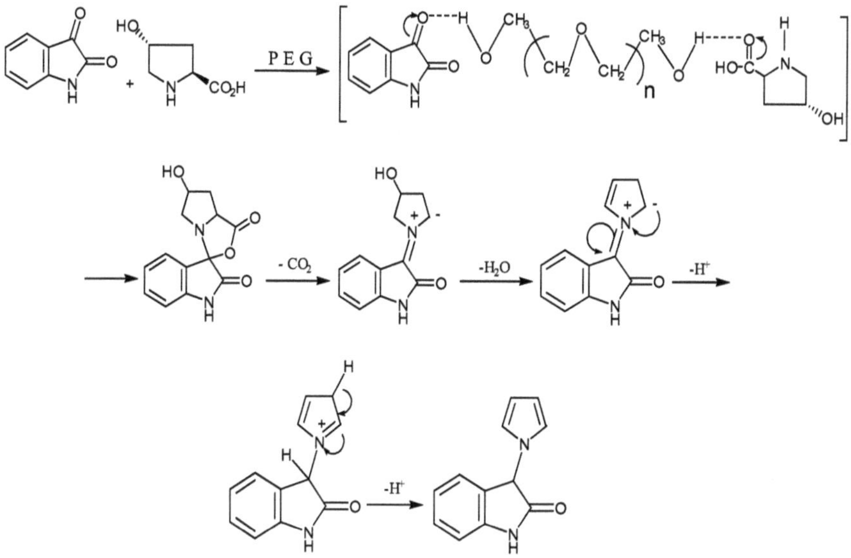

2(a-i)

References

1. (a) G.W. Gribble; J. Chem. Soc., Perkin Trans.1 (2000) 1045. (b) T.A. Gilchrist; J. Chem. Soc., Perkin Trans. 1 (1998) 615. (c) A. Nobuyoshi, O. Akihiko, M. Chikara and T.A. Gilchrist; J. Chem. Soc., Perkin Trans. 1 (1998) 615. (c) A. Nobuyoshi, O. Akihiko, M. Chikara, T. Tatsuya, O. Masami and S. Hiromitsu; J. Med. Chem. 42 (1999) 2946, (d) P.S. Baran, J.M. Richter and D.W. Lin; Angew. Chem. Int. Ed. 44 (2005) 609. (e) M. Torok, M. Abid and S.C. Mhadgut, Torok; Biochemistry 45 (2006) 5377.
2. (a) P. Kohling, A.M. Schmidt and P. Eibracht; Org. Lett. 5 (2003) 3213. (b) S. Yang and W.A. Denny; J. Org. Chem. 67 (2002) 8958. (c) S. Caron and E. Vazquez; J. Org. Chem. 68 (2003) 4140. (d) B. Jiang and C.J. Yang, Wang; J. Org. Chem. 67 (2002) 1396. (e) G. Bringmann, S. Tasler and H. Endress, Muhlbacher; J. Chem. Commun. (2001) 761.
3. Young-Won Chin, Song Won Lim, Seok-Ho Kim, Dong-Yun Shin, Young-Ger-Suh, Yang-Bae Kim, Young Choong Kim and Jinwoong Kim; Bioorg. Med. Chem. Let. 13 (2003) 79–81.
4. S. Funayama, K. Yoshida, C. Konno and H. Hikino; Tetrahedron Lett. 21 (1980) 1355.
5. J. Yamahara, M. Kim, T. Sawada and H. Fujimura; Shoyakugaku Zasshi 18 (1964) 33.
6. R. Dabur, A.K. Chhillar, V. Yadav, P.K. Kamal, J. Gupta and G.L. Sharma; J. Med. Microbiol. 54 (2005) 549–52.
7. (a) R.J. Sundberg, In Comprehensive Heterocyclic Chemistry, W.B. Clive, G.W.H. Cheeseman, Eds., Pergamon; *Oxford*, (1984) Vol. 4, pp. 313–368. (b) W. Seble, H.Y. Bryant, L.B. Stephen; J. Am. Chem. Soc., 120 (1998) 6621. (c) W. Seble, H.Y. Bryant, L.B. Stephen; J. Am. Chem. Soc., 121 (1999) 10251. (d) K. Miyashita, K. Kondoh, K. Tsuchiya, et al.; J. Chem. Soc., Perkin Trans. 1 (1996) 1261.
8. A. Hasaninejad, M.R. Zare, Mohammadizadeh and Z. Karami; J. Iran. Chem. Soc., Vol. 6, No. 1, March (2009) 153–158.
9. Shiho et al.; J. Am. Chem. Soc., 82 (1960) 4044–4054.
10. D.P. Singh, S.K. Deivedi, S.R. Hashim and R.G. Singhal; Pharmaceuticals. 3 (2010) 2416–2425.
11. P. Kohling, A.M. Schmidt and P. Eibracht; Org. Lett. 5 (2003) 3213.
12. T.J. Dickerson, N.N. Reddy and K.D. Jamda; Chem. Rev. 102 (2002) 3325.

71
Phytoremediation Potential of Induced Cd Toxicity in *Trigonella Foenum-Graecum* L. and *Vigna Mungo* L. by Neem Plants parts

R. Perveen, S. Faizan, S. A. Tiyagi and S. Kausar
Environmental Physiology Laboratory, Department of Botany,
Aligarh Muslim University, Aligarh (U. P.)-202002, India.
Email: rubina.perveen@gmail.com

Abstract

Plants can be used to remove heavy metal from the environment and hence reduce the toxic effect of heavy metal on health of all living organism. Cadmium is a non-essential heavy metal that does not have any metabolic use and can be harmful even at low concentrations. Fenugreek (Trigonella foenum-graecum L.) and Blackgram (Vigna mungo L.) are two important pulse crops grown in India to eradicate the malnutrition and hunger from developing country like India. Plants were grown in pots treated with 60 mg Cd kg^{-1} *soil as* $CdCl_2$ *caused significant damage to both the crops. The toxicity was significantly reduced when these plants earlier treated with Cd with different plants parts of neem. The changes in plant height, fresh as well as dry weights, percent pollen fertility, number of pods per plant, total chlorophyll content and nitrate reductase activity were observed. The highest improvement was noted in those plants treated with neem fruits alone while the minimum in those plants treated with Cd, neem leaf respectively. The effect was more pronounced in fenugreek than blackgram.*

Introduction

Fenugreek (*Trigonella foenum graecum* L.) and Blackgram (*Vigna mungo* L.) are important pulse crops. These have immense value and a good source of vitamins, proteins and essential oils. These crops are very important for agricultural economy and ability to increase soil fertility. Plant pathogens cause serve losses in productivity of fenugreek and blackgram [1, 2]. Cadmium (Cd) is a common metal pollutant introduced into the environment through industrial activities, sewage sludge application and commercial phosphorus fertilizers and subsequently become a part of the food chain [3]. Therefore, it is very essential to manage pathogens and reduced phytotoxicity of Cd in order to produce more plant biomass and grain of improved quality. This management objective can be achieved with the help of chemical fertilizers, broad spectrum pesticides and organic amendment of plant origin [4]. Bioorganic organic farming is a new concept that recently introduced in agriculture with an eye to stop applying the enormous amounts of agro chemicals that led to serve environmental and health troubles. The conjoint application of botanicals as organic manures is a suggested alternative way to replace the chemical fertilizations and obtain reasonable yield quality and high quality. This not only to reduce the phytotoxicity of Cd but increased the soil fertility. The present investigation aims at studying the different parts of neem plant and Cd individually as well as concomitantly in relation to growth parameters of fenugreek and blackgram.

Materials and Methods

The pot experiments were performed in randomized block during Rabi and Kharif season of (2010) in the net house of Department of Botany, Aligarh Muslim University, to investigate the efficacy of some botanicals such as Fenugreek (*Trigonella foenum graecum* L.) and Blackgram (*Vigna mungo* L.) respectively. The aim of these experiments were to study the toxicity of Cd alone and in combination with neem plant parts such as neem leaves and neem fruits.

M.M. Srivastava, L.D. Khemani, S. Srivastava, *Chemistry of Phytopotentials: Health, Energy and Environmental Perspectives*, DOI:10.1007/978–3-642–23394-4_71, © Springer-Verlag Berlin Heidelberg 2012

The seeds of *Trigonella foenum- graecum* L., and *Vigna mungo* L., were obtained from Indian Agricultural Research Institute, New Delhi. The seeds were surface sterilized with dilute solution of sodium hypochloride to prevent any fungal contamination. The 10 seeds of *Trigonella foenum-graecum* L. and *Vigna mungo* L., were sown is earthen pots containing 4 kg steam sterilized soil – manure mixture. Thinning was done after one week of seed germination and retains five plants per pot. The concentrations of Cd in the form of (60 mg Kg^{-1} of soil) $CdCl_2$, Cd + neem leaves, Cd+ neem fruits @ 110 kg N/ha alone and in combination were added before sowing of seeds as per inoculation scheme in Table 1 and 2. The plants with no added heavy metals served as control. The treatments were given at a temperature of 22±2 °C before 1 day of sowing. There were five replicates of each treatment. Necessary watering and weeding were done whenever required. The plants were kept at glasshouse benches in the randomized manner.

The experiments were terminated 120 days after seed germinations. Plant height, fresh as well as dry weights, percent pollen fertility, number of pods per plant, total chlorophyll content and nitrate reductase activity were measured. Pollen fertility (%) was estimated by the method of Brown [5], using stainability of pollen grains in 1 % acetocarmine solution.

Chlorophyll Estimation

The chlorophyll content in the fresh leaf was measured by Mackinney [6], 1 g of finely cut fresh leaves was ground to fine pulp using mortar and pestle after pouring 20 cm^3 of 80 % acetone. The mixture was centrifuged at 5,000 rpm for 5 minutes. The supernatant was collected in 100 cm^3 volumetric flask. The residue was washed three times, using 80 % acetone. Each washing was collected in the same volumetric flask and volume was made up to the mark, using 80 % acetone. The absorbance was read at 645 and 663 nm on spectrophotometer. The chlorophyll content present in the extract (mg kg^{-1} tissue) was calculated by following equation:

Chlorophyll a kg^{-1} tissue =

$$12.7\,(A_{663}) - 4.68\,(A_{663}) \times \frac{V}{1000 \times W}$$

Chlorophyll b kg^{-1} tissue =

$$20.2\,(A_{645}) - 18.02\,(A_{663}) \times \frac{V}{1000 \times W}$$

Total chlorophyll in fresh tissue =

$$20.2\,(A_{645}) + 8.02\,(A_{663}) \times \frac{V}{1000 \times W}$$

A = absorbance at specific wavelength
V = final volume of chlorophyll extract in 80 % acetone
W = fresh mass of tissue, used for extraction

NR Activity Estimation

The activity of nitrate reductase was measured by Jaworski [7]. The leaves were cut into small pieces (1 cm^2). 0.2 g of these chopped leaves weighed and transferred to plastic vials. To each vial 2.5cm^3 of phosphate buffer pH 7.5 and 0.5 cm of potassium nitrate solution was added followed by the addition of 2.5 cm^3 of 5 % isopropanol. These vials were incubated in BOD incubator ± 2°C in dark 0.4 cm^3 of incubated mixture was taken in a test tube to which 0.3 cm^3 each of sulfanilamide solution and NED-HCl was added. The test tube was left for 20 minutes, for maximum colour development. The mixture was diluted to 5 cm^3 with DDW. The absorbance was read at 450 nm on spectrophotometer. A blank was run simultaneously with each sample. Standard curve was plotted by using known graded concentration of $NaNO_2$ (Sodium nitrite) solution. The absorbance of each sample was compared with that on the calibration curve and nitrate reductase activity (n mg^{-1} h^{-1}) was noted on fresh mass basis.

Results and Discussion

The data presented in Table 1 and 2 clearly revealed that soil application of some botanicals and Cd alone as well as in combination with neem leaves and neem fruits significantly improved the plant growth, percent pollen fertility, and chlorophyll content in all treatments as compared to the untreated control. The highest improvement was noted with neem fruits and lowest was recorded with Cd + neem leaves and Cd alone respectively. The application of neem plant parts to soil is beneficial effects on soil nutrient, soil physical condition, soil biological activities and improvement of crop.

The improvement was more prominent in growth parameters of *Trigonella foenum –graecum* than *Vigna mungo*. The improvement may be due to reduction in plant pathogens and due to their manurial effect. Incorporation of botanicals increased microbial activity are known to bring about increased conversion of N to nitrate form [8]. Which in turn appears to be responsible for stimulation of nitrate reductase activity. Similarly chlorophyll content was also increased by amendments with their botanicals. Ahmad *et al.*, [9] also observed increased chlorophyll content due to application of organic matter. The application of Cd caused much reduction in all the growth parameters including NR activity and chlorophyll content of both the crop plants. Our results are in conformity with those of Mobin and Khan [10], who observed reduction in phytosynthetic content even at low concentration. Arduini *et al.*, [11] and Khan *et al.*, [12] also noticed reduction in plant growth and enzymes activity in the presence of Cd inoculated plants. However, addition of neem plants parts reduced the toxicity of Cd when inoculated concomitantly. This is an agreement with those of Braek *et al.*, [13] and Wang and Wood [14] when incorporated with algae and *Azolla* with heavy metals respectively.

The researcher outcome suggests the aptness of botanicals for management of diseases caused by fungi, insects and plant- parasitic nematodes along with affluence of soil health if the crop is managed under the umbrella of organic farming. The economics of using botanicals also works out to be adequate in high value commodity in pulse crops. Thus, the present finding strongly advocate the array of botanicals in general and neem in particular for pest management as well as

Table 1: Impact of Neem plant parts and Cd Inoculation on growth of *Trigonella foenum graecum* L (Each value is an average of 5 replicates)

Treatments	Plant height (cm)	Plant fresh weight (gm)	Plant dry weight (gm)	Pollen fertility (%)	Number of fruits/plants	Chlorophyll content (mgg^{-1})	NR Activity (μ mole NO_2-h^{-1}g^{1}fersh weight)
Control	24.84	10.63	3.54	96.74	24.08	4.327	0.373
Neem leaf	31.67	18.18	6.06	98.02	30.02	7.460	0.643
Neem fruit	**34.93**	**19.74**	**6.58**	**99.01**	**33.90**	**8.164**	**0.703**
Cd	**6.16**	**3.51**	**1.17**	**31.92**	**9.90**	**1.43**	**0.113**
Cd+ Neem leaf	18.67	7.86	4.16	55.34	14.68	2.50	0.403
Cd+ Neem fruit	21.62	8.50	4.75	59.24	16.08	2.14	0.470
C.D. (*P*=0.05)	2.39	1.27	0.59	5.44	2.23	0.128	0.037

Table 2: Impact of Neem plant parts and Cd Inoculation on growth of *Vigna mungo* L (Each value is an average of 5 replicates)

Treatments	Plant height (cm)	Plant fresh weight (gm)	Plant dry weight (gm)	Pollen fertility (%)	Number of fruits/plants	Chlorophyll content (mgg^{-1})	NR Activity (μ mole NO_2-h^{-1}g^{1}fersh weight)
Control	45.00	4.57	1.52	93.07	18.98	3.567	0.342
Neem leaf	63.34	8.39	2.80	95.09	24.67	4.386	0.538
Neem fruit	**70.85**	**8.46**	**2.82**	**96.70**	**27.05**	**4.793**	**0.647**
Cd	**30.04**	**1.51**	**0.50**	**40.20**	**07.09**	**1.660**	**0.114**
Cd+ Neem leaf	34.14	5.39	1.80	73.06	13.87	2.197	0.260
Cd+ Neem fruit	37.60	6.26	2.09	77.80	14.98	2.980	0.300
C.D. (*P*= 0.05)	3.83	0.54	0.26	4.37	1.77	0.136	0.040

reduction in phytotoxicity in an organic agriculture. Organic pulses are considered valuable and it is also believed that post- harvest losses of such vegetables are usually lower as to conventionally grown pulses. Such types of produce are free from any health hazard and also safe for environment.

References

1. S.A. Tiyagi, M. Parveen, A.A. Khan, and R. Kumar; Ann Appl Biol 118 (1991a), 106–107.
2. S.A. Tiyagi, M. Parveen, A.A. Khan and R. Kumar; Ann Appl Biol 118 (1991 b), 106–107.
3. G.J. Warner, Adv. Agron 51(1993), 173–212.
4. S.A.Tiyagi, M.M. Alam; Bioresour Technol 51 (1995), 233–239.
5. Brown, G.T., 1949. Pollen slide studies, Springfied, Illinios, USA: Charles C. Thomas publication.
6. G. Mackinney; J. Biol. Chem., 40 (1941), 315–322.
7. E.G. Jaworski; Biochem. And Biophys. Res. Comm., 43(1971), 1274–1279.
8. H.B. Gunner; Nature 197(1963) 1127–1128.
9. A. Ahmad S.A. Tiyagi M.M. Alam; Science Khyber, 3(1990), 165–170.
10. M. Mobin, and N.A. Khan; J. Plant Physiol., 164(2007.), 601–610.
11. I. Arduini, A. Masoni, M. Mariotti and L. Ercoli; Environ. Exp. Bot., 52 (2004), 89–100.
12. N.A. Khan, I. Ahmad, S. Singh and R. Nazar; World J. Agric. Sci., 2(2006), 223–226.
13. G.S. Braek, D. Malnes and A. Jansen; J. exp. Mar. Biol, Ecol,42 (1980): 39–54
14. H.K. Wang and J.M. Wood, Environ. Sci. Tech., 18 (2) (1984)). 106–109

72
Functionalized MCM-41 Type Sorbents for Heavy Metals in Water: Preparation and Characterization

S. Vashishtha, R. P. Singh and H. Kulshreshtha
Department of Chemistry, School of Chemical Sciences,
St. John's College, Agra
Email: vashishtha.shalini6@gmail.com

Abstract

Four novel sorbents were prepared by introducing functional groups as –NH_2-SH and –COOH into MCM-41 in various positions. The sorbents were characterized for BET surface area, pore size distribution, pore volume, surface functionality and surface group density. Multicomponent studies were also conducted. After sorption the metal-coated sorbents were tested to investigate the bonding mechanism of the metal ion using FTIR and XRD. The capacity was tested by batch process for Cadmium, Chromium, Lead and Zinc to study the effects of the different functional groups and their stereo specificity. The data were analyzed using various isotherms, kinetic and thermodynamic equations for the design of adsorption systems.

Introduction

Porous inorganic materials have been studied extensively for many years due to their utility as high surface area catalyst and sorption media. A significant breakthrough in the chemistry of material was accomplished with the synthesis of MCM-41. This new family of mesoporous material possesses uniform and large porous (2–10 nm), a large area over 1000m^2/g and a large adsorption capacity[1]. As a silica-based mesoporous material, M41S would be attractive as adsorbents, supports for catalysts, and hosts in inclusion chemistry, however it has several limitations due to the lattice of silica. From this view point, a number of groups have successfully incorporated various metal ions such as Al, Ti, V, Ga and B into the silica frame work of MCM-41. Metal oxides with porous structure have been synthesized by using surface directing agent similar to the templates used in the synthesis of M41S[2]. Some efforts have been made to obtain porous heteropoly acids, but none was accomplished due to the low thermal stability of the heteropoly anion[3, 4]. Layered mesostructured materials of $(C_{19}H_{42}N)_6(H_2W_{12}O_{40})[C_{12}H_{25}N(CH_3)_3]_6(H_2W_{12}O_{40})$ are of great interested because their walls are made of cluster anion in contrast to the amorphous walls of the other materials. The oxides of W, V, Nb and Mo precursors react with cationic surfactants, often forms similar cluster ion salts with lamellar structures. The aim of this work was to synthesis surfactant salts of phosphomolybdic acids (PMA) and to treat them with tetraethyl orthosilicate (TEOS) in a "sol-gel" reaction in order to obtain a three dimensional tunnel structure.

Material and Methods

The solution of surfactant was prepared by dissolving $CH_3(CH_2)_{15}N(CH_3)_3Br$ (CTAB=cetyltrimethyl ammonium bromide) in the distilled water. An appropriate amount of H_7 $[P(Mo_2O_7)_6].xH_2O$ (PMA) was added with vigorous stirring. After being stirred at room temperature for 1 h, the resulting mixture was allowed to react at 100 °C for 24 h. The resultant yellow solid product was filtered, washed with distilled water, and dried in air. Silicate was introduced into the salt by reacting 1 g of the cluster/surfactant salt with excess TEOS at 150 °C for 24 h in a Teflon-lined autoclave. The product was collected by filtration, washed with 95 % ethanol, and dried in air. CTAB was removed by stirring in 150 ml of 1M HCL/EtOH at room temperature for 4 h. The product was collected by filtration, washed with ethanol, dried in air and modified by using chemical moiety to introduce special selective

M.M. Srivastava, L.D. Khemani, S. Srivastava, *Chemistry of Phytopotentials: Health, Energy and Environmental Perspectives*, DOI:10.1007/978–3-642–23394-4_72, © Springer-Verlag Berlin Heidelberg 2012

surface functional groups,[5–7] in particular nonfunctionalized, -NH_2, -SH and -COOH to get four novel adsorbents A, B, C and D respectively.

The sorbents were characterized[8] for BET surface area, pore size distribution, pore volume, surface functionality and surface group density. After sorption the metal-coated sorbents were tested to investigate the bonding mechanism of the metal ion using FTIR and X-ray diffraction (XRD). The capacity was tested by batch process for Cadmium, Chromium, Lead and Zinc to study the effects of the different functional groups and their stereospecificity.[9] Single and multicomponent studies were also conducted. The data were analyzed using various isotherms, kinetic and thermodynamic equations for the design of adsorption systems.[10]

Results and Discussion

The IR spectrum of $(C_{19}H_{42}N)_3(PMo_{12}O_{40})$ as compared with those of PMA and the TEOS treated sample as shown in Fig. 1. The bands in the IR spectrum $(C_{19}H_{42}N)_3(PMo_{12}O_{40})$ correspond well to those of PMA, showing the typical features of Keggin anions. The four adsorption bands at 1062, 957, 881, 800 cm^{-1} are assigned to the v(P-O), v(Mo-O_t) (O_t refers to the terminal oxygens), v(Mo-O_c-Mo) (O_c refers to the corner oxygens) and v(Mo-O_e-Mo) (O_e refers to be edge oxygens) respectively. The red shift of the bands of $(C_{19}H_{42}N)_3(PMo_{12}O_{40})$ compared with those of PMA suggests some interaction between the cluster anions and the cationic headgroups of the surfactant. The IR spectrum of the TEOS-treated sample exhibits five absorption bands in the region between 1100 and 500 cm^{-1}.

It is different from the IR spectra of the PMA and $(C_{19}H_{42}N)_3(PMo_{12}O_{40})$.The peaks of heteropoly acid near 800 and 880 cm^{-1} are overlapped by the strong absorption peaks of silicate xerogel, which are near 800 cm^{-1}. Only the peak near 960 cm^{-1} can show the existence of the heteropoly acid. After the extraction of the template with HCL/EtOH, the IR spectrum corresponds well to the product before the extraction, showing the structure remains intact.

The XRD pattern of the $(C_{19}H_{42}N)_3(PMo_{12}O_{40})$ is shown in Fig. 2, and the (hkl) indexes are also indicated with the corresponding diffraction peaks. There are several sharp peaks at low angles (less than 10°), while the XRD pattern only shows broad lumps at angles greater than 10°. These features are again reminiscent of MCM-41 which shows sharp low-angle lines and diffuse high-angle lines, indicative of long-range ordering of channels but a locally glassy aluminosilicate framework. However, the peaks are different from those for MCM-41 but similar to those for the layer compounds at low angles which indicates that the salt has the layer structure. The interlamellar distance can be calculated from the 001 index as d_{001}=3.00 nm. On the basis of CPK modeling, a fully extended CTA ion has a length of 30A°; completely coiled up it extends~15 A°. As implies that the hydrocarbon tails penetrate each other, as might be expected.

The XRD patterns of the TEOS-treated sample and the solvent-extracted sample indicate that the peak near 2.9° change to 3.4° for the TEOS-treated sample, which means the interlamellar distance decreases when treated with TEOS. The strength of the peaks at low angles becomes less sharp than the salt. It is in contrast to the increase in the surface area of the material. We can deduce that the addition of Si

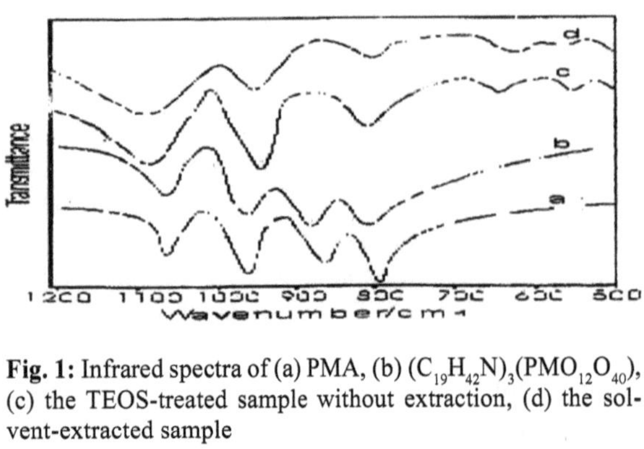

Fig. 1: Infrared spectra of (a) PMA, (b) $(C_{19}H_{42}N)_3(PMO_{12}O_{40})$, (c) the TEOS-treated sample without extraction, (d) the solvent-extracted sample

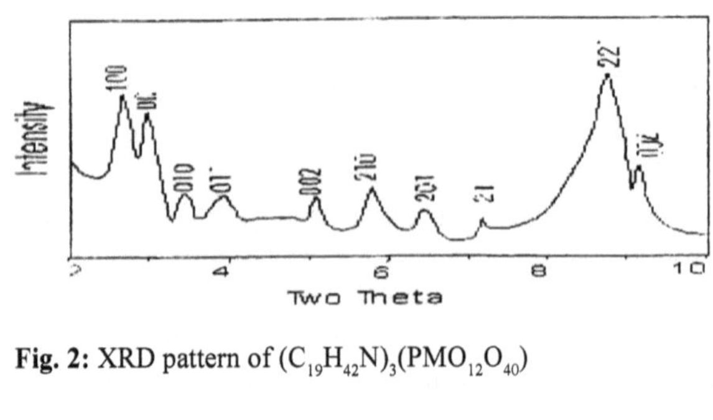

Fig. 2: XRD pattern of $(C_{19}H_{42}N)_3(PMO_{12}O_{40})$

leads to the bending of the layers, and at last results in the transformation of the lamellar salt into a microporous structured compound through the interaction between the Surfactant/heteropoly-acid and the silicate.

With the presence of the surfactant, the TEOS-treated sample is nonporous (0.05 m^2/g). After partial removal of the template, the surface area increases to 140 m^2/g. The bands around 3000 and 1500 cm^{-1} assigned to the stretching and bending vibrations of C-H of the surfactants are observed in the IR spectrum, indicating that some surfactant is still present and the extraction is uncompleted. Even in the elementary analysis the component of the element N decreases from 1.02% to 0.77%. If HCL/EtOH extraction is complete, the BET surface area may be more than 140 m^2/g.

The effects of various parameters affecting the adsorption such as initial toxic concentration, sorbent dose, contact time, pH and temperature were determined in the batch process. A, B, C and D could remove toxic ions in the order Pb(II)> Zn(II)> Cd(II)> Cr(VI). The order of toxic removal capacities for these adsorbents was found to be: D>C>B>A as per removal % and isotherm data. Adsorption decreased with rise in toxic concentration but increased with increase in sorbent dose and contact time. The optimum pH and contact time were different for different toxics. Cr (VI) as anion showed much lower removal than other metals because most of these sorbents including D has more negative or cation exchanging sites than positive or anion exchanging ones.

Conclusions

The porous structures with high surface area are obtained after the surfactant template is partially removed from TEOS-treated phosphomolybdate salts by solvent extraction. IR and XRD data indicate that silicate is indeed incorporated in the salt structures and the heteropoly structure remained intact. The hydrolysis product of TEOS is the SiO_2 xerogel. These highly porous MCM-41 based sorbents can be used as sorbents for heavy metals. They are also expected to act as catalysts, ion exchangers, demulsifiers *etc.* Their new applications need further research.

References

1. X.Bao, X.S. Zhao, X. Li and J. Li; Applied Surface Sc.; 237 (2004) 380–386.
2. X. Bao, X.S. Zhao, X. Li, P.A. Chia and J. Li; J Phys Chem B.; 108 (2004) 4684–4689.
3. A. Lapkin, B. Bozkaya, T. Mays, L. Borello, K. Edler and B. Crittenden; Catalysis Today. 81(2003) 611–621.
4. X.Y. Bao and X.S. Zhao; J Phys Chem B. 109 (2005) 10727–10736.
5. A.S.M. Chong, X.S. Zhao, A.T. Kustedjo and S.Z. Qiao; Microporous and Mesoporous Materials. 72(2004) 33–42.
6. R. Compostrini, M. Isehia, G. Carturan and L. Armelao; J Sol-Gel Sc Tech. 23(2002) 107–117.
7. C. Yang; Chinese Chem Lett. 14 (2003)96–99.
8. P. Delmelle, F. Villieras and M. Pelletier; Bull Volcanol. 67(2005)160 -169.
9. A. Bibby and L. Mercier. Chem. Mater. 14(2002) 1591–1597.
10. L. Bois, A. Bonhomme, A. Ribes, B. Pais G. Raffin and F. Tessier; Colloids and Surface A: Physicochemical and Engineering Aspects. 221(2003) 221–230.

73
Photocatalytic Degradation of Oxalic Acid in Water by the Synthesized Cu-TiO_2 Nanocomposites

Azad Kumar[1], A. Kumar[1] and R. Shrivastav[2]
[1]Department of Chemistry, Agra College, Agra- 282002
[2]Department of Chemistry Dayalbagh educational Institute, Dayalbagh-282110
Email id: kumarazad20@gmail.com

Abstract

TiO_2 is the most commonly used photo catalyst because of its high oxidation power, stability and non toxicity. Cu-TiO_2 nanocomposites were prepared using the solution impregnation method. After characterization for crystalline phase and particle size by XRD analysis, both the commercially procured TiO_2 and synthesized Cu-TiO_2 nanocomposites were used as photo catalyst in the photo- degradation of Carboxylic Acids (Oxalic Acid). The degradation of oxalic acid in the presence of pure TiO_2 and synthesized Cu-TiO_2 was done. The effective photo-degradation was found in case of oxalic acid in the presence of Cu-TiO_2 as compared to pure TiO_2.

Introduction

In many recent reports, the TiO_2 photo catalytic characteristics are to be greatly enhanced, which is largely due to the advent of nanotechnology and the progress made in the synthesis and characterization of nanostructured Titania. Nanoparticles of TiO_2 are mainly of interest for their, electrical optical and chemical properties.[1-8] At nanoscale not only the surface area of TiO_2 particles increases, but also it inhibits other effects on optical properties and size quantization.[9-12] An increased rate of photocatalytic reactions, when nanostructure Titania is used as photocatalysts, has been thus attributed to the increase in the redox potential as the size decrease.[13-15] TiO_2 is a excellent photocatalyst because the Photodegradation process occurs under ambient conditions, the oxidation of most organic substrates to CO_2 is complete and it is inexpensive and remain quite stable in contact with different substrates[16-17].

Methodology

Synthesis and Characterization

In this method suitable quantity of commercially obtained TiO_2 was dispersed in an alcoholic solution of Copper Acetate. The dispersion was agitated continuously for 4 hours at the elevated temp. (Just below the Boiling Point of Alcohol). After treatment the residue was sintered for 1 hour in air at 400°C. The obtain material was characterized by XRD for the phase and particle size determination.[18]

Photodegradation

In this study using TiO_2 and synthesized Cu-TiO_2 nanocomposites as photocatalyst. The photocatalytic degradation of oxalic acid was investigated. A solution of oxalic acid in water was prepared and in this solution suitable quantity of Photocatalyst was dispersed. The dispersion (reaction medium) was subjected to UV visible illumination for varying duration and residual concentration was determined by spectrophotometrically.[19-21]

Results and Discussion

In the XRD analysis shown in the figure1and 2 the observed peak is analyze phase whereas major peaks at 2θ. Angle 25.2, 37.2, 48.3 and 55.4 corresponds to Anatase phase whereas major peak at 2θ angle.

26.9, 28.2, 42.6 and 54.2 indicates the presence of rutile phase, and in Cu- TiO_2 decrease in peak inten-

M.M. Srivastava, L.D. Khemani, S. Srivastava, *Chemistry of Phytopotentials: Health, Energy and Environmental Perspectives*, DOI:10.1007/978-3-642-23394-4_73,

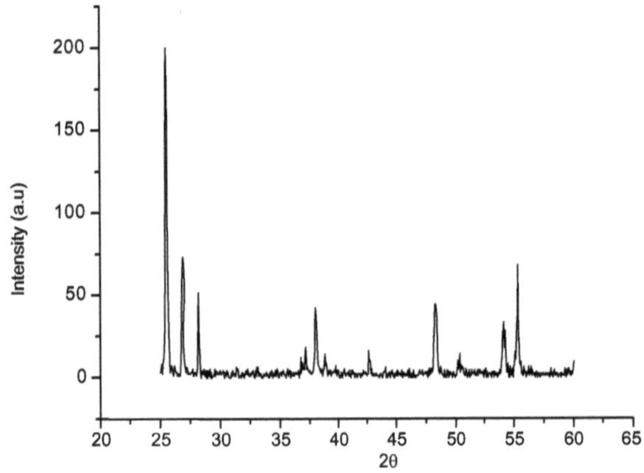

Fig. 1: Observed XRD pattern of TiO_2 (Merck)

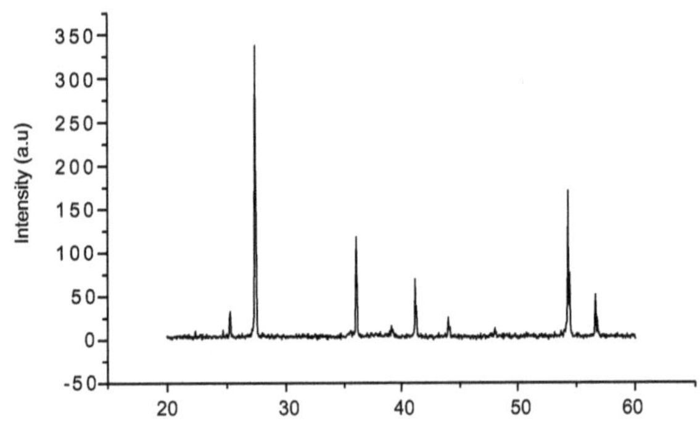

Fig. 2: Observed XRD pattern of Cu-TiO_2

Table 1: Photodegradation of oxalic acid using commercially and prepared photocatalyst

Reaction Time (min)	Oxalic acid concentration $\times 10^3$ (M)		
	†	*	**
0	5.2	5.2	5.2
60	5.2	4.6	1.9
120	5.2	4.1	1.4
180	5.2	3.9	1.2
240	5.2	3.1	0.8

*Without Photocatalyst
**Using TiO_2 as Photocatalyst
***Using Cu- TiO_2 as Photocatalyst

sity compared to TiO_2 but ever the observed some originally observed TiO_2 peaks.

The obtained yield of prepared nanocomposites of Cu- TiO_2 was more than 90% of the expected theoretical yield. This is due to decrease in crystallinity grain fragmentation and partial amorphization and particle size of TiO_2 and Cu- TiO_2 was 72 and 16 nanometer respectively.

Photodegradation of Oxalic Acid

The photodegradation of different concentration of oxalic acid are shown in Table 1. The prominent degradation of oxalic acid was found in two hours study in presence of Cu- TiO_2 in comparison to the commercially obtained TiO_2.

Conclusion

In this study, photocatalytic degradation of Carboxylic Acid namely oxalic acid was investigated. The measured values of residual concentration of Carboxylic Acid (oxalic acid) in the reaction mixture at different times of illumination (or reaction) have been shown in Table 1 and Fig. 1, 2. It is clear from the results shown that both TiO_2 and Cu- TiO_2 are proving as an effective photocatalyst for the degradation. However Cu- TiO_2 seems to be more effective as photocatalyst for the degradation of Oxalic Acid.

Reference

1. C . Xu, K. Richard, L.G Mc Mohan and S.V.M Khan; Applied catalysis B. environment ,64 (2006) 312-317.
2. L.C. Chem, Y.C. Ho, W.S. Guo, C.M. Huang and T.C. Pan; Electrochemica Acta. 54 (2009) 3884-3891.
3. A. Fuji Shima; Nature 238 (1972) 37-38.
4. H. Sia, Z. Zheng, Zhavh, L. Zhang and Z. Zou; Material Research Bulletin 44 (2009)1312-1316.
5. C. Karuna Karan and R. Dhanalakshmi; solar energy materials and solar cells. 92 (2008) 1315-1321.
6. R. Khan and T.J. Kim ; J. Hazardous materials 165 (2009) 1243-1247.
7. C.Y. Kuo and H. Lin; J. Hazardous materials 165 (2009)1243-1247.
8. C.Y. Kuo; J. Hazardous material 163(2009) 239-244.
9. A. Muller and A.K. Cheetham; The Chemistry of nanomaterial synthesis properties and applications (2004).
10. J.W. Shi, S.H. Chem, S.M.Wang, Wup and G.H. Xu; J. molecular catalysis a chem, 303(2009)141-147.
11. S. Kaur and V. Singh; J. Hazardous materials 141(1) (2007) 230-236.
12. D. Beydoun and R. Amal; J. Phys Chem B 104 (18) (2000) 4387-4396.
13. B. Balamurugan and B.R.Mehta; Thin solid films, 396 (2001) 90-36.

14. B. Damardji , H. Khalaf , L. Duclaux and B. David;Applied clay science 44 (2009) 201-205.
15. A. Gary and C.L. Epling; Chemosphere 46 (2002) 561-570.
16. S.C. Ameta, P.B. Punjabi, P. Rao and B. Single; J. Indian chem. Soc. 77 (2000) 157-160.
17. D.R Askland ;The science and Engineering of materials, 3rd Edition Chapman & Hall, London, 854 (1996).
18. H. Yoshida, Y. Lu, H. Nakayama and M. Hirothashi; J. Alloys and compounds 475 (2009) 383-386.
19. M. Saquib and M. Muneer; Dyes and Pigments 56(1) (2003) 37-49.
20. B.D. Cullity and S.R. Stock ;Elements of x-ray diffraction 3rd edition new Jersey, Prentice-Hall Inc. (2001).
21. K.Vinodgopal, D.E. Wynkoop and P.V. Kamat; Environ. Sci Tech . 30 (1996) 1660-1666.

74
Assessment of Insecticidal Properties of Some Plant Oils against *Spodoptera Litura* (Fab.)

P. Bhatt and R. P. Srivastava
[1]Bioactive Plant Natural product laboratory, Department of Entomology 'College of Agriculture, G.B Pant university of Agriculture and Technology, Pantnagar-263145, Uttarakhand, India
Email: priyangp30695@gmail.com

Abstract

Contact toxicity of three plant essential oils was evaluated against six days old larvae of Spodoptera litura (Fab), at 10,000ppm and 20,000ppm. Vetiveria zizanioides oil was found to be most toxic at 2% concentration giving 93.33% mortality 12 hours after exposure followed by Jatropha curcas oil (63.33%) and (60.00%) at 2 and 1% respectively. At 48hours after exposure khas oil at 2% gave maximum mortality (100%) followed by J. curcas oil (73.33%) and (63.33%) mortality at 2 and 1% respectively. Hedychium spicatum oil did not show any significant larval mortality.

Introduction

The move toward green chemistry processes and the continuing need for developing new crop protection tools with novel modes of action makes discovery and commercialization of natural products as green pesticides an attractive and profitable pursuit that is commanding attention. Essential oils are defined as any volatile oils that have strong aromatic components and that give distinctive odor, flavor or scent to a plant. These are the by-products of plant metabolism and are commonly referred to as volatile plant secondary metabolites. Plant essential oils are produced commercially from several botanical sources, therefore, justifying their placement under "green pesticides [1]. *Spodoptera litura* is an economically important pest in India, China and Japan causing considerable economic loss to many vegetables and field crops. [2] It is one of the economically important and polyphagous pest on field and horticultural crops [3] and causes economic loss of crops from 25.8–100 per cent [4]. It has a wide range of host, feeding on 112 species worldwide, of which 40 species are known from India [5]. The pest is resistant to most of the commonly used pesticides. Thus a study was done utilizing oils as pesticides so that there is no harmful effect on the environment.

Material and Methods

An experiment was conducted to test the residue contact toxicity of essential oils against six days old larvae of *Spodoptera litura.* The oils (*Hedychium spicatum, Jatropha curcas* and *Vetiveria zizanioides*)were obtained in their pure form from Medicinal plants Research and Development Centre. One and two percent dilutions were made in acetone. Residue contact method was followed as given by [6,7]. One ml of each concentration was coated as a thin film in both the lids of petri dish (diameter: 9 cm). The solvent was allowed to dry for one hour. After evaporation of the solvent ten six days old larvae of *S. litura* were given contact exposure for thirty minutes. In control the larvae were exposed to acetone alone. Three replications were maintained. Thereafter the larvae were transformed to rearing boxes with fresh castor leaves. The data on mortality was recorded after 12, 24 and 48 hours. The data obtained was analyzed in Complete Randomized Design(CRD).

Results and Discussion

The data on mortality count is presented in Table 1. At 2% and 12 hour interval *V. zizanioides* caused 93.3%

M.M. Srivastava, L. D. Khemani, S. Srivastava, *Chemistry of Phytopotentials: Health, Energy and Environmental Perspectives*, DOI:10.1007/978–3-642–23394-4_74, © Springer-Verlag Berlin Heidelberg 2012

Table 1: Effect of Plant Oils on Mortality count of *S. litura*

Plant oils	Dose (%)	Mortality 12HAE	24HAE	48HAE	% Corrected mortality
H. spicatum	2 1	6.6 10	16.6 10	16.66 13.3	16.66 13.33
J. curcas	2 1	63.33 60.00	73.33 63.33	73.33 63.33	73.33 63.33
V. zizanioides	2 1	93.33 36.66	100 36.66	100 40	100 40
Control					
Cd at 5% Sem± F value		31.27 10.31 **	32.64 10.76 **	35.22 11.61 **	

Table 2: Effect of Plant Oils on Feeding Behavior of *S. litura*

Plant oils	Dose (%)	Feeding 12HAE	24HAE	Reduced Feeding over control 12HAE	24HAE
H. spicatum	2 1	663.33 638.33	3062.00 3423.33	64.07 65.42	24.02 15.06
J. curcas	2 1	162.00 265.00	791.00 1032.66	91.22 85.64	80.37 74.37
V. zizanioides	2 1	4.66 872.00	0.00 1637.33	99.74 52.77	100 59.37
Control		1846.33	4030.33		
Cd at 5% Sem± F value		561.86 185.25 **	1165.77 384.37 **		

mortality followed by J. curcas (63.33%). At 24 hours interval *V. zizanoides* showed complete 100% mortality followed by *J. curcas* (73.33%) At 1% concentration *J. curcas* was most toxic (63.33%) followed by *V. zizanioides* (40%). *H. spicatum* did not show any significant effect on larvae. Recent investigations indicate that some chemical constituents of these oils interfere with the octopaminergic nervous system in insects. As this target site is not shared with mammals, most essential oil chemicals are relatively non-toxic to mammals and fish in toxicological tests, and meet the criteria for "reduced risk" pesticides [1]. The oils at their both concentration were effective in determining the rate of survival and feeding of *S. litura*. Maximum reduction in feeding over control was observed in *V. zizanioides* followed by *J. curcas* (Table 2). The results presented in the study will be useful to determine new strategies for pest control based on natural products.

References

1. O. Koul, S. Walia and G. S. Dhaliwal; Biopestic. Int. (2008) 63–64.
2. N. Ferry, M.G. Edwards, J.A. Gatehouse and A.M.R. Gatehouse. Curr. opin. Biotechnol (2004) 1551–161.
3. B.C. Dhir, H.K Mohapatra and B. Senapati. Indian J. Plant Protect. (1992) 215–217.
4. S.M. Murthy, M. Thippaiah and M.S. Kitturmntath. Insect Environment. (2006) 84–85.
5. M.G. Paulraj, Ph.D Thesis. Bharathidasn University. Trichy.Tamil Nadu India.
6. R.P. Srivastava and P. Proksch; Entomologia Generalis (1991) 265–274.
7. V. Vedamati. Thesis, (2004) G.B Pant University of Agriculture and Technology, pantnagar.137

75
Mentha Arvensis Assisted Synthesis of Silver from Silver Nitrate

S.K Shamna, S. Ananda Babu and H. Gurumallesh Prabu
Department of Industrial Chemistry, School of Chemistry, Alagappa University, Karaikudi
Email: hgprabu2010@gmail.com

Abstract

We report biosynthesis of silver using crude aqueous extract of Mentha arvensis at room temperature. The use of crude aqueous extract of Mentha arvensis as reducing agent is the first report on the synthesis of silver. XRD study suggests the formation of silver. UV-visible spectrum shows characteristic surface plasmon resonance (SPR) absorption peak of silver at 450 nm. FT-IR has emerged as a major tool to predict involvement of phytochemicals for the reduction of silver nitrate to silver. SEM analysis reveals the morphology of silver crystals and EDX spectrum confirms the formation of silver.

Introduction

Nanoscience has been established as a new interdisciplinary science. It can be defined as a whole knowledge on fundamental properties of nano-size object [1]. In particular, silver nanoparticles (AgNPs) have been the subject of substantial research in a present decade [2]. Silver nanoparticles were successfully synthesized in the natural polymeric matrix. Silver nitrate, gelatin, glucose and sodium hydroxide have been used as silver precursor, stabilizer, reducing agent and accelerator reagent respectively. Compared with other synthetic methods, greener method is rapid and simple to use [3]. The green synthesis of AgNPs involves three main steps, which must be evaluated based on green chemistry perspectives including, (i) selection of solvent medium, (ii) selection of the environmentally benign reducing agent and (iii) selection of non-toxic substances for the AgNPs stability [4]. Extracts from bio-organisms may act both as reducing and capping agents in AgNPs synthesis. The reduction of Ag^+ ions by combinations of biomolecules found in the extracts such as enzymes, proteins, amino acids, polysaccharides and vitamins is environmentally benign, yet chemically complex. An extensive volume of literature reports successful AgNPs synthesis using bioorganic compounds [5]. Polyoxometalates have the potential of synthesizing AgNPs because they are soluble in water and have the capability of undergoing stepwise, multi-electron redox reactions without disturbing their structure [6]. AgNPs can be successfully synthesized by using a variety of irradiation methods [7]. Biological syntheses of Gold and Silver nanoparticles of various shapes using different plant materials were reported. This is a simple, cost-effective, stable and reproducible aqueous room temperature synthesis method to obtain a self-assembly of Au and Ag nanoparticles. The size and shape of the nanoparticles are modulated by the varying the ratio of the metal salt and extract in the reaction medium. Variation of pH on the reaction medium gives silver nanoparticles of different shapes [8]. In the present investigation, we have examined the plant *Mentha arvensis* (common name is mint), a herbaceous perennial plant for the synthesis of Ag. This plant is famous for its high medicinal value and non-toxic.

Experimental

Mentha arvensis leaves were purchased from Karaikudi vegetable market. Silver nitrate was purchased from Merck and used as such. Buffer pH tablets were purchased from Ranbaxy and used as such. TKA–LAB Reinst system was used for the collection of high pure water. For the synthesis of silver nanoparticles, three types of extraction methods with *Mentha arvensis* leaf were carried out; (i) Extraction at room tempera-

M.M. Srivastava, L. D. Khemani, S. Srivastava, *Chemistry of Phytopotentials: Health, Energy and Environmental Perspectives*, DOI:10.1007/978–3-642–23394-4_75,

ture: 30 g of washed mint leaves were finely crushed and stirred with 90 ml de-ionized water at room temperature for 15 minutes and filtered, (ii) Extraction at 60 °C: 30 g of washed mint leaves were finely crushed and stirred with 90 ml de-ionized water at 60 °C by Soxhlet apparatus for 15 minutes and then filtered and (iii) Extraction at 100 °C: 30 g of washed mint leaves were finely crushed and stirred with 90 ml de-ionized water at 100 °C by Soxhlet apparatus for 15 minutes and then filtered. The filtrates obtained from the above said methods were used as reducing as well as stabilizing agent for the synthesis of Ag.

To a thoroughly clean beaker, 2.5 ml of *Mentha arvensis* leaf extract (reducing agent), 2.5 ml of aqueous silver nitrate (0.01M) solution and 1 ml of buffer were added. The beaker was placed on a magnetic stirrer and stirred well up to the formation of silver. The pH of the reaction was varied by three different conditions like acidic (pH-4), neutral (pH-7) and basic (pH-9). The reaction was regularly monitored by UV-Vis absorption spectra (surface Plasmon resonance) of the reaction mixture. UV-Vis spectrum was recorded using Jasco-V-530 Spectrophotometer. The FT-IR spectra were recorded using Perkin Elmer Spectrum RX1 FT-IR Spectrophotometer. The XRD analysis was performed using XPERT PANALYTIC. The particle size and morphology was determined by SEM HITACHI S300H coupled with EDX.

Results and Discussion

UV-Vis spectra of the reaction between Ag ions and extract of *Mentha arvensis* at room temperature and at 60 °C did not show corresponding UV-Vis absorption signals at pH 7 and 9. Representative absorption signal was observed only at pH 4 for the extract done at 100 °C. When the stirring was continued for 2 h, the solution color turned from light brown to dark brown. The change in color might be due to the reduction of silver nitrate. The characteristic SPR band was detected around 450 nm (Fig. 1) that could be due to the formation of silver.

The XRD result shows three distinct diffraction peaks at 38°, 44° and 77°, which are indexed for the planes 111, 200 and 311 respectively of the face centered cubic silver (Fig. 2). This data was matched with the database of joint committee on powder diffraction standards (JCPDS) file No. 89.3722 for silver. Thus, the formation of silver is confirmed. The additional peaks obtained could be due to the bio-inorganic compounds and protein matters present in the extract (as evidenced from the intensity of the Bragg reflections of strong X-ray scattering points in the crystalline phase arise from proteins in the nanoparticles synthesis). The average grain size of the silver particles synthesized by this greener method was calculated using Scherr's formula and observed as 90 nm. [9].

FT-IR has emerged as a major tool to predict involvement of functional groups. Individual FT-IR spectra were recorded for the crude extract as well as

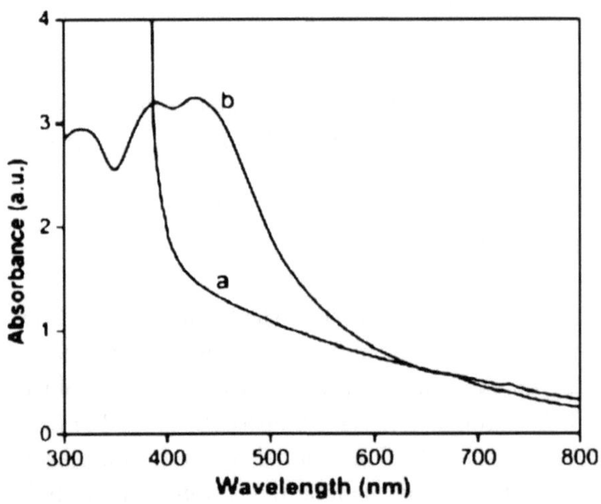

Fig. 1: UV-Vis spectra on the effect of time in the formation of silver using *Mentha arvensis* leaf extracted at 100 °C and at pH 4 : **(a)** before the reaction **(b)** after 2 h

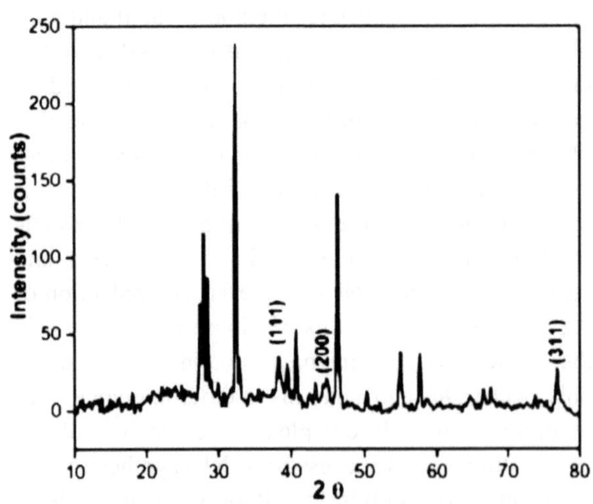

Fig. 2: XRD pattern of silver synthesized using *Mentha arvensis* extracted at 100 °C and stirred for 2 h

for Ag particles with extract. There was a variation in the FT-IR region about 1043 cm^{-1}(crude extract) when compared to (extract with Ag), which could be due to the reason that synthesized Ag particles chelate with (-C-C-) group to prevent agglomeration of silver particles. The peak observed at 1639 cm^{-1}(crude extract) was attributed to C=O stretching. These results indicated that the crude mint extract could act as potential source for the reduction of $AgNO_3$ to Ag^0. It was reported earlier that the carbonyl groups present in the extract are involved in the reduction and stabilization in the synthesis of Ag.

SEM analysis of synthesized silver particles was performed. Silver particles are distinctly seen from the SEM image (Fig. 3). The size of the silver nanoparticles obtained was in the range about 1000 nm.

The EDX energy dispersive spectrum reveals the peak of silver nanoparticles at ~3KeV (Fig. 4). Thus the presence of silver is confirmed. There is also a strong peak for Cu in the EDX spectrum, which is due to the use of copper plate in sample preparation. Other elemental signals reported in the spectrum are possibly due to the phytochemical elements present in the *Mentha arvensis* extract.

Fig. 3: SEM image of synthesized silver by *Mentha arvensis* leaf extract

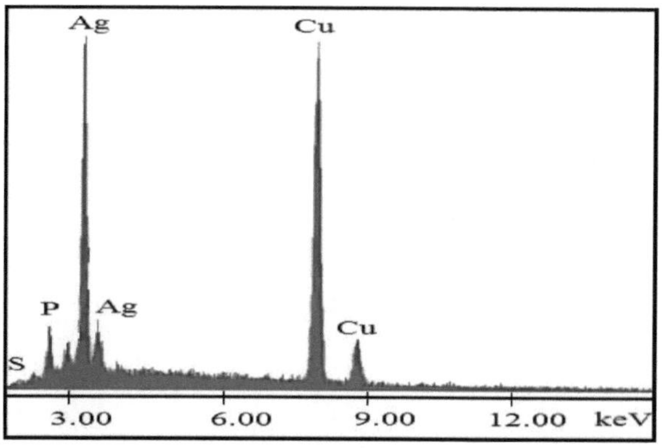

Fig. 4: EDX spectrum of synthesized silver by ***Mentha arvensis*** leaf extract.

Conclusion

The bio-reduction of aqueous Ag^+ ions by the mint leaf *(Mentha arvensis)* extract has led to the formation of silver particles. Mint leaf extract acted both as reducing as well as stabilizing agent. This method is a green chemistry approach towards the synthesis of silver particles. This is the first report on the synthesis of silver particles using mint (*Mentha arvensis*) leaf extract.

References

1. P. Williams, E. Keshavarz-Moore and P. Dunnil; J. Biotechnol. 48 (1996) 259–267.
2. Y. Yin, L. Zhi-Yuan, Z. Ziyi, G. Byron and V. Sagar; J. Mater Chem. (2002) 522.
3. M. Darroudi, M. Bin Ahmad, A. Halim Abdullah, N. Azowa Ibrahim, and Kamyar Shameli; Int. J. Mol. Sci. 11 (2010) 3898–3905.
4. P. Raveendran, J. Fu and S.L. Wallen; J. Am Chem Soc. 13940 (2003) 125.
5. J. Xie, J.Y. Lee, D.I.C. Wang and Y.P. Ting; ACS Nano. 429 (2007) 1.
6. A. Troupis, A. Hiskia, E. Papaconstantinou; Angew Chem Int Ed. 191(2002) 41.
7. J.P. Abid, A.W. Wark, P.F. Brevet and H.H. Girault; Chem Commun. (2002) 792.
8. P. Daizy; Physica E. 42 (2010) 1417–1424.
9. S.S. Shankar, A. Rai, A. Ahmad and M. Sastry; J. Colloid interface sci. 275 (2004) 496–502.

76

Synthesis of Colloidal Iridium Nanoparticles and Their Role as Catalyst in Homogeneous Catalysis – An Approach to Green Chemistry

A. Goel and S. Sharma
Department of Chemistry, Kanya Gurukul Mahavidyalaya,
Gurukul Kangri University, Jawalapur, Hardwar (U. K.), INDIA
Email: anjaligoel10@gmail.com

Abstract

The formation of colloidal iridium nanoparticles has been carried out by the chemical reduction of H_2IrCl_6 with methanol using polyvinyl-pyrrolidone as protecting agent. The synthesized iridium nanoparticles were characterized by IR, UV-vis Spectrophotometer, XRD, TEM and XPS methods of analysis. The catalysis by iridium nanoparticles was studied kinetically using the oxidation of amino acid – hexacyanoferrate(III) (abbreviated as AA & HCF(III) respectively) reaction in alkaline medium. The colloidal iridium nanoparticles were found to be more effective catalyst for oxidation of AA. The catalytic system could be recycled and reused without loss of activity.

Introduction

Colloidal transition metal nanoparticles are recently receiving increased attention since they hold promise for use as advanced materials with new electronic, magnetic, optical, and thermal properties[1]. In addition to their interesting physical properties exhibited due to quantum size effect, they also have applications in catalysis due to their large surface area and special morphologies. With the improved developments in nanochemistry, it is now possible to prepare soluble analogues of heterogeneous catalysts, which might have properties intermediate between those of the bulk metal and single metal-particle (homogeneous) catalysts. The nanoparticle-based catalytic systems generally exhibit superior catalytic activities than the corresponding bulk materials[2–9]. The recycling of both ligand and metal is highly desirable for reducing raw material costs and engineering a greener process via limiting the amount of waste chemicals for disposal. In this study, we have synthesized colloidal iridium nanoparticles by chemical reduction method. The catalysis by these iridium nanoparticles was studied kinetically using the oxidation of some amino acids (like glycine and alanine)- hexacyanoferrate(III) reaction in alkaline medium.

Experimental Section

Chemicals

All the reagents used were of AR grade. All solutions were prepared with doubly distilled water. $K_3[Fe(CN)_6]$ was recrystallized before use. Its solutions were kept in amber colored bottles to prevent photodecomposition. Catalyst was prepared by the reduction of H_2IrCl_6 (Merck). Polyvinyl-pyrrolidone (SRL) and Methanol (Merck) was used as stabilizing agent and reducing agent respectively. KCl (Merck) and NaOH (Merck) were used to provide the required ionic strength and alkalinity respectively.

Preparation of Colloidal Iridium Nanoparticles

In the present work colloidal iridium nanoparticles were synthesized by reduction of dihydro-hexachloroiridic acid (Merck) with methanol using polyvinyl-pyrrolidone as protecting agent. PVP-stabilized nanoparticles were obtained. These nanoparticles were dried at 50°C temperature and analyzed by UV-vis spectrophotometer, IR, XRD, XPS and TEM.

M.M. Srivastava, L. D. Khemani, S. Srivastava, *Chemistry of Phytopotentials: Health, Energy and Environmental Perspectives*, DOI:10.1007/978–3-642–23394-4_76, © Springer-Verlag Berlin Heidelberg 2012

Characterization of the Iridium Nanoparticles

Typical transmittance FTIR spectra of the PVP stabilized iridium nanoparticles and the pure PVP employed in the synthesis are shown in Fig. 1a and 1b respectively. The peaks were identified by comparing with literature value [10, 11]. Fig 1(a) shows that there is no peak due to precursor under the region 2550 cm^{-1}. The peak at 1690 (Fig. 1a) is due to red shift of resonance peak of pure PVP at 1714 cm^{-1} (Fig. 1b) indicating the interaction of >c=o group of PVP with nanoparticles. The appearance of a new weak peak at 2005 cm^{-1} due to iridium (PVP- stabilized) shows formation of iridium nanoparticles from H_2IrCl_6. UV-vis Spectrophotometer is a convenient technique for monitoring the progress of metal colloid formation. A PVP- H_2IrCl_6- methanol-water solution kept in an oil bath was sampled at different time and then the samples were characterized by UV-vis Spectrophotometer. As can be seen from Fig. 2 $IrCl_6^{-2}$ shows an absorption peak at 487 nm, the intensity of this peak decreases with the passage of time. After about 15 minutes the absorption peak starts disappearing indicating the reduction of $IrCl_6^{-2}$. After 25 min no peak was observed showing that $IrCl_6^{-2}$ has completely reduced. XRD was performed on the dry powders obtained after the evaporation of solvent using Bruker Axs D-8 Advance diffractometer with a scan rate 1^0 min^{-1} and a Cu K_α X-ray source ($\lambda = 0.154$ nm).

The XRD analysis (Figure 3) shows the presence of very small particles. The diffraction peaks were assigned as correspond to the iridium colloidal particles at 2-theta 34.4, 35.3, 38, 40 and 42 for iridium and iridium oxide. Analysis of the peaks broadening using the Scherrer equation gives an estimate of particle diameter. The X-ray diffractograms of these metallic particles show broad peaks characterization of materials with a small size.

Nanoparticles were also characterized by Transmission Electron Microscope (Philips). Fig. 4 shows a representative TEM image of iridium colloidal particles. It appears that the Ir (nano) particles are sepa-

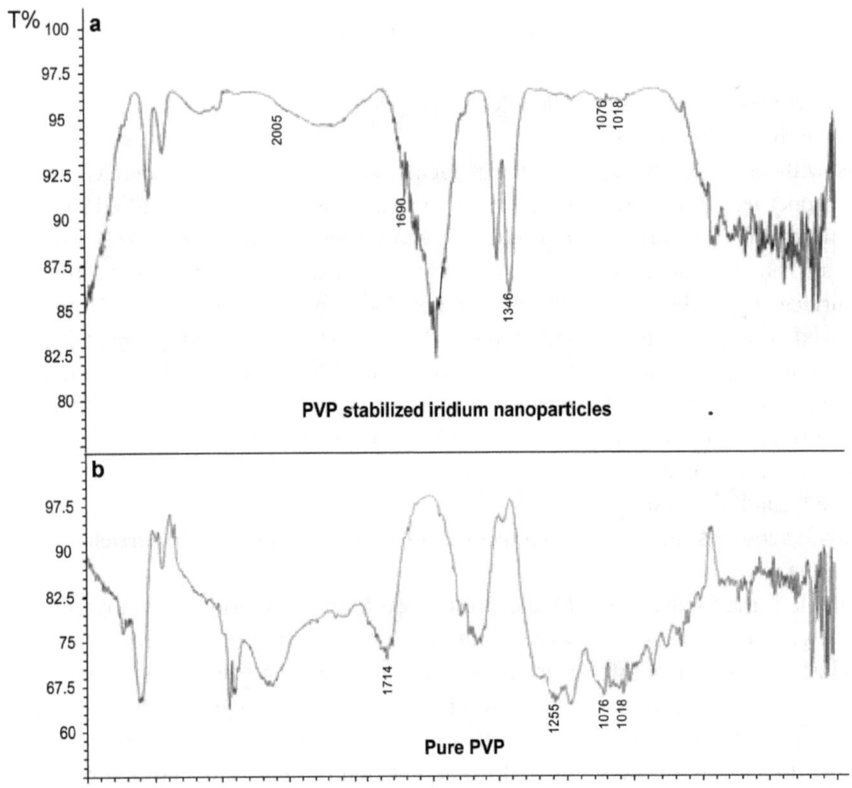

Fig. 1: FT-IR spectra of (a) PVP- stabilized iridium nanoparticles (b) Pure PVP

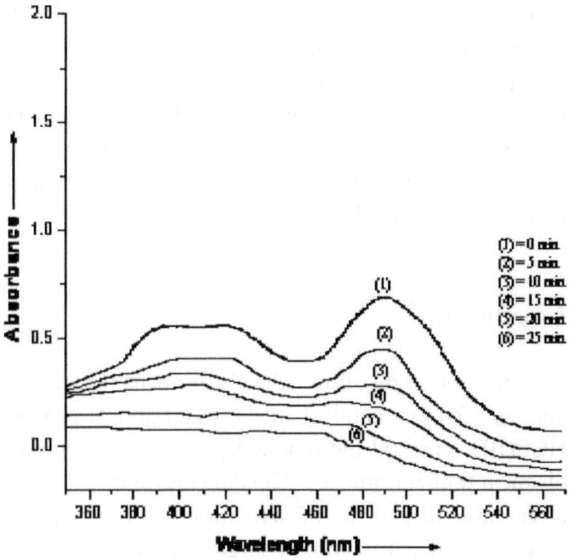

Fig. 2: Ultraviolet absorption spectra during the formation of iridium nanoparticles in the PVP- $H_2Ir\ Cl_6$ –methanol – water system after different irradiation time

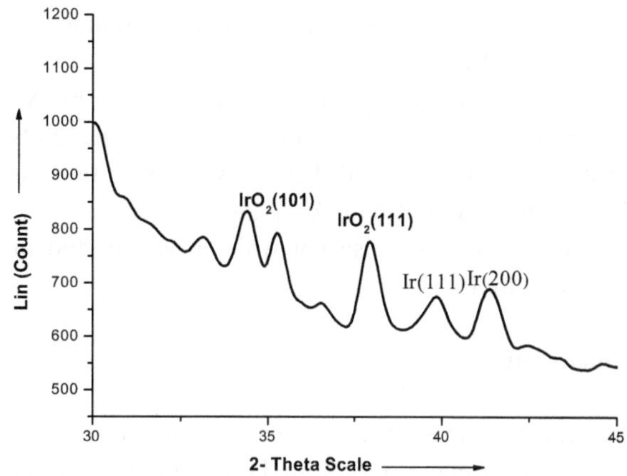

Fig. 3: XRD pattern of Ir (nano) particles

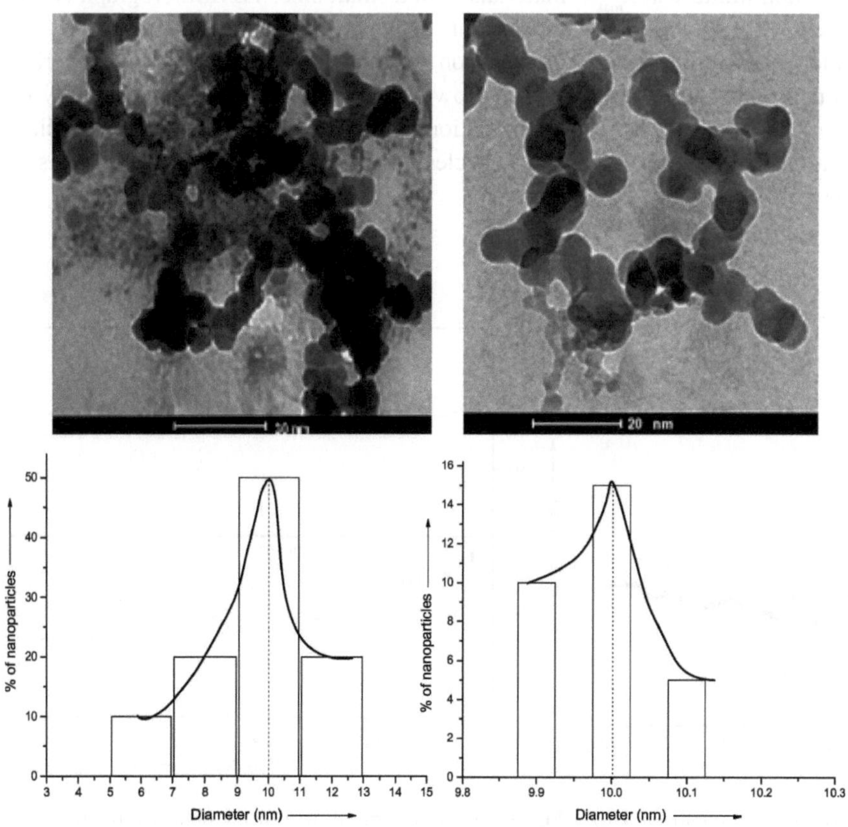

Fig. 4: TEM micrograph and corresponding particle size distribution histogram of iridium Nanoparticles

rated with no agglomeration tendency. Representative TEM micrograph shows that individual iridium nanoparticle is similar to spherical. The particle size distributions obtained from TEM images were fairly narrow and an average particle size was estimated to be around 10 nm.

The produced nanoparticles were analyzed by X-ray photoelectron spectroscopy (Fig. 5). The XPS spectrum shows typical Ir(0), IrO_2 absorption at 60.7eV, 65.0eV and 63.0eV, 67.5eV for $f_{7/2}$ and $f_{5/2}$. This absorption is consistent with those reported in the literature [12,13].

Kinetic Study

The kinetic measurements were performed at constant ionic strength of 0.50 mol dm^{-3} and at constant temperature 35.0 ±0.1°C. The requisite amount of each reactant was thermostatted at 35°C to attain thermal equilibrium. The appropriate quantities of reactants were mixed in a 100 ml iodine flask. The progress of reaction was measured by injecting a solution of amino acid into the reaction mixture at λ_{max} 420 nm corresponding to the $[Fe(CN)_6]^{-3}$. It was verified that there is negligible interference from other species present in the reaction mixture at this wavelength.

The kinetic study reveals that the oxidation of AA by HCF(III) in aqueous alkaline medium catalyzed by Ir-nano are same as catalyzed by bulk iridium [14,15]. The comparisons of results show that Ir-nano is more effective catalyst as compared to bulk iridium. They have the advantage of being recyclable and reusable but the recycled particles are agglomerated (Fig. 6). The catalyst was isolated via centrifugation and analyzed by XRD and TEM.

Conclusion

The present study shows that iridium nanoparticles acts as effective catalyst in the oxidation of AA by HCF(III) ions and the reaction proceeds through an electron transfer process. The data shows larger negative value of entropy of activation and a low value of energy of activation. A large –ive value of $\Delta S^{\#}$ shows that the complex formed is more ordered and polar. Iridium nanoparticles, which were prepared by reduction of H_2IrCl_6, remain stable after several months. The X- ray diffractograms of these metallic particles show broad peaks characterization of materials with a small size. TEM micrograph shows that iridium nanoparticles were separate by no agglomeration tendency and are similar to spherical with narrow size distribution. The kinetic study on the oxidation of AA by HCF(III) shows that iridium nanoparticles act as good catalyst in the above said reaction.

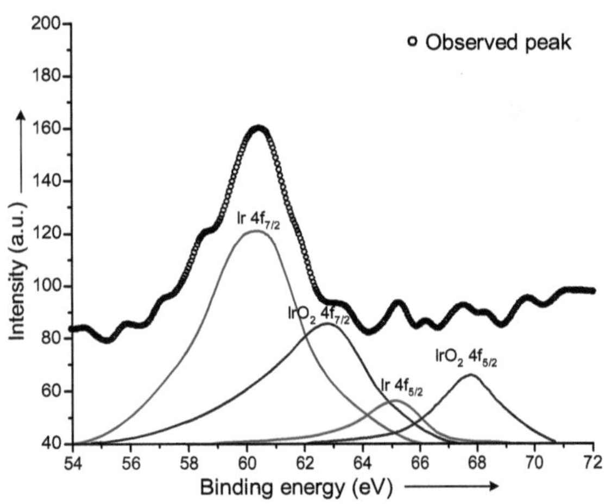

Fig. 5: X- ray photoelectron spectra of the PVP stabilized Ir(0) and IrO_2 nanoparticles

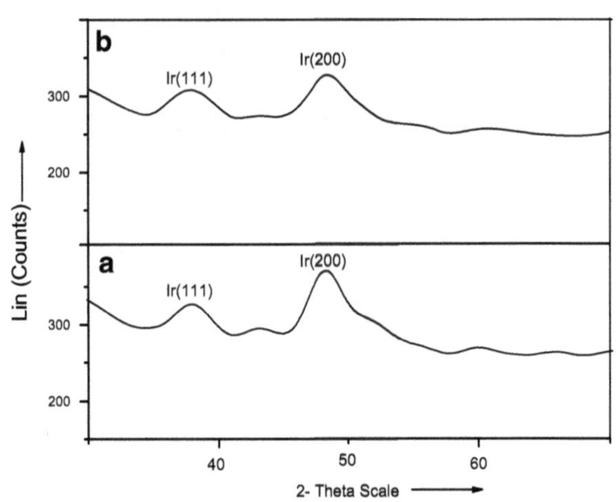

Fig. 6: XRD pattern of Ir(0)-nano obtained by recovery. **(a)** For recovered Ir-nano after completion of reaction with glycine **(b)** For recovered Ir-nano after the completion of reaction with alanine

Acknowledgment

The authors thank to Dr. Amish Josh, National Physical Laboratory, Delhi for providing facility of XPS analysis.

References

1. (a) V. L.Colvin, M. C. Schlamp and A. P. Alivisatos; Nature, 370 (1994) 354. (b) R. P. Andres et al. Nanostructured Materials Promise to Advance Range of Technologies. Chem. Eng. News, 1992, Nov. 18
2. V. Mevellec, A. Roucoux, E. Ramirez, K. Philippot and B. Chaudret; Adv. Syn. Catal. 346(1) (2004) 72–76
3. (a) L.N.Lewis, N. Lewis, J. Am. Chem. Soc., 1986, *108*, 7228. (b) L.N. Lewis, R. Uriarte, J. Organometallics, 9 (1990) 621
4. H. Hirai, H. Wakabayashi, and M. Komiyama; Bull. Chem. Soc. Jpn., 59 (1986) 545
5. M. Spiro; Catal. Today 17(1993) 517
6. G.Maayan and R.Neumann; Cat. Letter. 123(1–2) (2008) 41–45
7. C.W. Yen, M.L. Lin, A. Wang and S.A.Chen; J. Phys. Chem. C, 113(41) (2009) 17831–17839
8. V. Kiran, T. Ravikumar and A. K. Shukala; J. Electrochem. Soc. 157(8) (2010) 1201–1208
9. B.P. Fabio, R.O.Elaine, V.M. Oliveira and B.N. Fabio; Catal. Lett. 110(3–4) (2006)261–267
10. V. Mevellec, A. Roucoux , E. Ramirez and K. Philipoot; Adv. Synth. Catal. 346(2004)72–76
11. Y. Zhu, K. Chenyan, P.T. Ang and S.H. Narayan; Inorg. Chem. 47(13) (2008)5756–5761
12. S.P. In, S.K. Min, Y.K. Kung and S.L. Jae; Adv. Synth. Catal. 349(2007) 2039–2047
13. F. Bernardi, J.D. Scholten, G.H. Fecher and J. Dupont; Chem. Phys. Lett. 479(1–3) (2009) 113–116
14. A.Goel, Shakunj and S.Sharma; Int. J. Chem. Sci., 2008, 6(4), 1891–1899
15. A.Goel and S. Sharma; *Transition Met. Chem.*, 35(2010) 549–554

77

Toxic Level Heavy Metal Contamination of Road Side Medicinal Plants in Agra Region

J. Gautam[1], M. K. Pal[1], A. Singh[2], E. Tiwari[1] and B. Singh[1]
[1]Department of Chemistry, Institute of Basic Sciences, Dr. B. R. Ambedkar University, Khandari, Agra-282002
[2]Department of Chemistry, Hindustan College of Science and Technology, Farah, Mathura, India
Email: gjaiswar@gmail.com

Abstract

The heavy metals Cd, Co, Fe and Mn were determined in 9 samples of medicinal plants taken from road-side area of Agra, Firozabad and Shikohabad situated on NH-2 which are exposed with pollution. They were analyzed for heavy metal contamination by Atomic Absorption Spectroscopy (AAS). The results of AAS shows that the maximum concentration of Fe (154.4ppm) was found in Azadirachta indica (Agra) and the minimum concentration of Cd (0.000ppm) was found in Azadirachta indica (Agra and Shikohabad), Aloe-barbadensis (Agra and Shikohabad) and Eucalyptus globulus (Shikohabad). The phytochemical investigation were performed which show the presence of Alkaloids, Glycosides, Anthraquinones, Phenols and Flavonoids content in the medicinal plants which shows the maximum contents of phenols was found in Eucalyptus globulus (Agra, Firozabad and Shikohabad) and the minimum content of flavonoids was found in Azadirachta indica and Eucalyptus globulus located at Agra, Firozabad and Shikohabad.

Introduction

Environmental contamination of air, water, soil and food has become threat to continuous existence of biological life. This problem is being aggravated under ever increasing pressure on impacts of exponentially increasing population and pollution. Besides this compulsions of attaining higher degree of economic sufficiency and progress have inevitably necessitated the continuous exploration of industries which led to environmental pollution [1]. The heavy metal entry in to the biological system is increased now a day, due to their availability in the polluted environment and therefore accumulation in various plants parts has been increased in most of the plants parts. When contaminated plants parts are consumed as food to treat various diseases, besides accumulation, cause adverse effects in patients [2]. It is therefore necessary to check the crude drugs thoroughly for the heavy metal contamination before they are being released in to the market for selling. Medicinal plants can be easily contaminated with heavy metals from the environment (soil, water and air) during growth and the manufacturing processes when the ready-made products are produced [3]. Additional sources of heavy metal contamination are rainfall, atmospheric dust, plant protective agents and fertilizers [4–6]. The level of essential elements in plants is conditional, the content being affected by the geochemical characteristic of the soil and by the ability of plants to selectively accumulate some of these elements [7].

In the present study we determined the level of toxic heavy metals in road-side medicinal plants of Agra region. The concentration of heavy metals in medicinal plants was determined by atomic absorption spectroscopy (AAS) and phytochemical investigation was done by various tests for Alkaloids, Glycosides, Anthraquinones, Phenols and Flavonoids.

Experimental

All important chemicals such as Conc. HCl, H_2SO_4, HNO_3, $HClO_4$ and Glacial Acetic Acid were procured from Merck, Mumbai and were used as received. The organic solvents like Methanol, Benzene, Ammonia and Ferric chloride were obtained from Qualigens and they were used after purification by standard methods.

DOI:10.1007/978–3-642–23394-4_77,

Magnesium turnings were also obtained from Qualigens. The solvent acetone was obtained from Merck and used after double distillation and kept in a cool and dry place.

Collection of Plant Material

Azadirachta indica (Neem), *Aloe-barbadensis* (Aloe-vera) and *Eucalyptus globulus* (Eucalyptus) was collected from different geographical locations. The plants were sampled from areas, where they grow naturally. The environmental conditions under which the plants grow, the soil composition, and pH of the soil were recorded.

Sampling Areas

Spot 1 (Firozabad)
In this area the plant is exposed to both soil and air pollution. The samples were collected from roadside at a distance of 3 km from National Highways.

Spot 2 (Shikohabad)
This area is situated at a distance of 22 km from Spot 1. The sample were collected from roadside at a distance of 2 km. from National Highways.

Spot 3 (Agra)
This area is located at a distance of 62 km from spot 2. The samples were collected from roadside at a distance of 2 km. from National Highways.

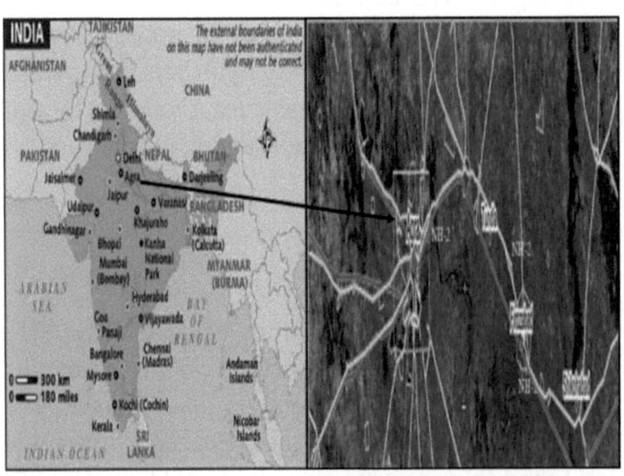

Fig. 1: Site Map

Sample Preparation, Extraction and Drying of Samples

The leaves of *Azadirachta indica* (Neem) and *Eucalyptus globulus* (Eucalyptus) were air dried under shade for one week, and *Aloe-barbadensis* (Aloe-vera) were air dried under shade for two weeks, and then powdered with the help of waring blender. Twenty-five grams of the powder was filled in the thimble and extracted successively with acetone using a Soxhlet extractor for 12 hours. All the extracts were concentrated using rotary vacuum evaporator and preserved in airtight bottle for further use. All the extracts were subjected for further analysis.

Extraction Method Development

Plant material and solvent were agitated with an orbital shaker at 115 rpm for three days at room temperature. The supernatant was drained and the residue was rinsed with 100 ml. extraction solvent (acetone) for 48 hours. The pooled extracts were filtered through Whatman filter paper and concentrated at 60°C using a rotary evaporator. The weight of the powders of *Azadirachta indica* (Neem), *Aloe-barbadensis* (Aloe-vera), and *Eucalyptus globulus* (Eucalyptus) were recorded. This process is given by Alkire et al [8].

Atomic Absorption Spectroscopy Study

AAS-ECIL 4141A spectrometer with pneumatic nebulization and mono-element lamps with hollow cathode was used for heavy metal contamination analysis.

Analysis of Phytochemicals

The obtained extracts were subjected to the following phytochemical screening tests according to Ateya, 1975 [9].

Test of Alkaloids
2 ml of plant extract corresponding to 2.5 gm of individual plant species was evaporated to dryness and the residue was heated on a boiling water bath with 5 ml of 2N HCl. After cooling the mixture was filtered and treated with few drops of Mayer's reagent composition (1.358 gm $HgCl_2$ and 5 gm KI dissolve in 100 ml distilled water). The sample was then observed for the presence of turbidity/ or precipitation.

Test of Anthraquinones (Borntreger's Test)

5 gm of plant extract was shaken with 10 ml of benzene. This was filtered and 50 ml of 10 % ammonia solution was added to the filtrate, the mixture was again shaken. Violet colors in ammoniacal phase confirm the presence of anthraquinones.

Test of Flavonoids

The solvent extract (5 ml corresponding to 1 gm of the individual plant species) was treated with a few drops of conc. HCl and 0.5 gm of magnesium turnings. The presence of flavonoids was indicated if pink or magenta-red color developed within 3 minutes.

Tests of Glycosides

0.5 gm. of solvent extract of individual plant species was dissolved in 2.0 ml of glacial acetic acid containing 1 drop of $FeCl_3$ solution. This was then underlaid with 1.0 ml of conc. H_2SO_4. A brown ring obtained at the interface indicated the presence of glycosides.

Tests of Phenol

Solvent plant extract and few drops of neutral ferric chloride solution 5 %, intense color developed indicated the presence of phenols.

Results and Discussion

Atomic Absorption Spectroscopy Study

The results of the atomic absorption spectroscopy of cadmium (Cd), Cobalt (Co), Manganese (Mn) and Iron (Fe) are shown in figures. The concentration of heavy metals not only varies from taxon to taxon but also within the same taxon in different organs [10] of the same plant grown under similar or in different environmental conditions [11]. However, in the present experiment the distribution of heavy metals in polluted, less-polluted and non-polluted area plants have shown variations. Cadmium concentration ranged be-

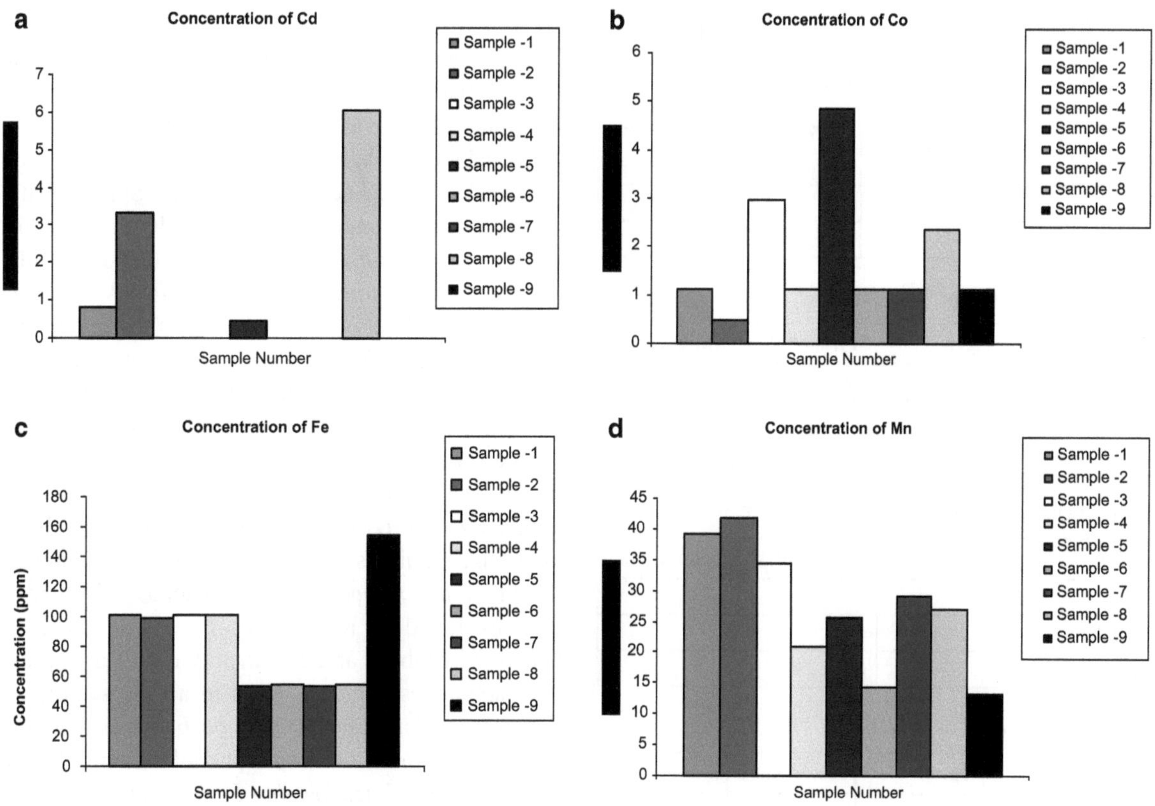

Fig. 2: AAS analysis: (a) Cd, (b) Co, (c) Fe and (d) Mn. Sample-1=*Eucalyptus globules* (Firozabad), Sample-2=*Eucalyptus globules* (Agra), Sample-3=*Aloe-barbadensis* (Shikohabad), Sample-4=*Aloe-barbadensis* (Agra), Sample-5=*Aloe-barbadensis* (Firozabad), Sample-6=*Azadirachta indica* (Shikohabad), Sample-7=*Eucalyptus globules* (Shikohabad), Sample-8 = *Azadirachta indica* (Firozabad) and Sample-9= *Azadirachta indica* (Agra)

tween 6.040 ppm as maximum in Azadirachta Indica (Firozabad) and 0.00 ppm as minimum in *Azadirachta Indica* (Shikohabad), *Azadirachta Indica* (Agra), *Aloe-barbadensis* (shikohabad), *Aloe-barbadensis* (Agra), and *Eucalyptus globules* (shikohabad). Cobalt element ranged between 4.840 ppm as the maximum in *Aloe-barbadensis* (Firozabad) and 0.480 ppm as minimum in *Eucalyptus globules* (Agra). Manganese element ranged between 41.76 ppm as the maximum in *Eucalyptus globules* (Agra) and 13.20 ppm as minimum in *Azadirachta indica* (Agra). Iron element ranged between 154.4 ppm as the maximum in *Azadirachta indica* (Agra) and 52.95 ppm as minimum in *Eucalyptus globules* (Shikohabad) and *Aloe barbadensis* (Firozabad).

Table 1: Observation of alkaloids

Trees	Agra	Firozabad	Shikohabad
Neem	+,+	+	+,+
Eucalyptus	+	+	+
Aloe-Vera	+	+	–

Table 2: Observation of Anthraquinones

Trees	Agra	Firozabad	Shikohabad
Neem	–	–	–
Eucalyptus	+	+	+
Aloe-Vera	+,+	+	+

Table 3: Observation of Flavonoids

Trees	Agra	Firozabad	Shikohabad
Neem	–	–	–
Eucalyptus	–	–	–
Aloe-Vera	+	+	+

Table 4: Observation of Glycosides

Trees	Agra	Firozabad	Shikohabad
Neem	+	–	+
Eucalyptus	–	+	+
Aloe-Vera	+	+	+

Table 5: Observation of Phenol

Trees	Agra	Firozabad	Shikohabad
Neem	+	+	+
Eucalyptus	+,+	+,+	+,+
Aloe-Vera	–	–	–

Heavy metal content of Cd, Co, Mn and Fe are increased in all the plant species collected from different polluted areas beyond the permissible limits assigned by WHO.

Phytochemical Study

Table 1–5 summarizes the quantitative determination of phytochemical constituents Alkaloids, Anthraquinones, Flavonoids, Glycosides and Phenols of *Azadirachta indica, Eucalyptus globules and Aloe-barbadensis* collected from Agra, Firozabad and Shikohabad respectively. The result shows the maximum contents of phenols was found in *Eucalyptus globulus* (Agra, Firozabad and Shikohabad) and the minimum content of flavonoid was found in *Azadirachta indica* and *Eucalyptus globulus* located at Agra, Firozabad and Shikohabad.

Conclusion

The concentration of Cadmium ranged between 6.040 ppm as maximum in *Azadirachta Indica* (Firozabad) and 0.00 ppm as minimum in *Azadirachta Indica* (Shikohabad and Agra), *Aloe-barbadensis* (Shikohabad and Agra) and *Eucalyptus globules* (Shikohabad). Cobalt element ranged between 4.840 ppm as the maximum in *Aloe-barbadensis* (Firozabad) and 0.480 ppm as minimum in *Eucalyptus globules* (Agra). Iron element ranged between 154.4 ppm as the maximum in *Azadirachta Indica* (Agra) and 52.95 ppm as minimum in *Eucalyptus globules* (Shikohabad) and *Aloe-barbadensis* (Firozabad). Manganese element ranged between 41.76 ppm as the maximum in *Eucalyptus globules* (Agra) and 13.20 ppm as minimum in *Azadirachta Indica* (Agra). The phytochemical analysis shows that alkaloids contents found in all the samples except *Aloe-barbadensis* (Shikohabad), glycosides contents found in all samples except *Azadirachta Indica* (Firozabad) and *Eucalyptus globules* (Agra), Anthraquinones contents found in all the sample of *Eucalyptus globules* and *Aloe-barbadensis* but not found in *Azadirachta Indica*, phenol contents found in all samples of *Azadirachta Indica* and *Eucalyptus globules* but not present in *Aloe-barbadensis* and flavonoids study reveals that it found only in samples of *Aloe-barbadensis*.

References

1. J.U. Lagerweff and A.W. Specht; Env. Sci. Technical. 4 (1970) 583.
2. J.U. Lagerweff; Soil Science Soiety of America, Medison, Wisconsin, (1972) pp. 619.
3. K. Chan, Chemosphere, 52 (2003) 1361.
4. R.M. Harrison and M.B. Chirgawi; EnViron. 83, (1989) 13.
5. J. Lars, British Medicinal Bulletin, 68 (2003) 167.
6. R.P. Djingova, G. Kovacheva and B. Wagner; Sci Total EnViron. 308 (2003) 235.
7. M.M. Al-Alawi and K.L. Mandiwana, J. Hazardous Material, 148 (2007) 43.
8. B.H. Alkire Erne, F. Ozanam and J.N. Chazalviel, J. Phys. Chem., 2 (1999) 68.
9. B.G. Ateya, B.M. Abo-Elkhair and I.A. Abdel-Hamid, Corrosion Science, 16 (1975) 163.
10. M.K. John and C.J. Laerhoven; Environ. Pollut. 10 (1976) 163.
11. R.Y.P. Bhatia, M. Prabhakar and R.K. Verma; J. Ind. Aced. Foren. Sci. 26(2) (1987) 20.

78
Biochemical Characteristics of Aerosol at a Suburban Site

Ranjit Kumar[1], K. M. Kumari[2], Vineeta Diwakar[3] and J. N. Srivastava[3]
[1]Department of Chemistry, Technical College,
[2]Department of Chemistry, [3]Department of Botany, Faculty of Science
Dayalbagh Educational Institute (Deemed University)
Dayalbagh, Agra-282110
Email: rkschem@rediffmail.com,

Abstract

Bio-aerosol plays very important role in climate change, rain patterns and public health. The concentration of airspora varied with the change in meteorological conditions as well as with the change in surrounding localities. Present study deals with biochemical study of aerosol at a suburban site in Agra. The maximum fungal colonies were recorded during the month of October and lowest were recorded in month of June. Aspergillus Species were dominant amongst isolated fungi. Protein estimation gives surrogate information about biological components of aerosol. The level of protein was in relation with presence of biological organism and debris in the aerosol.

Introduction

Air pollution has worsened the health status of residents of both developed and developing countries. Research in the past decades confirms that outdoor air pollution contributes to morbidity and mortality. Aerosol plays significant role in health problems due to its biotic and abiotic constituents. Although the biological mechanisms are not fully understood, state of art epidemiological studies have found consistent and coherent association between air pollution and various health outcomes.

Biotic components or bioaerosol is a group of organic aerosols ranging form 10 nm to 100 μm that are either alive, carry living organisms or are released form living organisms (viz., bacteria, fungi, virus, pollen, cell, cell debris and biofilms). Bioaerosols, as the name implies, have a biological origin. The surfaces of living and dead plants are important natural sources of airborne bacteria and fungal spores. Wind, waves and even rain may aerosolize microorganisms from natural accumulations of water. Bioaerosols occur naturally everywhere from the middle of the ocean to the middle of the arctic. They consist of particles such as bacterial spores and cells, viruses, fungal spores, protozoa, pollen, fragments of insects and skin scales. These include both fine and coarse particles. Viruses are among the smallest bioaerosol particles; some species are only a few tens of nanometers in size. On the other end of the size spectrum, pollen grains can be over 100 μm in diameter. As with other types of aerosols, bioaerosol concentrations vary significantly from location to location. Typical outdoor concentrations of airborne bacteria range from 100 to 1000 cfu/m^3 (cfu = colony forming units). Bioaerosol concentrations as high as 10^{10} cfu/m^3 may occur in environments such as textile mills. The presence of various types of bioaerosol in indoor air, in the troposphere and even in the stratosphere has been established [1]. Although biological components contributes 20 % load of aerosol but the biological fraction of ambient particulate matter can potentially cause significant health effects [2–5] and in the presence of chemical constituents its effects get exacerbated. Many numbers of diseases are caused by biological components viz. different microorganism, fungal spores, debris suspended in the atmosphere, hence biological characterization along with chemical constituents of aerosol is needed and performed in the present study.

M.M. Srivastava, L.D. Khemani, S. Srivastava, *Chemistry of Phytopotentials: Health, Energy and Environmental Perspectives*, DOI:10.1007/978–3-642–23394-4_78, © Springer-Verlag Berlin Heidelberg 2012

Materials and Methods

Sampling Site

The sampling site is Dayalbagh, a suburban site in Agra (27° 10′ N, 78° 05′ E) which is located in north central India, 200 km south east of Delhi. Two thirds of its peripheral boundaries (SE, W and NW) are surrounded by the Thar Desert of Rajasthan and is therefore a semi arid area. The soil type is a mixture of sand and loam, containing excess of salts. Agriculture is the major activity. The study site is about 10 km away from the industrial areas.

Sample Collection

To achieve the proposed study aerosol samples were collected using Glass fiber filter as a collecting surfaces. Predesiccated, preweighed and sterile filter was used and exposed in the atmosphere for aerosol collection. Samples were collected for the month of May-October (summer and monsoon season).

Analysis

After the sampling, filter was used for biological and chemical characterization. Filter paper was dissolved in 100 ml deionized and sterile water and filtrate was used for further analysis.

Fungal Analysis

One ml of filtrate was inoculated in SDA medium contained in sterile Petri plates at 27 °C for three days. After incubation period, the colonies appeared on SDA plates were counted and subculture until pure fungi were obtained. Fungal identification was carried out on the basis of shape, size and color of the colonies and compared with standard description on Fungi [6].

Chemical Analysis

Chemical analysis was carried out by protein estimation using 1 ml filtrate solution by Lowry method [7].

Results and Discussion

A total of 25 fungal types including 11 species were recorded. Figure 1 shows the monthly average data of predominant fungal colonies recorded and identified. The maximum fungal colonies were recorded during the month of October and lowest in the month of June. The high concentration of fungal counts in the month of October may be due to moderate rainfall, medium temperature and high humidity which are suitable of microbial growth. The minimum concentration in the month of June may be due to high temperature which is disastrous for fungal survival. Plate 1 shows fungi isolated from air. Total 11 species viz., *Aspergillus niger, A. flavus, A fumigates, Curvularia, Fusarium, Helminthosporium, Mucor, Penicillium, Rhizoporus, Trichothicium, Verticillium, Pleospora, Alternaria* were isolated.

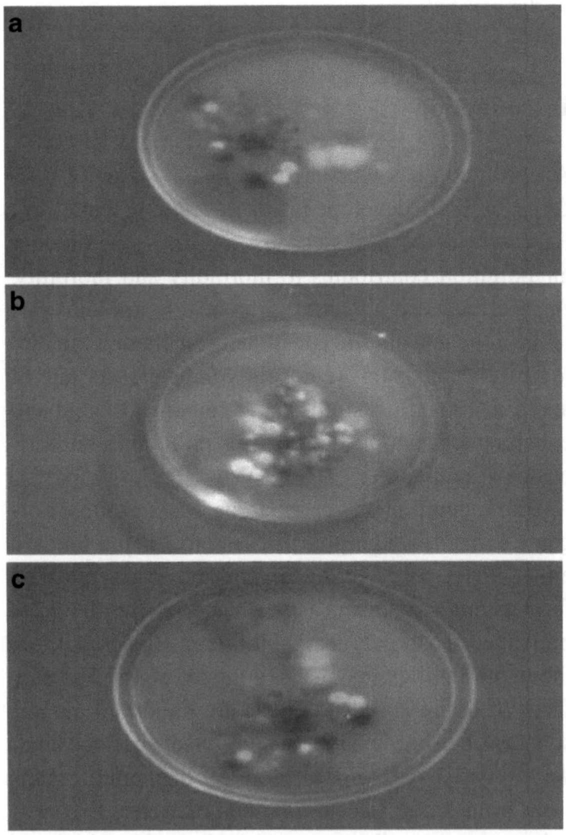

Plate 1: **(a)** *Aspergillus niger, A. flavus, Curvularia, Penicillium, Mucor, Pyricularia.* **(b)** *A. niger, A. fumigatus, Mucor, Penicillium, Verticillium, Alternaria, Fusarium, Cladosporium.* **(c)** *A. niger, Curvularia, Fusarium, Helminthosporium, Penicillium, Trichothicium*

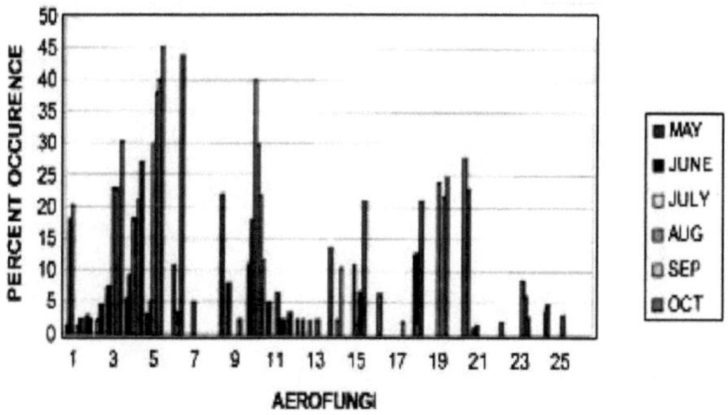

Fig. 1: Average monthly percent occurrence of dominant aero fungi at Agra.

Protein Estimation

Protein was estimated quantitatively. Protein estimation from exposed filter paper samples gives us information about the presence of biological organisms and debris in the aerosol. Figure 2 presents level of soluble protein in the filtrate of aerosol samples. Highest quantity of soluble protein was recorded during the month of October and lowest in the month of May.

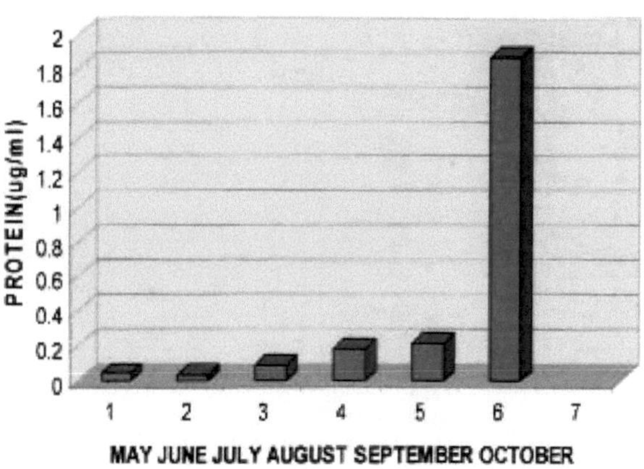

Fig. 2: Protein estimation

In the month of October soluble protein was found to be 1.869 µg/ml while in the month of May it was found to be 0.35 µg/ml. The protein concentration is directly related to the concentration of microorganism in the aerosol.

Conclusion

Ambient aerosol samples were collected at Dayalbagh, a suburban site of Agra region from May to October. Biological and chemical characterizations of aerosols were carried out. The concentration of airspora was highest in October and lowest in June. Atmospheric conditions viz., temperature, relative humidity, rainfall etc. plays very important role. The major fungal genera and species were found to be *Aspergillus, Mucor, Penicillium, and Fusarium* while *Aspergillus niger* was found to be dominant type. The level of protein was highest in October. It is akin to concentration of airspora. Protein gives surrogate information about the presence of biological organisms.

References

1. P. Ariya, and M. Amyot; New directions: the role of bioaerosols in atmospheric chemistry and physics; Atmos. Env. 38(2004) 123.
2. Schneider et al.; A coccidioidomycosis outbreak following the Northridge. Calif, Earthquake; Journal of American Medical Association 277 (1997) 904.
3. A.G. Miguel, G.R. Cass, M.M. Glovsky, J. Weiss; Allergens in paved road dust and airborne particles; Environ. Sc. and Tech. 33 (1999) 4159.
4. Boreson et al.; Atmos. Env. 38 (2004) 6029.
5. Viana et al.; characterizing exposure to PM aerosols for an epidemiological study; Atmos. Environ. 42 (2008) 1552.
6. M.P. Ellis and J. Ellis; Microfungi on land plants, An identification handbook, International book and periodical supply service, New Delhi India (1985).
7. O.H. Lowry et al.; The protein content determination; J. Biol. Chem. 193 (1951) 265.

79
Green Nanotechnology for Bioremediation of Toxic Metals from Waste Water

A. Kardam, K. R. Raj and S. Srivastava
Department of Chemistry, Faculty of Science, Dayalbagh Educational Institute, Agra 282110
Email: abhishekkrdm@gmail.com

Abstract

The present piece of work reports the preparation and characterization of nano cellulosic fibers (NCF) derived from Moringa Oleifera seed pods for decontamination of toxic metals. AFM image and SEM micrograph point out their long rod like elongated nano fibrillated morphology. These cellulosic fibers were characterized by FTIR & XRD analysis. The average grain size, calculated from the Debye Scherer equation was found to be 10 nm. The nano remediation batch experiments for cadmium and lead metal ions showed that prepared NCF (0.500 gm) posse's removal efficiency of 81.7 % and 84.5 % for Cd (II) and Pb (II) from initial concentration of 25 mg/l compared to control native Moringa Oleifera experiments of biomass dose (2gm).

Introduction

Appropriate integration of environmental chemistry and nanomaterials science creates significant breakthrough in a wide variety of environmental technologies as nano remediation. The opportunities and challenges of using nanomaterials in the purification of surface water, groundwater and industrial wastewater streams is a matter of continuing concern. Application of nanoparticles for the removal of heavy metals has come up as an interesting area of research. The ability to control, manipulate and design materials on the nano scale to remediate contaminants simultaneously avoiding environmental hazardous is a major challenge of 21st century i. e., *Green Nanotechnology*. Nanoparticles exhibit good adsorption efficiency especially due to higher surface area and greater active sites for interaction with metallic species [1–3]. The variety of inorganic nano structured materials has been explored for remediation but found to be associated with toxicity issues recently [4]. Unfortunately, not many of the most recent developments in the nature are able to satisfy the core concept of sustainability. One way to address issues related to sustainability is to incorporate renewable materials of miniaturized elements [5] of natural origin.

The use of organo nano structured material from pulp or other agricultural waste biomasses provides an important dimension to the environmental treatment technologies. Cellulose constitutes the most abundant and renewable polymer resource available worldwide and comprises of repeating β-D-glucopyranose units covalently linked through acetal functions between the OH groups of C_4 and C_1 carbon atoms providing highly suitable sorption characteristic properties of hydrophilicity, chirality and reactivity. In continuation of our work on biosorption of toxic metals using agricultural waste [6–9], the present piece of work reports the preparation and characterization of nano cellulosic fibers (NCF) derived from *Moringa Oleifera* seed pods for decontamination of toxic metals.

Experimental

Extraction and Preparation of Cellulose Nano fibers Physico-chemical treatment method

The *Moringa oleifera* seed pods were obtained in the nearby area of Dayalbagh Educational Institute, Agra and their seeds were extracted from the pods. Then the native pods without seeds were washed with distilled water several times and were dried in an oven at 80°C for 24 h. Then they were chopped to an approximate length of 5–10 mm and finally crushed into small fibers with the help of a mixer.

M.M. Srivastava, L. D. Khemani, S. Srivastava, *Chemistry of Phytopotentials: Health, Energy and Environmental Perspectives*, DOI:10.1007/978-3-642-23394-4_79,

Fibers were soaked into the sodium hydroxide solution of 0.5 M for 2 h and then washed several times with distilled water. After then, bleaching treatment with a sodium chlorite solution (pH 4) for 1 h at 50 °C was performed to remove the remained lignin and then washed with distilled water repeatedly. The pretreated pulp was hydrolyzed by 30 % 1M HCL + 70 % H_2SO_4 for 3h at 70 °C, and then washed with distilled water repeatedly. The acid treatment hydrolyzed the hemicelluloses and pectin by breaking down the polysaccharides to simple sugars and hence released the nano cellulose fibers. Then these fibers were dried and used for the sorption studies.

Microscopic Analysis

Scanning Electron Microscopy (SEM)

The morphology of nano cellulosic fibers were investigated by scanning electron microscopy (SEM) using a Leo Scanning Electron Microscope (LEO-400) instrument.

Atomic Force Microscopy (AFM)

Surface topographic image was taken by employing Atomic Force Microscope (Nano surf Easy-Scan, Switzerland; Version 1.8). The set point force was fixed at 20 μN for all the images, which were obtained for 256 × 256 data points for each scan size 5 × 5 μm. A drop of diluted nano fibers aqueous suspension (sonicated) was allowed to dry on optical glass substrate at room temperature and analyzed subsequently.

Fourier Transform Infrared Spectroscopy (FTIR)

FTIR analysis in solid phase was performed using a Shimadzu 8400 Fourier Transform Infrared spectroscopy. Spectra of the fiber before and after chemical treatment were recorded.

X-ray Diffraction (XRD)

X-ray diffraction pattern of samples was recorded by glancing angle X-ray diffractometer (Bruker AXS D8 Advance, Germany) using Cu Ka radiation at 40 kV and 30 mA. Scattered radiation was detected in the range 2Θ = 5–40°, at a speed of 2°/min.

Sorption Studies

Sorption studies were carried out in order to evaluate the sorption capacity of prepared Cellulose nano fibers and in comparison to the native *moringa oleifera* pod fibers.

Batch experiments (triplicates) were performed in clean air-conditioned environmental laboratory with Cd (II) and Pb (II) metal solutions separately. After proper pH adjustments [pH 6.5], a known quantity of biomaterial mixture was added and finally metal-bearing suspension was kept under stirring until the equilibrium conditions are reached. After shaking, suspension was allowed to settle down. The residual biomaterial sorbed with metal ions was filtered. Filtrate was collected and subjected for metal ions estimation using Flame Atomic Absorption Spectrometer [Perkin Elmer-2380]. Percent metal sorption by the sorbent was computed using the equation: $(C_o\text{-}C_e)/C_o \times 100$ where 'C_o' and 'C_e' are the initial and final concentration of metal ions in the solution.

The detail of the methodology remained the same as in our earlier publications [6–9].

Metal loaded biosorbent obtained from sorption experiment was transferred to Erlenmeyer flask and shaken with HNO_3 (0.5M, 50 ml) acid for 40 min. The filtrate was analyzed for number of regeneration cycles of biomass.

Results and Discussion

The Cellulose nano fibers extracted from native Moringa oleifera pod fibers were approximately weight about 40.5 – 42.8 % by w/w.

Morphological and Structural Characterization of Cellulose Nano Fibers

AFM image (Fig. 1) and SEM micrograph (Fig. 2) point out their long rod like elongated nano fibrillated network with high length (micrometer scale) and low diameter (>50 nm). However, individual nanoparticles cannot be distinguished.

The X-ray diffraction pattern (Fig. 3) display two well-defined characteristic peaks of cellulose at 2Θ = 12.5° and 2Θ = 22.5°. The average grain size, calculated from the Debye Scherer equation was found to be 10 nm.

FT-IR spectroscopy is a nondestructive method for studying the physico-chemical properties of lignocellulosic materials. The FT-IR spectra of untreated (Native) and chemically treated *Moringa oleifera* pod fi-

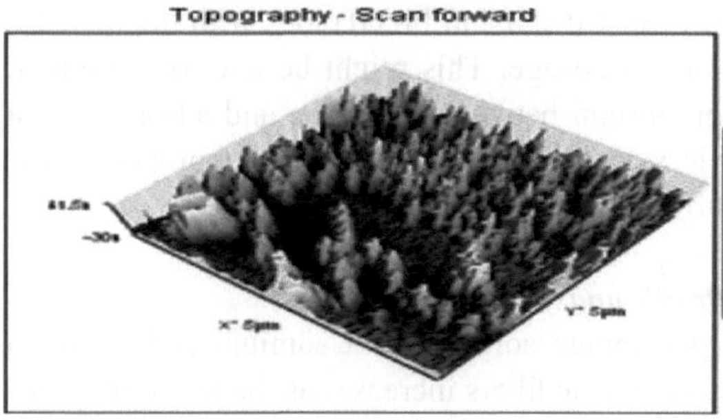

Fig. 1: AFM image of Cellulose nano fibers

Fig. 2: SEM micrograph of cellulose nano fibers

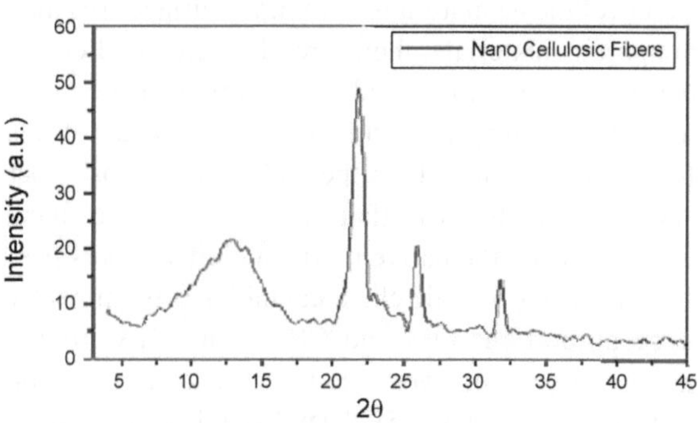

Fig. 3: XRD diffractogram of cellulose nano fibers

bers are shown in Fig. 4. Both the spectra are dominated by the peaks at 3,408 and 1,060 cm-1 that correspond to the stretching vibrations of O–H in cellulose and C–O in hemicelluloses and cellulose, respectively. The peak at 1,640 cm-1 in both samples is indicative of the water molecules and C = O bonds of hemicelluloses [10]. The variation of this peak in the treated sample in comparison with other native fibers spectra indicates the partial removal of hemicellulose. The peaks at 1507 and 1436 cm-1 in the untreated *moringa oleifera* pod fibers represent the aromatic C=C

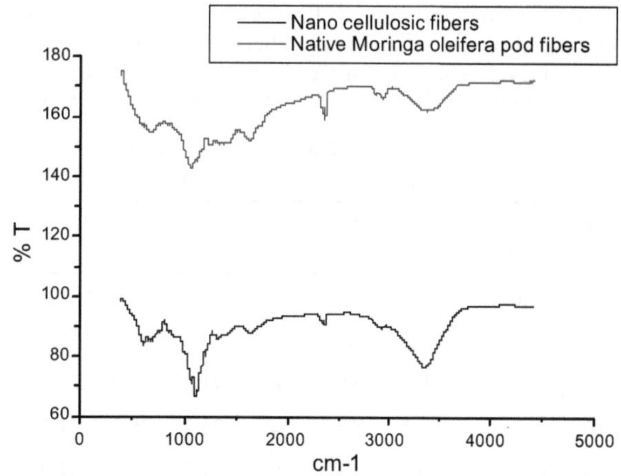

Fig. 4: FT-IR spectra of the untreated (Native) and chemically treated *Moringa oleifera* pod fibers

stretch of aromatic rings of lignin [11–13]. The intensity of these peaks decreased in the chemically treated fibers, which was attributed to the partial removal of lignin. The increase of the band at 896 cm-1 in the chemically treated fibers indicates the typical structure of cellulose. The absorbency at 1238 cm-1 is associated with the C–H, O–H or CH2 bending frequencies; and absorbencies at 1058 and 896 cm-1 are related to the C–O stretching and C–H rocking vibrations of cellulose.

Sorption Studies

Sorption studies lead to the standardization of the optimum conditions nano cellulosic fibers as: Cd (II) concentration (25 mg/l), contact time (40 min) and volume (200 ml) at pH 6.5 for maximum Cd (II) and Pb (II) removal i. e. 81.4 % and 84.7 % respectively.

It was observed that the % sorption efficiency of the nano fiber media were significantly (>2 folds) higher than the native *Moringa oleifera* pod fibers. This is mainly due to the high surface area to mass ratios of Cellulosic nano fibers which can greatly enhance the adsorption capacities. After the chemical treatment of the native Moringa oleifera fibers there is successive decrease in the lignin content this comprises of many aromatic ring linkage compounds. The physico chemical treatment (acid hydrolysis with H_2SO_4 + HCL) of the native fibers also increases the negative charge on the surface of the nano fibers

Table 1: represents Sorption efficiency (%) of native moringa oleifera fibers and cellulose nano fibers for Cd (II) and Pb (II) ions removal as function of metal concentration and biomass dosage at volume (200 ml), contact time (40 min) and pH 6.5

	Moringa oleifera pod fibers			**Nano cellulosic fibers**		
Biomass dosage (gm)	**0.5**	**1**	**2**	**0.5**	**1**	**2**
Cd (II) Conc.						
1 ppm	30.3	40.2	49.2	63.8	66.1	65.2
5 ppm	38.4	51.2	58.7	71.8	73.2	73.1
10 ppm	44.5	57.1	64.7	74.7	75.7	74.2
25 ppm	49.5	63.2	68.4	81.7	82.1	81.9

	Moringa oleifera pod fibers			**Nano cellulosic fibers**		
Biomass dosage (gm)	**0.5**	**1**	**2**	**0.5**	**1**	**2**
Pb (II) Conc.						
1 ppm	37.3	40.2	51.3	69.2	71.1	69.2
5 ppm	43.4	47.2	59.9	74.3	75.2	73.1
10 ppm	48.5	63.1	68.6	79.8	80.7	79.2
25 ppm	51.5	67.2	**73.8**	84.5	86.1	81.9

which enables the electro static interaction with the cationic metal species of Cd (II) and Pb (II) ions.

Effect of Metal Concentration

Percent sorption of Cd (II) and Pb (II) ions increased with the increase of metal concentration finally reaching to the optimum level for Cadmium and lead at 25 mg/cm^3 respectively. Later, an increase in initial concentration decreased the percentage binding. These observations can be explained by the fact that at low concentrations, the ratio of sorptive surface area to the metal ions available is high and thus, there is a greater chance for metal removal. When metal ion concentrations are increased, binding sites become more quickly saturated as the amount of biomass concentration remained constant.

Effect of Biomaterial Dosage

Percent sorption of Cd (II) and Pb (II) ions increased with the increase of biomass dosage in case of native fibers (Moringa oleifera fibers). However there is no sufficient increase in the % sorption efficiency of NCF for Cd (II) and Pb (II) ions with the increase in biomass dosage. This might be due to attainment of equilibrium between adsorbate and adsorbent at a very less dose of 0.5 g due to the novel properties exhibit by the nanomaterials.

pH Profile and Metal Binding

The percentage sorption of Cadmium and Lead on Nano cellulosic fibers increases as the pH of the solution increased from 4.5 to 8.5. The pH profile for the Cd (II) and Pb (II) sorption on nano cellulose fibers shows that metal sorption is a function of pH, exhibiting maximum sorption at pH 6.5. There was no significant difference in sorption behavior with further increase in pH up to 7.5. Investigation on pH variation beyond 7.5 yielded an apparent increase in sorption up to pH 8.5, which might be due to the precipitation carryover of Cd (II) and Pb (II) ions which starts at pH 7.5 [14]. Metal precipitation interferes and is undistinguishable from sorption phenomenon at pH 7.5. In fact, biosorption of Cadmium and lead occurs in the pH range 4.5 to 7.5 only. Based on our experimental findings and pertinent information available on the relevant topic, we synthesize an appropriate mechanism for metal binding to the nano cellulose fibers. Lignocellulosic materials have more surface area than nonporous materials; therefore, they are good candidates for sorption material. At the initial stage, metal ions are sorbed by some interaction and van der Waals forces between metals in the bath and cellulose fiber surface. In addition to that the cellulose nano fibers extracted from the native moringa oleifera fibers possesses various organic chemical moieties prominently large proportions OH^- and COO- entities. Carboxyl groups in cellulose are able to act as sources of ion exchange capacities [15]. They have the ability to interact with cationic metal ions and are likely to be active sites for the sorption of metal ions by involving electrostatic attraction between negatively charged groups of carboxylic acids and metallic cations.

Conclusion

The proposed bionanomaterial based remediation is likely to provide new avenues as cost effective, environment friendly, fast pre-treatment step for the decontamination of toxic metals from waste water as green nanotechnology.

Acknowledgements

The authors gratefully acknowledge Professor V.G Das and Professor L. D. Khemani, Director and Head of the Department of Chemistry, Faculty of Science, Dayalbagh Educational Institute, Dayalbagh, Agra for providing the necessary research facilities.

References

1. V. L. Colvin; Nature Biotech. 10 (2003) 1166–1170.
2. A. S. Nair and T. Pradeep, Applied Nanoscience, (2004) 59–63.
3. M. S. Diallo, S. Christie, P. Swaminathan, J. H. Johnson and W. A. Goddard. Environ. Sci. Technol., 39 (2005) 1366–1377.
4. B. Karn, T. Kuiken, M. Otto, Environ Health Perspect, 117 (12) (2009) 1823–1831.
5. B. Wang and M. Sain, BioResources 2 (3), (2007) 371–388.
6. A. Kardam, P. Goyal, J. K. Arora, K. R. Raj and S. Srivastava. Nat Acad Sci Lett. 32 (5–6) (2009) 179–181.
7. A. Kardam, K. R. Raj, J. K. Arora and S. Srivastava. Journal of Water Resource and Protection. 2 (2010) 339–344.
8. K. R. Raj, A. Kardam, J. K. Arora and S. Srivastava. Journal of Water Resource and Protection. 2 (2010) 332–339.
9. A. Kardam, K. R. Raj, J. K. Arora and S. Srivastava. Journal of instrumentation society of India, 40 (3) (2010) 175–176.
10. X. F. Sun, R. C. Sun, P. Fowler, M. S. Baird, Carbohydr Polym, 55 (2004) 379–391.
11. S. Panthapulakkal, A. Zereshkian, M. Sain, Bioresour. Technol, 97 (2006) 265–272.
12. B. Xiao, X.F Sun, R.C Sun, Polym. Degrad. Stab, 74 (2001) 307–319.
13. X. F. Sun, F. Xu, R. C. Sun, P. Fowler, M. S. Baird, Carbohydr. Res. 340 (2005) 97–106.
14. M. Iqbal, A. Saeed, N. Akhtar, Bioresource Technol., 81(2) (2002) 151–153.
15. B. L. Browning, Methods of Wood Chemistry. Interscience Publisher: New York, 1967.

80
Phyto Conservation: Folk Literature, Mythology and Religion to its Aid

M. R. Bhatnagar
Department of Sanskrit, Faculty of Arts, Dayalbagh Educational Institute, Dayalbagh, Agra- 282110
Email: anilb38@ gmail.com

Abstract

Plants have profoundly influenced the culture and civilization of India. Plants, an integral part of Indian life have had great impact on Folk literature, Mythology and Religion. The present study was undertaken to make a preliminary survey of the role played by Folk literature, Mythology and Religion in Phyto- conservation. The study reveals that on ceremonial functions, Folk songs, and Mythology and Folk tales several plants are associated with Gods and Goddesses, worshipped and considered as their abode. The sacred groves present in Western Ghats of India are vegetation preserved by tribals on religious grounds.

Introduction

India has been bestowed with a rich variety of Phyto diversity. All activities of life in this land centre around vegetation. Hence, reference of vegetation are closely woven with Folk literature, Mythology, folk songs etc. Religious practices of worship are replete with references of plants. Plants are worshipped and several Gods and Goddesses are associated with them. Plants like Bel and Vatare considered temple trees and held sacred and symbolic of certain Hindu Gods. To appeal to the human mind and bring in Phyto- conservation plants have been associated with Gods and Goddesses and a religious angle has been introduced since ancient times.

Mythology

These are ancient tales related to Gods and Godesses. Bel is an important sacred tree in Hindu mythology. The three leaflets resembling the Trishul or the Trident are considered symbolic of creation, preservation and destruction, the powers attributed to Lord Shiva. During the great festival of Durga Puja, Durga is invoked to descend to the earth through the newly sprouting Bel tree. The Sanskrit invocation reads.

"Abahayami Debitang Mrinmaya Sriphala Preete" Which underlines her fondness for Bel fruits, while the twings and the leaves are used in the performance of Homa, i. e. oblation to the fire God. The tree itself is spared the axe in view of its sacred status, indirectly conserving the species.

Vat is held in regard by the Hindus and is worshipped in many places in India. Sometimes the plant is associated with Hindu gods or demons. Such deities are believed to perch on its branches. Often the large trunk of the tree is surrounded by numerous stones all smeared with vermilion and worshipeed as symbols of Lord Shiva. The tribal people also consider the tree sacred and as an abode of their Gods.

Mango is considered sacred both by Hindu and Buddhists. Lord Buddha used to repose under a grove of mango trees. The Hindus consider the tree a transformation of God Prajapati, Lord of creations [1].

Tulsi is very sacred to Hindus and is worshipped. The term Haripriya denotes Tulsi which means dear to Lord Vishnu. According to Folk belief an evil spirit cannot come to a place where tulsi is planted.

Nariel or Coconut is considered sacred among the Hindus. The fruits are believed to fulfill one's desire and so it is offered to Gods. A whole green coconut with its stalk is an essential thing in Hindu religious ceremonies. Regarding the origin of the plant, there

M.M. Srivastava, L. D. Khemani, S. Srivastava, *Chemistry of Phytopotentials: Health, Energy and Environmental Perspectives*, DOI:10.1007/978–3–642–23394–4_80, © Springer-Verlag Berlin Heidelberg 2012

is a legend which says, "Rishi Vishwamitra sent king Trishanku from earth to heaven by his super human power obtained by long austerities. Indra got annoyed over this and sent back the king from heaven to earth that was again stopped by the Rishiputting a pole under him which ultimately became the coconut tree [2]. The plant is not cut by the Hindus; neither uses the wood for fuel because the plant is considered the seat of Lord Narayana.

Folk songs

Folk songs in worship of plants are considered. Folk songs in worship of Peepal are sung in Uttar Pradesh. Peepal is considered in Kumaon as a tree of the land where several rishis meditated and in Tibet as a tree of the land where Buddha meditated. Banyan tree as the abode of Shankar or Savitri; others consider that Brahma resides in its roots, Vishnu in its trunk and Shanker in its branches. Folk songs in praise of Sal, Aak, plantain, Saj, mango, Aonla, bamboo. Neem, Ashok, basil and Amaltas are sung believing that these plants are the abode of several Gods and Goddesses. In Bhojpuri and Magahi folk songs, the leaves of lotus are considered as the bed of Gaura-Parvati and in Bengal as the bed of Saraswati [3]. Bengal leads in display of love through ceremonial functions. The Bel tree is worshipped in Bengal and the twig is considered as a token of good crop. In Himachal the new crop of Maize is celebrated with fair called Minjar. During the 'Daur Durva Worship' festival in Kumaon, barley, wheat, gram, mustard and maize are the centre of the songs of worship. The festival of Basil is celebrated in Mysore. In Bundelkhand at the time of the festival of Mamulian, girls decorate the spiny green branches of Babul with colourful flowers. They offer various fruits to the tree for fulfilment of wishes. In Mandsor district of Madhya Pradesh a festival for opium is celebrated which is cultivated in this area [4].

The tribals consider cutting the Peepal and Banyan tree more sinful then killing a saint. In Uttar Pradesh, even folk songs for children prohibit the cutting of Peepal, banyan and sandal trees. In Kumaon, cutting of timber after sunset is believed to attract illness for children. [5]It is believed that using Peepal wood and Bamboos for burning in Yagya can destroy the whole family. Tribal women worship the fig, old clothes are placed on trees particularly, Jujuba to get wishes fulfilled.

Folk Tales

In folk tales there are numerous references of plants being associated with Gods and Godesses. According to one tale, Aonla was first born from Brahma's tear of love. Other trees and men were created after that. Regarding Champa and Jasmine it is said that two loving children were buried by their step mother in a heap of rubbish and here two trees grew- the Champa and Jasmine. The tree Arjuna is said to have been born of the two sons of Kubair after saint Narad cursed them. The banyan tree was created by Shiva. It is said that when the demon Bhasmasur wanted to destroy Shiva and marry Parvati, Shankara (Shiva) took the form of Banyan tree and stood firm. It is, therefore, said that banyan tree can never be destroyed. It is said that once Parvati became angry with Vishnu and Brahma and cursed them; the result was that first Peepal tree was created from the body of Vishnu and the first Palash tree from the body of Brahma [6] According to some tales, men have been born from trees and plants. Shiva is said to have married a twiner on a Peepal tree. Later, a black obstinate son was born from this twiner. The Kol and Bhil tribes consider themselves to be the progeny of this person born of a twiner.

Sacred Groves

Forest patches religiously preserved for the well being of society are known as "Sacred groves' or Dev-Rai or 'Dev- Ratiat' [7]. Felling trees or even removal of dead wood from such groves is banned [8]. These are located in remote tribal areas along the Western Ghats in India.

These forest patches preserved on religious grounds are indicators of type of vegetation that once existed along these hilly terrains, long before the dawn of modern civilization.

The practice of dedicating groves to deities is common in our country. All forms of vegetation in such a sacred groove including shrubs and climbers are supposed to be under the protection of the reigning deity of that grove and removal of even a small twig/is taboo [9]. These are ancient natural sanctuaries where plants and animals are afforded protection through the grace of some deity. These deities of primitive nature are often called Mother Goddess, are in the form of unshaped stone lumps smeared with red lead, mostly lying under tall trees or in open sky. They are namely Kalkia, Shirkai and Waghjai Taboo's like breaking wood may lead to serious 'illness or death has lead to preservation in these scared groves.

Folklores play important role in confirming such beliefs. It has helped the tribal population in preserving their traditional customs, rituals, ceremonies and a way of forest life. They also point out rewards and blessings for good behavior or act and heavy punishment for the atheist or infidel. They generally demand animal or human sacrifice which is symbolically offered now. The human sacrifice has been replaced by slaughtering a cattle or fowl. There are much folklore on reprisal of forest spirits for unauthorized hunting in scared grooves.

Information on location, area and associated deity for 233 sacred groves of various districts of state of Maharashtra has been worked out. The total area reported comes to 3,570 hectares. When complete information is worked out the total area under sacred grooves will be between 5,000 to 10,000 hectares or more.

Most of the folklores and beliefs of tribal culture have kept sacred groves unmolested till now. Due to rapid changes in civilization these beliefs and taboos are becoming less effective now.

Conclusion

Thus, from berry ancient times Indian folk life has not only been including trees, plants and flowers as members of their own family, but has also found in them the image of God [10] . It is for this reason that songs, tales and other expressions have been replete with deep affection for Phyto-diversity and have lead to their conservation.

Appendix

Botanical names of common names used

Aak	*Calatropis sp.*
Amaltas	*Cassia fistula*
Aonla	*Emblicaofficinalis*
Arjun	*Terminaliaarjuna*
Ashok	*Saracaasoca*
Bamboo	*Bambusa sp.*
Bel	*A eglemarmelos*
Champa	*Micheliachampa*
Jasmine	*Jasminium sp.*
Lotus	*Nelumbo sp.*
Opium	*Papaversomnifera*
Neem	*Azadirachtaindica*
Palash	*Buteamonosperma*
Peepal	*Ficusreligiosa*
Plantain	*Musa paradisiaca*
Saj	*Terminalia sp.*
Vat	*Ficusbengalensis*

References

1. M.S. Randhawa, All India Fine Art and Craft Society, New-Delhi (1964).
2. H.H. Wilson, Vishnu Puran, Punthi Pustak, Calcutta (1961).
3. S. Aryani, Magahi Lok Sahitya, Hindi Sahitya Sansar, Patna (1965).
4. R.D. Sankrityayan and K.D. Upadhyaya, Hindi Sahityaka BirhalItihas, Hindi Ka Lok Sahitya Nagri Pracharini Sabha, Varanasi (1960).
5. Anonymous, Uttar Pradesh Kelok Geet- Shakun Geet, Information Division, U.P. Govt Lucknow (1965).
6. S. Chaturvedi, Hamari Lok Kathayen, Sasta Sahitya Mandal, New Delhi (1965).
7. D.D. Kosambi Myth and Reality: Studies in the Formation of Indian Culture, Popular Press, Bombay (1962).
8. M. Gadgil and V.D. Vartak, The Sacred Groves of Western Ghats in India, Economic Botany vol. 30, (1976) p 152–160.
9. V.D. Vartakand M. Gadgil and D. Rahti, Proceedings of 60 Indian Science Congress, (1973) p 341.
10. S.C. Jain Kavya Men Padap Pushp, M.P. Prakashan Samiti, Bhopal, (1958).

MIX
Papier aus verantwortungsvollen Quellen
Paper from responsible sources
FSC® C105338
FSC
www.fsc.org

If you have any concerns about our products,
you can contact us on
ProductSafety@springernature.com

In case Publisher is established outside the EU,
the EU authorized representative is:
Springer Nature Customer Service Center GmbH
Europaplatz 3, 69115 Heidelberg, Germany

Printed by Libri Plureos GmbH
in Hamburg, Germany